M000159972

Brigham Young University
Harold B Lee Library

WITHDRAWN
SSP
6/20/2024

REPORTS ON ASTRONOMY

TRANSACTIONS OF THE
INTERNATIONAL ASTRONOMICAL UNION
VOLUME XXVIIA

COVER ILLUSTRATION: THE VEIL NEBULA

A wide-field image of the Veil Nebula, made as a colour composite of individual exposures from the Digitized Sky Survey 2. The field of view is $4°.2 \times 4°.2$.

The Veil Nebula, also known as the Cygnus Loop, is a large, bright (integrated $m_v = 7.4$) supernova remnant in the constellation Cygnus. The source supernova exploded some 5 000 to 10 000 yr ago, and the remnants have since expanded (presently $v \approx 180\,\mathrm{km\,s^{-1}}$) to cover an area of $3°.5$, over six times the size of the full Moon. Its distance $d = 540^{+100}_{-80}$ pc or around 1750 lyr (Blair *et al.* 2005).

It was discovered on 5 September 1784 by William Herschel. He described the western end of the nebula as *". . . Extended; passes thro' 52 Cygni . . . near 2 degree in length . . ."* and described the eastern end as *". . . Branching nebulosity . . . The following part divides into several streams uniting again towards the south . . ."*.

The nebula shows up very well in Hα and in several forbidden lines like [O III]. When resolved, some parts of the image appear to be rope-like filaments. The standard explanation is that the shock waves are so thin, less than one part in 50 000 of the radius, that the shell is only visible when viewed exactly edge-on, giving the shell the appearance of a filament. Undulations in the surface of the shell lead to multiple filamentary images, which appear to be intertwined.

The brighter segments of the nebula are listed in the New General Catalog under the designations NGC 6960, NGC 6979, NGC 6992, and NGC 6995. The easiest segment to find is NGC 6960 (the Witch's Broom Nebula), which runs through the 4.2 mag foreground star 52 Cygni. NGC 6979 – the central portion of the Veil Complex – is Pickering's Wedge, or Pickering's Triangular Wisp. This segment of nebulosity was discovered on Harvard College Observatory photographic plates by Williamina P. Fleming (1857–1911), but credit went to her supervisor Edward C. Pickering, as was the custom of the day in the 19th and early-20th century, and it is named after Pickering as a result.

References:
Blair, W. P., Sankrit, R., & Raymond, J. C. 2005, *AJ*, 129, 2268
IAU Division VIII Supernova Working Group, private communication
Mat Drummen, Sonnenborgh Observatory, private communication
Wikipedia <en.wikipedia.org/wiki/Veil_Nebula>

Credit: NASA; ESA; the Hubble Heritage (STScI/AURA)-ESA/Hubble Collaboration; and the Digitized Sky Survey 2 <www.spacetelescope.org/images/html/heic0712g.html>.

Acknowledgments: J. Hester (Arizona State University) and Davide De Martin (ESA/Hubble).

TRANSACTIONS OF THE
INTERNATIONAL ASTRONOMICAL UNION

2009 EDITORIAL BOARD

Chairman

I. F. CORBETT, IAU Assistant General Secretary
European Southern Observatory
Karel-Schwarzschild-Strasse 2
D-85748 Garching-bei-München
Germany
icorbett@eso.org

Advisors

K. A. VAN DER HUCHT, IAU General Secretary,
SRON Netherlands Institute for Space Research, Utrecht, the Netherlands
E. J. DE GEUS, *Dynamic Systems Intelligence B.V., Assen, the Netherlands*
U. GROTHKOPF, *European Southern Observatory, Germany*
M. C. STOREY, *Australia Telescope National Facility, Australia*

INTERNATIONAL ASTRONOMICAL UNION

UNION ASTRONOMIQUE INTERNATIONALE

International Astronomical Union

REPORTS

ON

ASTRONOMY

TRANSACTIONS
OF THE
INTERNATIONAL ASTRONOMICAL UNION
VOLUME XXVIIA

Edited by

KAREL A. VAN DER HUCHT
General Secretary of the Union

CAMBRIDGE
UNIVERSITY PRESS

HAROLD B. LEE LIBRARY
BRIGHAM YOUNG UNIVERSITY
PROVO, UTAH

CAMBRIDGE UNIVERSITY PRESS
The Edinburgh Building, Cambridge CB2 8RU, United Kingdom
32 Avenue of the Americas, New York, NY 10013-2473, USA
477 Williamstown Road, Port Melbourne, VIC 3207, Australia
Ruiz de Alarcón 13, 28014 Madrid, Spain
Dock House, The Waterfront, Cape Town 8001, South Africa

© International Astronomical Union 2009

This book is in copyright. Subject to statutory exception
and to the provisions of relevant collective licensing agreements,
no reproduction of any part may take place without
the written permission of the International Astronomical Union.

First published 2009

Printed in the United Kingdom at the University Press, Cambridge

Typeset in System LaTeX 2_ε

A catalogue record for this book is available from the British Library

Library of Congress Cataloguing in Publication data

ISBN 9780 521 856058 hardback
ISSN 1743-9213

Table of Contents

Preface

As input for the IAU XXVII General Assembly in Rio de Janeiro, Brazil, 3–14 August 2009, the General Secretary presents here to the IAU membership at large the *Reports on Astronomy 2006–2009, Transactions XXVIIA of the International Astronomical Union.*

This is the third and last volume of the triennial trilogy by your General Secretary 2006–2009, which also comprises the *Highlights of Astronomy* (science proceedings of the IAU XXVI General Assembly, Prague, August 2006), and the (administrative) *Proceedings of the XXVI General Assembly, Prague 2006, Transactions XXVIB of the International Astronomical Union.*

The individual triennial reports in this volume have been authored by the presidents c.q. chairpersons of the IAU Divisions/Commissions and Working/Program Groups. They present the activities of their IAU scientific bodies over the period August 2006 to November 2008. Earlier (annual) reports of these scientific bodies have been published in the IAU *Information Bulletins* Nos. 99 (January 2007) through 102 (July 2008).

In the absence of strict guidelines, the triennial reports presented in this volume display a great deal of diversity. While some authors simply present a list of conferences in their astronomical field during the triennium, others also list the literature in their field, and others, again, describe what actually happened in their scientific bodies, or provide a combination of the foregoing. At the same time, the triennial reports may reveal how these authors see their jobs as IAU Division/Commission president or Working/Program Group chair.

In any case, the present volume constitutes essential reading for all those interested in how the IAU goes about its business. I hope that it will inspire the astronomical community to reflect upon the role of the IAU and make plans for the future of this organization, which stands for what we all have in common: an unquenchable thirst for astronomical knowledge and desire to share our knowledge with each other and the world at large.

It is my pleasure to thank the IAU Division/Commission presidents and Working/Program Group chairpersons who have provided their triennial reports to me.

Karel A. van der Hucht
IAU General Secretary
Paris / Utrecht, 27 November 2008

Transactions IAU, Volume XXVIIA
Reports on Astronomy 2006–2009
Karel A. van der Hucht, ed.

© 2009 International Astronomical Union
doi:10.1017/S1743921308025222

DIVISION I   FUNDAMENTAL ASTRONOMY

ASTRONOMIE FONDAMENTALE

Division I provides a focus for astronomers studying a wide range of problems related to fundamental physical phenomena such as time, the inertial reference frame, positions and proper motions of celestial objects and precise dynamical computation of the motions of bodies in stellar or planetary systems in the Universe.

PRESIDENT	Jan Vondrák
VICE-PRESIDENT	Dennis D. McCarthy
PAST PRESIDENT	Toshio Fukushima
BOARD	Aleksander Brzezinski, Joseph A. Burns,
	Pascale Defraigne, Dafydd Wyn Evans,
	Toshio Fukushima, George H. Kaplan,
	Sergei A. Klioner, Zoran Knežević,
	Irina I. Kumkova, Chopo Ma,
	Richard N. Manchester, Gérard Petit

DIVISION I COMMISSIONS

Commission 4	Ephemerides
Commission 7	Celestial Mechanics and Dynamical Astronomy
Commission 8	Astrometry
Commission 19	Rotation of the Earth
Commission 31	Time
Commission 52	Relativity in Fundamental Astronomy

DIVISION I WORKING GROUPS

Division I WG	Second Realization of International Celestial Reference Frame
Division I WG	Numerical Standards in Fundamental Astronomy
Division I WG	Astrometry by Small Ground-Based Telescopes

INTER-DIVISION WORKING GROUPS

Division I-III WG	Cartographic Coordinates and Rotational Elements
Division I-III WG	Natural Satellites

TRIENNIAL REPORT 2006–2009

1. Introduction

The structure of Division I changed substantially since the IAU XXVI General Assembly (Prague, August 2006), reflecting the progress made in all astronomical disciplines covered by the Division. The former Working Group on *General Relativity in Celestial*

Mechanics, Astrometry and Metrology was promoted to the new Commission on *General Relativity in Fundamental Astronomy*, other Working Groups were either disbanded, if their tasks were accomplished (*Re-definition of Coordinated Universal Time, Precession and the Ecliptic*), or re-organized into new ones with re-defined tasks (*Future Developments in Ground-Based Astrometry, Nomenclature for Fundamental Astronomy*). Thus, three new Working Groups were created (*Second Realization of International Celestial Reference Frame; Numerical Standards in Fundamental Astronomy; Astrometry by Small Ground-Based Telescopes*).

The Division members at its business meeting in Prague also discussed and approved the proposal by the President, Toshio Fukushima, that the future Organizing Committee of Division I should consist of the Presidents and Vice-Presidents of all Commissions pertaining to the Division. This decision is reflected in the present composition of the OC.

2. Developments within the past triennium

Many scientific problems of Division I were discussed at Joint Discussion 16 (*Nomenclature, precession and new models in fundamental astronomy*), held during the IAU XXVI GA in Prague. Three resolutions, prepared by the Division and discussed at JD16 (Precession Theory and Definition of the Ecliptic; Supplement to the IAU 2000 Resolutions on reference systems; and Re-definition of Barycentric Dynamical Time, TDB) were endorsed by the General Assembly.

An important scientific discussion forum for the Division was established by organizing the series of symposia, officially held outside the IAU, Journées Systèmes de Référence Spatio-temporels. These meetings, originally French national ones that gradually became fully international, are held annually since 1988 in a European country. The Journées 2006 was not held because the JD16 at IAU GA in Prague had a similar theme. In September 2007 it was held at Meudon, France, with the sub-title *The Celestial Reference Frame for the Future* and in September 2008 in Dresden, Germany, subtitled *Astrometry, Geodynamics and Astronomical Reference Systems*.

One of the milestones for all Division members was IAU Symposium No. 248 *A Giant Step: from Milli- to Micro-arcsecond Astrometry*, held in October 2007 in Shanghai, China. The meeting was timed to coincide with the 10th anniversary of the release of the Hipparcos and Tycho Catalogues. It gathered scientists from all Commissions pertaining to Division I to discuss the problems connected with achieving higher astrometric accuracy (both in optical and radio wavelengths) and it touched not only observation techniques but also modelling, including general relativity aspects. Astrophysicists from other Divisions who are using astrometry and its results for their scientific interpretations also participated.

Members of Commission 4 concentrated mainly on developing software and calculating the ephemerides for the general use by astronomers, navigators, etc. .... Progress was made in the gradual implementation of IAU resolutions adopted in 2000, 2003 and 2006 into the contents of most major ephemeris books. Some observatories provide electronic access to ephemerides and significant astronomical phenomena and also to relevant software. A dominant role is played by the Jet Propulsion Laboratory (JPL) that traditionally produces very accurate planetary ephemerides, based on recent astronomical observations, gathered also from NASA and ESA space missions. Since 2006 also IMCCE (Observatoire de Paris) produces numerical ephemerides, INPOP, with comparable accuracy.

The problems of celestial mechanics and dynamical astronomy reach a very wide field across the whole of astronomy. The research, conducted by Commission 7, was concentrated on the problems of architectures of extra-solar planetary systems, interactions between galactic bars and halos in disc galaxies, spin-orbit dynamics in the solar system (with stress on the motion of Mercury), and binary solar system objects.

Commission 8 concentrated mainly on problems of space astrometry. A new reduction of the *Hipparcos* data, with much higher accuracy of parallaxes, was obtained, and significant efforts were devoted to the preparation of the ESA astrometric mission *Gaia*, now approved to be launched in 2011. It is also probable that Japan will launch, after 2010, an infrared astrometric satellite *JASMINE*. Other important activities were aimed at preparing new ground-based astrometric instrumentation and improving the reduction methods, realization and densification of the celestial reference frame, improving the positions and proper motions of observed stars and measuring precise trigonometric parallaxes of nearby and high proper-motion stars. Precise positions of Solar System bodies (asteroids, natural planetary satellites) were also intensively observed.

Research concerning the rotation of the Earth was made by the members of Commission 19, both from practical (measurement of Earth Orientation Parameters) and theoretical (studies of exchange of angular momentum between the solid Earth and its fluid parts) points of view. New observational techniques appeared (ring laser, new generation of VLBI with small antennas) and global circulation models of geophysical fluids were developed, including satellite gravity data from *GRACE*. As the new terrestrial reference frame (ITRF 2005) was derived, Earth Orientation Parameters since 1984 were reprocessed. The commission closely co-operated with the International Earth Rotation and Reference Systems Service (IERS) on updating the IERS Conventions, which provide a set of astronomical/geodetical constants and fundamental procedures. The practical use in space navigation and elsewhere requires also very precise prediction of Earth's orientation in space, so the necessary procedures were developed and inter-compared.

Time and its different scales is an important part of astronomy, namely as an independent variable for the description of all dynamical systems. Originally defined by Earth rotation, then by the motion of the Earth around the Sun, now by atomic clocks, and maybe by pulsars in future, the time and its relativistic aspects in different reference frames, and also its worldwide measurement and coordination, poses a lot of questions considered by Commission 31. International Atomic Time (TAI) provides a fundamental scale from which other scales are derived; it is now based on about 350 different clocks working at 65 laboratories all over the world. The possible change of definition of Coordinated Universal Time (UTC) in the future was discussed intensively in cooperation with other unions, mainly the International Telecommunication Union (ITU). Pulsars with very stable millisecond periods seem to be promising candidates for the future definition of time in astronomy. Joint Discussion 6 on *Time and astronomy*, to be held during the IAU XXVII General Assembly in Rio de Janeiro in August 2009, will provide a possibility of discussing these problems in detail.

With the increasing accuracy of observations and modelling of positions and motions of celestial bodies, the importance of relativistic effects are growing. The new Commission 52 on *General Relativity in Fundamental Astronomy* is very active in solving these tasks. Namely such problems as the impact of relativity on astronomical reference frames, relativistic modelling of observations, astronomical tests of general relativity, relativity in astrodynamics and space navigation, relativistic time scales and astronomical constants and units were studied. IAU Symposium No 261 (*Relativity in fundamental astronomy – dynamics, reference frames, and data analysis*), to be held at Virginia Beach, USA, in

April/May 2009, will provide a wide international forum to discuss these questions in detail.

The International Celestial Reference Frame (ICRF) was defined in 1997 by the adopted positions of 212 extragalactic radio sources. With many new VLBI observations gathered since then, it is now possible to obtain a better and denser realization of the ICRF, considering also temporal changes of source structures. The work on the second realization (ICRF-2) was done in close cooperation with the IERS and the International VLBI Service (IVS).

Further important progress was achieved in updating the current best estimates of astronomical constants, in collaboration with the IERS. This work will probably lead to a proposal of a new IAU System of Constants in 2009 (see the report of WG on *Numerical Standards of Fundamental Astronomy* for more details).

Small ground-based telescopes (roughly up to 2m diameter) were used, in many places of the world, to do high-precision astrometry of stars and solar system bodies. Most of these telescopes are now equipped with CCD detectors. During the last triennium, accurate positions and/or proper motions of some 100 million stars were obtained and thousands of asteroids, with concentration on Near Earth Objects, were observed. Small telescopes were also very active in observing other solar system bodies, such as satellites of the planets and their mutual phenomena. Thanks to a joint Franco-Chinese cooperation, a first spring school on astrometry *Observational Campaign of Solar System Bodies* was held in Beijing, China, in April 2008.

The Working Group on *Cartographic Coordinates and Rotational Elements* published its (2006) triennial report in Celestial Mechanics and Dynamical Astronomy (Seidelmann *et al.*, CeMDA 98, 155). Important changes included corrections to or updates to the recommended models for the rotation of the Sun, orientation of the Moon, orientation and size and shape of several Saturnian satellites, and orientation of Pluto and Charon. The recommended models for the size and orientation of several asteroids and comets were updated or given for the first time. The Working Group is starting to address changes needed for the next report and will present a draft for discussion and completion at the next IAU General Assembly.

The Working Group on *Natural Planetary Satellites* concentrated on gathering both historic and new astrometric observations. Based on them, theories of motion, masses, gravity fields, orbits and their evolution were derived for many satellites of planets and asteroids of the solar system.

3. Closing remarks

It is necessary to add that individual Commissions and Working Groups within Division I are not isolated; they mutually interact, cooperate and often organize meetings of common interest. This summary is meant only to show the most important progress achieved in fundamental astronomy during the last triennium, according to a subjective selection by the Division President. It can by no means cover all rapidly evolving fields of research that all components (Commissions and Working Groups) of the Division are filling. The detailed reports that follow give a more exhaustive image of what has been done.

Jan Vondrák
president of the Division

Transactions IAU, Volume XXVIIA
Reports on Astronomy 2006–2009
Karel A. van der Hucht, ed.

© 2009 International Astronomical Union
doi:10.1017/S1743921308025234

COMMISSION 4

EPHEMERIDES

ÉPHÉMÉRIDES

PRESIDENT	Toshio Fukushima
VICE-PRESIDENT	George H. Kaplan
PAST PRESIDENT	George A. Krasinsky
ORGANIZING COMMITTEE	Jean Eudes Arlot, John A. Bangert, Catherine Y. Hohenkerk, George A. Krasinsky, Martin Lara, Elena V. Pitjeva, Sean E. Urban, Jan Vondrák

TRIENNIAL REPORT 2006–2009

1. Jet Propulsion Laboratory, Caltech, USA

JPL planetary ephemeris development has been very active assimilating measurements from current planetary missions and supporting future missions. The NASA *Mars Science Laboratory* (*MSL*) mission with launch in 2009 requires knowledge of the Earth and Mars ephemerides with 30 m accuracy. By comparison, the accuracy of the Mars ephemeris in the widely used DE405 ephemeris was about 3 km. Meeting the *MSL* needs requires an ongoing program of range and very-long baseline interferometry measurements of Mars orbiting spacecraft. The JPL ephemeris DE421 was released three months before the landing of the Phoenix mission on Mars, and has met the 300 m requirement. Continued measurements are planned to support the *MSL* landing.

Measurements are now routinely made using several NASA and ESA spacecraft in orbit about Mars. VLBI measurements using DSN and ESA tracking stations have been supplemented by use of the Very Long Baseline Array for specific measurement campaigns. Currently the planetary ephemeris is aligned to the ICRF with an accuracy of 0.25 mas based on the VLBI measurements. VLBI and ranging measurements to the ESA Venus Express spacecraft have resulted in a significant improvement in the Venus ephemeris. Measurements of the *Cassini* spacecraft since 2005 have resulted in a significant improvement in the Saturn ephemeris. Measurements from the *MESSENGER* spacecraft encounters with Mercury are expected in the next several years. After a lapse of several years, lunar laser ranging are being included again in the JPL ephemeris development, to support planned lunar missions.

The JPL ephemeris coordinates time is not the TCB time scale recently adopted by the IAU. Instead the time scale is consistent with the earlier JPL coordinate time, and is nominally the same as TDB as it has now been redefined by the IAU as a linear function of TCB. Because of potential harm to NASA spacecraft, there is no plan to change the coordinate time scale for ephemerides to be used with NASA missions.

(*a*) Staff: E. Myles Standish retired in 2007 after 35 yr of leading JPL planetary ephemeris development. William Folkner is currently responsible.

(b) JPL planetary ephemerides are made available in ASCII 'export' format and in SPICE binary format via anonymous FTP from `<ftp://ssd.jpl.nasa.gov/pub/eph/planets/>`

(c) The observational data used in fitting the JPL planetary ephemerides is currently being updated, and available at `<iau-comm4.jpl.nasa.gov/plan-eph-data/>`

(d) Documentation on the latest JPL planetary ephemerides is in the form of memoranda available via anonymous FTP from `<ftp://ssd.jpl.nasa.gov/pub/eph/planets/ioms>`

2. Astronomical Applications Department, US Naval Observatory

This report covers activity in the Nautical Almanac Office (NAO) and its parent organization, the Astronomical Applications Department.

(a) Publications.

Publication of *The Astronomical Almanac* and *The Astronomical Almanac Online*, *The Nautical Almanac*, *The* (US) *Air Almanac*, and *Astronomical Phenomena* continued as a joint activity between Her Majesty's Nautical Almanac Office of the United Kingdom and the NAO. *The Astronomical Almanac* for 2009, released in January 2008, fully implements the resolutions adopted by the IAU in 2006, both within the tabular data and explanatory text. *The Air Almanac* for 2009, released in June 2008, is now available exclusively as an electronic publication on CD-ROM. U.S. Naval Observatory Circular 179, *The IAU Resolutions on Astronomical Reference Systems, Time Scales, and Earth Rotation Models: Explanation and Implementation*, was published on-line and in print form in October 2005. Work was underway on a major revision of *The Explanatory Supplement to the Astronomical Almanac*, in collaboration with P.K. Seidelmann (Univ. of Virginia) and numerous contributors.

(b) Software.

An update of the *Multiyear Interactive Computer Almanac*, MICA version 2.1, was completed and released in December 2006. The software is available in two editions for computers running Microsoft Windows and Apple Mac OS operating systems. A new version of the Naval Observatory Vector Astrometry Subroutines (NOVAS) that implements relevant IAU resolutions adopted in 1997 through 2006 was essentially completed. The software will be available in both Fortran and C editions. A major redesign of the Astronomical Applications Department web site `<aa.usno.navy.mil/>`, which included several new data services, was launched in September 2007. Usage of the web site varied from about 0.5 to 1.3 million visits per month.

(c) Research.

An active research program in positional and dynamical astronomy is underway within the department. Research topics included new methods of celestial navigation, determination of asteroid masses, and the theory of bodily tides.

3. National Astronomical Observatory and Japan Hydrographic and Oceanographic Department Japan

Annually National Astronomical Observatory of Japan (NAOJ) publishes the 'Calendar and Ephemeris', a basic almanac designed for astronomical observers, teachers, and citizens. From the 2009 edition, NAOJ not only implemented the New Precession Formula adopted by IAU in 2006, but also enhanced its volume, for example, almost doubled its size and its number of pages. This is the first step toward a development of a full-scale ephemeris. In addition, NAOJ has added more tools to its web site `<www.nao.ac.jp/koyomi/>`, such as local prediction of the Solar/Lunar Eclipse and the Transit of Mercury/Venus, and also established a web site for mobile phones.

The Japan Hydrographic and Oceanographic Department (JHOD) will finish publishing 'Japanese Ephemeris' by 2010 edition and plans to publish only Nautical Almanac and Abridged Nautical Almanac. JHOD has also decided to terminate International Lunar Occultation Centre (ILOC) activities, which will be taken over by the International Occultation Timing Association (IOTA) in March, 2009.

4. HM Nautical Almanac Office, UK Hydrographic Office

After having gone through a review by the UK Ministry of Defense (MoD) which established a continuing requirement for the services and publications of HM Nautical Almanac Office, it became part of the UK Hydrographic Office (UKHO), a UK government trading fund and part of the UK MoD, on 1 April 2006. Two staff members are now based at the UKHO in Taunton, Somerset, while one works off site at our previous host, the Rutherford Appleton Laboratory in Oxfordshire. Commercial viability is still important to HMNAO's operation, however, the requirements of both SOLAS (Safety of Life at Sea) and the Royal Navy take precedence.

Joint publications with the US Naval Observatory, The Nautical Almanac (NP 314), The Astronomical Almanac (AsA), its companion the AsA Online and Astronomical Phenomena, have been produced on schedule. Material throughout the AsA, but most noticeably in Section B, has been produced in accordance with all the recommendations and resolutions of the IAU General Assemblies up to and including 2006 including the Celestial Intermediate Reference System, the Earth rotation angle, IAU 2000A nutation and IAU 2006 precession. Effort has also gone into the expansion of web services provided by the office. Our general web site has been revamped, `<www.hmnao.com>`, and our Crescent MoonWatch, `<www.crescentmoonwatch.org>`, and Eclipse web sites, `<www.eclipse.org.uk>`, are under further development. UKHO versions of Rapid Sight Reduction Tables (NP 303), NavPac and Compact Data 2006-2010 (DP 330), The UK Air Almanac (AP 1602) and The Star Almanac for Land Surveyors (NP 321) have been published by UKHO and made available through their distributor network.

Despite the formal relocation of HMNAO to UKHO in 2006, the office is still expending significant effort in furthering its integration with its new operating environment.

HMNAO staff have also played an active role in the Division I WG on Nomenclature for Fundamental Astronomy and are involved with the WG on *Numerical Standards in Fundamental Astronomy*. Continued participation on the board of SOFA has been supplemented by HMNAO's provision and maintenance of the SOFA web site, `<iau-sofa.hmnao.com>`. HMNAO continues to use SOFA software in the production of its publications.

5. Institute of Applied Astronomy, Russia

(*a*) Fundamental ephemerides.
During the years 2005–2008 the regular publication of *The Russian Astronomical Yearbook* is continued. Planetary and lunar ephemerides are based on numerical model EPM-2004 available to outside users via `<ftp://quasar.ipa.nw.ru/incoming/EPM2004>`.
Ephemerides for planetary configurations, eclipses and occultations are updated and located at `<quasar.ipa.nw.ru/PAGE/EDITION/RUS/rusnew.htm>`. The ephemerides of

the Moon (as Tchebyshov polynomials) and the mutual phenomena in the system of the Galilean satellites of Jupiter are on the same Web site.

(*b*) Special ephemerides.

The Naval Astronomical Yearbook (annual issues for 2006–2009) and biennial *The Nautical Astronomical Almanac* (issues 2007–2008, 2009–2010) have been published. The basic purpose of producing the Almanac is to increase its applicability without essential increase of its volume and to give the same accuracy as NAY does. The explanation and part of auxiliary tables are given in both Russian and English versions.

(*c*) Software.

Constructing numerical dynamical models, fitting the ephemerides to observations, as well as preparation of the ephemerides for publishing are carried out in the framework of the universal program package ERA (Ephemerides for Research in Astronomy). The unified Windows/Linux as well as the DOS versions of the package ERA which are available via anonymous FTP, <ftp://quasar.ipa.nw.ru/incoming/ERA>, are under development. The first electronic version of *The Personal Astronomical Yearbook (PersAY)* has been constructed. It is intended for calculation of the ephemerides published in the Astronomical Yearbook, including the topocentric ephemerides for any observer. The system PersAY is implemented as the Win32 application on the basis of the package ERA. The first version of PersAY for interval 2000–2015 based the fundamental ephemerides DE405/LE405 and EPM2004 is available on <ftp://quasar.ipa.nw.ru/pub/PERSAY/persay.zip>. The electronic system the *Navigator* is in progress. It is intended for solution of basic naval astronavigating problems by the mode of remote access.

(*d*) Research work.

The updated Ephemerides of Planets and the Moon – EPM2008 – have been constructed by the simultaneous numerical integration of the equations of motion of the major planets, the Moon, the Sun, 301 biggest asteroids, 21 trans-Neptunian objects and the lunar physical libration accounting for the perturbations due to the solar oblateness and the massive ring of small asteroids. The parameters of EPM2008 have been fitted to lunar laser ranging measurements 1970–2008 as well as to planet and spacecraft observations 1900– 2007 of different types. The numerical ephemerides of the main satellites of the outer planets have been constructed and fitted to modern photographic and CCD observations. These ephemerides are used for improving the ephemerides of their parent planets, in which connection the ephemerides of the Galilean satellites are used for publication in the Russian Astronomical Yearbook.

6. Institut de Mécanique Céleste et de Calcul des Éphémérides, Observatoire de Paris, France

During the last three years, various works on ephemerides have been pursued at the Institut de mécanique céleste et de calcul des éphémérides (IMCCE). IMCCE, as institute of the Paris Observatory, is in charge to compute the official French ephemerides on behalf of the Bureau des longitudes. Therefore, besides scientific researches in the domains of theoretical celestial mechanics, astrometry and planetology, the research teams of IMCCE perform activities in dynamics and applied celestial mechanics with the goal to provide accurate ephemerides. Besides, the Ephemerides Service of IMCCE works for the improvement of the ephemerides books, softwares, web server and facilities.

(*a*) The new dynamical models.

Several new dynamical models have been developed for the planets and the natural satellites. The development of the new numerical planetary ephemerides named INPOP (Intégration Numérique Planétaire de l'Observatoire de Paris) has been carried on. A new version named INPOP06 has been published (Fienga *et al.* 2007; Fienga *et al.* 2008). The motion of the eight planets, Pluto and the Moon are modelized and fitted on the most accurate observations including tracking data from *Mars Global Surveyor* and *Mars Odyssey*. The accuracy is comparable to the accuracy of recent versions of the JPL DE414 ephemerides and of the EPM2004 ephemerides. With the goal to provide accurate ephemerides for the study of insolation quantities and paleoclimates, planetary dynamical models on very long term are also investigated by (Laskar 2007). Dynamical models of several planetary satellites systems have been developed or improved. The model of satellite motion named NOE (Numerical Orbit and ephemerides) takes into account various gravitational effects down to the small ones. Ephemerides are obtained by recomposition of quasi-periodic Fourier series issued from a frequency analysis of a numerical model coupled with digital filtering treatments. The motion of the Galilean satellites L1 has been modelized by (Lainey *et al.* 2006) and has been used to predict the next season of mutual events of the Galilean satellites by (Arlot 2008). Similarly, different models have been used by (Arlot & Thuillot 2008) to predict the next season of mutual events of the Saturnian satellites and of the eclipses by Saturn itself. NOE has also been applied to the development of new ephemerides of the Uranian satellites named LA06/NOE-7-06. An improved version, LA07/NOE-7-07 is now available on the ftp server of IMCCE `<ftp://ftp.imcce.fr/pub/ephem/satel/la06/>`. New ephemerides of the Martian satellites Phobos and Deimos (NOE-4-07) that are fitted to observations from 1877 to 2005 and include recent spacecraft observations by *Mars Global Surveyor* and *Mars Express* have been published by (Lainey *et al.* 2007). In an other work based on the use of the onboard camera (Super Resolution Camera) of the *Mars Express* space probe, an accurate dynamical model of the Martian satellites has been developed in a fruitful collaboration with the Royal Observatory of Belgium (Rosenblatt *et al.* 2008). This work combines observations of the moons positions from a spacecraft and from the Earth and this allows to assess the real accuracy of the spacecraft orbit. A collaboration with the Sternberg State Astronomical Institute of Moscow led to the development of the dynamical modeling of different satellites and the providing of ephemerides. Thus, ephemerides of Phoebe has been provided by (Emelyanov 2007). The problem of the propagation of errors in the dynamical model related to the fit of observations, and the sampling of these observations, is investigated by (Desmars *et al.* 2007). They showed some different statistical methods, re-sampling of observations, which allow a better estimation of the extrapolated accuracy in the future. During the last years, the model of the meteoroid streams developed by Vaubaillon has been successfully applied in international collaborations to the analysis of various observations and to the prediction of the date of the Earth encounter and their activity (Jenniskens *et al.* 2008).

(*b*) Ephemerides books.
IMCCE provides yearly ephemerides on behalf of Bureau des longitudes. Several books related to various Solar System objects and at different levels of accuracy are published. Several changes have been done in these books during the last years. The yearly book of ephemerides of high precision titled 'Connaissance des temps' has been recently transformed and revitalized by the introduction tables (instead of Chebychev coefficients) and many scientific texts upon constants, timescale, reference systems and transformations of coordinates. An ephemerides software allowing the computation of topocentric coordinates, rises and sets, is provided on a CD accompanying the book. On these last years we have introduced in this book the new planetary model INPOP06 and the new satellites models: L1 for the Galilean satellites and NOE for the Martian and the Uranian satellites. Three booklets are published to supplement the main ephemerides 'Connaissance des temps' and are guides for observers. They are titled 'Suppléments à la Connaissance

des temps' and concern the natural satellites. The first one gives graphic configurations and dates of the phenomena of the Galilean satellites. The second one gives the graphic configurations of the eight first satellites of Saturn. The third one gives positional ephemerides of several faint satellites of Jupiter and Saturn. A second yearly book titled 'Guide de données astronomiques - Annuaire du Bureau des longitudes' gives medium precision data. Data for the Sun, the Moon, the planets are given, but also ephemerides for bright comets and asteroids, stellar occultations by the asteroids and the Moon, phenomena of the Galilean satellites and other various phenomena. A scientific booklet is included in this book. Each year, a new topic is stated by some specialists. For the navigation, IMCCE publishes every year a nautical almanac, titles 'Ephémérides nautiques' and ephemerides for air navigation in the 'Ephémérides aéronautiques'.

(c) Electronic ephemerides.

Ephemerides on-line are available at the address <www.imcce.fr/>, the web site of IMCCE. The main improvement for these electronic ephemerides concerns the setting up of web services with the objectives to provide 'self-defined' data and interoperable services. Several ephemerides are now being transformed in order to become interoprable. This work is fully installed in the Virtual Observatory framework (Thuillot et al. 2006). The software labeled SkyBoT (Sky Bodies Tracker), has been developed in collaboration with the Centre de données de Strasbourg in France (CDS) and has been recently improved. It deals with a large and regularly updated database of ephemerides of the small Solar System Bodies to facilitate their identification on a 60 yr period. At the present time all the asteroids, the planets and 33 satellites are available. This software is well adapted to the data mining and has been implemented in other softwares thanks to the information which can be accessed at <www.imcce.fr/webservices/skybot/>. SkyBoT is also available through the Aladin sky atlas of the CDS (Berthier et al. 2006). The next improvement will be to introduce more satellites and the comets on the basis of the orbital elements (Rocher 2008) already available in the VizieR catalog of the CDS accessible at <webviz.u-strasbg.fr/viz-bin/VizieR?-source=B/comets>.

7. Real Observatorio de la Armada, Spain

The Real Observatorio de la Armada (ROA) is responsible for the publication of 'Efemérides Astron ómicas', a national Almanac with a similar layout to *The Astronomical Almanac*, which in a near future will incorporate the IAU recommendations regarding the ICRF. The ephemerides are computed from JPL DE405/LE405 fundamental ephemerides and USNO/AE98. The publication also incorporates ephemerides of the satellites of Jupiter, Uranus, and Neptune that are kindly provided by NAO (USNO). In addition, the publication includes information of apparent places of 194 stars, and prediction of lunar occultations for six Spanish sites.

Besides, ROA regularly issues a nautical almanac, in printed and electronic format, a booklet with the more relevant astronomical phenomena for the Sun, Moon and planets, and maintains a web site with interactive ephemerides computation and other astronomical information <www.roa.es/>.

8. Astronomical Institute, Czech Republic

The Institute, in close cooperation with the Observatory and Planetarium of Prague, issues every year an astronomical yearbook of about 250 pages for amateur astronomers 'Hvězdářská ročenka' (in Czech), with a limited precision. The yearbook is based on the VSOP82 ephemerides and it contains the ephemerides of the Sun, Moon, planets and their satellites, asteroids, comets, meteoric streams and variable stars. Many phenomena (such as eclipses, occultations, conjunctions and oppositions, etc., ...) are also included. All calculations are made at the Astronomical Institute, still in the 'old' system (i.e., using equinox, IAU1976 precession and IAU1980 nutation). We suppose to follow other 'big' ephemeris books such as The Nautical Almanac, The Astronomical Almanac, Connaissance des Temps, Astronomicheskii Ezhegodnik, etc., ... , when they decide on the new format of published data. In addition, the prediction of lunar occultations is provided also for four Romanian observatories and published in Romanian astronomical yearbook.

Toshio Fukushima
president of the Commission

Transactions IAU, Volume XXVIIA
Reports on Astronomy 2006–2009
Karel A. van der Hucht, ed.

© 2009 International Astronomical Union
doi:10.1017/S1743921308025246

COMMISSION 7

CELESTIAL MECHANICS AND DYNAMICAL ASTRONOMY
MÉCHANIQUE CÉLESTE ET ASTRONOMIE DYNAMIQUE

PRESIDENT   Joseph A. Burns
VICE-PRESIDENT   Zoran Knežević
PAST PRESIDENT   Andrea Milani
SECRETARY   David Vokrouhlický
ORGANIZING COMMITTEE   Evangelia Athanassoula,
                 Christian Beaugé, Bálint Érdi,
                 Anne Lemaitre, Andrzej J. Maciejewski,
                 Renu Malhotra, Alessandro Morbidelli,
                 Stanton J. Peale, Miloš Šidlichovský,
                 Ji-Lin Zhou

TRIENNIAL REPORT 2006–2009

1. Bars and bar-halo interactions in disc galaxies
(Evangelia Athanassoula & Albert Bosma)

The interplay of the disc and the dark halo resonances governs the secular evolution of disc galaxies, and the properties of their bar component (Athanassoula 2002). Martinez-Valpuesta *et al.* (2006), Ceverino & Klypin (2007) and Athanassoula (2007b) confirm and extend this work. Ceverino & Klypin (2007) calculate the orbital frequencies of each particle over the whole temporal evolution, and thus find much broader frequency peaks. In all cases, it is the same resonances that come into play, and, as in Athanassoula 2002, the angular momentum is emitted by near-resonant material in the bar region and absorbed by near-resonant material in the halo and the outer disc. The relative importance of each resonance, however, varies from one case to another. Furthermore, the second and third of the above mentioned studies examine the location of resonant orbits in configuration space and find compatible results.

Bar orbits can become vertically unstable, giving rise to a buckling and a subsequent thickening of part of the bar, which takes on a boxy or peanut shape. This results in a weakening of the bar strength. Martinez-Valpuesta *et al.* (2006) find that, following the initial buckling, the bar resumes its growth from deep inside the corotation radius and follows the ultraharmonic radius thereafter. Athanassoula & Martinez-Valpuesta (2008) find that the strength of the bar and of the peanut correlate and that stronger peanuts form in simulations that have experienced two or more bucklings. Finally, Athanassoula (2008) links the strength and the properties of the bar with those of the halo.

The halo, responding to the bar, forms a bar itself, called the halobar or the dark matter bar. Its properties, noted already in Athanassoula (2005), were studied by Berentzen & Shlosman (2006), Colin *et al.* 2006 and, more extensively, by Athanassoula (2007a). It lags the disc bar by only a few degrees at all radii and the difference between the two bar phases increases with distance from the centre. The two bars turn with roughly the same pattern speed. The length of the halo bar can be estimated by its phase difference with the disc bar and/or from the radius at which the halo shape turns from prolate to oblate. The inner parts of the halo rotate, but considerably less than the disc component.

Angular-momentum transfer within a galaxy could lessen or remove the central cusp in the halo mass distribution. Following previous work, Weinberg & Katz (2007a,b) (WK) and Sellwood (2006, 2008) consider this process using a simple model with a rigid bar in a galaxy with a live halo and no disc. Their results disagree. WK claim that about 10^8 particles are necessary for such simulations, while Sellwood finds converging results for particle numbers above a mere 10^5, and argues that the two studies differ because WK do not take into account resonant broadening due to the time dependence of the perturbation. Both parties agree that a very strong bar is necessary to remove a cusp. Fully self-consistent simulations (e.g., Colin *et al.* 2006) show that, as expected, the inwards concentration of the disc baryonic material during the secular evolution pulls the halo material inwards, thus increasing its central density.

Maciejewski & Athanassoula (2007, 2008) study the orbital structure in double-barred galaxies and show that the parent orbits are double-frequency orbits that do not close in any reference frame, but map onto closed curves called loops. Debattista & Shen (2007) and Shen & Debattista (2007) report on simulations showing that the inner bars pulsate, with their amplitude and pattern speed oscillating as they rotate through the primary bars. Heller *et al.* (2007) show that double bars can form naturally in *ab initio* simulations. The system evolves through successive dynamical couplings and decouplings, forcing gas inwards, and settles in a state of resonant coupling.

Manos & Athanassoula (2008) measure the fraction of chaotic orbits in various barred galaxy models, to see how this depends on parameters such as the bar strength and pattern speed, and outline the regions which are mainly populated by regular/chaotic orbits, guiding future observational studies. Voglis *et al.* (2007) use a combination of two different methods to measure chaos in self-consistent simulations.

Athanassoula & Beaton (2006) compare four fiducial N-body models with M31 observations, in particular isodensity shapes and radial light profiles in the near infrared. They argue that M31 has a sizable bar and constrain its length, strength and properties. The vertically thin part of the bar extends considerably further than the boxy bulge and gives information on the type of orbits that may compose it.

Tiret & Combes (2007a, b) compare bar formation in disc galaxies with dark matter and in disc galaxies following the MOND prescription for gravity. In the latter, bar formation is faster. They also show that gas speeds up the evolution.

Since cosmological simulations predict triaxial, rather than spherical, haloes, recently attention has been given to the evolution of discs embedded in such haloes. Berentzen & Shlosman (2006) (see also Berentzen *et al.* 2006) grow a disc in a triaxial halo, which responds adiabatically and, provided the disc is sufficiently massive, can become axisymmetric. In the *ab initio* simulations of Heller *et al.* (2007) the halo forms triaxial shapes and triggers what could be a first generation of bars. The evolution of the system is largely influenced by chaos introduced by the interaction of multiple nonaxisymmetric components (halo, oval disc, inner and outer bar). The halo does not tumble but its triaxiality evolves with time.

2. Architectures of extrasolar planetary systems (Eric B. Ford)

Radial-velocity observations dominate the observational constraints on the masses and orbital properties of exoplanets (Butler *et al.* 2006). The distribution of masses and orbital elements (Cumming *et al.* 2008) can be compared to parameterized theoretical models (Alibert *et al.* 2005; Armitage 2007; Ida & Lin 2008). Experience continues to demonstrate the value of observers publishing radial velocity data sets, as independent analyses often provide more detailed understanding of the full range of allowed orbital configurations and plausible formation scenarios. Independent analyses have revealed qualitatively different orbital solutions (Gozdziewski *et al.* 2007, 2008a; Gozdziewski & Konacki 2006; Beauge *et al.* 2008; Short *et al.* 2008). Refinements in computational techniques for Bayesian parameter estimation make rigorous Bayesian analyses routine for single planet systems and well-constrained multiple planet systems with negligible planet-planet interactions (Ford 2006; Gregory 2007a). Further research is needed into Bayesian model comparisons for establishing the significance of planet detections and orbital properties (Ford & Gregory 2007; Ford 2008; Gregory 2007b).

Computational limitations still leave the details of planetary migration in a gas disk uncertain, but mass growth and interactions between planets may explain the survival of planetary systems (Chambers 2006; Matsumura *et al.* 2007; Thommes *et al.* 2007; Morbidelli *et al.* 2008). Some models predict that migration will frequently place planets in mean-motion resonances (Kley *et al.* 2005; Beauge *et al.* 2006; Cresswell & Nelson 2008; Mandell *et al.* 2007; Crida *et al.* 2008), while other studies disagree (Quillen 2006; Lee *et al.* 2008; Thommes *et al.* 2008b; Adams *et al.* 2008). This distinction may be valuable in discriminating between migration models.

The eccentricity distribution of exoplanets remains a key observational constraint. Despite progress towards understanding disk- induced eccentricity growth (D'Angelo *et al.* 2006; Ogilvie & Lubow 2006; Cresswell *et al.* 2007; Britsch *et al.* 2008; Moorhead & Adams 2008), it is unclear whether single planets may emerge on eccentric orbits. N-body simulations exploring the late stages of evolution of dynamically active planetary systems suggest that late-stage planet-scattering may explain much of the observed eccentricity distribution, but the abundance of low-eccentricity planets suggests that disks might provide dissipation following the epoch of planet-scattering (Chatterjee *et al.* 2008; Ford & Rasio 2008; Juric & Tremaine 2008) and has inspired simulations of planet scattering in the presence of a disk (Moorhead & Adams 2005; Raymond *et al.* 2006; Kokubo & Ida 2007; Zhou *et al.* 2007; Chatterjee *et al.* 2008; Morishima *et al.* 2008; Thommes *et al.* 2008a). While correlations between eccentricity and other parameters could provide clues regarding planet-formation processes (Ribas & Miralda-Escude 2007; Ford & Rasio 2008), effects of uncertainties in orbital elements (Ford 2006) and measurement biases (Shen & Turner 2008) need to be better understood.

Many planetary systems exhibit significant secular eccentricity evolution (Barnes & Greenberg 2006; Libert & Henrard 2006) which can place constraints on their orbital histories and additional planets (Ford *et al.* 2005; Adams & Laughlin 2006; Sandor *et al.* 2007). Several multiple planet systems appear to be strongly influenced by mean-motion resonances. Assuming long-term dynamical stability (Lee *et al.* 2006; Gozdziewski *et al.* 2007, 2008; Michtchenko *et al.* 2008) implies that several systems are indeed participating in resonances. The secular evolution of these systems raises the possibility of diverse formation scenarios (Sandor & Kley 2006; Sandor *et al.* 2007). The 1:1 mean motion resonance has attracted particular attention due to viable alternative fits to radial velocity data (Gozdziewski *et al.* 2006), but observations already provide strong constraints on Trojans of transitting planets (Ford & Gaudi 2006; Croll *et al.* 2007).

A few planetary systems have inspired detailed studies thanks to complementary observations. *Spitzer* detections of debris disks around planet-host stars constrain formation histories (Alibert *et al.* 2006; Moro-Martin 2007). The combination of radial velocity and transit data constrain planets' physical properties (Torres *et al.* 2008). Spectroscopic measurements during transit constrain the inclination between the stellar spin axis and a planet's orbital angular momentum. Several systems appear to be well aligned (Winn *et al.* 2006; Wolf *et al.* 2006), but a few appear misaligned (Hebrard *et al.* 2008). While disk migration produces well-aligned systems, misaligned systems could arise due to either planet scattering (Chatterjee *et al.* 2008; Nagasawa *et al.* 2008) or interactions with a distant companion (Fabrycky & Tremaine 2007; Wu *et al.* 2007).

Looking forward, several detection techniques are closing in on terrestrial-mass planets. Measuring their frequency and orbital properties will test theoretical predictions for the distribution of masses and orbits of terrestrial planets (Veras & Armitage 2006; Ford *et al.* 2008; Raymond *et al.* 2008; Thommes *et al.* 2008a). Of particular interest for dynamicists, measurements of transit times are already sufficiently precise to detect Earth-mass planets (Agol *et al.* 2005; Holman & Murray 2005). Dynamicists are expected to play a key role in the interpretation of complex transit-timing signatures.

3. Solar System binaries: observations, characterizations, formations
(Daniel J. Scheeres)

The large percentage of binaries among the asteroid and trans-Neptunian populations allows us to infer that small bodies readily form binary or multiple component systems. A detailed review of binary systems appeared recently (Richardson and Walsh 2006), so my short summary will focus on the latest developments related to observation, characterization, and the formation and evolution of these binary systems.

Observations:
The number of observed binary systems in all realms of the solar system has increased to the point where it is feasible to draw statistical inferences about the density, shapes, strength and angular momentum content in these systems. The population of Kuiper Belt Binaries has been reviewed in Noll *et al.* (2008), the small Main Belt and Near-Earth binaries were summarized by Pravec *et al.* (2006, 2007), and the larger Main Belt binaries were described by Marchis *et al.* (2008a, 2008b). Statistics of how many NEA bodies are binaries remain relatively unchanged from Margot (2002), while those of small MB binaries have been rising with increased observations (Pravec *et al.* 2006). A number of multiple-body systems have been discovered in the Kuiper and Main Belts (Weaver *et al.* 2006, Marchis *et al.* 2005) as well as in the NEA population (Nolan *et al.* 2008). In addition to orbital binaries, contact binary bodies, which may represent a population of "failed binaries," have also been observed to be common among observed bodies (Benner *et al.* 2006, Kaasalainen *et al.* 2002, Mann *et al.* 2007, Lacerda *et al.* 2007). The asteroid Itokawa was described as a contact binary in Demura *et al.* (2006) and, if spun to fission, would form a highly unstable but gravitationally bound binary that would have a tendency to re-impact (Scheeres *et al.* 2007). There is also a recently discovered population of 'common origin' asteroids in the Main Belt (Vokrouhlicky and Nesvorny 2008), which have similar orbits today but had a predicted past close passage to one another; they may represent binary systems disrupted either by internal dynamics or by exogenous non-gravitational or gravitational perturbations.

Characterizations:
Densities of the Kuiper Belt population are $\sim$ 2–3 g/cm^3 for larger bodies and $\sim$ 1 g/cm^3 for smaller bodies, indicating compositional trends with size (Noll *et al.* 2008).

Low-density bodies have also been found among the Trojans with Patroclus' density being $\sim 0.8\,\mathrm{g/cm^3}$ (Marchis $et\ al.$ 2006) while those of NEA binaries lie between 1.3 and $2.1\,\mathrm{g/cm^3}$ (Pravec $et\ al.$ 2006). The best characterized NEA binary system to date is 1999 KW4 with detailed shape, spin and orbit information (Ostro $et\ al.$ 2006; Scheeres $et\ al.$ 2006; Fahnestock & Scheeres 2008a). Results include: a determination that the primary spins at its surface disruption limit, meaning that loose particles at the equator are close to or at orbital speeds; a significant density disparity between the two components, with the secondary being 40% denser than the primary; a precise characterization of the proto-typical 'spheroidal' primary shape and 'ellipsoidal' secondary shape which has been a noted feature of NEA binary systems (Pravec $et\ al.$ 2006). These results show that the detailed structure of NEA binaries contain significant inhomogeneities in shape and mass distribution.

Some researchers have used classical Jacobi, Roche & Darwin ellipsoids to understand the likely shapes and, in some cases, densities of bodies (Mann $et\ al.$ 2007, Lacerda $et\ al.$ 2007, Descamps & Marchis 2008). Holsapple and Michel (2008) and Sharma $et\ al.$ (2006) have developed extensions of classical theories that incorporate the effect of friction and strength to model aggregates in an averaged sense. Holsapple's studies show that larger bodies may follow classical results while smaller bodies will be dominated by material strength. Relevant to these studies are results from Pravec & Harris (2007) and Descamps & Marchis (2008) which show that the total angular momentum content of binary systems is consistent with the stability limits of rotating bodies. At the smaller size scales Scheeres (2007, 2008) has studied the mechanics of rigid bodies in contact, finding that they can change relative orientation or fission into binary systems as the rotation rate of the body evolves.

Formation and evolution:
Different formation mechanisms appear to be at play in different regimes of the solar system. For the Kuiper Belt, research has focused on 3-body effects and a variety of detailed mechanisms have been proposed (Goldreich $et\ al.$ 2002; Funato $et\ al.$ 2004; Lee $et\ al.$ 2007; Schlichting & Sari 2008); however a clear consensus of which mechanism dominates is yet to be achieved (Richardson & Walsh 2006). In the Main Belt, theories have generally focused on formation in impact events (Merline $et\ al.$ 2002), although recent speculation for smaller binary systems has focused on spin fission. Impact-created binaries now seem relatively well understood (Durda $et\ al.$ 2004) with observational support for the different classes of binaries formed by such cataclysmic events among the larger binaries in the Main Belt (Richardson & Walsh 2006). Until recently the favored formation mechanism for NEA binaries was via tidal flybys (Walsh & Richardson 2006). However a statistical analysis by Scheeres $et\ al.$ (2004) indicated that the disruption rate of asteroids due to close planetary flybys was inconsistent with the observed statistics and detailed simulations of binary creation and destruction by Walsh and Richardson (2008) could not produce the observed binary statistics of the NEA binary population. With the validation of the YORP effect (Lowry $et\ al.$ 2007; Taylor $et\ al.$ 2007; Kaasalainen $et\ al.$ 2007), attention has focused on fission from YORP spin-up as a mechanism for creating binaries and controlling some aspects of the shapes of asteroids in general (Scheeres 2007; Pravec & Harris 2007; Walsh $et\ al.$ 2008). This mechanism should also extend into the Main Belt and can explain some of the smaller binary systems there (Pravec $et\ al.$ 2006).

Evolution of binary systems provides life-time limits and constrains the production rate of these systems. Classical evolutionary mechanisms that focus on angular momentum transfer via tidal friction are still the probable mechanism for larger bodies, although great uncertainty remains in the appropriate empirical constants controlling this mechanism (Margot *et al.* 2002; Goldreich & Sari 2007). Among the small NEA and Main Belt binary populations, new mechanisms for evolution involving non-gravitational effects have been proposed. The Binary YORP (BYORP) effect (Cuk & Burns 2005) consists of non-gravitational forces acting on a binary secondary, causing the orbit to expand or contract with binary lifetimes predicted to be as short as 10^5 years (Cuk 2007), orders of magnitude faster than tidal evolution. More recently hypothesized is YORP induced expansion, introduced as an explanation for why the 1999 KW4 primary was at its spin disruption limit (Scheeres *et al.* 2006), where continued YORP spin-up of the primary can cause material to be lofted and angular momentum to be transferred to the orbit. Expansion times for this effect may be an order of magnitude faster than tidal evolution (Fahnestock & Scheeres 2008b; Harris, *et al.* 2008). As neither of these evolutionary mechanisms has been validated as of yet for small binaries, additional research and observations are needed. For any of these effects acting on NEA, close planetary flybys continue to remain an important contributor, essentially randomizing these more systematic effects (Walsh & Richardson 2008). However, MB binaries should be shielded from these randomizing effects and thus may represent a cleaner picture of binary evolution.

4. Spin-obit dynamics (Anne Lemaitre)

The field of spin-orbit dynamics has been renewed recently through theoretical, analytical and observational papers, owing to several space missions, and increases in the precision required for all the motions. The bodies of any mass can no longer be considered as point masses and the explanation of their present state needs very long-term integrations, including migrations and disk interactions, as well as the linking of orbital and rotational contributions. Recent contributions concern of course Mercury, but also the giant planets, the Moon, the Galilean and Saturnian satellites and asteroids.

The gravity field coefficients J_2 and C_{22} of Mercury are known with an uncertainty of 30% and 50%. The two space missions scheduled to Mercury (*Messenger* and *Bepi-Colombo*) should lead to a better knowledge of the gravitational potential, and through a precise observation of the rotation, of the interior of the planet (like the existence of a liquid core).

First, the probability for Mercury to be captured in a 3:2 spin-orbit resonance was calculated to be 52% by Correia & Laskar (2004), who explored a large variety of scenarios and initial conditions, and introduced planetary perturbations of the orbital and rotational motions.

Second, Mercury should theoretically be in a Cassini state, which means geometrically that its orbit pole, spin vector and normal to the inertial planes are coplanar; the dynamical interpretation is that all the free (also called proper) librations, depending on the initial conditions of the rotation, should have disappeared with time. Peale (2005) showed that core-mantle friction or tidal effects would drive the spin to its equilibrium value in less than 1 My and that even the excitation due to a potential impact should have been completely damped in less than 1 yr. Analytical rigid-body models (D'Hoedt & Lemaitre 2004) calculated the values of the three free frequencies of Mercury's rotational motion (of the libration in longitude, i.e., the resonant spin-orbit angle, of the node commensurability and of the wobble) as functions of C_{22}, and C (the third moment of inertia) in a rigid body, or C_m (the third moment of inertia of the mantle) for a liquid-core hypothesis.

The Cassini state is presently characterized by a small obliquity of about 2 arcminutes, mainly due to the precession of the orbital node and dependent on the present values of the eccentricity and the inclination. Numerical investigations, performed by Yseboodt & Margot (2006) and including planetary secular perturbations, have shown that the planet should follow the Cassini state, even if its position is time-evolving ; this behavior can also be explained by an adiabatic model developed by Peale (2006) and generalized to two degrees of freedom by D'Hoedt & Lemaitre (2008).

Following theoretical approaches, the short periodic orbital terms (88 days) induce forced oscillations on the libration in longitude, with amplitudes proportional to C_{22}/C_m. Observations of radar speckle patterns by Margot & et al. (2007) established that the planet occupied a Cassini state with an obliquity of 2.11 arcminutes and that the largest amplitudes of the oscillations in longitude were of 35.8 arcseconds; this value, even with the uncertainty of the value of C_{22} clearly shows a difference between C_m and C, which means that the mantle and the core are decoupled, and confirms the existence of a molten core.

The next step was to include planetary perturbations on Mercury's orbital motion (in addition to the 88 days forced libration). The contributions of Peale, Yseboodt & Margot (2007) and of Dufey, Lemaitre & Rambaux (2008) agree for the main terms (due to Venus and Jupiter orbital motions) but show divergences for the smaller contributions, due to the Earth or Saturn, for example. An interesting resonance, between the spin-orbit resonant frequency and the orbital motion of Jupiter, was mentioned in both papers.

The first flyby of Mercury by the probe Messenger in January 2008 (Solomon et al. 2008) should give much more precise values of J_2 and C_{22}, allowing a check on the different scenarios and hypotheses.

Saturn's obliquity (26.73°) also received new attention; the reason for such a high value (Jupiter's obliquity is only 3.12°) is not only explained by the obliquity of the core. A new mechanism was proposed by Ward & Hamilton (2004), based on a 1:1 commensurability between Saturn's spin-axis precession rate and the ν_{18} secular frequency of Neptune's orbital node, confirmed numerically by Hamilton & Ward (2004). The authors discussed different scenarios of circumplanetary disk dispersal, showing, for all the giant planets, how the sweeping of secular resonances could be followed by adiabatic tools and result in the increase of several degrees of the initial small obliquities. The lunar spin-axis dynamics was also revisited. Wisdom (2006) showed how the introduction of a non-constant orbital inclination and a non-uniform regression of the orbital node complicated the dynamics, by the introduction of new resonances and the apparition of local chaotic regions.

Following the models developed in the seventies to understand the Moon's libration, Henrard (2005a) proposed a three-dimensional rigid model of rotation for the Galilean satellites Europa and Io (Henrard 2005c), and estimated the values of the so-called free (or proper) frequencies. Thanks to the injection of the synthetic orbital theory of Layney, Arlot & Vienne (2004) in the model, Henrard (2005b) computed the main forced contributions induced by the other satellites' gravitational effects. The case of Io is particularly complex, due to its proximity to Jupiter and its capture in the Laplacian resonance with Europa and Ganymede. Its rotational behavior is far from a uniform oscillation around a Cassini equilibrium. Moreover Io's volcanic activity shows that a rigid-body model is surely not adequate. The last contribution of Henrard (2008) was to develop an analytical theory (based on the idea of Poincare 1910) to take into account the presence of a liquid core contained in a cavity filled by a non-viscid fluid of constant density; the presence of this supplementary degree of freedom (the spin of the core) in the problem, introducing possible new commensurabilities between the frequencies, may lead to chaotic motion in the vicinity of the Cassini state, for some values of the unknown core parameters.

The space mission *Cassini-Huygens* provided new data for Titan, showing in particular its slight super-synchronous rotation (+0.004%). Observers (Stiles *et al.* 2008 and Lorenz *et al.* (2008) interpreted it as a potential signature of the presence of an internal ocean, that would dissociate the rotation of Titan's crust from this of the core. Using Henrard's model, Noyelles, Lemaitre & Vienne (2008) calculated analytically the three proper frequencies of Titan's rotation and the main forced terms; in a second paper Noyelles (2008) showed that a dynamical forcing of the wobble frequency could have affected the measurement of Titan's spin rate, and could also be at the origin of the non-exact synchronism.

The minor planet dynamics, on very long periods of time, is strongly dependent on the thermal (Yarkovski) forces and their spin-vector evolution is influenced by the YORP torque, in particular, as shown by Nesvorný, & Vokrouhlický (2007), in the chronology and understanding of the asteroid families – like Erigone, Massalia, Merxia or Astrid – which likely results from catastrophic collisions, as identified in backward integrations.

Joseph A. Burns
president of the Commission

References

Section 1
Athanassoula, E. 2002, *ApJ*, 104, 340
Athanassoula, E. 2005, *Cel. Mech. & Dyn. Astr.*, 91, 9
Athanassoula, E. 2007a, *MNRAS*, 377, 1569
Athanassoula, E. 2007b, in: R. S. de Jong (ed.) *Island Universes, Astrophysics and Space Science Proceedings* (Dordrecht: Springer), p. 195
Athanassoula, E. 2008, in: M. Bureau, E. Athanassoula, & B. Barbuy (eds.) *Galactic Bulges, IAU Symposium No. 245* (Cambridge: CUP), p. 93
Athanassoula, E. & Beaton, R. L. 2006, *MNRAS*, 370, 1499
Athanassoula, E. & Martinez-Valpuesta, I. 2008, *MNRAS* (submitted)
Berentzen, I. & Shlosman, I. 2006a, *ApJ*, 648, 807
Berentzen, I. & Shlosman, I. 2006b, *ApJ*, 637, 582
Ceverino, D. & Klypin, A. 2007, *MNRAS*, 379, 1155
Colin, P., Valenzuela, O., & Klypin, A. 2006, *ApJ*, 644, 687
Debattista, V. P. & Shen, J. 2007, *ApJ*, 654, L127
Heller, C., Shlosman, I., & Athanassoula, E. 2007, *ApJ*, 657, L65 and *ApJ*, 671, 226
Maciejewski, W. & Athanassoula, E. 2007, *MNRAS*, 380, 999, and 2008, *MNRAS*, 389, 545
Manos, T. & Athanassoula, E. 2008, *MNRAS*, 2008arXiv0806.3563B
Martinez-Valpuesta, I., Shlosman, I., & Heller, C. 2006, *ApJ*, 637, 214
Sellwood, J. A. 2006, *ApJ* 637, 567 and 2008, *ApJ*, 639, 868
Shen, J., & Debattista, V. P. 2007, *ApJ*, 690, 758
Tiret, O. & Combes, F. 2007a, *A&A*, 464, 517 and 2007b, *A&A*, 483, 719
Voglis, N., Harsoula, M., & Contopoulos, G. 2007, *MNRAS*, 381, 575
Weinberg, M. & Katz, N. 2007, *MNRAS*, 375, 425 & 460

Section 2
Adams, F. C., Laughlin, G., & Bloch, A. M. 2008, *ApJ* 683, 1117
Agol, E., Steffen, J., Sari, R., & Clarkson, W. 2005, *MNRAS*, 359, 567
Alibert, Y., Mordasini, C., Benz, W., & Winisdoerffer, C. 2005, *A&A*, 434, 343
Alibert, Y., *et al.* 2006, *A&A*, 455, L25

Armitage, P. J. 2007, *ApJ*, 665, 1381

Barnes, R. & Greenberg, R. 2006, *ApJ*, 652, L53

Beaugé, C., Giuppone, C. A., Ferraz-Mello, S., & Michtchenko, T. A. 2008, *MNRAS*, 385, 2151

Beaugé, C., Michtchenko, T. A., & Ferraz-Mello, S. 2006, *MNRAS*, 365, 1160

Britsch, M., Clarke, C. J., & Lodato, G. 2008, *MNRAS*, 385, 1067

Butler, R. P., *et al.* 2006, *ApJ*, 646, 505

Chambers, J. E. 2006, *ApJ*, 652, L133

Chatterjee, S., Ford, E. B., Matsumura, S., & Rasio, F. A. 2008, *ApJ*, 686, 580

Cresswell, P., Dirksen, G., Kley, W., & Nelson, R. P. 2007, *A&A*, 473, 329

Cresswell, P. & Nelson, R. P. 2008, *A&A*, 482, 677

Crida, A., Sándor, Z., & Kley, W. 2008, *A&A*, 483, 325

Croll, B., *et al.* 2007, *ApJ*, 658, 1328

Cumming, A., Butler, R. P., Marcy, G. W., Vogt, S. S., Wright, J. T., & Fischer, D. A. 2008, *PASP*, 120, 531

D'Angelo, G., Lubow, S. H., & Bate, M. R. 2006, *ApJ*, 652, 1698

Fabrycky, D. & Tremaine, S. 2007, *ApJ*, 669, 1298

Ford, E. B. 2006, *ApJ*, 642, 505

Ford, E. B. 2008, *AJ*, 135, 1008

Ford, E. B. & Gaudi, B. S. 2006, *ApJ*, 652, L137

Ford, E. B. & Gregory, P. C. 2007, *ASPC*, 371, 189

Ford, E. B. & Rasio, F. A. 2006, *ApJ*, 638, L45

Ford, E. B. & Rasio, F. A. 2008, *ApJ*, 686, 621

Ford, E. B., Lystad, V., & Rasio, F. A. 2005, *Nature*, 434, 873

Ford, E. B., Quinn, S. N., & Veras, D. 2008, *ApJ*, 678, 1407

Goździewski, K. & Konacki, M. 2006, *ApJ*, 647, 573

Goździewski, K., Maciejewski, A. J., & Migaszewski, C. 2007, *ApJ*, 657, 546

Goździewski, K., Migaszewski, C., & Konacki, M. 2008b, *MNRAS*, 385, 957

Gregory, P. C. 2007b, *MNRAS*, 374, 1321

Gregory, P. C. 2007a, *MNRAS*, 381, 1607

Hébrard, G., Bouchy, F., Pont, F., *et al.* 2008, *A&A*, 488, 763

Holman, M. J. & Murray, N. W. 2005, *Sci*, 307, 1288

Ida, S. & Lin, D. N. C. 2008a, *ApJ*, 673, 487

Juric, M. & Tremaine, S. 2008, *ApJ*, 686, 603

Kley, W., Lee, M. H., Murray, N., & Peale, S. J. 2005, *A&A*, 437, 727

Kokubo, E. & Ida, S. 2007, *ApJ*, 671, 2082

Lee, A. T., Thommes, E. W., & Rasio, F. A. 2008, *arXiv* 801, 1926

Lee, M. H., Butler, R. P., Fischer, D. A., Marcy, G. W., & Vogt, S. S. 2006, *ApJ*, 641, 1178

Libert, A.-S. & Henrard, J. 2006, *Icarus* 183, 186

Mandell, A. M., Raymond, S. N., & Sigurdsson, S. 2007, *ApJ*, 660, 823

Matsumura, S., Pudritz, R. E., & Thommes, E. W. 2007, *ApJ*, 660, 1609

Michtchenko, T. A., Beaugé, C., & Ferraz-Mello, S. 2008, *MNRAS*, 387, 747

Moorhead, A. V. & Adams, F. C. 2008, *Icarus*, 193, 475

Moorhead, A. V. & Adams, F. C. 2005, *Icarus*, 178, 517

Morbidelli, A., Crida, A., Masset, F., & Nelson, R. P. 2008, *A&A*, 478, 929

Morishima, R., Schmidt, M. W., Stadel, J., & Moore, B. 2008, *ApJ*, 685, 1247

Moro-Martín, A., *et al.* 2007, *ApJ*, 668, 1165

Nagasawa, M., Ida, S., & Bessho, T. 2008, *ApJ*, 678, 498

Ogilvie, G. I. & Lubow, S. H. 2006, *MNRAS*, 370, 784

Quillen, A. C. 2006, *MNRAS*, 365, 1367

Raymond, S. N., Barnes, R., & Mandell, A. M. 2008, *MNRAS*, 384, 663

Raymond, S. N., Quinn, T., & Lunine, J. I. 2006, *Icarus*, 183, 265

Ribas, I. & Miralda-Escudé, J. 2007, *A&A*, 464, 779

Sándor, Z., & Kley, W. 2006, *A&A*, 451, L31

Sándor, Z., Kley, W., & Klagyivik, P. 2007, *A&A*, 472, 981

Shen, Y. & Turner, E. L. 2008, *ApJ*, 685, 553

Short, D., Windmiller, G., & Orosz, J. A. 2008, *MNRAS*, 386, L43

Thommes, E., Nagasawa, M., & Lin, D. N. C. 2008a, *ApJ*, 676, 728

Thommes, E. W., Bryden, G., Wu, Y., & Rasio, F. A. 2008b, *ApJ*, 675, 1538

Thommes, E. W., Nilsson, L., & Murray, N. 2007, *ApJ*, 656, L25

Torres, G., Winn, J. N., & Holman, M. J. 2008, *ApJ*, 677, 1324

Veras, D. & Armitage, P. J. 2006, *ApJ*, 645, 1509

Winn, J. N., *et al.* 2006, *ApJ*, 653, L69

Wolf, A. S., Laughlin, G., Henry, G. W., Fischer, D. A., Marcy, G., Butler, P., & Vogt, S. 2007, *ApJ*, 667, 549

Wu, Y., Murray, N. W., & Ramsahai, J. M. 2007, *ApJ*, 670, 820

Zhou, J.-L., Lin, D. N. C., & Sun, Y.-S. 2007, *ApJ*, 666, 423

Section 3

Benner, L. A. M., Nolan, M. C., Ostro, S. J., *et al.* 2006, *Icarus*, 182, 474

Cuk, M. 2007, *ApJ*, 659, L57

Cuk, M. & Burns, J. A. 2005, *Icarus*, 176, 418

Demura, H., Kobayashi, S., Nemoto, E., *et al.* 2006, *Science*, 312, 1347

Descamps, P. & Marchis, F. 2008, *Icarus*, 193, 74

Durda, D. D., Bottke, W. F., Enke, B. L., *et al.* 2004, *Icarus*, 170, 243

Fahnestock, E. G. & Scheeres, D. J. 2008a, *Icarus*, 194, 410

Fahnestock, E. G. & Scheeres, D. J. 2008b *BAAS*, 40, #2.02

Funato, Y., Makino, J., Hut, P., *et al.* 2004, *Nature*, 427, 518

Goldreich, P., Lithwick, Y., & Sari, R. 2002, *Nature*, 420, 643

Goldreich, P. & Sari, R. 2007, *arXiv* 0712.0446

Harris, A. W., Fahnestock, E. G., & Pravec, P. 2008, *BAAS-DDA*, 40, #14.01

Holsapple, K. A. & Michel, P. 2008, *Icarus*, 193, 283

Kaasalainen, M., Durech, J., Warner, B. D., *et al.* 2007, *Nature*, 446, 420

Lacerda, P. & Jewitt, D. C. 2007, *AJ*, 133, 1393

Lee, E. A., Astakhov, S. A., & Farrelly, D. 2007, *MNRAS*, 379, 229

Lowry, S. C., Fitzsimmons, A., Pravec, P., *et al.* 2007, *Science*, 316, 272

Mann, R. K., Jewitt, D., & Lacerda, P. 2007, *AJ*, 134, 1133

Marchis, F., Descamps, P., Hestroffer, D., & Berthier, J. 2005, *Nature*, 436, 822

Marchis, F., Hestroffer, D., Descamps, P., *et al.* 2006, *Nature*, 439, 565

Marchis, F., Descamps, P., Baek, M., *et al.* 2008a, *Icarus*, 196, 97

Marchis, F., Descamps, P., Berthier, J., *et al.* 2008b, *Icarus*, 195, 295

Margot, J. L., Nolan, M. C., Benner, L. A. M., *et al.* 2002, *Science*, 296, 1445

Merline, W. J., Weidenschilling, S. J., Durda, D. D., *et al.* 2002, *in Asteroids III*, W. F. Bottke Jr. *et al.* (eds.), U. Arizona Press, Tucson, 289

Nolan, M. C., Howell E. S., Benner, L. A. M., *et al.* 2008, IAUC 8921 '(153591) 2001 SN 263' (Feb 2008)

Noll, K. S., Grundy, W. M., Chiang, E. I., *et al.* 2008 *in The Solar System Beyond Neptune*, M. A. Barucci *et al.* (eds.), U. Arizona Press, Tucson, 345

Ostro, S. J., Margot, J.-L., Benner, L. A. M., *et al.* 2006, *Science*, 314, 1276

Pravec, P. & Harris, A. W. 2007, *Icarus*, 190, 250

Pravec, P., Scheirich, P., Kusnirak, P., *et al.* 2006, *Icarus*, 181, 63

Richardson, D. C. & Walsh, K. J. 2006, *Ann. Rev. Earth & Planet. Sci.*, 34, 47

Scheeres, D. J. 2007, *Icarus*, 189, 370

Scheeres, D. J. 2008, *PSS*, in press

Scheeres, D. J., Marzari, F., & Rossi, A. 2004, *Icarus*, 170, 312

Scheeres, D. J., Fahnestock, E. G., Ostro, S. J., *et al.* 2006, *Science*, 314, 1280

Scheeres, D. J., Abe, M, Yoshikawa, M., *et al.* 2007, *Icarus* 188, 425

Schlichting, H. E. & Sari, R. 2008, *ApJ*, 673, 1218

Sharma, I., Jenkins, J. T. & Burns, J. A. 2006, *Icarus*, 183, 312

Taylor, P. A., Margot, J.-L., Vokrouhlicky, D., *et al.* 2007, *Science*, 316, 274

Vokrouhlicky, D. & Nesvorny, D. 2008, *AJ*, 136, 280
Walsh, K. J. & Richardson, D. C. 2006, *Icarus*, 180, 201
Walsh, K. J. & Richardson, D. C. 2008, *Icarus*, 193, 553
Walsh, K. J., Richardson, D. C., & Michel, P. 2008, *Nature*, 454, 188
Weaver, H. A., Stern, S. A., Mutchler, M. J., *et al.* 2006, *Nature*, 439, 943

Section 4
Correia, A. & Laskar, J. 2004, *Nature*, 429, 848
D'Hoedt, S. & Lemaitre, A. 2004, *CM&DA*, 89, 267
D'Hoedt, S. & Lemaitre, A. 2008, *CM&DA*, 101,127
Dufey, J., Lemaitre, A., & Rambaux, N. 2008, *CM&DA*, 101,141
Hamilton, D. P. & Ward, W. R. 2004, *AJ*, 128, 2510
Henrard, J. 2005a, *CM&DA*, 91,131
Henrard, J. 2005b, *CM&DA*, 93,101
Henrard, J. 2005c, *Icarus*, 178,144
Henrard, J. 2008, *CM&DA*, 101, 1
Layney, V., Arlot, J. E., & Vienne 2004, *A&A*, 427, 371
Lorenz, R. D., Stiles, B. W., Kirk, R. L. *et al.* 2008, *Science*, 319, 1649
Margot, J. L., Peale, S. J., Jurgens, R. F., Slade, M. A., & Holin I.V. 2007, *Science*, 316, 710
Nesvorný, D. & Vokrouhlický, D. 2007, *AJ*, 134, 1750
Noyelles, B. 2008, *CM&DA*, 101, 13
Noyelles, B., Lemaitre, A., & Vienne, A. 2008, *A&A*, 478, 959
Peale, S. J. 2005, *Icarus*, 178, 4
Peale, S. J. 2006, *Icarus*, 181, 338
Peale, S. J., Yseboodt, M., & Margot, J. L. 2007, *Icarus*, 187, 365
Poincaré, H. 1910, *Bull. Astron.*, 27, 321
Solomon, S. C., McNutt, R. L., Watters, T. R., *et al.* 2008, *Science*, 321, 59
Stiles, B. W., Kirk, R. L., Lorenz, R. D., *et al.* 2008, *AJ*, 135, 1669
Ward, W. R. & Hamilton, D. P. 2004, *AJ*, 128, 2501
Wisdom, J. 2006, *AJ*, 131, 1864
Yseboodt, M. & Margot, J. L. 2006, *Icarus*, 181, 327

Transactions IAU, Volume XXVIIA
Reports on Astronomy 2006–2009
Karel A. van der Hucht, ed.

© 2009 International Astronomical Union
doi:10.1017/S1743921308025258

COMMISSION 8

ASTROMETRY
ASTROMÉTRIE

PRESIDENT   Irina I. Kumkova
VICE-PRESIDENT   Dafydd Wyn Evans
PAST PRESIDENT   Imants Platais
ORGANIZING COMMITTEE   Alexandre H. Andrei, Alain Fresneau, Petre P. Popescu, Ralf-Dieter Scholz, Mitsuru Soma, Norbert Zacharias, Zi Zhu

COMMISSION 8 WORKING GROUPS

Div. I / Commission 8 WG   Densification of the Optical Reference Frame

TRIENNIAL REPORT 2006–2009

1. Scientific highlights

Van Leeuwen has completed and published the new reduction of the *Hipparcos* data. Parallax accuracies have improved by up to a factor five for the brightest stars and correlations effectively removed.

Gaia is now fully funded and is well into the development and construction phase. This mission will revolutionize astrometry in much the same way that *Hipparcos* did. The launch is expected in late 2011.

IAU Symposium No. 248 'A Giant Step: From Milli- To Micro-Arcsecond Astrometry' was very successfully held in Shanghai, China PR, October 2007.

2. Instrumentation and reduction methods

BRAZIL: At Observatório Nacional, with FINEP funds, a state of the art CCD heliometer is being developed aiming to the continuation of the solar diameter measurements – first light was reached in 2008 and regular operation is due in 2009 (Andrei *et al.* 2006). Assafin and colleagues (2007) developed PRAIA, a fully automatized software for the astrometric and photometric treatment of CCD images. The Observatório Nacional, from 2007, leases 90 nights/yr at the ESO 2.2m Max Planck telescope enabling several long term projects. Specific projects are being pursued at the SOAR telescope, of which Brazil is co-proprietary.

CHINA PR: Wenjing Jin (Shanghai) reports that in order to reduce the cutoff height angle and improve the correction accuracy rigorous calculating formulae of the path-curving correction in the refraction delay of electromagnetic wave was derived. (Mao W. *et al.* 2008)

The observational astronomical refraction model toward the east, south, west, and north of the Yunnan Observatory was determined with a Lower Latitude Meridian Circle ($D = 40$ cm, $f = 5$ m). (Mao W. *et al.* 2007).

Chunlin Lu (Nanjing) reports that from October of 2006, the 1.0/1.2 m *NEOST* (Near Earth Objects Space Telescope) equipped with a 4096 × 4096 SI CCD detector was installed completely and began test observations. Due to fast optics and the high quantum efficiency of the CCD detector, the observational system can reach m = 22.46 in *B*-band with only a 40 s exposure, which enables an asteroid survey with great efficiency. (Zhao *et al.* 2008).

A method to improve binary orbits with a long-period is presented by using Hipparcos data and long-term data from ground-based observations. (Ren & Fu 2007).

Additionally, simulations have shown that the reliability of orbits derived from *Gaia* data is high only for binaries with periods < 8 yr otherwise the orbit should be derived from *Gaia* data and long-term ground-based data (Ren & Fu 2008).

Zhenghong Tang (Shanghai) reports a method of accurate calibration for LAMOST (Large Sky Area Multi-object Fiber Spectroscope Telescope). LAMOST is a reflecting Schmidt telescope, which will perform multi-object fiber spectroscopic observations. Four CCDs are installed at the plate edge to perform the guiding and calibration. The prototype of LAMOST (300 fibers) was adjusted in May 2007. The alignment accuracy is better than 0″.5 and the spectral lines are obtained from 97% fibers (Li *et al.* 2006, Yu *et al.* 2008). After a CCD camera (Model S1C-077) in drift scan mode was installed on the 25cm telescope of Shanghai Astronomical Observatory (SHAO), observations for GSS (Geo-Stationary Satellite) were carried out and internal positional precision is better than 0″.5 (Mao *et al.* 2007). Experiments with a rotating drift-scan mode was carried out in collaboration with Nikolaev Astronomical Observatory (NAO), Ukraine.

FRANCE: Arlot reports about a collaboration between IMCCE of Paris Observatory, USNO (Pascu), RBO (De Cuyper) for a new reduction of astrometric plates of the natural satellites of Jupiter, Saturn and Mars. Souchay reports about the reduction of mosaic CCD images obtained at the 3.6 m CFHT in order to derive coordinates of pulsars and QSOs in the optical at the mas level.

RUSSIA: Devyatkin (Pulkovo) reports the automatic telescope MTM-500M has been modernized for observations of Solar System bodies and is in operation at the Northern Caucasus station of the GAO RAS.

SPAIN: A new CCD camera with a Sloan *r'* bandpass filter has been bought for CMASF, to be delivered in June 2008. The CMASF will be operative again in autumn 2008. A collaboration between ROA, the Fabra Observatory and Barcelona University has undertaken the robotization of an old Baker Nunn camera. First light will take place in the second half of 2008. After testing the camera performace at ROA, it will be moved to the Montsec Observatory, northwest Spain.

UKRAINE: Shul'ga (Nikolaev) reports about the Fast Robotic Telescope of RI NAO for observations of NEAs, artificial satellites, and space debris on all orbits. Filonenko (Kharkov) reports that the DSLR Canon 300D camera has been installed at the 20/190 cm AZT-7 telescope for observations of small Solar System bodies up to 15 m. Kleshchonok (Kiev) reports that the automatic complex UNIT for fast photometry using a CCD camera Rolera MGI and telescope Celestron CGE 1400 has been developed for observation of stellar occultations by the Moon.

UK: The new reduction of the *Hipparcos* data was completed and presented as part of a book describing the new reduction and the proper use of the data in scientific applications (van Leeuwen 2007). Parallax accuracies have been improved by up to a factor 5 for stars near magnitude 3 to 4, and correlations in the underlying data have been reduced by more than a factor 10 to an insignificant level. This has greatly benefited the study of parallaxes and proper motions of open clusters, while for Cepheids the number of stars that can be incorporated in a parallax-based PL study has increased by a factor 4 (van Leeuwen, Feast, *et al.* 2007).

USA: The StarScan plate measure machine (USNO) completed measuring all applicable Black Birch, AGK2, and Hamburg Zone astrograph plates (Zacharias *et al.* 2008).

3. Space astrometry

Lindegren (Lund) reports on the European Space Agency's astrometry mission *Gaia*. This was proposed in 1994, included in the ESA Science Program in 2000, confirmed by the ESA Science Program Committee in 2006, and is now well into the development and construction phase with a target launch date in late 2011. The scientific case for *Gaia* rests on its ability to obtain accurate astrometric measurements for very large and complete (flux-limited) samples of stellar, extragalactic, and solar-system objects, and the matching collection of synoptic, multi-epoch photometric, and radial-velocity data. The scientific objectives include an extremely broad range of topics in Galactic and stellar astrophysics, solar-system astronomy, reference frame, and fundamental physics, as exemplified in the Proceedings of the symposium *The Three-Dimensional Universe with Gaia* (Turon *et al.* 2005). During its operational phase of five years, the *Gaia* mission will provide an all-sky optical astrometric and photometric survey completely to 20th magnitude. The scanning and measurement principle is similar to *Hipparcos*, but vastly more efficient and accurate thanks to the use of a large array of CCDs and bigger telescope (up to 1.4 m aperture). The expected sky-averaged parallax accuracy is 8 - $25\,\mu$ As for the brighter stars (<15 mag) and $\sim 300\,\mu$ as at 20 mag, with similar accuracies for the positions at mid-epoch (~ 2015) and the annual proper motion components. The photometric survey uses two slitless prism spectrographs with a resolution of $R \simeq 10$ - 30 in two wavelength bands, 330 - 680 nm and 640 - 1050 nm. The photometric accuracy is difficult to quantify but the instrument has been designed with a view to enable determination of relevant astrophysical parameters (T_{eff}, $\log g$, [Fe/H]) for a broad range of stars. Radial velocities at the 1 - 5 km s^{-1} accuracy level will be measured for the brighter stars using a slitless spectrograph operating in the Ca II triplet region (847 - 874 nm, $R = 11500$). Thus, for many millions of stars, *Gaia* is expected to provide full phase-space information, accurate individual distances, and complementary astrophysical data together with a systematic analysis of stellar variability. The satellite, including the scientific instruments, launch and mission operations, are fully funded by ESA. The prime industrial contractor for building the satellite and instruments is EADS Astrium. The production of the *Gaia* Catalogue is however the task of the scientific community and will be achieved by the *Gaia* Data Processing and Analysis Consortium (DPAC). The ESA web site <www.rssd.esa.int/Gaia> maintains up-to-date information about the status of the mission, news items, information sheets, meetings, etc.

BRAZIL: Andrei and Colleagues from Observatorio Nacional, Observatorio do Valongo, SYRTE/Observatoire de Paris, Observatoire de Bordeaux, and Observatorio do Porto develop the Initial Quasar Catalogue for the *Gaia* mission (Souchay *et al.* 2007).

JAPAN: Mitsuru Soma reported about the *JASMINE* and Nano-JASMINE projects. JASMINE is an astrometric mission that observes in an infrared Kw band (2.0 microns) (Gouda 2007, 2008; Yano 2008; Yamada 2008). This project is in development, with a target launch date around 2015 - 2020. It is designed to perform a survey towards the Galactic bulge with a single-beam telescope, determining positions and parallaxes accurate to 10 μas and proper motions to 10 μas/yr for stars brighter than Kw = 11 mag. *JASMINE* will observe about a few ten million stars brighter than the limiting magnitude

(Kw = 14 mag). *JASMINE* development proceeds mainly by staff at NAOJ, in collaboration with Japanese universities and engineers at JAXA. It should be noted that *JASMINE* has been identified as particularly significant for astrometry by the SOC of IAU Symposium No 248. *Nano-JASMINE* is planned to demonstrate the first space astrometry of Japan and to perform experiments for verification of some techniques and operations of *JASMINE* (Kobayashi 2008; Suganuma 2008; Yamauchi 2008). *Nano-JASMINE* has a size and weight of about $50 \, \mathrm{cm}^3$ and $20 \, \mathrm{kg}$. The telescope has a $5 \, \mathrm{cm}$ diameter with a focal length of about $1.7 \, \mathrm{m}$. The detector will be a $1\mathrm{k} \times 1\mathrm{k}$ CCD (z-band). A candidate orbit for *Nano-JASMINE* is a sun-synchronous orbit. The accuracy will be about $3 \, \mathrm{mas}$ at $z = 7.5$ mag for parallaxes and for proper motions $0.1 \, \mathrm{mas/yr}$ by combining the results with *Hipparcos* data. *Nano-JASMINE* will probably be launched by Cyclone-4 rocket in July 2010. The development is by NAOJ and Kyoto University in strong collaboration with Prof. Nakasuka and his group (Tokyo).

RUSSIA: Kiselev from Pulkovo reports on a new observational programme of visual-binary and multiple stars proposed for *Gaia* observations. The stars with a history of 20 - 40 years were chosen from the Pulkovo List (Kiselev 2007).

USA: Benedict *et al.* have used the *HST* Fine Guidance Sensors to carry out mas precision astrometry to determine parallaxes of Galactic Cepheids (Benedict 2007, AJ, 133, 1810) and AM CVn binary stars (Roelofs 2007, ApJ, 666, 1174). *HST*-FGS data was obtained to establish perturbation orbits due to planetary-mass companions to nearby stars like ϵ Eri (Benedict 2006, AJ, 132, 2206), the M-dwarf companion to HD 33636 (Bean 2007, AJ, 134, 749) v And and others (Benedict *et al.* 2008, 2009). *HST*-FGS observations continue for RR Lyr and Pop II Cepheid stars to calibrate a Pop II Period-Luminosity relationship. Once *HST* has its last repairs, Bendict *et al.* will measure the parallaxes of nine metal-poor stars to establish a Pop II main sequence and address distances and ages of Galactic globular clusters.

4. Reference frames

BRAZIL: Work on the extragalactic frame concentrates on the reconciliation between optical and radio positions (Camargo *et al.* at the SOAR, NTT, ESO 2.2 m), and, also, on the astrometric bearing of astrophysical quantities of quasars, as the morphology, the variability, the redshift, and color locus, as developed in theses by Melo and Antunes Filho.

CHINA PR: Wenjing Jin reports that a list of 173 candidate stable sources selected from ICRF-ext2 was proposed after comparing the solutions and performing statistical analysis. (Qiao *et al.* 2008). The precession and equinox motion correction were obtained from various samples of PPM and ACRS proper motion data. The results show that obvious systematic difference 1.5 mas/yr was derived from two catalogues mainly due to the internal systematic error of the FK5 proper motion (Zhu 2007).

FRANCE: Souchay reports at `<ftp://syrte.obspm.fr/pub/LQAC/LQAC_2008.ascii>` about the Large Quasar Astrometric Catalogue used to monitor the ICRS and maintain the ICRF. The increased number of observations of southern ICRF quasars will assess the astrometric quality of the QSOs. Capitaine investigated high-precision methods for locating the celestial intermediate pole (CIP) and origin (CIO) at a few microarcsec.

ROMANIA: Popescu reports on work carried out by the Astronomical Institute of the Romanian Academy in improving relative positions of reference stars around ICRF radio-sources. Observations have taken place since 2006, at Belogradchik Observatory, Bulgaria, in collaboration with the Bulgarian Academy. The 60 cm Zeiss telescope is used to observe ICRF defining and candidate northern sources with a 6x6 arcmin field. There are planned also observations of asteroids in the ICRF source fields. A pilot investigation has been published on the astrometry of 59 northern ICRF sources using significant approaches of asteroids to these sources (A&A 476, 989, 2007). For the ICRF objects, the (x,y) errors per source ranged from 50 mas to 10 mas. The RA and Dec reduction mean errors were 46 mas. Investigations into the orbits of these asteroid have been carried out. All astronomical (12989) plates observed since 1930 have been archived within the Wide-Field Plate Archive Program initiated by IAU Commission 9. 5000 of them have a preview stored in the WFPDB. The intention is to add the CCD images acquired for the last 5 yr and form the origin of the future Bucharest Virtual Observatory.

5. Positions and proper motions

BRAZIL: Teixeira and Colleagues from the Bordeaux Observatory (Rappaport 2006; Ducourant 2006; de Souza 2007) conducted position and proper motion studies based on CCD Meridian Circles observations.

CHINA PR: Li Chen (Shanghai) reports that the positions and absolute proper motions of stars in the regions of the young (2-4 Myr old) open clusters NGC 2244 (the central cluster in the Monoceros R2 association) and NG 6530 (the dominant cluster in the Sgr OB1 association) were determined based on photographic plate material obtained at Shanghai Astronomical Observatory, with time baselines of 34 and 87 yr, respectively. Membership probabilities and velocity dispersions for both clusters were estimated (Chen *et al.* 2007). Chunlin Lu (Nanjing) reports that 9 astrometric standard fields ($5° \times 7°$) for LAMOST have been primarily completed with NEOST at Xu Yi, an observing station of the Purple Mountain Observatory. These fields include two SDSS standard fields and seven open clusters, such as Hyades, Pleiades, Praesepe, NGC 2281, etc., in which six clusters are the targets of the WIYN project. The magnitude limit is 21.0 mag. The internal precision of position is $0''.07$ at the observational epoch and external accuracy is $0''.1$ by using comparison between 2 SDSS fields and SDSS standard fields in the ACR (Astrometric Calibration Regions) catalog. The systematic errors induced by CTE, variation of atmospheric Refraction, etc., especially the displacement of the tangent point of CCD plane in the observations with drift scan mode were investigated by comparison between catalogues. The results of comparison between ACR - CMC13 and UCAC2 indicate that there are significant magnitude equations in declination of ACR and CMC13, and waves with period approximated to the size of the FoV of CCD drift scanning exist in the difference of position in both of UCAC2 - ACR and UCAC2 - CMC13 (Jiang 2008). Zhanghong Tang (Shanghai) reports that the systematic errors of GSC 2.3 were investiagted by using the comparison between GSC2.3 and UCAC2. The results indicate that the systematic error at the plate edges is $0''.5$ and the dividing line at radius $2.5°$ around the plate center is clear as well as a magnitude equation with non-linear features of about $0''.2$ from peak to peak. Reprocessing the plates can improve the final positional accuracy at 20th magnitude to $0''.2$ (Tang *et al.* 2008).

FRANCE: Ducourant (<ftp://cdsarc.u-strasbg.fr/pub/cats/J/A+A/469/1221>) has archived the PM2000 and CDC2000 catalogues at CDS (I/300 and I/303) Fresneau (with Vaughan and Argyle) reports the archiving of the Sydney Observatory Galactic Survey with proper motion accuracies of 3 mas/yr for half a million stars to B = 14 mag. Analysis of 600 astrographic plates archived at Macquarie University compared to GSC 1.2 suggests the detection in the solar neighborhood of the interface between the extreme disk and the thin disk of the Galaxy along the Galactic meridian $l = 330°$. Guibert reports of the discovery of a nearby M9 dwarf detected in the DENIS database with a proper motion of 2.5 arcsec/yr.

GERMANY: Brosche (Bonn/Daun) reports that Brosche and Schwarz (Landau, now Siegen) have compared their results on the motions of K-type giants perpendicular to the Galactic plane with other data; the increase of the velocity dispersion with increasing distance was confirmed. This behavior leaves no room for essential dark matter in the disk. A more generalized treatment is foreseen for the future. Röser (Heidelberg) reports that Röser, Schilbach, Schwan, Kharchenko, Piskunov, & Scholz (2008) compiled a new astrometric catalogue of positions and proper motions in the ICRS, called PPM-Extended (PPMX) (arXiv:0806.1009). The catalogue contains about 18 million stars with limiting magnitude 15.2 in the magnitude system of GSC 1. Besides the astrometric information, J, H and K_s magnitudes (from 2MASS) are given for 99.8% of the stars, and, for the subset of stars, which are in Tycho-2, Johnson B and V magnitudes are published. The positions and proper motions are derived from a rigorous weighted least-squares solution from various source catalogues: Astrographic Catalogue, Tycho-2, GSC 1.2, UCAC2, CMC14 and 2MASS, etc. For the subset of 4.5 million stars contained in the Astrographic Catalogue the typical mean error of the proper motions is 2 mas/y, for the remaining stars with first epoch in GSC 1.2 it is 10 mas/y. The catalogue is available from the CDS.

JAPAN: Proper motions of SiO maser sources around the Galactic Center were obtained with VLBA (Oyama *et al.* 2008). These results are important for studying the character and the mass of the super-massive black hole which is believed to be located at the center of the Galaxy.

RUSSIA: Khrutskaya (Pulkovo) reports that new high accuracy proper motions of 38,600 stars in the Pulkovo areas with galaxies have been obtained for zone −5 to +40°. The new proper motion accuracy is 1 - 3 mas/y. Systematic errors of proper motions with stellar magnitude are detected in UCAC2 (Khruyskaya 2006). Kinematic study of proper motions in TYCHO-2 and UCAC2 showed the average rotation around Galactic axis Y to be -0.37 ± 0.04 mas/yr for the most distant stars ($d \simeq 900$ pc). The rotation is interpreted as the residual rotation of ICRS/Tycho-2 system with respect to the inertial coordinate system (Bobylev & Khovritchev 2006). A catalogue of 235 areas (ERS, the IAU main list) in declination zone −17° to +80° has been compiled. The positional accuracy of 21355 stars ($9 - 17^m$) is $0''.04 - 0''.10$. Only 10 650 stars with $\delta < +50°$ have proper motions from UCAC2. The comparison in 198 areas between ERS and UCAC2 has been fulfilled in the zone from −30° to +45° (Ryl'kov 2007). Coordinates of more than 12 500 stars for 78 areas with Galactic radio sources have been obtained.

SPAIN: ROA has continued observations in order to extend CMC14 because a small zone was not published in 2005 due to bad weather. Also the southern survey limit was increased −40°. A new catalogue (CMC15) will be published in 2009 - 2010 with the stars brighter than $r' = 17$ and declination between −40° and +50°. During the period September 2006 - June 2007, CMASF (Argentine) has continued CCD observations to produce a survey of stars with $V < 16$ for $-55° < \delta < +30°$. Unfortunately, the CCD camera has failed and it has not been possible to repair it. Thus a sub-catalogue with $-30° < \delta < 0°$ is planned to be published by the end of summer of 2008.

UKRAINE: Ivanov (Kiev) reports the compilation of a catalogue of 555 200 stars with high-proper motions (> 0.04 arcsec/yr) up to 16 mag in zone $-2.5° < \delta < +90°$. Fedorov (Kharkov) reports on an all-sky (excluding the Galactic center) absolute proper motion catalogue of 285 million stars ($12^m < B < 20^m$). The proper motions were derived from 2MASS and USNO-A2.0 positions. The absolute proper motions were calculated using about 1.45 million galaxies from 2MASS with mean error < 1 mas/yr. Fedorov (Kharkov) reports that the first version of XC1 catalogue of positions and proper motions of 856,421 stars around ICRF sources in the Northern hemisphere was integrated into the CDS (I/302 XC1). Basey (Odessa) reports about processing the catalogue of 4984 stars ($-3° < \delta < 3°$) in the FK5 system which has been observed at the meridian circle in 1931–1935. Pinigin (Nikolaev) reports on a collaboration with Pulkovo Observatory (Ryl'kov) on a compiled catalogue of 22000 reference stars ($10–16^m$) in 235 fields around extragalactic radio sources of the ICRF list with $-17° < \delta < +80°$.

USA: Platais (Baltimore, MD) reports on the Deep Astrometric Standard (DAS) initiative intended to calibrate the focal plane arrays of large telescopes (Platais 2006). First epoch imaging in all DAS fields has been obtained. The radio VLA observations of candidate compact sources are complete in three DAS fields and VLBI observations have been made in one field (Fey 2007). The Northern and Southern Proper Motion (NPM, SPM) data (Lick Observatories, Yale/San-Juan) were re-reduced in a joint Yale/USNO effort (SPM4, Girard 2008). The UCAC3 (Zacharias 2009) contains highly accurate positions and proper motions of about 80 million stars. The 20 cm USNO 'red lens' astrograph has been refurbished for a new all-sky observing program with a 440 million pixel camera. This 'U-mouse' astrometric survey will begin in 2009 from Cerro Tololo. For more details see the IAU WG on Densification of the Optical Reference Frame report and various contributions at IAUS 248. NOFS participates in the PanSTARRS and LSST projects (D.Monet). The FASST instrument continues to observe mainly solar system targets for JPL. The Navy Prototype Optical Interferometer (NPOI) successfully observed a selected number of bright Hipparcos stars in global astrometry mode (Hutter, 2008).

6. Trigonometric parallaxes – nearby and high proper-motion objects

BRAZIL: Penna and Colleagues, in collaboration with the Observatorio di Torino, conducted a long term parallax program for a large sample of up to 200 L and M cool dwarfs. Teixeira and colleagues (Ducourant, 2007 and 2008) concentrated the work on milli-arsecond parallaxes of specific nearby objects.

CHINA PR: Chunlin Lu (Nanjing) reports that a new reasonable weight assigning scheme and a three-piecewise continuous model were adopted to fit the data for 48 low-mass stars of the main sequence in the solar neighborhood. Taking into account the constraints from observational luminosity, the empirical mass-luminosity relation (MLR) in K, J, H and V bands, and mass-metallicity-luminosity relation (MMLR) in V band for low-mass stars are improved Xia, 2008).

GERMANY: Schilbach (Heidelberg) reports that Schilbach, Röser & Scholz measured trigonometric parallaxes of 10 ultra-cool subdwarf candidates. The observations extended over 3 yr and were carried out with the OMEGA2000 IR-camera on the 3.5 m telescope on Calar Alto (Spain). At the time of this writing, the final observations were being obtained. Preliminary reductions yielded precisions of the parallaxes between 1 and 2 mas. Due to a relatively large and deep field, the parallaxes can be referred directly to galaxies. Scholz (Potsdam) reports that Scholz, Kharchenko, Lodieu, & McCaughrean (2008) have discovered an extremely wide and very low-mass pair with a common proper motion. This pair consists of a late-type (M7) dwarf and an ultra cool subdwarf (sdM7) sharing exactly the same very large proper motion of about 860 mas/yr. The two stars are separated by about six degrees on the southern sky, corresponding to a projected physical separation of about 5 pc at the assumed common distance of about 50 pc. The large separation and the different metallicities of dwarfs and subdwarfs make a common formation scenario as a wide binary (later disrupted) improbable. It seems more likely that this wide pair is part of an old halo stream.

JAPAN: Hachisuka (2006) obtained trigonometric parallaxes of H_2O masers near the star-forming region W3(OH) with VLBA. Imai (2007) obtained a trigonometric parallax of an H_2O maser feature associated with the low-mass young stellar object IRAS 16293–2422 using VERA. Hirota (2007) obtained a trigonometric parallax of Orion KL with VERA. Honma (2007) performed high-precision astrometry of H_2O maser sources in the Galactic star-forming region S269 with VERA and successfully detected the smallest trigonometric parallax ever measured. Hirota (2008) obtained the trigonometric parallax of an H_2O maser feature associated with the young stellar object SVS 13 with VERA. Nakagawa (2008) have started trigonometric parallax measurements of the Galactic Mira variables, which are important to deduce the precise period-luminosity relation in the Galaxy.

USA: Gatewood (Allegheny Observatory) concludes the 100 year parallax program with new results on selected nearby stars and the Pleiades distances (Gatewood 2008). The Naval Observatory Flagstaff Station (NOFS) Parallax Program continues (Dahn, Harris 2006 - 2009).

7. Solar System

BRAZIL: V. Martins and colleagues, in collaboration with Sicardy's group, keep several programs on the astrometry, dynamics, and astrophysics of asteroids, natural satellites and TNOs, notably on the Pluto/Charon system (Sicardy 2006; Descamps 2007; Gulbis 2006).

CHINA PR: Chunlin Lu (Nanjing) reports that the PMOE planetary/lunar ephemeris framework was established in 2003, and has been improved in recent years. Various effects on the bodies in the solar system have been taken into consideration. The further improvement of PMOE 2003 by using the second post-Newtonian (2PN) theory is in progress. Based on the PMOE 2003 ephemeris framework, Li *et al.* have calculated the orbits of the *LISA*, *ASTROD I* and *ASTROD* spacecrafts, and proposed the methods for the orbital optimization of the *LISA* spacecrafts, cooperating with Gerhard *et al.* at ESA. Also the celestial phenomena and lunar phases in Xia Shang and Zhou dynasty in China (BC 2100-BC 771) have been calculated and published (Li 2008). Wenjing Jin (Shanghai) reports that it is the first time to observe the mutual events of Galilean satellites in China with the 1m telescope at Yunnan Observatory. The comparisons were made with the theoretical modes, Lieske's E5 and Lainey's L1. The accuracy of positions in RA and Dec are 103 and 88 mas for Lieske's theory and 74 and 80 mas for Lainey's theory (Peng & Noyelles 2007). In addition, some further reduction has also carried out by comparison of normal CCD observations and some mutual events observations. The results show a better agreement between internal and external observations (Peng *et al.* 2008) The next campaign of mutual events will be in 2009 for the Galilean and in 2010 for Saturnian satellites. 210 positions of Phoebe (S9) were obtained with the 1 m telescope at the Yunnan Observatory during the years 2003 - 2005, using a CCD image-overlapping calibration method proposed by Peng *et al.* After the observed positions of Phoebe were compared with its theoretical positions, computed by the new JPL ephemeredes DE405 and SAT199, the mean residuals (O-C) are $0''.21$ and $-0''.05$ in right ascension and declination, respectively, with a standard deviation of $0''.06$ (Peng & Zhang 2006). Kaixian Shen (Xian) reports that a total of 115 frames of Phoebe were obtained with the 1.56 m telescope at Shanghai in 2003 - 2004. A comparison with three high quality ephemeredes, including the JPL SAT185 by Jacobson was made. The standard deviations in right ascension and declination are $0''.058$ and $0''.078$ respectively (Qiao 2007). In order to enhance determination of the Phoebe's orbit, it is necessary that the observations cover the longest time span as possible, in this reduction the 686 Earth-based astrometric observations available from 1905 to 2004, including the 101 new CCD observations were used (Shen 2008). Previously there were less than 400 observations of Triton with an accuracy of better than $0''.15$. Recently the 943 astrometric observations of Triton with the 1.56 m telescope at Shanghai during the period of 1996-2006 were obtained and compared with the theoretical positions provided from JPL and IMCCE. The standard deviation of O-C is $0''.04$ (Qiao 2006).

RUSSIA: Devyatkin (Pulkovo) reports on observations of Solar System bodies taken at the ZA-320M of GAO RAS (Descamps 2008; Devyatkin 2008). Kiseleva reports that astrometric investigations of the Galilean satellites of Jupiter are completed using the 26″ Pulkovo refractor. Estimation of Galilean satellites motion theories gives accuracies up to $0''.10$ (Kiseleva 2008). The observation results have been included in the PHEMU03 catalogue (Arlot 2008). Absolute and relative ('satellite-satellite') astrometry using photographic observations (2006-2007) at the Pulkovo 26″ has shown the presence of errors in the theoretical motion of Saturn. Pulkovo has taken part in the International Program of mutual phenomena of Saturn and Uranus. In 2006-2007 CCD-observations of asteroids by the Normal astrograph at Pulkovo were started (Khrutskaya, 2006 and 2007). Databases of astrometrical catalogues and Solar System and double star observations from Pulkovo Observatory are available at <www.puldb.ru> (Khrutskaya 2007).

UKRAINE: Filonenko and Velichko (Kharkov) report on CCD observations of comets 17P/Holmes, C/2007 E2 Lovejoy, C/2007 F1 LONEOS, and 8P/Tuttle, processed using UCAC2 as a reference catalogue. Kleshchonok (Kiev) reports that observations of stellar occultations by the Moon using a portable TV complex are continuing. The catalogue of stellar occultations contains more then 120 events. Kazantseva (Kiev) reports about the first results in the network of synchronous television and visual observations of stellar occultations. Baransky (Kiev) reports that 2164 CCD observations of 144 comets and minor planets have been obtained at the 0.7 m reflector and reported to MPC (code 585). Pinigin and Hudkova (Nikolaev) report that CCD observations of selected asteroids were taken at the RTT-150 (Turkey) in 2004-2007 with cooperation with Turkish National Observatory TUBITAK (Aslan) and Kazan University, Russia (Gumerov). About 5 500 of topocentric positions of 68 asteroids and 517 positions of 17 NEAs have been obtained (2004-2007), with about 3 200 of them for 2004-2005 sent to the MPC (code A84). Ivantsov (Nikolaev) reports about new values of 21 asteroid masses, which have been determined by the dynamic method using ground-based CCD observations and based on DE405. The relative error of mass determination is less than 50% for 12 asteroids.

8. Open and globular clusters and the Galaxy

BRAZIL: Work on Galactic clusters continues by the use of high quality proper motions to determine cluster membership (Dias 2006).

CHINA PR: Li Chen (Shanghai) reports that based on a most complete open clusters sample with metallicity, age and distance data as well as kinematic information, some preliminary statistical analysis regarding the spatial and metallicity distributions of the Milky Way disk is presented. In particular, a radial abundance gradient of 0.058 dex/kpc was derived. (Chen 2008). The orbits and theoretical tidal radii for a sample of 45 Galactic globular clusters were calculated. It is found that an orbital phase dependence between

theoretical and observed tidal radii is evident (Wu 2008). Using published accurate observational data of radial velocities and proper motions of stars in the open cluster M11, the
distance of M11 was determined, which is in quite good agreement with the luminosity
distances of the cluster given by some authors (Zhao & Chen 2007). Beijing-Arizona-
Taiwan-Connecticut (BATC) multi-band photometric data were used to determine the
membership of open cluster M48. By comparing observed spectral energy distributions
of M48 stars with theoretical ones, membership probabilities of 750 stars with limiting
magnitude of 15.0 in the BATC c band were determined. 323 stars with membership
probabilities higher than 30% were considered to be candidate members of M48 (Wu
2006). Absolute proper motions and radial velocities of 202 open clusters in the solar
neighborhood, which can be used as the tracers of the Galactic disk, were employed
to investigate the kinematics of the Galaxy in the solar vicinity. The results derived
from the observational data of proper motions and radial velocities of a subgroup of
117 thin disk open clusters are: the mean heliocentric velocity components of the open
cluster system $(u_1, u_2, u_3) = (-16.1 \pm 1.0, -7.9 \pm 1.4, -10.4 \pm 1.5)$ km s^{-1}, the characteristic velocity dispersions $(\sigma_1, \sigma_2, \sigma_3) = -17.0 \pm 0.7, 12.2 \pm 0.9, 8.0 \pm 1.3)$ km s^{-1}, the
Oort constants$(A, B) = (14.8 \pm 1.0, -13.0 \pm 2.7)$ km s^{-1}kpc^{-1}, and the large-scale radial
motion parametersof the Galaxy $(C, D) = (1.5 \pm 0.7, -1.2 \pm 1.5)$ km s^{-1}kpc^{-1} (Zhao
2006). Zi Zhu (Nanjing) reports that 301 open clusters with complete spatial velocity
measurements and ages were selected to estimate the disk structure and kinematics of
the Milky Way. The distance of the Sun to the Galactic center was derived as $R_0 = 8.03$
± 0.70 kpc. The mean rotation velocity of the Milky Way was obtained as 235 ± 10 km/s.
Using a dynamic model for an assumed elliptical disk, a clear weak elliptical potential
of the disk with ellipticity of $\epsilon(R_0) = 0.060 \pm 0.012$ is detected, the Sun was found to be
near the minor axis with a displacement of $30° \pm 3°$. The motion of clusters is suggested
to be on an oval orbit other than the circular rotation (Zhu 2008). Stellar samples of
main sequence (MS) and horizontal branch stars were obtained with spectral types O-B,
A, F, G, K-M to calculate the scale height of the Galactic disk by using *Hipparcos* data.
The results indicate that the scale height is 103.1 ± 3.0 pc and 144.0 ± 10.0 pc from the
samples of O-B of MS stars and horizontal branch. The Sun is located at 15.2 ± 7.3 pc
and 3.5 ± 5.4 pc above the mean plane of the disk for two of the samples. (Kong & Zhu
2008). The Galactocentric distance by a pure kinematical model was determined from
two components of the Galactic thin disk: 1200 O-B stars and 270 Galactic open clusters.
An estimated value of $R_0 = 8.25 \pm 0.79$ kpc was derived from the former one while $R_0 =$
7.95 ± 0.62 kpc from latter one with a simple Oort-Lindblad model of Galactic rotation. A
direct comparison shows that the above results are in good agreement with the best value
suggest by Reid (Shen & Zhu 2007). Based on *Hipparcos* proper motions and available
radial velocity data of O-B stars, the local kinematical structure of the young disk population of ~1500 O-B stars, not including the Gould-belt stars, have been re-examined.
A systematic warping motion of the stars about the direction to the Galactic center has
been reconfirmed. A negative K-term implying a systematic contraction of stars in the
solar vicinity has been detected. (Zhu 2006).

UKRAINE: Kharchenko (Kiev) reports that the all-sky compiled catalogue of 2.5 million stars (ASCC-2.5) has been made and 650 Galactic open clusters were identified in
it (about 130 previously unknown objects). The cluster sample is complete within the
Galactic disk area of about 1 kpc radius, and apparent integrated magnitudes brighter
than $V = 8$. Rybka (Kiev) reports that the local velocity field of more than 53 000 Red
Clump giants within 1 kpc of the Sun were investigated using the three-dimensional
Ogorodnikov-Milne model. It was shown, that Galactic rotation of the stellar group with

$|b| < 30°$ has peculiarities, such as in values of the phase offset parameter $\phi = 6.9 \pm 0.6°$ and the contraction parameter $K = 6.3 \pm 1.1 \, \mathrm{km \, sec^{-1} \, kpc^{-1}}$.

USA: Platais (Baltimore, MD) reports on deep astrometry in the WIYN Open Cluster Study (WOCS) Program. A total of 645 old long-focus telescope plates have been scanned with the STScI GAMMA measuring machine, covering 54 open clusters. Catalogs of proper motions and positions are constructed for Blanco 1, IC 2391, M 67, NGC 6253, and the globular cluster ω Cen. Astrometry for the last three clusters is based entirely upon the CCD mosaic frames.

9. Education in astrometry

Zhanghong Tang (Shanghai) reports that the first Chinese-French Spring School on Astrometry 'Observational campaign of solar system bodies', was held in Beijing, April 7-13, 2008. 28 students and young astronomers participated. The purpose of organizing the spring school was to foster young students for their specialization. The lectures were given by astrometric scientists on fundamental astrometry, receptors, telescopes and images for astrometric purpose, astrometry through photometry and observational campaigns for phenomena, practical astrometry.

10. Symposia, colloquia, conferences

Journées 2007 *The Celestial Reference Frame for the Future*, Meudon, France, 17-19 September 2007

IAU Symposium No. 248, *A Giant Step: From Milli- To Micro-Arcsecond Astrometry*, Shanghai, China, 15-19 October 2007, W. Jin, I. Platais & M. A. C. Perryman (eds.) (Cambridge: CUP)

ADELA 2008 – *IV Meeting on Dynamical Astronomy in Latin-America*, Mexico City, 12-16 February 2008.

Acknowledgement

We thank the designated national representatives for their contribution to this report.

Irina I. Kumkova
president of the Commission

Transactions IAU, Volume XXVIIA
Reports on Astronomy 2006–2009
Karel A. van der Hucht, ed.

© 2009 International Astronomical Union
doi:10.1017/S174392130802526X

DIVISION I / COMMISSION 8 / WORKING GROUP
DENSIFICATION OF THE OPTICAL REFERENCE FRAME

CHAIR   Norbert Zacharias
PAST CHAIR   Imants Platais
MEMBERS   William F. van Altena, Beatrice Bucciarelli,
                 Thomas E. Corbin, Christine Ducourant,
                 Dafydd Wyn Evans, Ralph A. Gaume,
                 Irina I. Kumkova, Brian D. Mason,
                 François Mignard, David G. Monet,
                 Jose L. Muinos Haro, Jean Souchay,
                 Sean E. Urban

TRIENNIAL REPORT 2006–2009

1. Introduction

A continuation of this WG was voted for at the IAU GA 2006 in Prague. The International Celestial Reference Frame (ICRF) is defined by the positions of 212 distant quasars at radio wavelengths. The primary, optical reference frame is the Hipparcos Celestial Reference Frame (HCRF), which is the Hipparcos Catalog without astrometric 'problem' stars (in: H. Rickman (ed.) 2001, *Proceedings IAU XXIV General Assembly, Transactions IAU XXIVB* (San Francisco: ASP), Resolution B1.2). The Tycho-2 catalog with its 2.5 million brightest stars forms the first step in the densification of the optical reference frame. However, the limiting magnitude of about $V = 12$ of the Tycho-2 catalog is not sufficient for most applications in astronomy and the goal of this IAU Working Group is to further extend the grid of highly accurate positions and motions toward more and fainter stars. The web site of this WG is at <ad.usno.navy.mil/dens_wg/>.

2. Developments within the past triennium

A catalog of the approximately 113 000 known quasars was compiled and published (Souchay *et al.* 2008, A&A) with emphasis on astrometry. Cross referencing between radio, optical and IR data made it possible to significantly improve the positions of targets in traditional QSO catalogs and to derive interstellar extinction and absolute luminosities.

The Automatic Transit Circle (ATC) at La Palma, formerly known as CAMC, continues to operate under the new owner, the Institute of Astrophysics of the Canary Islands (IAC), managed by ROA. Declination zones between –15 and –40 are being observed. The San Fernando automated transit circle telescope continues observations from the Southern Hemisphere, and a catalog of stars to 16th mag in the 0 to –30 declination zone is in preparation (Muinos 2009).

Astrometric and photometric properties of the Guide Star Catalog II version 2.3, listing almost a billion objects to mag 21 in the J photographic band, have been extensively tested by the OATo and STScI teams against the SDSS, 2MASS and UCAC2 catalogs for the assessment of random and systematics errors. Their analysis has shown that a magnitude-dependent astrometric error of the order of 0.1 to 0.2 arcseconds is present (Bucciarelli *et al.* 2008, in IAU Symposium No. 248). A significant improvement in astrometric accuracy is expected when using fainter calibrators (Tang *et al.* 2008, in IAU Symposium No. 248) such as the imminent UCAC3 catalog. A complete review of the construction methods and properties of GSC 2.3 has been accepted for publication in the Astronomical Journal (Lasker, Lattanzi, McLean, *et al.* 2008).

The Astronomisches Rechen-Institut published the PPM-Extended (Roeser *et al.* 2008, A&A), a compiled catalog of over 18 million stars based on the GSC 1.2 and other available data to arrive at accurate proper motions for galactic dynamics studies.

Second-epoch observations for the Yale/San-Juan Southern Proper Motion (SPM) program have continued. CCD frames are obtained with the double astrograph at Cesco Observatory in El Leoncito, Argentina (Girard *et al.* 2008, in IAU Symposium No. 248). Sky coverage south of -20 degrees declination is essentially complete as of June 2008, with over 16850 pointings recorded and only 76 remaining. Proper-motion determinations for 6 globular clusters (Casetti-Dinescu *et al.* 2007, *AJ*, 134, 195) provided a testbed for the combination of the first-epoch photographic plate material with second-epoch CCD astrometry. Results were excellent and indicate that the proper motion uncertainties are dominated by the first-epoch material. Thus, the original plan to obtain 2-fold overlap with the CCD survey has been modified. Adopting a single-fold observing strategy has permitted the completion of the survey south of -20 degrees despite poor weather and recent mechanical problems with the astrograph dome. Lower-priority fields north of -20 degrees will continue, as will observations of some scientifically important fields where second-epoch plate material exists but for which the added precision and extended baseline of new CCD observations warrant a new epoch. Funds for these continued CCD observations have been included in a proposal to the National Science Foundation.

Work is well underway on the SPM4 catalog that will cover the sky south of -20 degrees. In addition to the reduction of the CCD frames, the PMM pixel data for all of the first-epoch plates (as well as the fraction of the sky covered by second-epoch plates) are being re-reduced in a collaboration with the U.S. Naval Observatory (USNO). It is anticipated that the SPM4 will be completed in Fall 2008, contain roughly 100 million stars and galaxies and be more than 90 percent complete to $V = 17.5$.

The SPM and Northern Proper Motion (NPM) first epoch data are also needed to derive proper motions of the faint stars in the USNO CCD Astrograph Catalog (UCAC) program. All scan data from the NPM and SPM plates were obtained from the Naval Observatory Flagstaff Station (NOFS) and processed through a modified StarScan pipeline (Zacharias *et al.* 2008, PASP, 120, 644) to arrive at global x,y data.

Extensive code development for the final reduction of UCAC has been completed and all 275,000 CCD frames pixel data were processed at the USNO in spring 2008. The complete all-sky UCAC3 catalog will contain positions and proper motions of about 80 million stars (8 to 16 mag) with improved astrometric and photometric accuracy. Completeness of UCAC3 will also be significantly higher than in UCAC2, including high proper motion stars, and several hundred thousand newly discovered double stars (Zacharias 2008, in IAU Symposium No. 248).

Optical design studies for the USNO Robotic Astrometric Telescope (URAT) were completed and the primary mirror obtained in spring 2008. This telescope features two full-size aperture (0.85 m) corrector lenses to support the secondary mirror without a spider construction. Currently no funds are identified to complete this new telescope, however the focal plane assembly for URAT has been funded. The world's largest monolithic CCD chip with 111 million pixels was manufactured in 2007 and tested at the USNO astrograph (Zacharias *et al.* 2007, SPIE 6690-08). The URAT focal plane consists of 4 of these detectors. For phase 1 of URAT (nicknamed 'U-mouse') this 4-shooter camera is being mounted on the re-furbished USNO 20 cm astrograph 'redlens'. First light is expected in March 2009 and a new sky survey will commence from Cerro Tololo. A single exposure covers 27 square degrees of sky and a 20-fold overlap of the entire Southern Hemisphere can be completed by 2011 to solve for positions, proper motions and parallax of stars in the 8 to 18 mag range in a 680 - 750 nm bandpass. The resulting star catalog will be on the ICRF, directly linking the *Hipparcos* stars and optical counterparts of compact extragalactic radio reference frame sources.

Platais and collaborators continued with the Deep Astrometric Standard (DAS) initiative (Platais *et al.* 2008, Commission 8 report in IAU Transactions XXVIB) to provide absolute positions and motions to 25th magnitude in selected areas.

3. Closing remarks

The major near future task will be to merge all relevant, astrometric, observed catalogs. This huge task will involve systematic error analysis and applying corrections to bring the data onto a common system, globally as well as locally. New observational data relevant for the densification of the optical reference frame are expected over the next couple of years from projects like U-mouse (U.S. Naval Observatory), PanSTARRS (Univ. of Hawaii and collaborators) and JMAPS (proposed micro-satellite, U.S. Naval Observatory). Even further into the future, the Space Interferometry Mission (SIM Planetquest) will be able to define positions and proper motions of individual celestial objects on the micro-arcsecond level. The *Gaia* (ESA) mission with expected catalog release near the year 2020 will define the optical reference frame at unprecedented densities and magnitude limits.

Norbert Zacharias
chair of the Working Group

Transactions IAU, Volume XXVIIA
Reports on Astronomy 2006–2009
Karel A. van der Hucht, ed.

© 2009 International Astronomical Union
doi:10.1017/S1743921308025271

COMMISSION 19

ROTATION OF THE EARTH
ROTATION DE LA TERRE

PRESIDENT	**Aleksander Brzezinski**
VICE-PRESIDENT	**Chopo Ma**
PAST PRESIDENT	**Véronique Dehant**
ORGANIZING COMMITTEE	**Pascale Defraigne, Jean O. Dickey, Cheng-Li Huang, Jean Souchay, Jan Vondrák, Patrick Charlot** (IVS representative), **Bernd Richter** (IERS representative), **Harald Schuh** (IAG representative)

TRIENNIAL REPORT 2006–2009

1. Introduction

The Commission supports and coordinates scientific investigations in the Earth rotation and related reference frames. Several changes had been introduced to the structure of Commission 19 since the IAU XXVI General Assembly in Prague, 2006. The Organizing Committee of Commission 19 has been substantially reduced. It consists now of six ex-officio members, the Commission president, vice-president, past president and representatives from the International Association of Geodesy (IAG), International Earth Rotation and Reference Systems Service (IERS), International VLBI Service for Geodesy and Astrometry (IVS), and five members at-large who are nominated by the OC, selected by the Commission members and elected by the IAU GA for a maximum of two terms. The modified terms of reference of Commission 19, the list of members and other details can be found at the Commission website <iau-comm19.cbk.waw.pl/>.

A brief description of the most important developments in the related field is given in Section 2, while Sections 3 - 5 contain the reports of cooperating services/institutions. The list of references comprise only the most important papers which have been cited in the report; an extended list of references provided by the members of Commission 19 will be posted at the Commission website.

2. Developments within the past triennium

The activities related to Commission 19 are mostly developed in the different institutions, at scientific meetings and in the WGs of Division I. The most important developments are described below.

(1) Establishment of the Global Geodetic Observing System (GGOS):
An important part of GGOS is related to the activity of Commission 19 and much has been done already in this field; for details see <www.ggos.org> and <geodesy.unr.edu/ggos/ggos2020/>.

(2) Developments of the observation techniques:

Ring laser gyroscope: an advantage of this emerging technique is that a single instrument is capable to determine the polar motion of the instantaneous rotation axis. However, the measurements are still not stable over periods longer than a few days. Therefore only the diurnal and subdiurnal variations can be estimated. Further progress has been attained in the analysis and interpretation of the ring laser measurements made in Wettzell.

VLBI 2010: a new VLBI system based on small antennas (10–12 m diameter) has been proposed by the IVS WG3. The following three performance goals have been identified: a) accuracies of 1 mm for site position and below 1 mm/year for velocity (TRF); b) continuous measurement of EOP; c) rapid generation and distribution of the IVS products.

(3) *GRACE*:

The *GRACE* observations allow the determination of the excitation functions, derived from changes in the gravity field. Such excitation functions can be compared with those derived from individual geophysical fluids, and also with the geodetically-determined polar motion excitations. The utility of *GRACE* fields in measuring polar motion excitation is now comparable to what can be obtained directly from the geophysical fluids.

(4) Development of the global circulation models of geophysical fluids:

Progress has been attained in modeling the atmospheric circulation (e.g., new re-analysis model ERA40; experimental models with hourly resolution) and the land hydrology (several new models become available, including those based on *GRACE* data).

(5) Reference frames and EOP:

The implementation of the IAU 2006 Resolutions related to Earth's orientation and reference systems has been prepared. IAU 2006 Resolution B1 adopted the P03 precession model as a replacement to the precession part of the IAU 2000A precession-nutation in order to be consistent with both dynamical theories and the IAU 2000 nutation (Hilton *et al.* 2006). EOP series have been reprocessed since 1984. Pole coordinates are now fully consistent with ITRF2005. The nutation offsets and UT1 are made consistent with the International Celestial Reference Frame (ICRF) through the IVS combined solution. Working groups have been established by the IAU, IERS and IVS with the goal of presenting the second realization of the ICRF at the IAU General Assembly in 2009.

(6) IERS Conventions:

Much has been done in order to present a consistent set of the IERS Conventions, in agreement with the current state of knowledge, and to have it actually put into practice by analysis centers. The IERS Conventions Center has worked on updating the IERS Conventions (2003), since 2005 with the help of an Advisory Board on IERS Conventions Update. A workshop on the IERS Conventions has been organized at the BIPM in September 2007.

(7) EOP predictions:

Predictions of the EOP are necessary for practical use, e.g., in space navigation. Currently, predictions of EOP are provided by the IERS Rapid Service/Prediction Center for Earth Orientation Parameters. The EOP Prediction Comparison Campaign has been organized between July 2005 an March 2008. The results have been presented at several international meetings including EGU General Assemblies 2007 and 2008, Journees 2007 and AGU Fall Meeting 2007, and will be published in *Journal of Geodesy*; see <www.cbk.waw.pl/EOP_PCC/> for details. Another organization involved in the improvement of the EOP predictions is the IERS Working Group on Prediction.

Two important review of works on Earth rotation variations were published in 2007 in *Treatise of Geophysics, Vol.3: Geodesy, Elsevier*, Chapter 3.11 – Long Period Variations by R. Gross, and Chapter 3.12 – Nearly Diurnal Variations by V. Dehant and P. Mathews.

3. Report of IERS (compiled by the IERS Central Bureau)

3.1. *Publications, websites, and workshops*

The following IERS publications and newsletters appeared between 2005 and 2008: J. Souchay and M. Feissel-Vernier (eds.), The International Celestial Reference System and Frame, 2006 (IERS Technical Note No. 34); IERS Annual Reports 2004, 2005, and 2006; IERS Bulletin A, B, C, and D (weekly to half-yearly); IERS Messages Nos. 76 to 131; ITRF Mail Nos. 63 to 76.

A new IERS Data and Information System (DIS) at the website <www.iers.org>, maintained by the Central Bureau, is running in operational mode since the end of 2005. It presents information related to the IERS and the topics of Earth rotation and reference systems. It provides tools for searching within the products (data and publications), working with the products and downloading them. The DIS provides also links to other servers, including about 20 websites run by other IERS components.

The following IERS Workshops were held: Combination, Potsdam, Germany, October 2005; Global Geophysical Fluids, San Francisco, USA, December 2006; Conventions, Sèvres, France, September 2007. The IERS also organized two GGOS Unified Analysis Workshops, at Monterey, USA, December 2007, and Vienna, Austria, April 2008. Abstracts and presentations of all these workshops are available at the IERS website.

3.2. *Activities of the IERS components*

From 2005 to 2008, the IERS Directing Board met twice each year. Summaries of the meetings are available in the Annual Reports and at the IERS website.

The Central Bureau concentrated its work on the development of the IERS Data and Information System. It also coordinated the work of the Directing Board and the IERS in general, organized workshops and issued publications.

The work of the Analysis Coordinator and of the Working Group on Combination focused on coordinating the Combination Pilot Project (CPP) for the generation of weekly inter-technique combination products, mainly EOP. Version 2.02 of the SINEX data format was developed. The major achievement of the CPP is that each of the Technique Centres (IGS, ILRS, IVS, except IDS) provides a combined solution based on the weekly SINEX files of the individual analysis centres on a routine basis.

The Rapid Service/Prediction Centre is responsible for providing Earth orientation parameters on a rapid turnaround basis and does so through the IERS Bulletin A and data sets made available through its website and ftp areas. For more information see report 5.12 below.

The IERS further increased the quality of their time series of Earth Orientation Parameters. Bulletin B and C04 series were recomputed and aligned to the EOP solution associated to the ITRF2005, resulting in the new EOP 05 C04 series. The system of Bulletin A was changed to match the system of this new series.

In October 2006, the ITRS Centre released the new ITRF2005 based on time series of station positions and Earth Orientation Parameters (Altamimi *et al.*, 2007).

Involvement by ICRS Centre personnel in the celestial reference frame VLBI program has continued, increasing the number of observations of ICRF quasars in the southern celestial hemisphere and continuing an extensive observing program in the northern hemisphere. This observing program will eventually result in a new realization of the ICRS, tentatively called ICRF 2 (see also the reports of Commission 8 and WG ICRF-2).

The Conventions Centre has made several modifications to the Conventions through the Conventions update web page including a complete rewrite of Chapter 9 (Tropospheric Model). Other chapters have been revised to make them more consistent with current IAU recommendations. A Conventions Workshop was held in September 2007 to prepare a new edition, expected to be produced in 2009.

The eight Special Bureaus of the Global Geophysical Fluids Centre provide data sets reflecting the mass transport of geophysical fluids, which are important for Earth rotation studies. A plan for improving the GGFC data and activities is being developed.

The activities of the ITRS Combination Centres at DGFI and at IGN concentrated on the computations for the ITRF2005 and on its evaluation. For information regarding the work of the Combination Research Centres see the reports 5.1, 5.5 and 5.11 below.

Three new working groups (WG on Prediction, WG on Site Survey and Co-location, IERS/IVS WG on the Second Realization of the ICRF) were established to support the work of the IERS product centres.

3.3. *Report of IVS (compiled by IVS Coordinating Center)*

The IVS continued to fulfill its role as a service within the IAU by providing necessary products for the densification and maintenance of the celestial reference frame as well as for the monitoring of Earth orientation parameters, in particular UT1 and nutation, with two 24-hr sessions per week.

UT1 Intensive measurements are continued on five days (Monday through Friday, Int1) on the baseline Wettzell (Germany) to Kokee Park (Hawaii, USA) and on weekend days (Saturday and Sunday, Int2) on the baseline Wettzell (Germany) to Tsukuba (Japan). In August 2007 a third Intensive series (Int3) started to fill the 36-hour gap in the data series between Int1 and Int2.

The VLBI 2010 Committee was formed to work on designing and implementing the next generation VLBI system. In November 2007 the first fringes were found with proof-of-concept hardware installed at the Goddard Geophysical and Astronomical Observatory.

For the second half of August 2008 a continuous VLBI campaign (CONT08) is being organized that will have 15 days of continuous observations on an 11-station network. With a recording rate of 512 Mbps and a time period that avoids the transitional months of March and October, CONT08 is expected to yield a high signal level in the atmospheric excitation functions for high temporal resolution Earth rotation investigations.

4. Report of SOFA (chair: Patrick Wallace)

The SOFA Review Board provides an authoritative set of fundamental-astronomy algorithms through a website (<www.iau-sofa.rl.ac.uk>) hosted by the UK Hydrographic Office and operated by Her Majesty's Nautical Almanac Office. SOFA's third issue of software was made in August 2007; it contained 161 Fortran routines (an increase of 40), including definitive implementations of the IAU 2006 precession models and related items, plus a 38-page explanatory document covering the Earth attitude algorithms. A fourth issue occurred in March 2008, comprising minor corrections and improvements. A second implementation of SOFA, this time using the C programming language, was completed in June 2008 and is being prepared for release. The Fortran SOFA already has some users in the aerospace industry, and the new C version is likely to increase interest. SOFA is increasingly coordinating its work with the IERS Conventions effort, with the object of eliminating unnecessary overlaps and evolving common standards.

5. Report of the individual institutions and national projects

5.1. *SYRTE Department of Paris Observatory*

Implementation of the IAU 2006 Resolutions: Various ways of forming the IAU 2000/2006 precession-nutation matrix have been discussed by Capitaine & Wallace (2006) and the corresponding precession-nutation procedures have been provided by Wallace & Capitaine (2006). Simplified implementations have also been proposed (Capitaine & Wallace 2008); these offer different compromises between size and precision and take advantage of the use of the celestial coordinates of the CIP instead of using the nutation angles in longitude and obliquity. The 'IAU 2006 Glossary' produced by the IAU WG on 'Nomenclature for Fundamental astronomy' (Capitaine *et al.* 2008a) was made available on the WG website (`<syrte.obspm.fr/iauWGnfa>`) providing the definitions corresponding to the IAU 2000 Resolutions and IAU 2006 Resolutions B1, B2 and B3. This includes, in particular the appropriate terminology for the pole, the Earth's angle of rotation, the longitude origins and the related reference systems and corresponding time scales.

Comparison between models and observations for the celestial motion of the CIP: A comparative study by Souchay *et al.* (2007) discussed the differences between the rigid Earth theory (REN 2000), the non-rigid Earth nutation theory (MHB 2000), and observational data. A detailed consideration by Mathews *et al.* (2007) of the ERA-2005 precession-nutation model that G. Krasinsky presented at the IAU 2006 General Assembly (JD16) made clear the fundamental differences in the main features of the IAU and the ERA-2005 precession-nutation models. This study showed severe deficiencies in the ERA-2005 model that were confirmed by Capitaine *et al.* (2008b) through further comparisons between the models as well as with the most recent VLBI observations.

The first step of the Descartes-nutation project devoted to the integration of the rotational equations developed as functions of the X, Y coordinates of the CIP in the geocentric celestial reference system (GCRS) were completed (Capitaine *et al.*, 2006) and the second step was started. A method has been developed by Zerhouni *et al.* (2008) for determining the GCRS CIP coordinates from the high accuracy LLR measurements over a 20-yr period.

Earth Rotation and geophysical excitation: Different studies have been conducted within the Earth orientation and Space geodesy group of the SYRTE Department, to study the relationship between the Earth rotation parameters (ERP) and the geophysical phenomena that affect Earth rotation. These concern the effects of the geophysical external fluids (Gambis *et al.* 2008; Seoane *et al.* 2008) and internal parts (Varga *et al.* 2007). Other effects include the solar activity (Chapanov & Gambis 2008). Observation of ERP are now available with an increased accuracy, i.e., 40 microarcseconds for pole components and 3 microseconds for UT1.

Journées 2007 on space and time reference systems: The Journées 2007 organized at Paris Observatory with the sub-title 'The Celestial Reference Frame for the future' included several sessions devoted to Earth rotation and reference systems.

5.2. *Royal Observatory of Belgium*

General objectives: The objectives of the work are to better understand and model the Earth rotation and orientation variations, and to study physical properties of the Earth's interior and the interaction between the solid Earth and the geophysical fluids. It is based on theoretical developments and on the analysis of data from Earth rotation monitoring and general circulation models of the atmosphere, ocean, and hydrosphere. We worked on the improvement of VLBI observations as well as of analytical and numerical Earth rotation models. The angular momentum budget of the complex system composed of the solid Earth, the core, the atmosphere, the ocean, the cryosphere, and the hydrosphere was studied, in order to better understand the dynamics of all components of Earth orientation.

Theory: We demonstrated that the contributions of Poisson terms in the tidal potential to long-periodic nutations are small but significant at the microarcsecond level and that the liquid core has an important contribution in that effect (Folgueira *et al.* 2007); (2) we computed the second-order torque on the tidal redistribution and the Earth's rotation (see Lambert and Mathews 2007); (3) we studied the electromagnetic coupling at the core-mantle boundary and its effects on the tides and nutation, using numerical integration; (4) we established an analytical method to compute the topographic coupling at the core-mantle boundary and its effects on nutations; (5) we showed that the atmosphere could globally excite the FCN but the atmospheric data could not reproduce the exact time variability of its amplitude (Lambert 2006); we estimated the time variable amplitude of the FCN from VLBI data and proposed our model for the IERS Conventions.

Observations and data interpretation: (6) precise VLBI-only EOP and reference frames determination is routinely done in collaboration with the Paris Observatory IVS Analysis Center; (7) analyzing VLBI data, we studied the influence of the analysis strategy on Earth EOP determination and showed that this can reach a few tens of microarcseconds in nutation amplitudes (Lambert *et al.*, 2008, Lambert and Dehant V., 2007); (8) we showed that using GPS-based determinations of the VLBI station positions could improve the determination of VLBI-only Earth rotation parameters; (9) we estimated time variable amplitude of FCN from VLBI data; our model showed that the atmosphere could globally excite the FCN; (10) using a Bayesian approach and VLBI observations in time domain, we estimated the limits within which the geophysical parameters may be expected to change, and to which extent the inner core parameters and precession may vary (Koot *et al.*, 2006 and 2008); we showed that the errors of the adopted parameter values were underestimated; (11) we estimated the FCN free mode and showed that both results are coherent; (12) we analyzed polar motion when the two main oscillations canceled each other (from November 2005 till February 2006); we explained the small centimeter level loops by atmospheric and oceanic contributions (Lambert *et al.*, 2006).

5.3. *Research unit 'Earth rotation and global dynamic processes' in Germany (chair: Jürgen Müller)*

In order to organize joint research activities in 'Earth rotation and global dynamic processes' in Germany, since the beginning of 2006 ten related sub-projects are supported by the German research funding organization DFG (Deutsche Forschungsgemeinschaft) in the frame of a research unit (Müller *et al.* 2005). Based on the general survey of Schuh *et al.* (2003), the main objective of this coordinated project is a comprehensive description and explanation of underlying physical phenomena contributing to variations of Earth rotation. Such an integral treatment of Earth rotation became possible by combining experts of observational techniques, data processing and analysis as well as in particular modelling. The research unit with participating scientists and institutions from geodesy, geophysics, meteorology, and oceanography will provide significant contributions to international programs such as GGOS and GMES (Global Monitoring for Environment and Security).

In close cooperation with the research unit a complex Earth system model is developed in a research project supported by DFG. The dynamical system model couples numerical models of the atmosphere, of ocean tides and circulation as well as of continental discharge considering consistent mass, energy and momentum fluxes between these near-surface subsystems of the Earth in order to allow for explanations and interpretations of geodetically observed variations of global parameters of the Earth. More information on the DFG Research Unit FOR584 can be obtained from www.erdrotation.de.

5.4. *Space Research Centre of the Polish Academy of Sciences*

Atmospheric and oceanic excitation of Earth rotation (Aleksander Brzeziński). A successful attempt has been made to retrieve the diurnal and semidiurnal components of polar motion and UT1 from analysis of the routine observations by VLBI (Brzeziński & Bolotin 2006). These components are expressed by a time series beginning in 1984, which can be interpreted by comparison with geophysical determinations of the excitation functions. Brzeziński (2007) developed and implemented the computational algorithm of the so-called geodetic excitation of nutation based on VLBI observations. Korbacz *et al.* (2008) performed detailed analysis of the new high resolution atmospheric and oceanic angular momentum data (AAM – from ERA40 reanalysis project, OAM – Ocean Model for Circulation and Tides forced with ERA-40 data). They estimated the corresponding perturbations in EOP and compared them to earlier results based on different AAM and OAM data sets. A separate part of this analysis was devoted to the atmospheric thermal tide S1 occurring at exactly the frequency of 1 cycle per solar day (Brzeziński, 2008).

Geophysical excitation functions of polar motion (Barbara Kołaczek & Jolanta Nastula). Gravimetric polar motion excitation functions computed from the data derived from the Gravity Recovery and Climate Experiment (GRACE) differ from the mass terms of geodetic observations and from the geophysical excitation functions of polar motion of seasonal time scales up to 20 mas. Thus gravimetric excitation functions need further improvements for studies of polar motion (Nastula *et al.* 2007; Nastula *et al.* 2008). Hydrological excitations of polar motion are not able to explain the residual excitations. The differences between amplitudes of the seasonal oscillations of hydrological excitation of the polar motion computed from various hydrological models are of the order of 5 mas (Nastula *et al.* 2006; Nastula *et al.* 2008). Study of the regional patterns of atmospheric variability was continued in a fine-resolution network of sectors over an extended 60-yr period at annual time scales. (Salstein & Nastula 2006). Spectral characteristic of polar motion along loops during 2005 - 2006 and 1999 - 2000 were studied by Nastula & Kołaczek (2007).

Modeling and prediction of the Earth Orientation Parameters (Wieslaw Kosek). The time frequency analyses methods were used to detect wideband oscillations in the EOP data, which were then used to construct the EOP model data. Prediction of such a model reveals the influence of the most energetic oscillations on the forecast errors of the EOP data (Kosek *et al.* 2006; 2008). The mean prediction errors for various prediction algorithms and techniques were calculated during the Earth Orientation Parameters Prediction Comparison Campaign (Kalarus *et al.* 2008). The combination of least-squares extrapolation and univariate or multivariate autoregressive prediction was used to predict UT1-UTC (Niedzielski & Kosek 2007) as well as local and global sea level anomaly data from TOPEX/Poseidon and Jason-1 altimetry (Niedzielski & Kosek 2005).

5.5. *Astronomical Institute, Academy of Sciences of the Czech Republic and Faculty of Civil Engineering, Czech Technical University, Prague*

The new combined astrometric catalogue EOC-3 (Earth Orientation Catalogue) has been provided by the combination of astrometric observations of latitude/universal time with astrometric catalogues (ARIHIP, TYCHO-2). The new catalog contains also quasiperiodic terms reflecting orbital motion of stars (Vondrák & Štefka 2007). The new solution of EOP from optical astrometry in 1899.7 - 1992.0 in the system of this new catalogue was derived by Vondrák *et al.* (2008). The atmospheric and oceanic excitation of free core nutation, the stability of its period and the quality factor were studied by Vondrák & Ron (2006, 2007, 2008).

A method of non-rigorous combination of different techniques to obtain simultaneously station coordinates and Earth orientation parameters was developed and tested. Basically, three approaches were studied - the short-term, e.g., monthly, and the long-term combinations (Pešek & Kostelecký 2006a), and, as a response to a call for producing weekly combinations, the method of non-rigorous combination was slightly modified to produce such very short-term solutions (Pešek & Kostelecký 2006b). We also started to use the new constraints to ensure the continuity and smoothness of EOP of the non-rigorous combination (Štefka & Pešek 2007; Štefka *et al.* 2008). A more effective algorithm for sparse systems from the GNU Gama package (<`www.gnu.org/software/gama`>) was implemented, decreasing the necessary computation time by one order. Other information on our activities is available at the web pages of the Center for Earth Dynamics Research (<`pecny.asu.cas.cz/cedr¿`>), joining five Czech institutions active in astronomy and geosciences research.

5.6. *Institute of Geodesy and Geophysics (IGG) of the Vienna University of Technology*

One main research area at IGG focuses on sub-daily and episodic variations of Earth rotation. Hourly polar motion and universal time was estimated in an optimal and consistent manner from Very Long Baseline Interferometry (VLBI) observations (Englich *et al.* 2008) and compared with results obtained from GPS. So far, no significant signal could be seen at the terdiurnal band. The relative Sagnac frequency variation of a new emerging technology, the ring laser gyroscope, was investigated for sub-daily signals of Earth rotation (Mendes Cerveira *et al.* 2008b). As it is envisaged to combine high-resolution ring laser data with VLBI observations, the ring laser gyroscope theory was carefully investigated (Mendes Cerveira *et al.* 2008a).

Uncertainties of the atmospheric excitation of Earth rotation were evaluated w.r.t. the diverse possible calculation options. Forecast data of the ECMWF was used to increase the temporal resolution (Boehm *et al.* 2008). The largest uncertainty arises from the adopted numerical weather model itself. An investigation was conducted on the improvement of a combination of nutation rates from GPS and GLONASS with nutation offsets from VLBI (Kudryashova *et al.* 2008). The potential benefit of including the future European satellite system GALILEO was studied, too.

5.7. *Central Astronomical Observatory at Pulkovo RAS (Pulkovo Observatory)*

The following topics were investigated: determination of the EOP from VLBI observations; Free Core Nutation (Malkin 2007; Malkin & Miller 2007); connection of the Earth rotation with other geophysical phenomena (Gorshkov 2007); assessment and improvement of the celestial reference frame (Sokolova & Malkin 2007; Malkin 2008). An IVS Analysis Center PUL was organized in 2006. Related info and results can also be found at <`www.gao.spb.ru/english/as/ac_vlbi/`>. A permanent GPS station PULK has been included in the EPN network and is used for the EOP determination in the Russian State EOP Service (in cooperation with NAVGEOCOM, Moscow). Group members participated in the several Working Groups of IAG, IERS and IVS.

5.8. *Institute of Applied Astronomy of the Russian Academy of Sciences (IAA)*

During the last three years IAA has continued to support the EOP Service including daily processing of VLBI, SLR and GPS observations. The CONT05 VLBI data have been analyzed to determine subdiurnal variations of EOP and other parameters (Finkelstein *et al.* 2006). Significant improvements to the AC software were maid. A new version of the VLBI data processing software QUASAR (Gubanov *et al.* 2007) is now used for global EOP + TRF + CRF solutions on regular basis. Quasar VLBI Network observations are processed to support GLONASS (Finkelstein *et al.* 2007). The period and phase of FCN data obtained from VLBI observations have been analyzed (Gubanov 2008). The analysis of radio source position time series has been performed in the framework of the IVS WG on ICRF (Kurdubov & Skurikhina 2008). The secular decrease of the Earth's dynamical ellipticity has been derived from the analysis of VLBI data (Krasinsky 2008).

5.9. *Sternberg Astronomical Institute of Moscow State University, Moscow*

A new program package ARIADNA for processing VLBI observations has been developed and tested. It is planned that beginning in July 2008 it will be used for the processing of VLBI observations. Research continued on the modeling VLBI delay both for ground and space-ground interferometers (Radioastron mission). This package was used to improve the data set of stable and potentially stable sources for the next radio ICRF. The theory of apparent motion of the 'stable' radio sources, based on the analysis of VLBI-derived time series of their coordinates, was developed. A new approach for the realization of the new radio ICRF was developed.

Five new GPS/GLONASS stationary points have been installed (two in Moscow and three in the North Caucasus region). GPS/GLONASS data are used both for monitoring the point motions and EOP estimation.

5.10. *Universities of Alicante and Valladolid, Spain*

The groups of the universities of Alicante and Valladolid (Spain) continued investigating the accurate modeling of the non-rigid Earth rotation. The efforts were focused on extending the analytical Hamiltonian theory up to the second order, both in the sense of perturbation theory and in the case of contributions previously disregarded. Some examples are illustrated in (Ferrándiz *et al.* 2007) and (Escapa *et al.* 2008). In addition, we participated in the IAU Division I WG on *Precession and the Ecliptic* (Hilton *et al.* 2006).

Theoretical studies on the rotation of other celestial bodies of the Solar System (Moon, Mercury, etc.) and on some fundamentals of the rotational motion have been initiated (e.g., Efroimsky & Escapa 2007; Ferrándiz & Barkin 2007; Barkin & Ferrándiz 2006), often in cooperation with scientists from other centers such as National Aeronautics and Space Administration (USA), National Astronomical Observatory of Japan, US Naval Observatory (USA), Royal Observatory of Belgium and Sternberg Astronomical Institute. (Russia).

These investigations have been partially supported by Spanish projects AYA2004-07970, AYA2007-67546, *Junta de Castilla y León* project VA070A07, and also by the Descartes Prize Nutation consortium.

5.11. *Jet Propulsion Laboratory (JPL), USA*

JPL continued to investigate Earth orientation variations including their excitation by earthquakes (Gross and Chao 2006), by atmospheric winds (Dickey *et al.* 2007; Gross *et al.* 2008a), and by ocean tides (Gross, 2008). The consistency of Earth orientation, gravity, and shape measurements was evaluated by comparing them both with each other and with models of surface geophysical fluids (Gross *et al.* 2008b). A review of the theory of Earth orientation variations, the techniques used to measure them, and their causative mechanisms was published (Gross 2007a). JPL also continued to support tracking and navigation of interplanetary spacecraft by acquiring and reducing VLBI, GPS and LLR data and by using a Kalman filter to both combine these with other Earth orientation measurements in order to produce optimal estimates of past variations in the Earth's orientation and to predict its future evolution (Gross 2007b).

5.12. *U.S. Naval Observatory (USNO), USA*

The USNO serves the community as: the IERS Rapid Service/ Prediction Center; the co-host of the IERS Conventions Center along with the Bureau International des Poids et Mesures; the co-host of the IERS ICRS Center along with the Paris Observatory; a VLBI Correlator Center, an Operations Center, an Analysis Center and Analysis Center for Source Structure for the IVS; and an Analysis Center for the IGS. USNO has improved the quality of its EOP combination by using improved data sets and reducing the smoothing applied to the data. The predictions have been improved by using a new polar motion prediction algorithm and improved AAM forecasts. USNO is preparing for the upcoming release of the next registered edition of the IERS Conventions, anticipated to be released in the next year. The VLBI correlator is undergoing a modernization of the control computer and the addition of playback units which promises to improve the processing of large VLBI sessions. USNO also plays a significant role in the generation of the ICRF-2. The quality of the USNO GPS solutions has improved through the use of better antenna phase center models.

5.13. *National Geospatial-Intelligence Agency (NGA), Basic and Applied Research Office, USA*

Several research efforts have been initiated to improve Earth Orientation (EO) prediction. A new research project was begun with the Naval Research Laboratory to improve the atmospheric and oceanic model data available for EO research. This effort is focused on improving our understanding of excitation mechanisms of polar motion and length of day variations and on advancing techniques used for the prediction of EO parameters. This has resulted in an improved understanding of the angular momentum transfers between the atmosphere, oceans, and land, and how these interactions affect the excitation of high frequency polar motion (Johnson 2008). Currently, NGA's University Research Initiative (NURI) grant program is sponsoring academic research projects to improve atmospheric data products and prediction methods available for near real-time EO predictions. T. Johnson, formerly at the USNO, is now a project scientist in the NGA, coordinating these efforts.

5.14. *Related researches in China*

National Astronomical Observatory of China (NAOC) and National University of San Juan of Argentina (NUSJ) set up a cooperative 60 cm SLR station (ID 7406) in San Juan. It has started routine observations from the beginning of 2006, and contributes to the ITRF, etc. The daytime tracking function and a co-located GPS station will be realized. In the NAOC group, the secular variation of the Earth's rotation is studied using ancient astronomical observations (Li, 2006), while the variation of the vertical is also studied using observations of time and latitude (Han *et al.* 2007; Ma 2007).

Shanghai Astronomical Observatory (SHAO): an artificial neural networks (ANN) method has been applied to predict EOPs (x, y & LOD). An operational prediction series of the AAM has been incorporated into the ANN model as an additional input in the real time prediction of the LOD, and the results show that the LOD prediction is significantly improved in comparison with that obtained by using LOD data only (Wang *et al.* 2008). The coupling between the magnetic field near the core-mantle-boundary and Earth nutation is discussed in a numerical integration approach, and it is shown that the

effect is one order of magnitude smaller than the gap of the out-of-phase part of the
−1.0-year nutation between the observations and theoretical nutation models (Huang
et al. 2007). The free oscillations of a 12-layers Earth model with rotation is studied by
the Galerkin method, while boundary conditions are treated with the Tau method. It
is shown that the Galerkin method is an alternative tool for such a study (Zhang et al.
2008). A new integrated formula to obtain the equilibrium figures of the Earth's interior
to third-order accuracy has been developed, in which both the direct and indirect contribution of the anti-symmetric crust layer are included, i.e., all non-zero order and odd
degree terms are included. Using this new potential theory and replacing the homogeneous outermost crust and oceanic layers in PREM with various real surface layers data,
the global dynamic flattening (H) is obtained and it is shown that the 1% difference
between H-PREM and H-obs can be removed by 2/3 (Liu & Huang 2008).

Wuhan Institute for Geodesy and Geophysics(WHIGG): the correlation between DLOD
and ENSO is studied, and it is shown that the maximum appears around three-five-year
timescales; weak correlation exists around the biennial timescales (Liu et al. 2005). The
concept of the normal Morlet wavelet transform (NMWT) is developed and applied to
polar motion (PM), and finds that the ocean always offsets the atmospheric effects on
annual PM and that the observed Chandler wobble has only one instantaneous frequency
all the time. (Liu et al. 2008).

Aleksander Brzeziński
president of the Commission

References

Altamimi, Z., *et al.* 2007, *J. Geophys. Res.*, Vol. 112, No.B09401
Barkin, Y. & Ferrándiz, J. M. 2007, in: A. Brzeziński, N. Capitaine, & B. Kołaczek (eds.), *Proc. Journées 2005*, Space Res. Centre PAS, Warsaw, p. 127
Boehm, J. *et al.* 2008, in: D. Behrend & K. Baver (eds.), *Proc. IVS 2008 General Meeting*, Russian Science Series, in press
Brzeziński, A. 2007, *J. Geodesy*, Vol. 81, p. 543
Brzeziński, A. 2008, in: N. Capitaine (ed.), *Proc. Journées 2007*, Obs. Paris, p. 180
Brzeziński, A. & Bolotin, S. 2006, in: A. Brzeziński, N. Capitaine, & B. Kołaczek (eds.), *Proc. Journées 2005*, Space Res. Centre PAS, Warsaw, p. 211
Capitaine, N. & Wallace, P. T. 2006, *A&A* 450, 855-872
Capitaine, N. & Wallace, P. T. 2008, *A&A*, 477, p. 277
Capitaine, N., Folgueira, M., & Souchay, J. 2006, *A&A* 445, p. 347
Capitaine, N., *et al.* 2008a, in: *Highlights of Astronomy*, Vol. 14, p. 474
Capitaine, N., *et al.* 2008b, in: D. Behrend & K. Baver (eds.), *Proc. IVS 2008 General Meeting*, Russian Science Series, in press
Chapanov, Y. & Gambis, D. 2008, in: N. Capitaine (ed.), *Proc. Journées 2007*, Obs. Paris, p. 206
Dehant, V. & Mathews, M.P. 2008, in: T. Herring & J. Schubert (eds.), *Treatise of Geophysics*, 3, 295
Dickey, J. O., Marcus, S. L., & Chin, T. M. 2007, *Geophys. Res. Letters*, 34, L17803
Efroimsky, M. & Escapa, A. 2007, *Celest. Mech. Dyn. Astr.*, 98, pp. 251-283
Englich, S., Heinkelmann, R., & Schuh, H. 2008, in: D. Behrend & K. Baver (eds.), *Proc. IVS 2008 General Meeting*, Russian Science Series, in press
Escapa, A., Getino, J., & Ferrándiz, J. M. 2008, in: N. Capitaine (ed.), *Proc. Journées 2007*, Obs. Paris, p. 113
Ferrándiz, J. M. & Barkin, Y. 2007, in: Anne Lemaitre (ed.), *Proc. International Workshop: Rotation of Celestial Bodies*, (Presses Universitaires de Namur), pp. 21-23
Ferrándiz, J. M., *et al.* 2007, in: A. Lemaitre (ed.), *Proc. International Workshop: Rotation of Celestial Bodies*, (Presses Universitaires de Namur), p. 9

Finkelstein, A., *et al.* 2006, *Proc. of IAA RAS*, No.14, 3, in Russian

Finkelstein, A., *et al.* (2007), *Astron. Letters*, Vol. 34, No.1, 59

Folgueira, M., Dehant, V., Lambert, S.B., & Rambaux, N. 2007, *A&A*, 469(3), 1197

Gambis, D., Salstein D., & Richard, J.-Y. 2008, in: N. Capitaine (ed.), *Proc. Journées 2007*, Obs. Paris, p. 210

Gorshkov, V. 2007, *Solar System Research*, 41, 65

Gross, R. S. 2007a, in T. A. Herring (ed.), *Geodesy, Treatise on Geophysics*, Vol. 3 Oxford: Elsevier, p. 239

Gross, R. S. 2007b, *Jet Propulsion Laboratory Publ.* 07-5, , Pasadena, CA, USA

Gross, R. S. 2008, *J. Geodesy*, submitted

Gross, R. S. & Chao, B. F. 2006, *Surv. Geophys.* 27(6), p. 615

Gross, R. S., de Viron, O., & van Dam, T. 2008a, in: N. Capitaine (ed.), *Proc. Journées 2007*, Obs. Paris, p. 212

Gross, R. S., *et al.* 2008b, in: F. Sanso (ed.), *IAG Symposium Series*, Vol. 133 New York: Springer-Verlag, in press

Gubanov, V. S. (2008), *Astron. Letters*, in press

Gubanov, V. S, Kurdubov, S. L., & Surkis, I. F. (2007), *Proc. of IAA RAS*, No.16, 61, in Russian

Han, Y. B., Hu, H., *et al.* 2007, *Earth, Moon, and Planets*, 100, 125

Hilton, J. L., *et al.* 2006, *Celest. Mech. Dyn. Astron.* 94, 351

Huang, C. L., Dehant, V., Liao, X. H., *et al.* 2007, *Highlights of Astronomy*, Vol. 14, p. 483

Johnson, T. J. 2008, *J. Geophys. Res.* vol. 113, No. B04407

Kalarus, M., Kosek, W., & Schuh, H. 2008, in: N. Capitaine (ed.), *Proc. Journées 2007*, Obs. Paris, p. 159

Koot, L., de Viron O., & Dehant V. 2006, *J. Geodesy*, 79, 663

Koot, L., Rivoldini, A., de Viron, O., & Dehant, V. 2008, *J. Geophys. Res.*, in press

Korbacz, A., Brzeziński, A, & Thomas, M. 2008, in: N. Capitaine (ed.), *Proc. Journées 2007*, Obs. Paris, p. 188

Kosek, W., Rzeszotko, A., & Popinski, W. 2006, in: A. Brzeziński, N. Capitaine, & B. Kołaczek (eds.), *Proc. Journées 2005*, Space Res. Centre PAS, p. 121

Kosek, W., Kalarus, M., & Niedzielski, T. 2008, in: N. Capitaine (ed.), *Proc. Journées 2007*, Obs. Paris, p. 165

Krasinsky, G. (2008), in: D. Behrend & K. Baver (eds.), *Proc. IVS 2008 General Meeting*, Russian Science Series, in press

Kudryashova, M., *et al.* 2008, in: D. Behrend & K. Baver (eds.), *Proc. IVS 2008 General Meeting*, Russian Science Series, in print

Kurdubov, S. & Skurikhina, E. 2008, in: D. Behrend & K. Baver (eds.), *Proc. IVS 2008 General Meeting*, Russian Science Series, in press

Lambert, S. B. 2006, *A&A*, 457, 717

Lambert, S. B. & Dehant, V. 2007, *A&A*, 469, 777

Lambert, S.B., & Mathews, P.M. 2006, *A&A*, 453, 363

Lambert, S., Bizouard, C., & Dehant, V. 2006, *Geophys. Res. Letters*, 33, L03303,

Lambert, S. B., Dehant, V., & Gontier, A.-M. 2008, *A&A*, 481(2), 535

Li, Y. 2006, *Chin. J. Astron. Astrophys.* 6(5), 629

Liu, Y. & Huang, C. L. 2008, in: J. W. Jin, I. Platais & M. A. C. Perryman (eds.) *Proc. IAU Symp. No. 248*, 403

Liu, L. T., Hsu, H. T., & Grafarend, E. W. 2005, *J. Geodyn.*, 39, 267

Liu, L. T., Hsu, H. T., & Grafarend, E. W. 2007, *J. Geophys. Res.*, Vol. 112, No.B08401

Ma, L. H. 2007, *The Observatory*, 127(1200), 364

Malkin, Z. 2007, *Solar System Research*, 41, 492

Malkin, Z. 2008, *J. Geodesy*, 82, 325

Malkin, Z. & Miller, N. 2007, in: J. Boehm, A. Pany, & H. Schuh (eds.), *Proc. 18th European VLBI for Geodesy and Astrometry Working Meeting*, Vienna, Austria, p. 93.

Mathews, P. M., Capitaine, N., & Dehant, V. 2007, 2007arXiv0710.0166M

Mendes Cerveira, P. J. *et al.* 2008, in: D. Behrend & K. Baver (eds.), *Proc. IVS 2008 General Meeting*, Russian Science Series, in print

Mendes Cerveira, P. J. *et al.* 2008b, *Advances in Geosciences*, in print

Müller, J., Kutterer, H., & Soffel M. 2005, in: N. Capitaine (ed.), *Proc. Journées 2004*, Obs. Paris, p. 121

Nastula, J. & Kołaczek, B. 2007, *Artificial Satellites*, Vol. 42, p. 1

Nastula, J., Kołaczek, B., & Popiński, W. 2006, in: A. Brzeziński, N. Capitaine, & B. Kołaczek (eds.), *Proc. Journées 2005*, Space Res. Centre PAS, p. 207

Nastula, J., Ponte, R. M., & Salstein, D. A. 2007, *Geophys. Res. Letters*, Vol. 34, L11306

Nastula, J., Salstein, D. A., & Kołaczek, B. 2008, in: N. Capitaine (ed.), *Proc. Journées 2007*, Obs. Paris, p. 220

Niedzielski, T. & Kosek, W. 2005, *Artificial Satellites*, 40 (3), p. 185

Niedzielski, T. & Kosek, W. 2008, *J. Geodesy*, 82, 83

Pešek, I. & Kostelecký, J. 2006a, *Studia Geoph. et Geod.*, 50, 537

Pešek, I. & Kostelecký, J. 2006b, in: L. Gerhátová & J. Hefty (eds.), *GPS+GLONASS+Galileo: nové obzory geodézie*, STU Bratislava, 51 (in Czech)

Salstein, D. A. & Nastula, J. 2006, in: A. Brzeziński, N. Capitaine, & B. Kołaczek (eds.), *Proc. Journées 2005*, Space Res. Centre PAS, Warsaw, p. 195

Schuh, H., *et al.* 2003, *Mitteilungen des Bundesamtes für Kartographie und Geodäsie*, Band 32, Frankfurt/Main

Seoane, L., Bizouard, C., & Gambis, D. 2008, in: N. Capitaine (ed.), *Proc. Journées 2007*, Obs. Paris, p. 196

Sokolova, J. & Malkin, Z. 2007, *A&A*, 474, 665

Souchay, J., Lambert, S., & Leponcin-Lafitte, C. 2007, *A&A*, 472, 681

Štefka, V. & Pešek, I. 2007 *Acta Geodyn. Geomater.*, 4 (148), 129

Štefka, V., Pešek, I., & Vondrák J. 2008, in: N. Capitaine (ed.) *Proc. Journées 2007*, Paris, 169

Varga, P., *et al.* 2007, *Acta Geod. Geoph.*, Vol. 42 (4), p. 433

Vondrák, J. & Ron, C. 2006, *Acta Geodyn. Geomater.*, 3 (143), 53

Vondrák, J. & Ron, C. 2007, *Acta Geodyn. Geomater.*, 4 (148), 121

Vondrák, J. & Ron, C. 2008, in: *Proc. Journées 2007*, Paris, 95

Vondrák, J., Ron, C., & Štefka, V. 2008, in: J. W. Jin, I. Platais & M. A. C. Perryman (eds.) *Proc. IAU Symp. No 248*, 89

Vondrák, J. & Štefka, V. 2007, *A&A*, 463, 783

Wallace, P. T. & Capitaine, N. 2006, *A&A* 459, 981; *A&A* 464, 793

Wang, Q. J., Liao, D. C., & Zhou, Y. H. 2008, *Chin. Sci. Bull.*, 53 (7), 969

Zerhouni, W., Capitaine, N., & Francou, G. 2008, in: N. Capitaine (ed.), *Proc. Journées 2007*, Obs. Paris, p. 123

Zhang, M., Seyed-Mahmoud, B., & Huang, C.L. 2008, in: J. W. Jin, I. Platais & M. A. C. Perryman (eds.), *Proc. IAU Symp.* No. 248, 409

Transactions IAU, Volume XXVIIA
Reports on Astronomy 2006–2009
Karel A. van der Hucht, ed.

© 2009 International Astronomical Union
doi:10.1017/S1743921308025283

COMMISSION 31

TIME
L'HEURE

PRESIDENT
VICE-PRESIDENT
PAST PRESIDENT
ORGANIZING COMMITTEE

Pascale Defraigne
Richard N. Manchester
Demetrios Matsakis
Mizuhiko Hosokawa, Sigfrido Leschiutta,
Demetrios Matsakis, Gérard Petit,
Zhai Zao-Cheng

TRIENNIAL REPORT 2006–2009

1. Introduction

The realization and dissemination of the international time scales is the responsibility of the section on Time, Frequency and Gravimetry of the BIPM. Commission 31 supports and coordinates investigations associated with Time definitions, realizations, astronomical data relevant to atomic timekeeping, such as pulsar data. The major developments achieved during the period 2005 - 2008 in that domain are reported here.

2. Consultative Committee for Time and Frequency

The CCTF held its 17th meeting on 14 and 15 September 2006. Following the discussions, six Recommendations were adopted and submitted to the Comité International des Poids et Mesures (CIPM). The list is the following:

Recommendation CCTF 1 (2006): Recommended values of standard frequencies for applications including the practical realization of the metre and secondary representations of the second;

Recommendation CCTF 2 (2006): Concerning secondary representations of the second;

Recommendation CCTF 3 (2006): Concerning the use of measurements of the International Atomic Time (TAI) scale unit; this Recommendation specifies the role of the Working Group on Primary Frequency Standards in evaluating new PFS and their use for the steering of TAI;

Recommendation CCTF 4 (2006): Concerning the use of Global Navigation Satellite System (GNSS) carrier phase techniques for time and frequency transfer in International Atomic Time (TAI); this Recommendation promotes the development of GNSS carrier phase techniques both for the laboratories equipment and for the BIPM to generate solutions and prepare their use in TAI computation.

Recommendation CCT 5 (2006): Improvement to Global Navigation Satellite System (GNSS) time transfer;

Recommendation CCTF 6 (2006): Coordination of the development of advanced time and frequency transfer techniques; this Recommendation promotes the activity of a recently created Working Group on Coordination of the Development of Advanced Time and Frequency Transfer Techniques, with the aim that time and frequency transfer techniques match the progresses in primary frequency standards.

Additionally discussed were the present form of UTC and the interest of preserving the leap second. The work of the ITU-R Special Rapporteur Group (SRG) was reviewed and a new CCTF working group was created to express a view of the CCTF regarding this issue.

3. Computation and dissemination of TAI and UTC

Reference time scales TAI and UTC have been computed regularly and published in the monthly Circular T. Definitive results have been available in the form of computer-readable files on the BIPM home page and on printed volumes of the Annual Report of the BIPM Time Section for 2005 (Volume 18) and of the BIPM Annual Report on Time Activities for 2006 and 2007, Volumes 19 and 20. Since January 1st, 2006, BIPM Time activities have been incorporated in the Time, frequency and gravimetry section, a new organization which is reflected in the BIPM publications on time.

Research concerning time-scale algorithms includes studies to improve the long-term stability of the free atomic time scale EAL and the accuracy of TAI. The number of laboratories participating to TAI steadily increases and is about 65 in 2008. The number of clocks participating in the ensemble time scale EAL also increases to about 350, among them more than 80% are either commercial caesium clocks of the HP5071A type or active, auto-tuned hydrogen masers. About 15% of the clocks reach maximum weight and they provide nearly 50% of the total weight. The medium-term stability of EAL, expressed in terms of an Allan deviation, is estimated to be 0.4×10^{-15} for averaging times of twenty to forty days over the period 2006 - 2008.

To characterize the accuracy of TAI, estimates are made of the relative departure, and its uncertainty, of the duration of the TAI scale interval from that of Terrestrial Time TT, as realized by primary frequency standards. Over 2006-2008, individual measurements of the TAI frequency have been provided by twelve primary frequency standards including nine Cs fountains. Detailed reports on the operation of primary standards are published in the Annual Report. Over 2006-2008, the global treatment of individual measurements has led to a relative departure of the duration of the TAI scale unit from that of TT ranging from 2×10^{-15} to 4×10^{-15}, with a standard uncertainty of order 0.5×10^{-15} to 1×10^{-15}. The steering procedure has been changed in 2006, to allow more frequent steerings in the aim of providing a more accurate TAI without impeding its stability.

4. The future of UTC

The ITU-R Special Rapporteur Group (SRG), created withing the ITU Working Party 7a to study the question ITU-R 236/7 'The Future of the UTC Time Scale', presented its report to the Working Party 7A in 2007. Several documents have been studied by the SRG, coming from countries members of the ITU, from organisations (like the BIPM), from laboratories providing with experiences during past leap seconds, etc, ... Due to the difficulties induced by a non-continuous UTC in some applications requiring a time scale without jumps (GNSS, tlcoms, newtworks, etc, ...), and as the IERS is able to provide a way to get UT1 in real time with a precision better than UTC, the present opinion is in favour of a new definition of UTC. There remains however still opposition from some national delegation. During its last meeting in April 2008, the WP7A decided to prepare a complete report of the content of the discussions since 2000; this report will be attached to the recommendation which should progress, after approbation by the Study Group 7, up to the World Radiocommunication Conference to be validated as official recommandation. The new definition could be in use 5 year after the approbation by the WRC, i.e., not before 2017.

5. Pulsars and time scales

Pulsars, especially those with periods in the millisecond range, are extraordinarily good clocks and have the potential to establish a new timescale. Pulse times of arrival can be determined with a precision of a few tens of nanoseconds at best, so an averaging interval of a year or more is required to obtain rate measurements to a precision comparable to that of atomic frequency standards. A pulsar timescale will therefore be useful only for detecting long-term variations in the atomic-clock-based timescales. However, a pulsar timescale is essentially independent of terrestrial effects, is based on an entirely different physical principle, namely rotation of a macroscopic body, and is potentially continuous for billions of years.

Just as existing timescales are based on a statistical average of data from an ensemble of atomic clocks, the best pulsar timescale will be based on measurements of an ensemble of pulsars. Significant progress has been made in the past few years in establishing such ensembles, known as pulsar timing arrays. The Parkes Pulsar Timing Array (PPTA) has been timing 20 millisecond pulsars every 2-3 weeks since 2004 (Manchester *et al.* 2006) and similar timing arrays are being established in Europe (EPTA) and in North America (NANOGrav). With such measurements, variations in atomic timescales over intervals of a month or two can be investigated. Techniques for analysis of such data are currently under development: Rodin (2006) uses a Wiener filter approach, whereas Zhong & Yang (2007) advocate the use of a wavelet transform method to isolate variations having different timescales.

6. Laboratory reports

USNO

To provide accurate and precise time, the USNO currently maintains an ensemble of 73 cesium-beam frequency standards and 32 cavity-tuned masers in three buildings in Washington, DC and at the Alternate Master Clock (AMC), at Schriever Air Force Base in Colorado. A non-operational cesium-based atomic fountain has reached stabilities of 1.E-15 at one day, and one rubidium-based fountain has reached stability better than 5.E-16 at two days. An improved timescale algorithm involving Kalman-based hourly characterization and steering of the Master Clock against the USNO ensemble was installed in 2004. Since 1 September 2006 UTC(USNO) stayed within 3 ns RMS of UTC. USNO has participated in Two Way Satellite Time Transfer (TWSTT) for the generation of TAI, and has calibrated its transatlantic timing link with the PTB in Germany on a biannual basis. It has participated as a contributor of clock information to the International GPS Service (IGS) via time-stable Ashtech Z123Ts at all three of its buildings, and contributes to two real-time GPS networks.

USNO serves as the timing reference for GPS, and has also been designated to serve in the role of measuring the GPS-Galileo Time Offset (GGTO). With the advent of GPS III, the stability and robustness requirements for the USNO Master Clock will become more stringent. In order to prepare for this, the USNO has embarked upon an ambitious campaign of infrastructure improvement. One aspect is the constructing of a new clock building, with a temperature stability spec of 0.1 degrees C and 3% relative humidity at each interior point, at all times. The second is the construction of six operational rubidium fountains, three of which will be located at the AMC and all of which are designed to operate continuously. We anticipate that the new building will be ready for operations in test-mode by 30 September 2008, along with two rubidium fountains. A third aspect is the design of new timescale algorithms which incorporated the superior performance of atomic fountains. Of equal importance is improved time transfer, and the USNO is therefore giving heavy emphasis to GPS receiver development, GPS calibration, and also to operational improvements for Two Way Satellite Time Transfer (TWSTT). While the design goals of these innovations are often expressed in terms of accuracy or precision, the explicit motivation behind many of the specifications is to guarantee robustness at the required levels.

Royal Observatory of Belgium

The ROB presently maintains 3 Cesium clocks HP5071A with standard tubes, and two active H-maser (CH1-75 and CH1-75A). The time scale UTC(ORB) is generated from the frequency of the active H-maser, of which the auto-tuning is performed using the second active maser CH1-75A. The steering of UTC(ORB) is performed only using the circular T, but UTC(ORB) is also monitored on a daily basis using a comparison with the IGS time scale and with several UTC(k)s produced by time laboratories collocated with IGS stations. This last comparison is done using a self-developed software named Atomium, using a combination of GPS code and carrier-phase measurements and the IGS products for satellite orbits and clocks. The ROB also developed during this triennium some methods for the combination of time transfer data (GNSS and TWSTFT).

LNE-SYRTE, Observatoire de Paris

The three atomic fountain clocks of the LNE-SYRTE have all reached an accuracy of a few times 10^{-16} and contribute regularly to UTC. Thanks to these references our atomic time scale now has a frequency accuracy comparable with that of TAI. Our strontium optical clock has reached an uncertainty of 2.6×10^{-15}, while our mercury optical clock project has advanced, with the achievement of laser cooling of mercury atoms and the beginning of cold atom spectroscopy. The compact cold atom caesium clock HORACE has reached a stability of 3.7×10^{-13} at 1 s. Our second two-way satellite time transfer system has been installed and tested and will soon begin regular comparisons with laboratories in the Asian region. Development of optical fibre frequency transfer links has continued, demonstrating a stability of 4×10^{-19} at 1 day over a link of 86 km. Our cold atom gyrometre has reached a sensitivity of 2.4×10^{-7} rad/s at 1 s and 10^{-8} rad/s in the long term, on a par with the best commercial fibre laser gyros, while our cold atom absolute gravimetre has reached a sensitivity of 1.7×10^{-7} m/s^2 at 1 s, which is competitive with the best mechanical absolute gravimetres. Other ongoing work includes the space clock project PHARAO/ACES, an atom-chip clock project, GPS receiver P-code calibration for Galileo, femtosecond laser combs and ultra-stable lasers.

7. Workshops, colloquia and conferences

Various meetings relating to the scope of Commission 31 were held. In addition to the usual meetings (see <www.astro.oma.be/IAU/COM31/>) of the time and frequency community , the fifth International Symposium on time scale algorithms, sponsored by the ROA, the USNO, the BIPM and the INRIM, was held in San Fernando (Spain) on 28-30 April 2008. A special issue of *Metrologia* will publish the Proceedings. Commission

31 also organizes a Joint Discussion at the Brazil IAU General Assembly in 2008, named 'Time and Astronomy'. This JD will provide a forum for discussion of recent work on UT1 determination as well as UT1 modeling and prediction, on the recent developments in precision pulsar timing and its application to time scales, planetary ephemerides, detection of gravitational waves and tests of gravitational theories and on the present realizations and performance of atomic time scales and time transfer techniques.

Pascale Defraigne
president of the Commission

References

Manchester, R. N. 2006, *ChJAA*, 6, suppl. 2, 139
Rodin, A. E. 2006, *ChJAA*, 6, Suppl. 2, 157
Zhong, C. X. & Yang, T. G. 2007, *ChJAA*, 31, 443

Transactions IAU, Volume XXVIIA
Reports on Astronomy 2006–2009
Karel A. van der Hucht, ed.

© 2009 International Astronomical Union
doi:10.1017/S1743921308025295

COMMISSION 52

RELATIVITY IN FUNDAMENTAL ASTRONOMY
RELATIVITÉ DANS
ASTRONOMIE FONDAMENTALE

PRESIDENT   Sergei A. Klioner
VICE-PRESIDENT   Gérard Petit
ORGANIZING COMMITTEE   Victor A. Brumberg, Nicole Capitaine,
Agnès Fienga, Toshio Fukushima,
Bernard R. Guinot, Cheng Huang,
Francois Mignard,
Kenneth P. Seidelmann,
Michael H. Soffel, Patrick T. Wallace

TRIENNIAL REPORT 2006 - 2009

1. The new Commission

The tremendous progress in technology which we have witnessed during the last 30 years has led to enormous improvements of observational accuracy in all disciplines of fundamental astronomy. Relativity has been becoming increasingly important for modeling and interpretation of high accuracy astronomical observations during at least these 30 years. It is clear that for current accuracy requirements astronomical problems have to be formulated within the framework of General Relativity Theory. Many high-precision astronomical techniques have already required the application of relativistic effects, which are several orders of magnitude larger than the technical accuracy of observations. In order to interpret the results of such observations, one has to construct involved relativistic models. Many current and planned observational projects can not achieve their goals if relativity is not taken into account properly. The future projects will require the introduction of higher-order relativistic effects. To make the relativistic models consistent with each other for different observational techniques, to formulate them in the simplest possible way for a given accuracy, and to formulate them in a language understandable for astronomers and engineers who have little knowledge of relativity are the challenges of a multidisciplinary research field called Applied Relativity.

The new IAU Commission 52 on *Relativity in Fundamental Astronomy* (RIFA) has been established during the IAU XXVI General Assembly (Prague, 2006) to centralize the efforts in the field of Applied Relativity and to provide an official forum for corresponding discussions. The general scientific goals of the Commission are:

- clarify geometrical and dynamical concepts of Fundamental Astronomy within a relativistic framework;
- provide adequate mathematical and physical formulations to be used in Fundamental Astronomy;

- deepen the understanding of the above results among astronomers and students in astronomy;
- promote research needed to accomplish these tasks.

2. The work of the Commission over the period September 2006 - June 2008

During this period the work of the Commission has been initiated. A web page containing all the information concerning the work of the Commission has been created and activated: <astro.geo.tu-dresden.de/RIFA>. The invitation to join the Commission has been widely distributed in the astronomical community.

Several scientific and educational projects have been initiated in 2007. The educational projects include compilation of a list of open problems in the field of applied relativity as well as a list of frequently asked questions. The Relativistic Glossary for astronomers has been also created and is available on the web page of the Commission. These three tools are intended to serve the broad astronomical community. The corresponding documents are expected to be updated and enriched in the future by the Commission itself and by future Working Groups of the Commission.

The following three scientific topics were identified by the Commission as important to discuss in the years to come:

1. 'Units of measurements' for astronomical quantities in the relativistic context

   In the literature (including very recent papers) one can find different units used in precise work: 'TDB units', 'TCB units', 'TT units' along with 'SI units'. The co-existence of these units is related to the relativistic scaling of time and space coordinates. On the other hand, the IAU 1991 resolutions clearly state that only SI units without any additional relativistic scaling should be used for all astronomical quantities (astronomical units like AU are not meant here). A balanced approach to this issue should be suggested and discussed. This would help to unify the notations and numerical values of astronomical constants throughout the literature. As a material for discussions Klioner (2008) has published a concise review of the problem of relativistic scaling of astronomical quantities.

2. 'Astronomical units' in the relativistic framework

   It is known that a significant freedom exists in the definition of the system of astronomical units in the framework of relativity. This freedom has been discussed in a number of recent publications, but up to now no standard choice has been agreed upon. Moreover, the complexity of relativistic modification of the current system of astronomical units together with the fact that the original reasons for astronomical units are no longer important for current practice naturally invokes a discussion on a possible simplification of the system of astronomical units. This question is clearly a delicate one since it concerns many parts of astronomy. Nevertheless, it seems to be right time to discuss these issues at the level of IAU Commissions and Working Groups.

3. Time-dependent ecliptic in the GCRS

   An improved definition of ecliptic adopted by the IAU XXVI General Assembly is given in the Barycentric Celestial Reference System. On the other hand, theories of Earth rotation for which ecliptic plays an important role should be defined in the GCRS. Therefore, a GCRS ecliptic has to be discussed. It has to be clarified if relativistic effects can affect the definition of the ecliptic at some perceptible level of accuracy.

Besides this activity, the Commission has decided to propose an IAU Symposium on *Relativity in Fundamental Astronomy: Dynamics, Reference Frames, and Data Analysis* in 2009. This proposal has been accepted as IAU Symposium No. 261. It has the goal to overview and summarize the progress that Applied Relativity has made during the past quarter of a century (the first and only IAU Symposium devoted to that field, namely IAU Symposium No. 114 on *Relativity in celestial mechanics and astrometry* has been held in StPetersburg, Russia, in 1985) and to develop the basis for the future of this discipline. Principal topics of the IAU Symposium No. 261 include

- astronomical reference frames in the relativistic framework;
- relativistic modelling of observational data;
- astronomical tests of relativity;
- relativistic dynamical modelling;
- relativity in astrodynamics and space navigation;
- modern observational techniques in fundamental astronomy;
- time measurement and time scales; and
- astronomical constants and units of measurements.

3. Plans for the period July 2008 - July 2009

The first attempt to discuss the issue of 'TDB units' (the first item in the list above) has been undertaken. To this end an ad hoc discussion forum, a so-called task team, has been initiated. The work of the task team 'TDB units' is planned to be finished within one year from now. Conclusions concerning the choice of wording and semantics when speaking about units of measurements will be made available afterwards.

IAU Symposium No. 261 on *Relativity in Fundamental Astronomy: Dynamics, Reference Frames, and Data Analysis* will be held from 25 April till 1 May 2009 in Virginia Beach, USA. The Proceedings of this Symposium are expected to be ready by the next IAU General Assembly and are intended to serve as a reference for the field of Applied Relativity for the years to come.

4. Recent developments in Applied Relativity

Not pretending to completeness, let us give a list of recent interesting developments in the field of Applied Relativity.

Like many other research fields, General Relativity in general and Applied Relativity in particular has many old ideas which have been abandoned or insufficiently developed in the past. It is certainly useful to recall and critically re-consider some of these ideas which seem to be forgotten undeservedly. One such idea was recently reiterated by Brumberg (2007) who demonstrated that in accordance to the old idea of Infeld the variational principle of the Einstein field equations may be used to derive the commonly employed Einstein-Infeld-Hoffman equations of motion from the linearized metric. Although this approach does not seem to lead to new practical results in the first post-Newtonian approximation, it might be interesting to apply this idea to simplify the derivation of the equations of motion of a system of N arbitrary shaped bodies in higher post-Newtonian approximations.

Recent re-definition of TDB adopted by the IAU XXVI General Assembly has created the possibility to define the time scales of new solar system ephemerides in full consistency with General Relativity. The ephemeris groups at JPL and in Paris Observatory have positively reacted on this possibility (Fienga, Manche & Laskar 2007; Folkner 2007). This fosters the hope that in the nearest future the solar system ephemerides will become consistently 4-dimensional as required by the nature of General Relativity.

Earth rotation is the only astronomical phenomenon which is observed with very high accuracy, but modelled in a Newtonian way. Although a number of attempts to estimate and calculate the relativistic effects in Earth rotation have been undertaken, no consistent theory has appeared until now. At least two projects have been recently started to improve the situation. Brumberg & Simon (2007) consider the formally Newtonian equations of rotational motion with all quantities relativistically transformed into dynamically non-rotating version of the GCRS. Klioner, Soffel & Le Poncin-Lafitte (2008) have started to develop the fully post-Newtonian theory of Earth rotation using numerical integration of the post-Newtonian equations of rotational motion.

A series of recent papers (Kouba 2004; Larson *et al.* 2007) consider additional relativistic effects in the GPS model. Although the relativistic model for GPS is relatively simple, the technical complexity of the GPS observations makes it difficult for experts in relativity to go beyond the simplest model. For this reason every step towards understanding of higher-order effects is important and potentially has a significant impact on the future high-accuracy applications of GPS, GLONASS and Galileo. A detailed review on Relativity in Geodesy has been published by Müller, Soffel & Klioner (2007).

The investigation of relativistic light propagation in the gravitational field of non-spherical, rotating and moving bodies has been continued. Kopeikin & Makarov (2007) have constructed a compact model for light deflection of sources observed close to giant planets of the solar system. This model will be used in the interpretation of the future high-accuracy experiments. Le Poncin-Lafitte & Teyssandier (2008) have developed an algorithmic procedure enabling one to determine explicitly the light ray connecting any two points located at a finite distance in the gravitational field of an isolated axisymmetric body.

Considering the possibility to have very accurate clock (stable and accurate at the level of $10^{-16} - 10^{-17}$) in space in the near future, special care must be taken in practical relativistic modelling of such clocks. Duchayne, Mercier & Wolf (2007) have investigated the relativistic modelling of high-accuracy clock on board of an Earth satellite, and time and frequency comparison with such a clock. The requirements for the accuracy of orbital data have been also investigated.

Another important research area is planning and verifying astronomical tests of General Relativity and alternative theories of gravity. Bertotti, Ashby & Iess (2008) have discussed the effect of the motion of the Sun on the light travel time in experiments such as Cassini relativity experiment. After an analysis of NASA's Orbit Determination Program (ODP) the authors confirm the claimed accuracy of the Cassini experiment (Bertotti, Iess & Tortora, 2003). Considering the anticipated accuracy of the future relativistic experiments it is important to go beyond the standard least-square fits and to provide realistic errors of the achieved estimates of relativistic parameters. A sophisticated statistical analysis of the relativistic experiments with ESA's mission BepiColombo (Milani *et al.* 2002) has been performed by Ashby, Bender & Wahr (2007).

ESA's second space astrometry mission, *Gaia* is progressing towards its launch in 2011. The unprecedented accuracy of *Gaia* (up to a few microarcsecond) makes it especially difficult to keep the whole data processing chain fully compatible with General Relativity. Significant efforts are being made in this area by the *Gaia* scientific community. A detailed description of the relativistic models will be made available to the public in due time.

Finally, the work on the improvement of the relativistic formulations and semantics of the IERS Conventions has been continued. One can hope that the wording in the next version of the IERS Conventions will be more consistent from the point of view of General Relativity.

5. Closing remarks

Applied Relativity is a multidisciplinary research field. Progress here requires dedicated efforts both from the side of theoretical work and from the side of practical implementation of relativistic concepts and ideas into every-day astronomical practice. It is clearly a challenge to combine knowledge in theoretical general relativity and in practical observational techniques and modelling of complex astronomical phenomena. IAU Commission 52 allows us to organize a dynamic and fruitful exchange between the experts in these areas, and thus give a direction to future developments in the area.

Sergei A. Klioner
president of the Commission

References

Ashby, N., Bender, P. L., & Wahr, J. M. 2007, *Phys. Rev. D*, 75, 022001
Bertotti, B., Ashby, N., & Iess, L. 2008, *Class. Quan. Grav.*, 25, 045013
Bertotti, B., Iess, L., & Tortora, P. 2003, *Nature*, 425 374
Brumberg, V. A. 2007, *Cel. Mech. Dyn. Astr.*, 99, 245
Brumberg, V. A. & Simon, J. -L. 2007, *Notes scientifique et techniques de l'insitut de méchanique céleste*, S088
Duchayne, L., Mercier, F., & Wolf, P. 2007, in press [arXiv:0708.2387v2]
Fienga, A., Manche, H., & Laskar, J. 2007, private communication
Folkner, W. 2007, private communication
Klioner, S. A. 2008, *A&A*, 478, 951
Klioner, S. A., Soffel, M., & Le Poncin-Lafitte, Chr. 2008, in: N. Capitaine (ed.), *The Celestial Reference Frame for the Future*, Proc. of Journeés, 2007, Observatoire de Paris, 139
Kopeikin, S. M. & Makarov, V. V. 2007, *Phys. Rev. D*, 75, 062002
Kouba, J. 2004, *GPS Solutions*, 8, 170
Milani, A., *et al.* 2002 *Phys. Rev. D*, 66, 1
Larson, K. M., Ashby, N., Hackman, C., & Bertiger, W. 2007, *Metrologia*, 44, 484
Le Poncin-Lafitte, Chr., & Teyssandier, P. 2008, *Phys. Rev. D*, 77, 044029
Müller, J., Soffel, M., & Klioner, S. A. 2007, *J. Geod.*, 82, 133

Transactions IAU, Volume XXVIIA
Reports on Astronomy 2006–2009
Karel A. van der Hucht, ed.

© 2009 International Astronomical Union
doi:10.1017/S1743921308025301

DIVISION I / WORKING GROUP
NUMERICAL STANDARDS OF FUNDAMENTAL ASTRONOMY

CHAIR   Brian J. Luzum

MEMBERS   Nicole Capitaine, Agnès Fienga,
William M. Folkner, Toshio Fukushima,
James L. Hilton, Catherine Y. Hohenkerk,
George A. Krasinsky, Gérard Petit,
Elena V. Pitjeva, Michael H. Soffel,
Patrick T. Wallace

TRIENNIAL REPORT 2006 - 2009

1. Introduction

The IAU Working Group (WG) on *Numerical Standards for Fundamental Astronomy* has been tasked with updating the IAU Current Best Estimates (CBEs), conforming with the IAU Resolutions, IERS Conventions and Système International d'Unités whenever possible. As part of its effort to achieve this, the WG is working in close cooperation with IAU Commissions 4 and 52, the IERS, and the BIPM Consultative Committee for Units.

This is the third IAU WG to be tasked with producing CBEs and is adding to the legacy of the two previous WGs. The first Sub-group on Numerical Standards of the IAU WG on *Astronomical Standards* was headed by E. M. Standish and the WG report (Standish 1995) established the rules which are still used today. For instance, this group decided on the two-tiered approach to the astronomical constants that we are currently using and also created the first CBEs for a list of IAU constants.

This work was continued by T. Fukushima and his IAU WG on *Astronomical Standards* (Fukushima 2000; Fukushima 2003). Many of the updates concerned work on constants in a general relativistic framework and improved estimates of the precession constant. This revised list of CBEs is the current IAU CBEs.

The excellent work of both these WGs has helped to establish the precedent and allows us to improve incrementally the values for which there are now better estimates.

2. Changes since the last Current Best Estimates

In addition to the need to update the CBEs because of improved estimates, there have also been significant changes that impact the IAU CBEs. Since the IAU CBEs were adopted, the IERS Conventions 2003, a document widely used by the astronomical and geodetic communities, has been produced. This reference contains estimates of many of the constants included in the IAU CBEs.

One significant change was the adoption of a new precession model with IAU 2006 Resolution B1. This resolution accepted the conclusion of the IAU Division I Working

Group on *Precession and the Ecliptic* (Hilton *et al.* 2006) and adopted the P03 precession theory of Capitaine *et al.* (2003). This resolution also replaced the terms 'lunisolar precession' and 'planetary precession' with 'precession of the equator' and 'precession of the ecliptic'. Another change is the redefinition of Barycentric Dynamical Time (TDB) that occurred with the adoption of IAU 2006 Resolution B3.

These resolutions fundamentally alter the status of the associated constants. For instance, the general precession found in the IAU CBEs is no longer the appropriate quantity to describe precession. It should be replaced with either the rate of precession of the equator in longitude (or of the Celestial Intermediate Pole), or a number of precession quantities or expressions. The resolutions also changed the status of the constant L_B to a defining constant.

3. Changes to the Current Best Estimates

The current WG started where the previous IAU WG tasked with providing CBEs left off, by using the existing IAU CBEs as the starting draft. From this starting point, the WG has proceeded to update the CBEs based on internationally adopted values and recent research. Some examples are adopting values:

- from the Committee on Data for Science and Technology (CODATA) 2006;
- recommended by the IAU 2006 Resolutions;
- based on recent research to modify most of the celestial object masses and adding

masses for Ceres, Pallas, Vesta, and Eris. These improvements are possible due to years of high-precision observations of spacecraft as they near planets and their satellites.

The changes to the CBEs serve two significant purposes. First, they keep the IAU CBEs consistent, where possible, with international standards. They also keep the constants consistent with the most accurate estimates currently available.

To date, nine additional constants have been added to the list of CBEs, one constant has been superseded by another, one constant has been removed from the list and the numerical values for nine additional constants have been replaced by more current values. The draft version of the CBEs is available at <maia.usno.navy.mil/NSFA/CBE.html>.

4. Additional concerns

In addition to updating the list of CBEs, the WG is addressing the larger issues surrounding the adoption of IAU CBEs. These include the mechanism to keep the CBEs current and the way in which these constants will be provided, the procedure to document the theoretical context of the constants, and whether the IAU should revise its current list of adopted constants to correspond with the new list of CBEs.

The mechanism for maintaining the IAU CBEs has been discussed and, to date, three options have been considered. The current method is for the IAU to form a WG when it believes that the CBEs need to be updated. This method has worked in the past and there is no reason to believe that it would not work in the future. Another possibility would be to enlist the aid of the IERS Conventions Center to maintain the IAU CBEs. Since the IERS is an IAU service organization and it already has a mechanism in place to maintain CBEs, it is possible that the IERS Conventions could be used to maintain the IAU CBEs as well. In addition to providing a tested process of maintaining CBEs, this would also ensure that the IAU and IERS use consistent CBEs. The biggest problem with this method is that the user communities and the areas of research of the IERS and the IAU are slightly different and there is a possibility that these differences could be problematic. A third option is to create a permanent WG within IAU Commission 4 that would maintain the list of CBEs. One potential problem with this option is that in the past, the IAU has been reluctant to allow WGs to exist indefinitely. More details will need to be obtained to determine the potential status of a WG either within an IAU Commission or Division.

The numerical values for the CBEs are not numbers that exist in isolation; they are defined fully within the theoretical context in which they are estimated. The WG will need to account for this in the presentation of the CBEs by making the theoretical underpinnings apparent to the users of the CBEs. The level to which this is done and the method of achieving this are still under consideration.

Electronic information is likely to play an important role in achieving the proper level of documentation for the CBEs. The extent to which electronic information is used is also a topic that is still under consideration. Electronic media can be used as either a primary source for defining information or a secondary source, providing supplemental information.

Considering the possibility that a new IAU System of Constants will be proposed, there has been a brief discussion of the level of accuracy necessary for this system. In order to provide an appropriate response, it is necessary to understand the users and appreciate the uses of the system of constants. There has been preliminary discussion within the Working Group regarding the possibility of proposing both a list of CBEs for the users with the highest accuracy needs and a new IAU System of Constants with a reduced accuracy for a larger community of users. This issue is still under discussion.

5. Update of IAU system of constants

There are now significant differences between the CBEs and the current IAU system of constants. This is due to both increasing accuracy of estimates and to changes in astronomical theory. As a result of this, there is a consensus to recommend to the 2009 IAU General Assembly that the IAU System of Constants be updated. There is also a consensus to recommend that the IAU should seek a mechanism to periodically update the value of the astronomical constants and CBEs in the future.

Brian J. Luzum
chair of the Working Group

References

Capitaine, N., Wallace, P. T., & Chapront, J. 2003, *A&A*, 412, 567

Fukushima, T. 2000, in: K. J. Johnston, D. D. McCarthy, B. J. Luzum, & G. Kaplan, (eds.) Proc. IAU Colloquium 180, Washington, DC, USA 27-31 March 2000, 417

Fukushima, T. 2003, in: H. Rickman (ed.), *Highlights of Astronomy*, 13, 107

Hilton, J. L., Capitaine, N., Chapront, J., Ferrandiz, J. M., Fienga, A., Fukushima, T., Getino, J., Mathews, P., Simon, J.-L., Soffel, M., Vondrak, J., Wallace, P., & Williams, J. 2006, *Celest. Mech. Dyn. Astr.*, 94, 351

Standish, E. M. 1995, in I. Appenzeller (ed.), *Highlights of Astronomy*, 12, 180

Transactions IAU, Volume XXVIIA
Reports on Astronomy 2006–2009
Karel A. van der Hucht, ed.

© 2009 International Astronomical Union
doi:10.1017/S1743921308025313

DIVISION I / WORKING GROUP
ASTROMETRY BY SMALL GROUND-BASED TELESCOPES

CHAIR   William Thuillot
VICE-CHAIR   Magdalena Stavinschi
MEMBERS   Alexander H. Andrei, Jean-Eudes Arlot,
                   Marcelo Assafin, N. Bazey,
                   George A. Gontcharov, Rustem Gumerov,
                   Jin Wenjing, Jose L. Muinos Haro,
                   Panayiotis Niarchos, Jose Pereira Osório,
                   Dan Pascu, Thierry Pauwels,
                   Gennadiy I. Pinigin, Y. Prostyuk,
                   A. Pugliano, Theodore J. Rafferty,
                   Jane L. Russell, Vladimir V. Rylkov,
                   M. Luisa Sanchez, Alexander A. Shulga,
                   Jean Souchay, Zhenghong Tang,
                   Ramachrisna Teixeira, Arthur R. Upgren,
                   William F. van Altena, Roberto Vieira Martins
                   Norbert Zacharias

TRIENNIAL REPORT 2006 - 2009

1. Introduction

At the IAU XXVI General Assembly in 2006, the Division I decided to create the Working Group on *Astrometry by Small Ground-Based Telescopes* (WG-ASGBT). Its scientic goals are to foster the follow-up of small bodies detected by the large surveys including the NEOs; to set-up a dedicated observation network for the follow-up of objects which will be detected by *Gaia*; to contribute to the observation campaigns of the mutual events of natural satellites, stellar occultations, and binary asteroids; and to encourage teaching astrometry for the next generation. The present report gives the main activities carried out in these areas with small telescopes (diameter less than 2 m).

2. Research developments within the past triennium

Several members of the working group report on recent activities in astrometry by small ground-based telescopes carried out by themselves or their team. The Working Group has presented a part of these activities during IAU Symposium No. 248 in Shanghai in 2007 (Thuillot *et al.* 2008).

N. Zacharias (USNO, Washington, USA) reports that the Northern and Southern Proper Motion (NPM, SPM) data (Lick Observatories, Yale/San-Juan) were re-reduced in a joined Yale/USNO effort. The UCAC3 is completed for a 2009 release of higly accurate positions and proper motions of about 80 million stars. The 20 cm USNO 'red lens' astrograph has been refurbished for a new all-sky observing program with a 440 million pixel camera. This 'U-mouse' astrometric survey will begin in 2009 from Cerro Tololo. For more details please see the IAU WG on *Densification of the Reference Frame* triennial report in this volume and Zacharias (2008, in IAU Symposium No 248).

Wenjing Jin (Shanghai Astronomical Observatory, China) reports on various astrometric activities in China. The peculiar following works were published. The 1.0/1.2 m NEOST (Near Earth Objects Space Telescope) equipped with a 4096 × 4096 SI CCD detector was installed completely and began to test observation. More than 188 new asteroids have been found including an Apollo-type NEO - 2007 JW2. A new Jupiter-family comet, P/2007 S1, was discovered (Zhao *et al.* 2008).

A collaboration with Nikolaev Astronomical Observatory (NAO), Ukraine, used the 25 cm telescope of Shanghai Astronomical Observatory equipped with a CCD camera S1C-077 in drift scan mode. Observations for GSS (Geo-Stationary Satellite) were carried out (Mao *et al.* 2007).

Astrometric observations of Triton have been done at the Shanghai 1.56 m telescope. Comparison with the theoretical positions provided from JPL and IMCCE showed a standard deviation of $0''.04$ of the O-C (Qiao *et al.* 2006).

Astrometric observations of Phoebe has been done at the 1 m telescope of the Yunnan Observatory and analyzed. Comparison the JPL ephemerides DE405 and SAT199 showed mean residuals of 210 mas and −50 mas in right ascension and declination, respectively, and a standard deviation of 6 mas (Peng & Zhang 2006).

Observations of the same satellite has been done at the Shanghai 1.56 m telescope. Comparison with JPL ephemerides show standard deviations in right ascension and declination are 58 mas and 7 mas respectively (Qiao *et al.* 2007).

Mutual events of the Galilean satellites has been observed with the 1 m telescope at Yunnan Observatory and analyzed. Astrometric positions were deduced and the comparisons were made with the theoretical models, Lieske's E5 and Lainey's L1. The accuracy of positions in RA and DEC are 103 and 88 mas for Lieske's theory while 74 and 80 mas for Lainey's theory (Peng & Noyelles 2007; Peng *et al.* 2008).

At IMCCE, Paris Observatory, France, improvement of the dynamical models and researches for the detection of small suspected effects (tidal effects, effect of the Yarkovsky force, . . .) on the motion of the Solar System bodies are performed. Highly accurate observational data are required and small telescopes are generally well adapted for this goal. Mutual events of the Uranian satellites, which occur every 42 yr and had never been observed until this occurrence, were predicted by (Arlot *et al.* 2006) and observed through an international campaign. The analysis is still in progress, only preliminary results are published at that time (Birlan *et al.* 2008).

As a complement of a program of detection of satellites of asteroids by large telescopes, observations by small telescopes are very useful such as those of mutual events. Such events have been predicted, surveyed and observed thanks to international campaigns: Antiope in 2005 (Descamps *et al.* 2007a), Patroclus in 2006 and 2007 (Berthier *et al.* 2007), Kalliope in 2007 (Descamps *et al.* 2007b). Orbital parameters and morphological characterization of these systems have been deduced.

An analysis of observations of planets and natural satellites performed at the meridian transit circle of Bordeaux Observatory, France, during the period 1997 - 2007 has been done. The observations of the planetary satellites provide pseudo observations of the planet itself. The results show the interest of continuing this type of observation (Arlot *et al.* 2008).

Researchers of IMCCE are involved in the preparation of the *Gaia* mission for the program related to the Solar System objects. A ground-based network of observers is being organized in order to perform follow-up observations of the Solar System bodies which could be detected by *Gaia* but difficult or impossible to re-observe by the satellite itself (Mignard *et al.* 2007).

M. Assafin (Obs. di Valongo, Rio de Janeiro, Brazil) reports on the activity of the Brazilian astrometric group. Astrometry of natural satellites is performed at the 1.6 m and 0.6 m of the Itajuba National Observatory, in particular for Jupiter (Elara, Himalia, Pasiphae, Carme), Saturn (Titan, Phoebe, Iapetus, Hiperion), Uranus (all main five ones), Neptune (Triton, Nereid). This group gives also an important contribution to the international campaigns of observation of the mutual events of the satellites lead by IMCCE, and recently for the Uranian satellites. It is also strongly involved in the campaign of observation of the mutual events of the binary asteroids (Descamps *et al.* 2007a).

Through a collaboration with B. Sicardy (Observatoire de Paris) predictions of stellar occultations by Pluto, TNOs or natural satellites are periodically made on the basis of the Brazilian astrometic observations. The improvement of these predictions thanks to last minute astrometry helps to choose the location of the mobile telescopes. The occultation event by Pluto on 14 July 2007 was successfuly observed on the 1.6 m telescope by this group (Sicardy *et al.* 2007). In a collaboration with astronomers from the Bucharest astronomical institute, Romania, the observations of ICRF sources have been performed at the 0.6 m Zeiss telescope at Belogradchik Observatory, Bulgaria (Assafin *et al.* 2007).

Th. Pauwels (ROB, Brussels, Belgium) describes the activity of the astrometry group of the Royal Observatory of Belgium. It performs several observations at the Uccle Schmidt telescope (0.85 cm aperture, 2 m focal length): 6687 positions of asteroids and 16 positions of comets were published in the MPCs (Observers P. De Cat, H. Debehogne, R. Desmet, E. Elst, A. Jonckheere, T. Pauwels, K. Puttemans). 41 positions of asteroids and 16 positions of comets were published in the MPECs (Observers P. De Cat and T. Pauwels). 28 occultations of stars by minor planets were attempted, of which 1 positive event. 8 light curves were obtained of mutual occultations and eclipses of an asteroid and its satellite (Berthier *et al.* 2007).

G. Pinigin, in collaboration with A. Shulga, A. Ivantsov and L.Hudkova, reports on observations and analysis done at the Research Institute National Astronomical Observatory (RI NAO, Nikolaev, Ukraine) with robotic or automatic telescopes. Fast Robotic Telescope FRT ($D = 300$ mm, $F = 1500$ mm)in a multichannel mode with rotary platform is used for observations of NEAs, artificial satellites and space debris in all orbits. The Axial Meridian Cirle AMC ($D = 180$ mm, $F = 2480$ mm), a modern CCD robotic telescope, has been upgraded with the CCD computer control and remote access. It is used for observations of stars and small Solar System bodies up to 17.5 mag. The telescope AZT-8 ($D = 700$ mm, F = 285 mm) is used for CCD astronomical research of objects up to 19 mag. in the near-Earth space since 2006 (agreement between Nat. Cent. Space Control and RI NAO).

G. Pinigin also reports on the international joint project between Turkish National Observatory TUBITAK (Z. Aslan), Kazan University, Russia (R. Gumerov) and NAO (Ukraine) on the observations of the small Solar System bodies at the TUBITAK RTT150 telescope ($D = 1500$ mm, F/8). The positional and photometric observations of asteroids down to 20 mag were made there in 2004 - 2008, and part of them was sent to the MPC. According to the astrometric precision obtained (50 mas), this telescope is well adapted for the *Gaia* follow-up program. New masses for 21 asteroids were determined by the dynamical method using modern ground-based CCD observations of the selected asteroids and dynamical model of asteroid motions, based on the DE405 (Aslan *et al.* 2007, Ivantsov 2007).

CCD observations of 200 fields around extragaclactic radiosources obtained in 2000-2003 with the RTT150 (TUG, Turkey) and 1 m telescope of Yunnan Observatory (Shanghai Astronomical Observatory, China) were used for collaborative research on refinement of linking optical-radio reference frames in Nikolaev with the reference 2MASS catalog.

W. van Altena (Yale Univ., New Haven, USA) reports on the second-epoch observations for the Yale/San-Juan Southern Proper Motion (SPM) program which will consist of approximately 100 million absolute proper motions from magnitude 5 to 17.5, magnitudes and colors in the blue and visual passbands. CCD frames are obtained with the 51 cm double astrograph at Cesco Observatory in El Leoncito, Argentina. Sky coverage south of –20 degrees declination is essentially complete as of June 2008.

Proper-motion determinations for six globular clusters (Casetti-Dinescu *et al.* 2007) provided a testbed for the combination of the first-epoch plate material with second-epoch CCD astrometry. Work is well underway on the next SPM catalog that will cover the sky south of -20 degrees. It is anticipated that the SPM4 will contain roughly 100 million stars and galaxies and be more than 90 percent complete to $V = 17.5$.

3. Education

Thanks to a joint initiative by astronomers from France (IMCCE) and China (Shanghai observatory, Beijing Planetarium, Jinan University of Guangzhou), a first Chinese-French Spring School on Astrometry 'observational campaign of solar system bodies', was held in Beijing, 7–13 April, 2008. The purpose was to foster students and young researchers to do researches in astrometry and to study new astrometric methods. 28 students and young astronomers attended this spring school. The lectures were given on fundamental astrometry, receptors, telescopes and images for astrometric purpose, astrometry through photometry and observational campaigns for phenomena, practical astrometry, space astrometry. In 2008, a census is organized by the working group in order to identify the universities and observatories where astrometry is teached and to know the topics of the main cursus delivered there. The results will be posted on the working group web page <www.imcce.fr/astrom/>.

William Thuillot
chair of the Working Group

References

Arlot, J.-E., Lainey, V., & Thuillot, W. 2006, *A&A*, 456, 1173
Arlot, J.-E., Dourneau, G., & Lecampion, J.-F. 2007, *A&A*, 484, 869
Aslan, Z., Gumerov, R., *et al.* 2007, in: O. Demircan, S. O. Selam & B. Albayrak (eds.), *Solar and Stellar Physics Through Eclipses*, ASP-CS, 370, 52
Assafin, M., Nedelcu, D. A., Popescu, P., *et al.* 2007, *A&A*, 476, 989
Berthier, J., Marchis, F., *et al.* 2007, *AAS*, Proc. DPS, 39, 35.05
Birlan, M., Nedelcu, D. A., *et al.* 2008, *Astron. Nach.*, 329, 567
Casetti-Dinescu, D. I., Girard, T. M., Herrera, D., *et al.* 2007, *AJ*, 134, 195
Descamps, P., Marchis, F., Michalowski, T., *et al.* 2007, *Icarus*, 187, 482
Descamps, P., Marchis, F., Pollock, J., *et al.* 2007, *P&SS*, 56, 1851
Ivantsov, A. V. 2007, *Kinematics and Physics of Celestial Bodies*, 23, 3, 108
Mao, Y. D., Tang, Z. H., Tao Jun, *et al.* 2007, *Acta Astronomica Sinica*, Vol. 48, No. 4, 475

Mignard, F., Cellino, A., Muinonen, K. *et al.* 2007, *EM&P*, 101, 97

Peng, Q. Y. & Zhang, Q. F. 2006, *MNRAS*, 366, 208

Peng, Q. Y. & Noyelles, B. 2007, *ChJAA*, Vol. 7, No. 2, 317

Peng, Q. Y., Emelyanov, N. V., *et al.* 2008, in: W. J. Jin, I. Platais & M. A. C. Perryman (eds.), *A Giant Step: from Milli- to Micro-arcsecond Astrometry*, Proc. IAU Symp. No. 248 (Cambridge: CUP), p. 114

Qiao, R. C., Tang, Z. H, Shen, K. X., Dourneau, G., *et al.* 2006, *A&A*, 454, 379

Qiao, R. C., Yan, Y. R., Shen, K. X., *et al.* 2007, *MNRAS*, 376, 1707

Sicardy, B., Widemann, T. *et al.* 2007, *AAS*, Proc. DPS, 39, 62.02

Thuillot, W., Stavinschi, M., Assafin, M., & the ABSGBT working group 2008, in: W. J. Jin, I. Platais & M. A. C. Perryman (eds.), *A Giant Step: from Milli- to Micro-arcsecond Astrometry*, Proc. IAU Symposium No. 248 (Cambridge: CUP), p. 286

Zhao, H. B.,Yao, J. S, & Lu, H. 2008, in: W. J. Jin, I. Platais & M. A. C. Perryman (eds.), *A Giant Step: from Milli- to Micro-arcsecond Astrometry*, Proc. IAU Symposium No. 248 (Cambridge: CUP), p. 505

Transactions IAU, Volume XXVIIA
Reports on Astronomy 2006–2009
Karel A. van der Hucht, ed.

© 2009 International Astronomical Union
doi:10.1017/S1743921308025325

DIVISIONS I-III / WORKING GROUP
CARTOGRAPHIC COORDINATES & ROTATIONAL ELEMENTS

CHAIR   Brent A. Archinal
PAST-CHAIR   P. Kenneth Seidelmann
MEMBERS   Michael F. A'Hearn, Edward L. Bowell,
                      Albert R. Conrad, Guy J. Consolmagno,
                      Dale P. Cruikshank, Daniel Hestroffer,
                      James L. Hilton, Gregory A. Neumann,
                      George A. Krasinsky, Jürgen Oberst,
                      Philip J. Stooke, Edward F. Tedesco,
                      David J. Tholen, Peter C. Thomas

TRIENNIAL REPORT 2006 - 2009

The IAU/IAG Working Group on *Cartographic Coordinates & Rotational Elements* published its (2006) triennial report containing current recommendations for models for solar system bodies (Seidelmann *et al.* 2007). P. Kenneth Seidelmann stepped down as chairperson and B. A. Archinal was elected chairperson at the Working Group business meeting that took place at the IAU XXVI General Assembly in Prague in 2006.

Important changes for the 2006 report includes corrections or updates to the recommended models for the rotation of the Sun, orientation of the Moon, orientation and size and shape of several Saturnian satellites, and orientation of Pluto and Charon. The recommended models for the size and orientation of several asteroids and comets were updated or given for the first time.

The Working Group is currently reformulating its membership and is starting to address changes needed for the next report. These will likely include changes for Mercury, due to recent Earth-based radar (Margot *et al.* 2007) and *MESSENGER* mission results, a new model for the Moon's orientation, a possible update to the Mars orientation model, and further updates for Saturnian Satellites due to recent *Cassini* mission results. An unusual orientation model will need to be developed for Titan where *Cassini*-RADAR observations have revealed a new pole position and periodic orientation changes apparently due to a sub-surface ocean (Stiles *et al.* 2008; Lorenz *et al.* 2008). A draft of the next report will be presented for discussion and completion at the IAU XXVII General Assembly in Rio de Janeiro, Brazil, in 2009.

Brent A. Archinal
chair of the Working Group

References

Margot, J. L., Peale, S. J., Jurgens, R. F., Slade, M. A., & Holin, I. V. 2007, *Science*, 316, 710
Lorenz, R. D., Stiles, B. W., Kirk, R. L., Allison, M. D., *et al.* 2008, *Science*, 318, 1649
Seidelmann, P. K., Archinal, B. A., A'Hearn, M. F., *et al.* 2007, *Cel. M&DA*, 98, 155
Stiles, B. W., Kirk, R. L., *et al.*, and the *Cassini*-RADAR Team 2008, *AJ*, 135, 1669

Transactions IAU, Volume XXVIIA
Reports on Astronomy 2006–2009
Karel A. van der Hucht, ed.

© 2009 International Astronomical Union
doi:10.1017/S1743921308025337

DIVISIONS I-III / WORKING GROUP
NATURAL PLANETARY SATELLITES

CHAIR   Jean-Eudes Arlot
MEMBERS   Kaare Aksnes, Carlo Blanco,
                     Nikolaj V. Emelianov, Robert A. Jacobson,
                     George A. Krasinsky, Jay H. Lieske,
                     Dan Pascu, Michel Rapaport,
                     Mitsuru Sôma, P. Kenneth Seidelmann,
                     Donald B. Taylor, Roberto Viera-Martins,
                     Gareth V. Williams

TRIENNIAL REPORT 2006 - 2009

1. Activities of the Working Group on Natural Planetary Satellites

The main goal of the Working Group was to gather astrometric observations made during the triennum as well as old observations not yet published in the data base. The WG encouraged the making of new observations. A Spring School was organized in China in order to teach the observational techniques of natural satellites to students and young astronomers. New theoretical models of the motion of the satellites and fit of the current models to new observations were used in order to make ephemerides of all the planetary satellites with tools useful for observations such as configurations. These ephemerides named MULTISAT are available at <www.imcce.fr/sat> or at <lnfm1.sai.msu.ru/neb/nss/nssephme.htm>.

Original ephemerides are also available on JPL's Horizons ephemerides and on MPC ephemerides for irregular satellites. A workshop has been held in Paris in November 2006 for organizing campaigns of observations. The problem of a standard format for the astrometric observations of the natural satellites raised and will have to be solved during the next triennum.

2. Selected works performed during the triennum

2.1. *The Martian satellites*

The data from MEX and MGS were analysed and information on the tidal dissipation within Mars were deduced from observation of the shadow of Phobos on Mars (Bills *et al.*, *J. Geophys. Res.*, 110, E7, 7004). Numerically integrated orbits for Phobos and Deimos were produced from fits to all observations including MEX and MGS (Lainey *et al.*, *A&A*, 465, 1075; Jacobson, LPI Contribution 1377). Jacobson's orbits were produced in support of the MRO Project and incorporated MRO imaging observations as well.

2.2. *The Galilean satellites*

Observations of the 1997 mutual events were published (Arlot *et al.*, *A&A*, 451, 733) and astrometric data were deduced from these photometric observations (Emelyanov &

Gilbert, *A&A*, 453, 1141). A new theory was published (Lainey *et al.*, *A&A*, 456, 783) and studies have been made on the tidal dissipation in Io (Lainey & Tobie, *Icarus*, 179, 85), on the rotation of Io and Europa (Henrard, *Icarus*, 178, 144; *Cel. M&DA*, 91, 131; 93, 101) and on the free and forced obliquities of the Galileans (Bills, *Icarus*, 175, 233).

2.3. *The inner satellites of Jupiter*

Several papers were published on Amalthea after the fly-by of Galileo on the gravity field (Weinwurm, *Adv. Sp. Res.*, 38, 2125) and on the density of Amalthea (Anderson *et al.*, *Science*, 308, 1291). Results were obtained from *Cassini* observations (Cooper *et al.*, *Icarus*, 181, 223).

2.4. *The outer satellites of Jupiter*

Observations were reported (Veiga, *A&A*, 453, 349). New orbits were calculated using all observations (Emelyanov, *A&A*, 435, 1173). The mass of Himalia was determined from perturbations on other satellites using ground-based observations (Emelyanov, *A&A*, 438, L33). Theoretical works were performed on an analytical theoretical model (Beauge' *et al.*, *AJ*, 131, p. 2299) and the dynamical evolution of this family of satellites (Christou, *Icarus*, 174, 215).

2.5. *The main satellites of Saturn*

The data of the Cassini probe provide many results on the gravity field, the shape (Jacobson *et al.*, *AJ*, 132, 711; 2520). Star occultations by Titan were reported (Sicardy *et al.*, *JGR*, 111, S91) as astrometric observations by the *HST* (French *et al.*, *PASP*, 118, 246). A new analysis technique for mutual events data was published (Ramirez *et al.*, *A&A*, 448, 1197) as a new image-processing technique for astrometry (Peng, *MNRAS*, 359, 15, 97).

2.6. *The faint inner satellites of Saturn*

The hypothetical satellites seen by Cassini led to study of their stability (Mourao *et al.*, *MNRAS*, 372, 1614). New orbits of the inner satellites have been deduced from *Cassini* and old observations (Spitale *et al.*, *AJ*, 132, 692; Jacobson *et al.*, *AJ*, 135, 261). The orbit of the newly discovered satellite Anthe was fit to *Cassini* data (Cooper *et al.*, *Icarus*, 195, 765).

2.7. *The outer satellites of Saturn*

A new orbit of Phoebe was calculated using all the available observations (Emelyanov, *A&A*, 473, 343). An analysis of albedo was made using Cassini data (Porco *et al.*, *Science*, 307, 1237) as spectrophotometry, useful to characterize the families of satellites and their formation (Buratti *et al.*, *Icarus*, 175, 490).

2.8. *The satellites of Uranus*

Astrometric observations of the main satellites were performed (Izmailov *et al.*, *Solar System Res.*, 41, 42), of Puck (Veiga & Bourget, *A&A*, 454, 683), mutual events were predicted (Arlot *et al.*, *A&A*, 456, 1173; Christou, *Icarus*, 178, 171)) and observed (Hidas *et al.*, *MNRAS*, 384, 38). Satellites U-12 to 17 have been named (CBET 323, 2005). New ephemerides for the main satellites were produced from fits to all but the mutual event data (Jacobson, *BAAS*, 39(3), 453).

2.9. *The satellites of Neptune*

Astrometric observations of Triton were performed (Qiao *et al.*, *MNRAS*, 376, 1707) as predictions of eclipses of Nereide (Mallama, *Icarus*, 187, 620). Constraints on the orbital evolution of Triton was published (Matija & Gladman, *ApJ*, 626, 113) and a study has been made of the disturbing function on the inner satellites (Yokoyama, *Adv. Sp. Res.*, 36, 569) followed by the modelling of the precession of the equator of Neptune (Nascimento *et al.*, 35th COSPAR Scientific Assembly, 18–25 July 2004, Paris, France).

New ephemerides for Triton, Nereid, and Proteus were produced by fitting all observations through the opposition of 2007 (Jacobson, *BAAS*, 40(2), 296). NOTE: Qiao's observations contain systematic errors and are unusable.

2.10. *The satellites of asteroids*

These objects are in fact binary or triple objects, the center of mass of the system being not inside the largest object. Nowadays a lot of binary (or triple) systems have been discovered. No data base of astrometric observations is available and no ephemeris is published.

2.11. *The rings*

The rings of the giant planets have been extensively observed by the space probes *Galileo* and *Cassini* and also by the *HST*. The dynamics are studied and the systems ring-moon have also been studied. The ring-moon system of Uranus has been studied (Showalter & Lissauer, *Science*, 311, 973; Gibbard *et al.*, Icarus, 174, 253).

2.12. *Miscellaneous*

The observability of the Natural Planetary Satellites has been explored by Tanga & Mignard (Proc. *Gaia* Symposium, Paris, 2004). The capture of irregular satellites of the giant planets has been studied (Nesvorny *et al.*, *AJ*, 133, 1962) as their chaotic behaviour (Mel'Nikov & Shevchenko, *Solar System Res.*, 39, 322).

Jean-Eudes Arlot
chair of the Working Group

Transactions IAU, Volume XXVIIA
Reports on Astronomy 2006–2009
Karel A. van der Hucht, ed.

© 2009 International Astronomical Union
doi:10.1017/S1743921308025349

DIVISION II SUN AND HELIOSPHERE

SOLEIL ET HELIOSPHERE

Division II of the IAU provides a forum for astronomers and astrophysicists studying a wide range of phenomena related to the structure, radiation and activity of the Sun, and its interaction with the Earth and the rest of the solar system. Division II encompasses three Commissions, 10, 12 and 49, and four Working Groups.

PRESIDENT	**Donald B. Melrose**
VICE-PRESIDENT	**Valentin Martínez Pillet**
PAST PRESIDENT	**David F. Webb**
SECRETARY	**Lidia van Driel-Gesztelyi**
BOARD	**Jean-Louis Bougeret, James A. Klimchuk**
	Alexander Kosovichev, Rudolf von Steiger

DIVISION II COMMISSIONS

Commission 10	Solar Activity
Commission 12	Solar Radiation and Structure
Commission 49	Interplanetary Plasma and Heliosphere

DIVISION II WORKING GROUPS

Division II WG	Solar Eclipses
Division II WG	Solar and Interplanetary Nomenclature
Division II WG	International Solar Data Access
Division II WG	International Collaboration on Space Weather

TRIENNIAL REPORT 2006-2009

1. Introduction

Solar physics is a distinct subdiscipline with strong overlaps with both astronomy and space physics. Compared with other branches of astronomy and astrophysics, there is an enormous amount of detailed data on the Sun and on various aspects of solar activity. Since the beginning of the space age, instruments on spacecraft have provided more and more detailed data with higher and higher resolution. The two spacecraft of NASA's *STEREO* mission, and the Japanese-led multi-national *Hinode* spacecraft were launched during the past triennium, and are providing yet a further advance in resolution of solar phenomena. During this period solar activity has been near minimum, and there is an expectation that these missions will provide important new insights into solar flares as the Sun becomes more active (see the report from Commission 10 for further details).

One opinion of the role of solar physics in the IAU was put by the prominent solar radiophysicist, Paul Wild, who died on May 10, 2008. With more than a touch of irony, in his opening address at the IAU General Assembly in Sydney in 1973, Wild said (Wild

1973): *"I have the feeling that to most astronomers the Sun is rather a nuisance. The reasons are quite complex. In the first place the Sun at once halves the astronomer's observing time from 24 to 12 hours, and then during most of the rest of the time it continues its perversity by illuminating the Moon. Furthermore I have met numerous astronomers who regard solar astronomy to be now, as always before, in a permanent state of decline – rather like Viennese music or English cricket. Nevertheless those who study the Sun and its planetary system occasionally make significant contributions. There were, for instance, Galileo and Newton who gave us mechanics and gravitation; Fraunhofer who gave us atomic spectra; Eddington and Bethe who pointed the way to nuclear energy; and Alfvén who gave us magneto-hydrodynamics. Perhaps the point to be recognized is that the Sun has more immediately to offer to physics rather than astronomy."*

The role of solar physics in initiating new fields of research of relevance to both physics and astronomy has continued since this was written. Wild's own research was one of the stimulants for the new field of plasma astrophysics, including collisionless shock waves, plasma emission processes and particle acceleration. Other areas of wider interest to physics and astronomy to which solar physics has made notable recent contributions include neutrino oscillations, astrophysical dynamos, and high-energy astrophysics.

2. Structure of the Division

Division II includes three Commissions, 10, 12 and 49, and four Working Groups. The current Commission presidents, elected at the Prague General Assembly, are James Klimchuk, Commission 10, Valentin Martínez Pillet, Commission 12, and Jean-Louis Bougeret, Commission 49. Also at the IAU XXVI General Assembly, in Prague, 2006, Lidia van Driel-Gesztelyi was elected the Division's first secretary.

Commission 10 *Solar Activity*, focuses on transient aspects of the Sun, including flares, prominence eruptions, coronal mass ejections (CMEs), particle acceleration, magnetic reconnection and topology, coronal loop heating, and shocks in the corona. Commission 12 *Solar Radiation and Structure*, emphasizes steady-state aspects of the Sun, including long-term irradiance, helioseismology, magnetic field generation, active regions, photosphere, and chromosphere. Commission 49 *Interplanetary Plasma and Heliosphere*, studies the solar wind, shocks and particle acceleration, both transient and steady-state, e.g., corotating, structures within the heliosphere, and the termination shock and boundary of the heliosphere. There can be considerable overlap among the Commissions, such as in the areas of magnetic activity, solar evolution, particle acceleration, and space weather.

The four working groups involve the topics of solar eclipses, solar and interplanetary nomenclature, access to solar data including virtual observatories, and space weather.

The organizing committee of Division II includes the president, vice-president, secretary and immediate past president of the Division, together with the presidents and vice-presidents of the three Commissions. The position of Divisional secretary had been discussed and agreed in principle at the Sydney General Assembly, but it was only at the Prague General Assembly that it was decided to elect a secretary from among the vice-presidents of the Commissions.

On the 50th anniversary of the *International Geophysical Year*, the IAU endorsed 2007 as the *International Heliophysical Year* (IHY). 'Heliophysical' is a broadening of the concept 'geophysical', extending the connections from the Earth to the Sun andinterplanetary space. IHY is taking advantage of the many near and deep-space missions/spacecraft launched over the past decade or so. IHY activities extend over the period 2007–2009, and are summarized on the website at URL <hy2007.org>.

IAU participation in the IHY has been coordinated through the Division, with the IHY subgroup chaired by Nat Gopalswamy, with David Webb the IAU representative for the IHY. IAU Symposium No. 257 on *Universal Heliophysical Processes* (September 2008, Greece) is a specific example of IAU involvement in IHY activities.

3. Working Groups

The four current Working Groups of the Division are *Solar Eclipses*, *Solar and Interplanetary Nomenclature*, *International Solar Data Access*, and *International Collaboration on Space Weather*.

3.1. *Solar Eclipses (Jay M. Pasachoff)*

The Working Group on *Solar Eclipses* is chaired by Jay M. Pasachoff (USA) and includes Iraida S. Kim (Russia), Hiroki Kurokawa (Japan), Jagdev Singh (India), Vojtech Rusin (Slovakia), Atila Ozguc (Turkey), Yihua Yan (China), Fred Espenak (USA), Jay Anderson (Canada, consultant on meteorology), Glenn Schneider (USA), and Michael Gill (UK, maintainer of Solar Eclipse Mailing List).

The WG has a web sites <www.eclipses.info> and <www.totalsolareclipse.net>. The WG has as its task the coordination of solar eclipse efforts, particularly making liaisons with customs and other officials of countries through which the path of totality passes and providing educational information about the safe observation of eclipses for the wide areas of the Earth in which total or partial eclipses are visible. The work is coordinated with that of the Program Group on Public Information at the Times of Eclipses of IAU Commission 46 on *Education and Development* (<www.eclipses.info>). Two members, Espenak and Anderson, produce the widely used NASA Technical Publications with eclipse paths and detailed information, available as hard copies or online, linked through www.eclipses.info or via the NASA Eclipse Web Site at <eclipse.gsfc.nasa.gov/>. The bulletins are available directly at URL <eclipse.gsfc.nasa.gov/SEpubs/bulletin.html>.

A notable past success was the distribution of material for tens of thousands of eye-protection filters after online consultation. A review article on eclipses will be published in Nature during 2009 as part of the *International Year of Astronomy*.

During the past triennium, following the Africa/Greece/Turkey/Asia eclipse (29 March 2006), total eclipses had a hiatus until the eclipses of 1 August 2008 and 22 July 2009. The 2008 total eclipse started in northern Canada and Greenland, crossed the extreme north (within $7°$ of latitude from the North pole), and descended through central Russia (including 2 minutes 20 seconds at the Siberian city of Novosibirsk), western Mongolia, and China. The 2009 total eclipse will start in India and cross Bangladesh before it enters China. It will cross Hangzhou and Shanghai while proceeding eastward, where it will pass just south of the main Japanese islands, crossing some small Japanese islands, before it reaches its maximum duration in the Pacific. As the longest eclipse in the saros, it passes 5 minutes 50 seconds on the centerline near Shanghai and reaches 6 minutes 39 seconds in the Pacific. The date of the 2009 eclipse clashed with the originally proposed start date of the IAU XXVII General Assembly in Rio de Janeiro, which is now scheduled to start less than two weeks after the eclipse.

The annular eclipse of 22 September 2006 was visible from Guyana, Suriname, French Guiana, adjacent Brazil, and the Atlantic Ocean, with partial phases visible from eastern South America, western and southern Africa, and part of Antarctica. Eclipse-knowledge-able professionals and amateurs gathered especially in Kourou, French Guiana. The annular eclipse of 7 February 2008 was visible from Antarctica, with one French amateur astronomer viewing it from the top of Antarctica's tallest mountain and with partial phases visible from eastern Australia and New Zealand. The 26 January 2009 annular eclipse will be visible from Indonesia, especially western Java and eastern Sumatra, with partial phases visible from southern Africa, southern India, southeast Asia, and western Australia. The annular eclipse of 20 May 2012 begins in southeastern China, touches the southeast coast of Japan, crosses the Pacific, and ends in the southwestern United States, from northern California through parts of Utah, Arizona, and New Mexico, and ending in northwestern Texas. Of course, partial phases will be widely visible to the sides of these paths of annularity.

Partial solar eclipses included 19 March 2007, visible widely over Asia east of India, as far east as western Japan and western Alaska, and as far south as all of Korea; and 11 September 2007, visible in South America except for its northern part. There will be no other partial eclipses through 2010.

The following triennium will include the total eclipse of 11 July 2010 visible from some French South Pacific islands and Chile's Easter Island, and ending over southern Chile and Argentina. It will also include the 15 January 2010 annular eclipse, visible from Africa, the southern tip of India, northern Sri Lanka, Myanmar, and China, with partial phases visible from eastern Europe through Asia except for Japan and past mid-Java. The following total eclipse will be on 13 November 2012, which will begin and cross land only in northern Australia, including the Cape York peninsula, cross the ocean north of New Zealand, and continue over the southern Pacific, ending in the ocean substantially west of Chile.

3.2. *Solar and Interplanetary Nomenclature (Edward W. Cliver)*

The Working Group on *Solar and Interplanetary Nomenclature* is chaired by Edward W. Cliver (USA) and includes Jean-Louis Bougeret (France), Hilary Cane (Australia), Takeo Kosugi (Japan; deceased, 26 November 2006), Sara Martin (USA), Reiner Schwenn (Germany), and Lidia van-Driel-Gesztelyi (France, UK, Hungary).

With the help of the broader community, the WG identifies terms used in solar and heliospheric physics that are thought to be in need of clarification, and then commissions topical experts to write essays reviewing the origins of terms and their current usage or misusage. The first six essays were published in Eos, the weekly publication of the American Geophysical Union, under 'The Last Word' heading. The most recent essay in the series, and the only one published from September 2006 - June 2008, appeared in Solar Physics, called, *The misnomer of 'post-flare loops'* (Švestka 2007). At present no additional articles are in the queue.

3.3. *International Solar Data Access (Robert D. Bentley)*

The Working Group *International Solar Data Access* is chaired by Robert D. Bentley (UK). Originally formed as a group intended only to cover the solar part of Division II, the WG has been extended to include heliophysics data sets needed to support Space Weather and related studies. Many of the members of the WG have discussed the idea of building a virtual observatory (VO) in heliophysics. In order to address science problems that span the disciplinary boundaries this would would provide enhanced access to solar and heliospheric data, and to magnetospheric and ionospheric data for planets with magetic fields and/or atmospheres. In June 2007, several members of the WG met at the *Virtual Observatories in Geosciences* (VOiG) Conference held in Denver, CO, USA; the VOiG Conference was an activity held within the context of the Electronic Geophysical Year (eGY). Also in June 2007 Bentley presented a paper at the

IHY meeting in Turin entitled 'Building a Virtual Observatory in Heliophysics' that discusses many of the issues related to metadata that need to be resolved if you are to conduct automated searches across the domains. The WG has a website at the URL: `<www.mssl.ucl.ac.uk/grid/iau/index.html>`.

3.4. *International Collaboration on Space Weather (David F. Webb)*

The Working Group for *International Collaboration on Space Weather* is chaired by David Webb (USA). The WG has as its main goal to help coordinate the many activities related to space weather at an international level. Its website is at the URL: http://www.iac.es/proyecto/iau_divii/IAU-DivII/main/spaceweather.php. The site currently includes the international activities of the *International Heliospheric Year* (IHY), the *International Living with a Star* (ILWS) program, the *Climate and Weather of the Sun-Earth System* (CAWSES) WG on *Sources of Geomagnetic Activity*, and *Space Weather* studies in China.

The *International Heliospheric Year* is an international program of scientific collaboration that during the time period 2007–2009, centered on 2008, the 50th anniversary of the *International Geophysical Year*. Nat Gopalswamy (USA) is the chair of the IHY subgroup within Division II. The IHY was considered to be of sufficient importance to the IAU to have its own IAU scientific representative, currently David F. Webb. The physical realm of the IHY encompasses all of the solar system out to the interstellar medium, representing a direct connection between in-situ and remote observations. The IHY working group helped identify national leaders for the IHY program. The IHY organization has an International Advisory Committee headed by Roger M. Bonnet (France) and an International Steering Committee headed by Joseph M. Davila (USA). Coordinators were appointed for eight regions of the world and ~ 60 countries have functioning national committees. Complete information on the IHY can be found at the main IHY URL: `<http://ihy2007.org>`. Four key activities are planned under the IHY program: science activities, the UN Basic Space Sciences (UNBSS) initiative, IGY Gold, and public outreach activities.

The CAWSES Working Group on *Sources of Geomagnetic Activity*, also chaired by Nat Gopalswamy (USA), has as its objectives to understand how solar events, such as CMEs and high speed streams, impact geospace by investigating the underlying science and developing prediction models and tools. Finally, the Working Group on Space Weather Studies in China is chaired by Jingxiu Wang (China) and is involved with many new initiatives on space weather.

4. New website

During the last triennium a new Division II web site was established, and is maintained at URL: `<www.iac.es/proyecto/iau_divii/IAU-DivII/main/index.php>`. The web site provides a description of the Division objectives, lists the members of the Organizing Committee, informs of the latest Division-wide news and upcoming events and provides links to the three Commissions and four Working Groups (as well as to the IAU main web page). Publicly distributable documentation related to the Division is made available through the website, including IAU Newsletter or Symposia selections when available. The web site also hosts a private section (requiring authentication) that allows the members of the OC of the Division (and Commissions when implemented) to retrieve submitted Symposium, JD, etc., proposals during the evaluation phase. Currently, following the IAU making available (mid 2008) updated member lists, Division-wide and Commission-wide email listings are being implemented.

5. IAU meetings

The Division has played a leading role in the following IAU meetings, that have been held since the last triennial report Webb *et al.* (2006):

IAU S233 on *Solar Activity and its Magnetic Origin* was held in Cairo, Egypt, March 31-April 3, 2006 (Bothmer & Hady). The symposium was timed to follow the total solar eclipse of March 29, 2006.

At the General Assembly in Prague in 2006, Division II was involved in three Joint Discussions and one Special Session:

- JD1 on *Cosmic Particle Acceleration - from Solar System to AGNs* (Karlický & Brown 2007).
- JD3 on *Solar Active Regions and 3D Magnetic Structure* (Choudhary & Sobotka 2007).
- JD8 on *Solar and Stellar Activity Cycles* (Kosovichev & Strassmeier 2007).
- SpS5 on *Astronomy for the Developing World*, subsession on the IHY.

Symposia since the IAU involving Division II were:

- IAU S247 on *Waves & Oscillations in the Solar Atmosphere: Heating and Magneto-Seismology* was held in Porlamar, Isla de Margarita, Venezuela, 17-22 September 2007 (Erdélyi & Mendoza-Briceño 2007).
- IAU S257 on *Universal Heliophysical Processes* was held in Ionnina, Greece, 15–19 September 2008. The proceedings are to be edited by Dave Webb and Nat Gopalswamy.

One further symposium is to be held soon after the deadline for the present report:
- IAU S259 on *Cosmic Magnetic Fields – from Planets, to Stars and Galaxies* in Puerto Santiago, Tenerife Spain, 3-7 November 2008. Contact: Klaus G. Strassmeier.

Donald B. Melrose
president of the Division

References

Bothmer, V. & Hady, A. A. (eds.) 2006, *Solar Activity and its Magnetic Origin*, Proc. IAU Symp. No. 233 (Cambridge: CUP)

Choudhary, D. P. & Sobotka, M. 2007, in: K. A. van der Hucht (ed.), *Highlights of Astronomy*, Vol. 14, 139 (Cambridge: CUP)

Erdélyi, R. & Mendoza-Briceño, C. A. (eds.) 2007, *Waves & Oscillations in the Solar Atmosphere: Heating & Magneto-Seismology* (Cambridge: CUP)

Karlický, M. & Brown, J. C. 2007, in: K. A. van der Hucht (ed.), *Highlights of Astronomy*, Vol. 14, 79 (Cambridge: CUP)

Kosovichev, A. G. & Strassmeier, K. G. 2007, in: K. A. van der Hucht (ed.), *Highlights of Astronomy*, Vol. 14, 271 (Cambridge: CUP)

Webb, D. F. 2007, in: O. Engvold (ed.), *Transactions IAU XXVB, Reports on Astronomy 2002-2005* (Cambridge: CUP), p. 69

Wild, J. P. 1973, in: G. Contopoulos (ed.), *Highlights of Astronomy*, Vol. 3, 3

Švestka, Z. 2007, *Solar Physics*, 246, 393

Transactions IAU, Volume XXVIIA
Reports on Astronomy 2006–2009
Karel A. van der Hucht, ed.

© 2009 International Astronomical Union
doi:10.1017/S1743921308025350

COMMISSION 10

SOLAR ACTIVITY
ACTIVITÉ SOLAIRE

PRESIDENT
VICE-PRESIDENT
SECRETARY
PAST PRESIDENT
ORGANIZING COMMITTEE

James A. Klimchuk
Lidia van Driel-Gesztelyi
Carolus J. Schrijver
Donald B. Melrose
Lyndsay Fletcher,
Natchimuthuk Gopalswamy,
Richard A. Harrison,
Cristina H. Mandrini, Hardi Peter,
Saku Tsuneta, Bojan Vršnak,
Jing-Xiu Wang

TRIENNIAL REPORT 2006 - 2009

1. Introduction

Commission 10 deals with solar activity in all of its forms, ranging from the smallest nanoflares to the largest coronal mass ejections. This report reviews scientific progress over the roughly two-year period ending in the middle of 2008. This has been an exciting time in solar physics, highlighted by the launches of the *Hinode* and *STEREO* missions late in 2006. The report is reasonably comprehensive, though it is far from exhaustive. Limited space prevents the inclusion of many significant results. The report is divided into the following sections: Photosphere and chromosphere; Transition region; Corona and coronal heating; Coronal jets; flares; Coronal mass ejection initiation; Global coronal waves and shocks; Coronal dimming; The link between low coronal CME signatures and magnetic clouds; Coronal mass ejections in the heliosphere; and Coronal mass ejections and space weather. Primary authorship is indicated at the beginning of each section.

2. Photosphere and chromosphere (C. J. Schrijver)

2.1. *Quiet-Sun field within the photosphere*

The Solar Optical Telescope (SOT) on board the *Hinode* spacecraft provides an unprecedented combination of spatial resolution and continuity of observations. The SpectroPolarimeter focal-plane instrumentation exploits that to measure the polarization signals from the photospheric plasma. Lites *et al.* (2008) and Ishikawa *et al.* (2008) show direct evidence that much of the magnetic field in the quiet-Sun photosphere is essentially horizontal to the solar surface. This observation is the direct confirmation of the existence of the weak field for which less direct evidence had been found by Harvey *et al.* (2007), and contributes to Hanle de-polarization effects discussed by, e.g., Trujillo Bueno *et al.* (2004).

The nearly vertical component is found primarily in the downflow network of the granular convection, corresponding to the well-known network field. The horizontal field is mostly found in the interior of the convective cells. Despite this significant preference for a separation by upflow and downflow domains, flux has been observed to also emerge already largely vertical even within the interior of the granular convective cells (Orozco-Suarez *et al.* 2008). And, perhaps not surprisingly, the conceptually expected evolutionary pattern of emerging flux is also seen: Centeno *et al.* (2007) report on observations in which field is seen to first surface nearly horizontally and subsequently – as it is advected to the downflow lanes – rights itself to be nearly vertical.

Lites *et al.* (2008) measure the mean flux density of the horizontal field to be about five times higher than that associated with the nearly vertical field component. Interestingly, radiative MHD simulations of near-surface stratified convection by Schuessler and Voegler (2008) show a very similar orientation-dependent ratio for the field. Steiner *et al.* (2008) reach a similar conclusion based on their numerical experiments: they argue that the granular upflows allow field to be stretched horizontally, being advected from over the cell centers only slowly in the stagnating, overshooting, upper-photospheric flows. Both studies support the conclusion that near-surface turbulent dynamo action significantly contributes to the internetwork photospheric field. A study by Abbett (2007) elucidates how such a turbulent-dynamo field would connect sub-photospheric and coronal layers through a complex and dynamic chromospheric layer in between; work by Isobe *et al.* (2008) explores numerically the frequent reconnective interactions expected with the overlying chromospheric canopy field, suggesting that this and the associated wave generation may have significant consequences for atmospheric heating and driving of the solar wind.

2.2. *The solar dynamo(s): global and local aspects*

Ephemeral bipolar regions are at the small end of the active region spectrum. Their properties over the solar cycle are an extension of those of their larger counterparts: they follow the general butterfly pattern, and have the proper preferential orientation of their dipole axes relative to the equator, but with a spread about the mean that increases towards the smaller bipoles. In this, they are a natural extension of the active region population. Where they were known to differ from the large regions is in the fact that they are the first to appear and last to fade for a given sunspot cycle.

Now, work by Hagenaar *et al.* (2008) uncovers another distinct property of ephemeral regions: the emergence frequency decreases with increasing local flux imbalance (consistent with findings by Abramenko *et al.* (2006) and Zhang *et al.* (2006) who differentiated only coronal hole regions from other quiet-Sun regions). Hagenaar *et al.* (2008) find that the rate of flux emergence is lower within strongly unipolar network regions by at least a factor of 3 relative to flux-balanced quiet Sun. One consequence of this is that because coronal holes overlie strong network regions, there are fewer ephemeral regions, and therefore fewer EUV or X-ray bright points within coronal holes.

The ephemeral-region population thus takes an interesting position in the study of solar magnetic activity: with the smallest-scale internetwork field perhaps largely generated by a local turbulent dynamo (Schuessler & Voegler 2008), and with the active regions associated with a global dynamo action, the ephemeral region population has signatures of both. Voegler & Schuessler (2007) show that local dynamo action can lead to a mixed-polarity field similar to the flux balanced very-quiet network field. It remains to be seen what such experiments predict in case there is a net flux imbalance, i.e., a background 'guide field': are fewer ephemeral regions generated, or does reconnection with the background guide field cause fewer of them to survive the rise to the surface (see discussion by Hagenaar *et al.* 2008).

2.3. *Emerging flux: observations and numerical experiments*

Observations made with the *Hinode*-SOT show unambiguously that magnetic flux bundles that form active regions do not emerge as simply curved arches, but rather as fragmented collections of undulating flux bundles. Each bundle likely crosses the photosphere one or more times between the extremes of the emerging region (e.g., Lites 2008). This is likely the result of the coupling to the near-surface convective motions, and the difficulty of relatively heavy sub-photospheric material to drain from the dipped field segments. Reconnection between neighboring supra-photospheric flux bundles could pinch off the sub-photospheric mass pockets, thus allowing the field to rise into the corona. Radiative MHD simulations of emerging flux by Cheung *et al.* (2008) support this interpretation: they show the 'serpentine' nature of the emerging flux, with characteristics that resemble the observed patterns of emerging flux, flux cancellation with associated downflows, convective collapse into strong-field flux concentrations, and photospheric bright points. Note that an example of field dipping into sub-photospheric layers is also discussed by Abbett (2007).

2.4. *Upper-chromospheric dynamics: spicules and waves*

Spectacularly sharp Ca II-H narrow-band filter observations made both with *Hinode*-SOT and the Swedish Vacuum Solar Telescope reveal ubiquitous jet-like features (called spicules or fibrils) above the solar limb. The relatively long-lived, broad population among these (discussed by De Pontieu *et al.* 2007a) appear to be caused by acoustic shock waves propagating upward from the photosphere. These shock waves cause the chromospheric material to undulate with almost perfectly parabolic height-time profiles, and saw-tooth velocity patterns. These shock-induced fibrils occur both in plages and in quiet Sun.

A more enigmatic phenomenon is formed by the much finer and more transient population of hair-like high extension of the chromospheric plasma discussed by De Pontieu *et al.* (2007b). Their origin remains subject to debate, but their transverse displacements point to the ubiquitous existence of Alfvén-like waves propagating into the corona. This is the most direct observational evidence for the existence of such waves reported to date. The estimated power suffices to heat the quiet-Sun corona and power the solar wind: these waves have amplitudes of 10-25 km s^{-1} for periods of 100-500 seconds. Alfvén-like waves with similar periods have also been observed for the first time in coronal loops, using the CoMP instrument (Tomczyk *et al.* 2007).

3. Transition region (H. Peter)

The transition region from the chromosphere to the corona, originally thought of as a simple thin onion shell-like layer, is a spatially and temporally highly complex part of the solar atmosphere. So far we are missing a unifying picture combining the numerous phenomena observed in emission lines formed from a couple of 10 000 K to several 100 000 K. Some of the aspects are re-interpreted by Judge (2008), who attempts to explain the transition region as being due to cross-field conduction of neutral atoms. The emission measure increasing towards low temperatures and the persistent redshifts are two of the major observational facts to be explained. A collection of one-dimensional transient models (Spadaro *et al.* 2006) and a three-dimensional MHD model (Peter *et al.* 2006) gave quantitative explanations for this. In agreement with the latter model, Doschek (2006) could show that the bulk of the (low) transition region emission originates from small cool loops. Using rocket imaging data, Patsourakos *et al.* (2007) give further direct observational evidence for the existence of such small loop-like structures dominating the transition region emission.

The magnetic field has to connect the transition region structures down to the chromosphere, and in case they are part of hot coronal elements also to the corona (Peter 2007). However, correlations between the transition region and the photosphere cannot be identified in a unique way (Sánchez Almeida et al. 2007). Larger features in the photosphere, such as moving magnetic features, might well leave an imprint in the outer atmosphere (Lin et al. 2006). In general, the connection from the chromosphere to the transition region is quite subtle and hard to identify in observations (Hansteen et al. 2007). A new way to investigate the relation between transient events in the transition region and the chromosphere was presented by Innes (2008). She studied the chromospheric emission of molecular hydrogen near 111.9 nm during microflaring events and proposes that the (coronal) energy deposition in the microflare also heats the chromosphere and thus affects the opacity for molecular hydrogen lines.

The energy balance, one part being the heating process, is largely determining the pressure of the transition region and thus implicitly also the mass loading. Combining various models and observations, Aschwanden et al. (2007) argue that the bulk part of the heating is located deep down, basically reflecting an exponential decay of the heating rate with height, on average. As speculated earlier, Tian et al. (2008a) could now show that the persistent blue-shifts in upper transition region lines are not due to the solar wind outflow, but due to mass loading of loops. Recent Doppler shift observations with the Extreme ultraviolet Imaging Spectrometer (EIS; Culhane et al. 2007a) on board Hinode indicate that the redshifts are due to radiative cooling and subsequent bulk downflows within the loops (Bradshaw 2008).

New investigations of coronal moss, i.e., the (upper) transition region footpoint areas of large hot loops, show that in moss regions the temperature is inversely related to the density (Tripathi et al. 2006). Using comparisons with models, Warren et al. (2008b) show that in order to understand this moss within the framework of a steady uniform heating model, one needs to assume that the moss plasma is not fully filling the volume. However, it remains to be seen if such a static model is applicable at all, because one might suspect a spatially varying heating rate (see above), and if the assumption of static moss is justified.

Motivated by direct magnetic field measurements in the corona indicating the presence of Alfvén waves (Tomczyk et al. 2007) and observations with SOT on Hinode above the limb, McIntosh et al. (2008) re-interpreted the widths of transition region lines across the solar disk. They conclude that the present observations are consistent with a line-of-sight superposition of Alfvénic disturbances in small-scale structures. How this relates to the new finding of Doschek et al. (2007) that the (non-thermal) line widths are largest not in the brightest parts of an active region but in dimmer regions adjacent to bright loops remains to be seen. Doschek et al. (2007) find broad lines related to potential outflow locations, so maybe this problem hints to different acceleration and heating mechanisms in open and closed field regions. Another new difference between (globally) open and closed field regions, was proposed by Tian et al. (2008b), who find evidence that the expansion of transition region structures is more rapid in the coronal holes as compared to the quiet Sun. Dolla & Solomon (2008) analyzed line widths above the limb in order to determine the (kinetic) temperatures of minor ions in presumably open field regions. They find the smallest mass-to-charge-ratio ions to be the hottest at a given height, but their analysis remains inconclusive with regard to supporting or disproving the proposed heating by ion-cyclotron resonances.

While being in orbit nearly 13 years now, the instruments on board SOHO, SUMER in particular, still give numerous new valuable results on the transition region. The EIS instrument on board *Hinode* covers wavelengths around 17–21 nm and 24–29 nm. This mainly includes emission lines formed from 1 to several MK, but also a small number of lines from the transition region, and allows good density diagnostics (Feldman *et al.* 2008). Given the spectral range, the main science topics are grouped around active region phenomena, while the transition region can also be investigated (Young *et al.* 2007). Besides these instruments, which will provide the main source for observations of transition region lines in the coming years, rocket experiments complement these data.

4. Corona and coronal heating (J. A. Klimchuk)

The past two years have seen considerable progress in understanding the magnetically-closed corona and how it is heated. This short report highlights just some of the important contributions. Much effort has been devoted to determining the properties of the heating – how it varies in time and space and whether it depends on physical parameters such as the strength of the magnetic field and the length of field lines. Some studies have concentrated on individual coronal loops, while others have addressed active regions as a whole. These efforts have both clarified some issues and raised new questions.

Let us first consider distinct, measurable loops. A short history is useful. For many years after the *Skylab* soft X-ray observations, it was thought that loops are static equilibrium structures maintained by steady heating. Then came the EUV observations from *SOHO*-EIT and *TRACE*. These revealed that warm (~ 1 MK) loops are much too dense for static equilibrium and have super-hydrostatic scale heights. Modeling efforts showed that the excess densities and large scale heights could be explained by impulsive heating. Because of their temperature response, EIT and *TRACE* are sensitive to the loops when they are cooling by radiation, well after the heating has ceased. The problem is that loops are observed to persist for longer than a cooling time, so a monolithic model is not viable. This led to the suggestion that loops are bundles of thin, unresolved strands. The observed high densities, large scale heights, and long lifetimes can all be explained if the strands are heated at different times by a storm of nanoflares. Since the strands are in different stages of cooling, a range of temperatures should be present within the loop bundle at any given time. In particular, there should be a small amount of very hot (> 5 MK) plasma. See Klimchuk (2006) for a discussion of these points and original references.

Whether loops are isothermal or multi-thermal has been intensely debated over the past several years. Double- and triple-filter observations from *TRACE* seem to suggest that the most narrow loops are isothermal (Aschwanden 2008). However, it has been demonstrated that many different thermal distributions, including ones that are broad, can reproduce the observed intensities, even with three filters (Schmelz *et al.* 2007a; Patsourakos & Klimchuk 2007; Noglik & Walsh 2007). Spectrometer observations provide far superior plasma diagnostics. The results here are mixed. Studies made with *SOHO*-CDS continue to find evidence for both isothermal loops and highly multi-thermal loops (Schmelz *et al.* 2007b; Cirtain *et al.* 2007), while studies made with the new EIS instrument on *Hinode* find that loops tend to be mildly multi-thermal (Ugarte-Urra *et al.* 2008; Warren *et al.* 2008a). Where temporal information is available, there is clear evidence that the loops are evolving, but the evolution is generally slower than expected for radiative cooling. Loop lifetimes are extremely important and require further investigation. A loop bundle will be only mildly multi-thermal if the storm of nanoflares is short-lived; however, the observed lifetime of the loop will then be correspondingly short. If a loop is observed to persist for much longer than a cooling time (López Fuentes *et al.* 2007), then its thermal distribution is expected to be broad. More work is needed on whether the lifetimes and thermal distributions of loops are consistent. Finally, Landi & Feldman (2008) have found that one particular active region is dominated by three distinct temperatures, which would greatly challenge our understanding if correct.

Modeling the plasma properties of whole active regions is a relatively new endeavor. In addition to providing valuable information on coronal heating, these research models are the forerunners of eventual operational models for nowcasting and forecasting the solar spectral irradiance. This is of great practical value, since the irradiance controls the dynamics, chemistry, and ionization state of the terrestrial upper atmosphere and thereby affects radio signal propagation, satellite drag, etc. Active region models based on static equilibrium are able to reproduce the observed soft X-ray emission reasonably well, but they fail, often miserably, at reproducing the EUV emission. The model corona is too faint in the EUV (there are no warm loops) and the moss at the transition region footpoints of hot loops is too bright. Winebarger *et al.* (2008) have demonstrated that better agreement can be obtained in the core of an active region by using a combination of flux tube expansion and filling factors near 10%. Filling factors of this magnitude have been measured in moss with EIS (Warren *et al.* 2008b). Small filling factors are consistent with the idea of unresolved loop strands. Reale *et al.* (2007) have developed a new multi-filter technique using *Hinode*-XRT data that reveals considerable thermal structure on small but resolvable scales.

Active region models based on impulsive heating are in much better agreement with observations than are static models. In particular, the predicted coronal EUV emission is greatly enhanced (Warren & Winebarger 2007; Patsourakos & Klimchuk 2008a). The predicted moss emission is still too bright, but these models assume constant cross section flux tubes, and expanding tubes will improve the agreement, as they do for static models. It is also likely that the brightness of the observed moss is diminished by spicules and possibly by other cool absorbing material.

Nanoflare models predict that very hot plasma should be present throughout the corona, albeit in very small quantities (Klimchuk *et al.* 2008; this paper presents a highly efficient IDL code for modeling dynamic loops and is available upon request). The intensities of very hot spectral lines are expected to be extremely faint, due to the small emission measures and possible also to ionization nonequilibrium effects (Bradshaw & Cargill 2006; Reale & Orlando 2008). Measurable quantities of very hot ($\sim 10\,\mathrm{MK}$) plasma have been detected outside of flares by the *CORONAS*-F spectroheliometers (Zhitnik *et al.* 2006), *RHESSI* (McTiernan 2008), and XRT (Siarkowski *et al.* 2008; Reale *et al.* 2008). The derived differential emission measure distributions, $\mathrm{DEM}(T)$, are consistent with the predictions of nanoflare models. The $\mathrm{DEM}(T)$ derived from EIS spectra for $T \leqslant 5\,\mathrm{MK}$ are also consistent with the predictions (Patsourakos & Klimchuk 2008b). Other tests of the nanoflare idea include emission line Doppler shifts, broadening, and wing enhancements that are associated with evaporating and condensing plasma (Patsourakos & Klimchuk 2006; Hara *et al.* 2008; Bradshaw 2008).

Information on the distribution of nanoflare energies can be inferred from the intensity fluctuations of observed loops (Parenti *et al.* 2006; Pauluhn & Solanki 2007; Parenti & Young 2008; Sarkar & Walsh 2008; Sakamoto *et al.* 2008; Bazarghan *et al.* 2008). Proper flares are known to have a power law energy distribution with an index < 2. Extrapolating to smaller energies implies that nanoflares cannot heat the corona, as first pointed out by Hudson (1991). However, it is now believed that the power law index for small events is > 2, though with a large uncertainty (Benz 2004; Pauluhn & Solanki 2007; Bazarghan *et al.* 2008). Furthermore, the subset of proper flares that are not associated with CMEs also have a power law index > 2 (Yashiro *et al.* 2006; see Section 6). Since the physics of eruptive events and non-eruptive events (nanoflares and confined flares) is likely to be much different, it is not surprising that they obey different power laws.

Thermal nonequilibrium, a phenomenon thought to be important for prominence formation (Karpen & Antiochos 2008), may also play an important role in ordinary loops. Loop equilibria do not exist for steady heating if the heating is highly concentrated low in the loop legs. Instead, cool condensations form and fall to the surface in a cyclical pattern that repeats on a time scale of hours. Resolvable condensations are indeed observed in active regions, but only in a small fraction of loops (Schrijver 2001). As a possible explanation for other EUV loops, Klimchuk & Karpen (2008) appeal to the multi-strand concept. The individual tiny condensations that occur within each strand will not be detected as long as the strands are out of phase. It is encouraging that the models predict excess densities similar to those of observed EUV loops. Mok *et al.* (2008) report thermal nonequilibrium behavior in their active region simulations. Hot loops form and cool, but without producing localized condensations.

We can summarize the state of understanding as follows. Much of the magnetically-closed corona is certainly *not* in static equilibrium, but much of it could be. A significant portion – perhaps the vast majority – is heated impulsively or is in thermal nonequilibrium, or some combination thereof. Most coronal heating mechanisms that have been proposed involve impulsive energy release (Klimchuk 2006; Uzdensky 2007; Cassak *et al.* 2008; Dahlburg *et al.* 2008; Rappazzo *et al.* 2008; Ugai 2008). It should be noted, however, that nanoflares that recur sufficiently frequently within the same flux strand (on a time scale much shorter than the cooling time) will produce quasi-static conditions. It is clear that more observational and theoretical work is required before the coronal heating problem will be solved.

5. Coronal jets (L. van Driel-Gesztelyi)

Coronal bright points are often observed to have jets – collimated transient ejections of hot plasma. *Hinode* (Kosugi *et al.* 2007) can now study the fine detail of jets which tend to occur preferentially inside coronal holes, which is consistent with reconnection taking place between the open magnetic field of the coronal hole and the closed loop field lines. Observations with the *Hinode*-XRT instrument (Golub *et al.* 2007) revealed that jets from polar coronal holes are more numerous than previously thought (60 jets day^{-1}, Savcheva *et al.* 2007; and even 10 jets hr^{-1}, Cirtain *et al.* 2007). The EIS instrument (Culhane *et al.* 2007a) allows direct measurement of the velocity of jets in the corona for the first time. The footpoints of the loops are seen to be red-shifted which is consistent with downflowing cooling plasma following reconnection. The (blue-shifted) jet is the dominant feature in velocity space but not in intensity (Kamio *et al.* 2007). Another new feature of jets is post-jet enhancement of cooler coronal lines observed by *Hinode*-EIS. This can be explained by the hot plasma in the jet not having sufficient velocity to leave the Sun and then falling back some minutes later (Culhane *et al.* 2007b).

Hinode-XRT observations of jets at the poles have shown mean velocities for jets of 160 km s^{-1} (Savcheva *et al.* 2007). Multiple velocity components were found in jets by Cirtain *et al.* (2007) in XRT polar coronal hole data: a spatio-temporal average of about 200 km s^{-1} as well as a much higher velocity measured at the beginning of each jet – with speeds reaching 800 km s^{-1}. Cirtain *et al.* (2007) interpret this early (and sometimes recurrent) fast flow as being due to plasma ejected at the Alfvén speed during the relaxation phase following magnetic reconnection. The mass flux supplied by about 10 jets per hour occurring in the two polar coronal holes was estimated to produce a net flux of 10^{12} protons m^{-2} s^{-1} which is only a factor of 10 less than the current estimates of the average solar wind flux. These small jets are providing a substantial amount of mass that is being carried into interplanetary space. A 3D numerical simulation has been carried out to compare with these observations (Moreno-Insertis *et al.* 2008) and is found to be consistent with several key observational aspects of polar jets such as their speeds and temperatures.

A study of the 3D morphology of jets became possible for the first time with stereoscopic observations by the *euvi*-SECCHI imagers (Howard *et al.* 2008) on board the twin *STEREO* spacecraft. The most important geometrical feature of the observed jets was found to be helical structures showing evidence of untwisting (Patsourakos *et al.* 2008). This is in agreement with the 3D model proposed by Pariat *et al.* (2008) with magnetic twist (untwisting) being the jet's driver.

6. Flares (L. Fletcher and J. Wang)

In this brief review we focus on progress in flare energy build-up and flare prediction, flare photospheric effects, high energy coronal sources, non-thermal particles, the flare-CME relationship, and recent advances with *Hinode*.

Where is the magnetic free energy stored in a flaring active region? Using the increasingly robust methods for extrapolating magnetic fields from vector magnetic field measurements, Schrijver *et al.* (2008) find evidence for pre-flare filamentary coronal currents located < 20 Mm above the photosphere and Regnier & Priest (2007) show that in a newly-emerged active region the free energy is concentrated within the first 50 Mm (in an older, decaying region it resides at higher altitudes). Horizontal shear flows close to the neutral line prior to large flares (Deng *et al.* 2006) confirm the concentration of free energy in a small spatial scale, and Schrijver (2007) finds that if the unsigned flux within 15 Mm of the polarity inversion line exceeds 2×10^{21} Mx a major flare will occur within a day. Though Leka & Barnes (2007) find that the probability of flaring has only a weak relationship to the state of the photospheric magnetic field at any time, single or synthesized magnetic parameters are being used with some success to quantify flare probability and productivity. Georgoulis & Rust (2007) introduce the effective connected magnetic field of an active region, finding that this exceeds 1600 and 2100 G for M- and X-class flares, respectively, at 95% probability. LaBonte *et al.* (2007) surveyed the helicity injection prior to X-class flares producing a CME, finding occurrence only if the peak helicity flux exceeds 6×10^{36} Mx s^{-1}. Cui *et al.* (2007) find that flare probability increases with active region complexity, non-potentiality, and length of polarity inversion line. Impulsive phase HXR sources are concentrated where the magnetic field is strong, and where the reconnection rate is high (Temmer *et al.* 2007; Jing *et al.* 2008; Liu *et al.* 2008).

The last three years have seen effort directed towards understanding the magnetic and seismic effects of flares near the photosphere. It is clear that the photospheric magnetic field changes abruptly and non-reversibly during the flare impulsive phase (see e.g. Sudol & Harvey 2005 for a recent survey). Rapid changes in sunspot structure have also been detected by Chen *et al.* (2007) in 40% of X-class flares, 17% of M flares, and 10% C flares, while Wang (2006) finds variations in magnetic gradient close to the polarity inversion line consistent with a sudden release of magnetic shear. The obvious future task is to analyze vector magnetograms to identify changes in the 'twist' component of the field.

Flare-generated seismic waves, discovered by Kosovichev & Zharkova (1998) and amply confirmed in Cycle 23, also show the flare's photospheric impact. Donea *et al.* (2006),

Kosovichev (2006, 2007) and Zharkova & Zharkov (2007) show that flare HXR sources and seismic sources correlate in space and time. Seismic sources are associated also with white light kernels, responsible for the majority of the flare radiated energy and strongly correlated with HXR sources (Fletcher *et al.* 2007) but the total acoustic energy is a small fraction of total flare energy (Donea *et al.* 2006). Nonetheless, looking at the cyclic variation of the total energy in the Sun's acoustic spectrum, Karoff & Kjeldsen (2008) propose that – analogous with earthquakes – flares may excite long-duration global oscillations.

The *RHESSI* mission has discovered several new types of flare coronal HXR sources, and we highlight here a hard-spectrum HXR source at least 150 Mm above the photosphere, with a nonthermal electron fraction of about 10% (Krucker *et al.* 2007b). Hard spectrum gamma-ray (200-800 keV) coronal sources have also been found, suggesting coronal electron trapping (Krucker *et al.* 2008). A soft-hard-soft spectral variation with time is present in some coronal sources as well as footpoints (Battaglia & Benz 2006), and this may be explicable by a combination of coronal trapping and stochastic electron acceleration (Grigis & Benz 2006). A curious observation with the Owens Valley Solar Telescope of terahertz emission may come from a compact coronal source of electrons at 800 keV, if the electrons are radiating in a volume with a magnetic flux density of 4.5 kG (Silva *et al.* 2007).

Discovering the origin and properties of the flare electron distribution continues to motivate advanced modeling and observations. Particle-in-cell simulations, used for some years in magnetospheric physics, have been harnessed to study acceleration in coronal magnetic islands produced by magnetic reconnection (Drake *et al.* 2006), and wave-particle distributions in current sheet and uniform magnetic field geometries (Karlicky & Barta 2007; Sakai *et al.* 2006). Flare Vlasov simulations are also being developed (e.g. Miteva *et al.* 2007; Lee *et al.* 2008). Detailed *RHESSI*-HXR spectroscopy has led to a new diagnostic for the flare electron angular distribution, based on photospheric HXR albedo (Kontar *et al.* 2006a). This diagnostic suggests that electron distributions might not be strongly downward-beamed in the chromosphere (Kontar *et al.* 2006b; Kasparova *et al.* 2007; though see Zharkova & Gordovskyy 2006 for an alternative explanation). Xu *et al.* (2008) studied *RHESSI* flares having an extended coronal source, finding evidence for an extended coronal accelerator. Looking at the larger coronal context, Temmer *et al.* (2008) show that peaks in the flare electron acceleration rate and in the CME acceleration rate are simultaneous within observational constraints (∼ 5 minutes). *RHESSI* and *WIND* observations suggest that the spectral indices of flare and interplanetary electrons are correlated, but in a way that is inconsistent with existing models for flare X-ray emission, and the number of escaping electrons is only about 1/500th of the number of electrons required to produce the chromospheric HXR flux (Krucker *et al.* 2007a).

However, outstanding questions about the total electron number and supply in solar flares has prompted various authors to suggest alternatives to the 'monolithic' coronal electron beam picture. For example Dauphin (2007), Gontikakis *et al.* (2007) examine acceleration in multiple, distributed coronal region sites, while Fletcher & Hudson (2008) investigate the transport of flare energy to the chromosphere in the form of the Poynting flux of large-scale Alfvénic pulses, in strong and low-lying coronal magnetic fields.

The relationship between flares and CMEs continues to be an important topic. The general consensus regarding the spatial correspondence between CME position angle and flare location in the pre-*SOHO* era was that the flare is located anywhere under the span of the CME (e.g., Harrison 2006). However, using 496 flare-CME pairs in the SOHO era, Yashiro *et al.* (2008) found that the offset between the flare position and the central position angle (CPA) of the associated CME has a Gaussian distribution centered on zero, meaning the flare is typically located radially below the CME leading edge. This finding suggests a closer flare-CME relationship as implied by the CSHKP eruption model. Many flares are not associated with CMEs. Yashiro *et al.* (2006) studied two sets of flares one with and the other without CMEs. The number of flares as a function of peak X-ray flux, fluence, and duration in both sets followed a power law. Interestingly, the power law index was > 2 for flares without CMEs, while < 2 for flares with CMEs. In flares without CMEs, the released energy seems to go entirely into heating, which suggests that nanoflares may contribute significantly to coronal heating (see Section 4).

The launch of *Hinode* promises significant advances in flare physics in the next cycle. Thus far there have only been a small number of well-observed large flares, but observations of small flares start to show the combined power of *RHESSI* and *Hinode*. For example, Hannah *et al.* (2008) find a microflare not conforming to the usual relationship between flare thermal and nonthermal emission, and Milligan *et al.* (2008) show evidence for hot downflowing plasma in the flare corona, not explained in any existing flare model. Observational evidence for a new kind of reconnection, called slip-running reconnection, has been found by Aulanier *et al.* (2007), and sub-arcsecond structure in the white light flare sources has been demonstrated by Isobe *et al.* (2007). We look forward to the continued operation of these instruments, and the theoretical advances that they will bring, in the rise of Cycle 24.

7. Coronal mass ejection initiation (L. van Driel-Gesztelyi)

In recent years our physical understanding of CMEs has evolved from cartoons inspired by observations to full-scale numerical 3D MHD simulations constrained by observed magnetic fields. Notably, there has been progress made in simulating CME initiation by flux rope instabilities as inspired by observed filament motions during eruption which frequently include helical twisting and writhing (e.g. Rust & LaBonte 2005; Green *et al.* 2007). Several of these simulations use the analytical model of a solar active region by Titov & Démoulin (1999) as initial condition. The model contains a current-carrying twisted flux rope that is held in equilibrium by an overlying magnetic arcade. The two instabilities considered as eruption drivers are the ideal MHD helical kink and torus instabilities. The helical kink instability sets in if a certain threshold of (flux rope) twist (~ 2.5 turns for line-tied flux ropes) is reached (e.g., Török & Kliem 2005). Above this threshold, twist becomes converted into writhe during the eruption, deforming the flux rope (or filament) into a helical kink shape. On the other hand, a current-carrying ring (or flux rope) situated in an external poloidal magnetic field (B_{ex}) is unstable against radial expansion when the Lorentz self-force or hoop force decreases more slowly with increasing ring radius than the stabilizing Lorentz force due to B_{ex}. Known as the torus instability, its possible role in solar eruptions has been examined by Kliem & Török (2006) and Isenberg & Forbes (2007). In solar eruptions, the torus instability does not require a pre-eruptive, highly twisted flux rope, but (i) a sufficiently steep poloidal field decrease with height above the photosphere and (ii) an (approximately) semi-circular flux rope shape. Both the helical kink and torus instabilities may be responsible for initiating and driving prominence/filament eruptions and thus CMEs. The magnetic field decrease with height above the filament was shown to be critical whether a confined eruption or a full eruption occurs as well as for determining the acceleration profile, corresponding to fast CMEs for rapid (field) decrease, as it is typical of active regions, and to slow CMEs for gentle decrease, as is typical of the quiet Sun (Török & Kliem 2005, 2007; Liu 2008). The latter means that CMEs from complex active regions with steep field gradients in the corona are more likely to give rise to fast CMEs – something that is indeed observed.

More complex CME initiation models involve multiple magnetic flux systems, such as in the magnetic break-out model (e.g. Antiochos *et al.* 1999, DeVore & Antiochos 2008). In this model, magnetic reconnection removes unstressed magnetic flux that overlies the highly stressed core field and this way allows the core field to erupt. The magnetic break-out model involves specific nullpoints and separatrices. A multi-polar configuration was also included in the updated catastrophe model (Lin & van Ballegooijen 2005), the flux cancellation model (Amari *et al.* 2007), and the MHD instability models (Török & Kliem 2007). In an attempt to test CME initiation models with special attention to the breakout, Ugarte-Urra *et al.* (2007) analyzed the magnetic topology of the source regions of 26 CME events using potential field extrapolations and *TRACE*-EUV observations. They found only seven events which could be interpreted in terms of the breakout model, while a larger number of events (12) could not be interpreted in those terms. The interpretation of the rest remained uncertain. On the other hand, the CME event analyzed by Williams *et al.* (2005) provided a good example to indicate that also a combination of several mechanisms, e.g. magnetic break-out and kink instability, can be at work in initiating CMEs.

8. Global coronal waves and shocks (B. Vršnak)

The research on globally propagating coronal disturbances (large-amplitude waves, shocks, and wave-like disturbances) continued to be very dynamic. Maybe the most prominent characteristic of the past triennium was an enhanced effort to combine detailed multi-wavelength observations with the theoretical background. The empirical research resulted in a number of new findings, leading to new ideas and interpretations, whereas the theoretical research provided a better understanding of physical processes governing the formation and propagation of global coronal disturbances.

For the first time the EUV signatures of a global coronal wave ('EIT waves') were measured at high cadence by *STEREO*-EUVI, related to the eruption of 2007 May 19 (Long *et al.* 2008; Veronig *et al.* 2008). Long *et al.* (2008) reported for the first time the wave signatures at 304 Å. Furthermore, they confirmed the idea by Warmuth *et al.* (2001) that velocities of EIT waves measured by *SOHO*-EIT are probably significantly underestimated due to the low cadence of the EIT instrument. Veronig *et al.* (2008) revealed reflection of the wavefront from the coronal hole boundary, indicating that the observed disturbance represents a freely-propagating MHD wave.

The data from pre-*STEREO* instruments continued to be exploited fruitfully. Mancuso & Avetta (2008) analyzed the UV-spectrum (*SOHO*-UVCS) of the 2002 July 23 coronal shock, and concluded that the plasma-to-magnetic pressure ratio β could be an important parameter in determining the effect of ion heating at collisionless shocks. Employing the extensive data on CMEs, solar energetic particle (SEP) events, and type II radio bursts during the *SOHO* era, Gopalswamy *et al.* (2008b) demonstrated that essentially all type II bursts in the decameter-hectometric (DH) wavelength range are associated with SEP events once the source location is taken into account. Shen *et al.* (2007) proposed a method to determine the shock Mach number by employing the CME kinematics, type II burst dynamic spectrum, and the extrapolated magnetic field. Analyzing one Moreton wave that spanned over almost 360°, Muhr *et al.* (2008) revealed two separate radiant points at opposite ends of the two-ribbon flare, indicating that the wave was driven by the CME expanding flanks. Veronig *et al.* (2006) found out that the Moreton/EIT wave segments where the front orientation is normal to the coronal hole boundary can intrude into the coronal hole up to 60–100 Mm.

Regarding the nature of global coronal disturbances, some new ideas appeared. For example, Attrill *et al.* (2007) attributed EIT waves to successive reconnections of CME flanks with coronal loops, which could explain the association of EIT 'waves' and shallow coronal dimmings which are formed behind the bright front. Wills-Davey *et al.* (2007) proposed that slow EIT waves are caused by MHD slow-mode soliton-like waves. Delannee *et al.* (2008) performed a 3D MHD simulation to show that EIT 'waves' could be a signature of a current shell formed around the erupting structure. Balasubramaniam *et al.* (2007) demonstrated that the visibility of Moreton waves increases when sweeping over filaments and filament channels, so they put forward the idea that a significant contribution to the Moreton-wave H_α signature might be coming from coronal material of enhanced density.

The question of the origin of coronal shocks and large amplitude waves continues to be one of the central topics in this field. The published studies showed a variety of results, some biased towards the CME-driven option, some favoring the flare-ignited scenario, and some finding arguments for small scale ejecta (e.g., Chen 2006; Pohjolainen & Lehtinen 2006; Shanmugaraju *et al.* 2006a,b; Subramanian & Ebenezer 2006; Cho *et al.*, 2007; Liu *et al.* 2007; Reiner *et al.* 2007; White 2007; Grechnev *et al.* 2008; Muhr *et al.* 2008; Veronig *et al.* 2008; Magdalenic *et al.* 2008; Mancuso & Avetta 2008; Pohjolainen 2008). To illustrate the current level of ambiguity in such studies, let us mention that for one well observed event two sets of authors came to diametrally opposite conclusions: Vršnak *et al.* (2006) favored a flare driver, whereas Dauphin *et al.* (2006) advocate a CME. The status of the 'CME/flare controversy' was reviewed recently by Vršnak & Cliver (2008).

Related to the formation and propagation of large-amplitude waves and shocks, a number of important theoretical papers were published. Pagano *et al.* (2007) investigated the role of magnetic fields and showed that a CME-driven wave propagates to longer distances in the absence of magnetic field than in the presence a weak open field. Ofman (2007) modeled the wave activity following a flare by launching a velocity pulse into a model active region and demonstrated that the resulting global oscillations are in good agreement with observations. Employing the photospheric magnetic field measurements, Liu *et al.* (2008) performed a 3D MHD simulation of a CME, and showed that the shock segment at the nose of the CME remains quasi-parallel most of the time. In the simulation of reconnection in a vertical current sheet, Barta *et al.* (2007) revealed the formation of large-amplitude waves associated with changes of the reconnection rate, which might explain flare-associated type II bursts in the wake of CMEs. Zic *et al.* (2008) developed an analytical MHD model describing the formation of large-amplitude waves by impulsively expanding 3D pistons. The model provides an estimate of the time/distance at which the shock should be formed, dependent on the source-surface acceleration, the terminal velocity, the initial source size, the ambient Alfvén speed, and plasma β.

Finally, it should be noted that a comprehensive review on coronal waves and shocks was published by Warmuth (2007). Gopalswamy (2006e) reviewed the relationship between CMEs and type II bursts, while Mann & Vršnak (2007) surveyed the relationship between CMEs, flares, coronal shocks, and particle acceleration.

9. Coronal dimming (R. Harrison and L. van Driel-Gesztelyi)

There is no strict definition of the phenomenon which we call coronal dimming. Most authors consider coronal dimming to be a depletion of extreme-UV (EUV) or X-ray emission from a large region of the corona, which is thought to be closely associated with coronal mass ejection (CME) activity. Clearly, understanding the onset phase of a CME is one of the key issues in solar physics today, so the study of such dimming activity could well be of critical importance. However, most of the literature deals with dimming in a rather hand-waving manner, with the emphasis on phenomenological studies and associations, no strict definitions of what constitutes a dimming event (e.g., the depth of the depletion in intensity, the size of the dimming region, etc.) and little in terms of a physical interpretation of the plasma characteristics of the dimming region. Having said that, some key studies are emerging which do tackle such issues head on, and with the advent of the new STEREO and Hinode spacecraft, along with the on-going SOHO and TRACE missions, as well as the up-coming SDO mission, we have many tools to address this area of research effectively.

Coronal dimming is not a newly discovered phenomenon; Rust and Hildner (1976) reported such an event using *Skylab* observations. More recently, from the late 1990s, dimming was reported using X-ray and EUV, imaging and spectroscopic data, from the *SOHO* and *Yohkoh* spacecraft (e.g. Sterling & Hudson 1997; Harrison 1997; Gopalswamy & Hanaoka 1998; Zarro *et al.* 1999; Harrison & Lyons 2000), and dimming has taken center-stage in the study of mass ejection onset in recent years (e.g., recent studies include Moore & Sterling 2007; Zhang *et al.* 2007; Reinard & Biesecker 2008). In many ways coronal dimming has become a well established phenomenon.

The majority of dimming reports involve EUV or X-ray imaging, and we have excellent tools aboard *SOHO, TRACE, STEREO* and *Hinode* to identify and study the topology and evolution of dimming regions. On the other hand, there are spectroscopic studies of dimming which are providing key plasma information, despite having limited fields of view or cadence. The combination of imaging and spectroscopy is essential, but it is worth stressing some of the spectroscopic studies because they stress the physical processes which are involved in the dimming and, perhaps, the CME onset process.

EUV spectroscopy has been used to confirm that the dimming process represents a loss of mass – i.e., it is a density depletion – rather than a change in temperature (Harrison & Lyons 2000; Harrison *et al.* 2003). Indeed, these studies have demonstrated the loss of between 4.3×10^{10} and 2.7×10^{14} kg, in each case consistent with the mass of an overlying, associated CME. If we are identifying the plasma which becomes (part of) the CME, then this is an exciting phenomenon; studies focusing on the properties of the dimming plasma, before, during and after the event, will be essential for understanding the CME onset (Harrison & Bewsher 2007).

Hudson *et al.* (1996) showed that the timescale of the dimming formation observed in *Yohkoh*-soft x-ray Telescope (SXT; Tsuneta *et al.* 1991) data is much faster than corresponding conductive and radiative cooling times. More recently, data obtained by the *Hinode*-Extreme ultra-violet Imaging spectrometer (EIS; Culhane *et al.* 2007) have shown detection of Doppler blue-shifted plasma outflows of velocity ≈ 40 km s^{-1} corresponding to a coronal dimming (Harra *et al.* 2007). This result confirms a similar finding (Harra & Sterling 2001) obtained with the *SOHO*/coronal Diagnostic spectrometer (CDS; Harrison *et al.* 1995). In addition, *SOHO*-CDS limb observations have been used to show the formation of a dimming region through the outward expansion of pre-CME EUV loops (Harrison & Bewsher, 2007), which is consistent with such blueshifts. Imada *et al.* (2007) find that *Hinode*-EIS data of a dimming shows a dependence of the outflow velocity on temperature, with hotter lines showing a stronger plasma outflow (up to almost 150 km s^{-1}. These works collectively support the primary interpretation of coronal dimmings as being due to plasma evacuation.

Statistical studies are becoming important in truly establishing the relationship with CMEs. Reinard & Biesecker (2008) have recently studied the properties of 96 dimming events, using EUV imaging, associated with CME activity. They confirmed earlier studies which showed that the dimming events could be long-lasting, ranging from 1 to 19 hours, and compared the size of the dimming regions to the associated CMEs. They also tracked the number of dimming pixels through each event and showed that the 'recovery' after the dimming often took the form of a two-part slope (plotted as dimming area vs. time).

Bewsher *et al.* (2008) have produced the first statistical and probability study of the dimming phenomenon using spectroscopy. They recognized that while we have associated CMEs and dimming, there has not been a thorough statistical study which can really identify the degree of that association, i.e., to put that relationship on a firm footing. Using spectroscopy, they also recognized the importance of studying this effect for different temperatures. They made use of over 200 runs of a specific campaign using the SOHO spacecraft with an automated procedure for identifying dimming.

Key results included the following: Up to 84% of the CMEs in the data period can be back-projected to dimming events – and this appears to confirm the association that we have been proposing. However, they also showed, as did other spectral studies, that the degree of dimming varies between temperatures from event to event. If different dimming events have different effects at different temperatures then this is a problem for monitoring such events with fixed-wavelength imagers.

Assuming that magnetic field lines of the CME are mostly rooted in the dimmings, several properties derived from the study of dimmings can be used to obtain information about the associated CME. Firstly, calculations of the emission measure and estimates of the volume of dimmings can give a proxy for the amount of plasma making up the CME mass (Sterling & Hudson 1997; Harrison & Lyons 2000; Harrison *et al.* 2003; Zhukov & Auchère 2004). Secondly, the spatial extent of coronal dimmings can give information regarding the angular extent of the associated CME (Thompson *et al.* 2000; Harrison *et al.* 2003; Attrill *et al.* 2007, van Driel-Gesztelyi *et al.* 2008). Thirdly, quantitative measurement of the magnetic flux through dimmings can be compared to the magnetic flux of modeled magnetic clouds (MC) at 1 AU (Webb *et al.* 2000; Mandrini *et al.* 2005; Attrill *et al.* 2006; Qiu *et al.* 2007), see Démoulin (2008) for a review. Fourth, studying the evolution of the dimmings, particularly during their recovery phase can give information about the evolution of the CME post-eruption (Attrill *et al.* 2006; Crooker & Webb 2006) providing proof for e.g. magnetic interaction between the expanding CME and open field lines of a neighboring coronal hole. Finally, study of the distribution of the dimmings, their order of formation and measurement of their magnetic flux contribution to the associated CME enabled Mandrini *et al.* (2007) to derive an understanding of the CME interaction with its surroundings in the low corona for the case of the complex 28 October 2003 event. They, building on the model proposed by Attrill *et al.* (2007), demonstrated that magnetic reconnection between field lines of the expanding CME with surrounding magnetic structures ranging from small- to large-scale (magnetic carpet, filament channel, active region) make some of the field lines of the CME 'step out' from the flaring source region. Magnetic reconnection is driven by the expansion of the CME core resulting from an over-pressure relative to the pressure in the CME's surroundings. This implies that the extent of the lower coronal signatures match the final angular width of the CME. Through this process, structures over a large-scale magnetic area become CME constituents (for a review see van Driel-Gesztelyi *et al.* 2008). From the wide-spread coronal dimming some additional mass is supplied to the CME.

Observations show that coronal dimmings recover whilst suprathermal uni- or bi-directional electron heat fluxes are still observed at 1 AU in the related ICME, indicating magnetic connection to the Sun. The questions why and how coronal dimmings disappear whilst the magnetic connectivity is maintained was investigated by Attrill *et al.* (2008) through the analysis of three CME-related dimming events. They demonstrated that dimmings observed in *SOHO*-EIT data recover not only by shrinking of their outer boundaries but also by internal brightenings. They show that the model developed in Fisk & Schwadron (2001) of interchange reconnections between 'open' magnetic field and small coronal loops is applicable to observations of dimming recovery. Attrill *et al.* (2008) demonstrate that this process disperses the concentration of 'open' magnetic field (forming the dimming) out into the surrounding quiet Sun, thus recovering the intensity of the dimmings whilst still maintaining the magnetic connectivity of the ejecta to the Sun.

Although this brief summary cannot report on all studies, it is clear that we have made progress very recently in putting the dimming phenomenon on a firm footing – the association is real – and we are making in-roads into studies of the plasma activities leading to the dimming/CME onset process. With the continuation of the *SOHO* mission, as well as *TRACE*, combined with the new *STEREO* and *Hinode* missions and the up-coming *SDO* mission, this is a topic which will receive much attention in the next few years.

10. The link between low-coronal CME signatures and magnetic clouds (C. Mandrini)

A major step to understanding the variability of the space environment is to link the sources of coronal mass ejections (CMEs) to their interplanetary counterparts, mainly magnetic clouds (MCs), a subset of interplanetary CMEs characterized by enhanced magnetic field strength when compared to ambient values, a coherent and large rotation of the magnetic field vector, and low proton temperature (Burlaga 1995). Identifying the solar sources and comparing qualitatively and quantitatively global characteristics and physical parameters both in the Sun and the interplanetary medium provide useful tools to constrain models in both environments.

Under the assumption that dimmings (see Section 9) at the Sun mark the position of ejected flux rope footpoints (Webb *et al.* 2000), the magnetic flux through these regions can be used as a proxy for the magnetic flux involved in the ejection and, thus, be compared to the magnetic flux in the associated interplanetary MC. Another proxy for the flux involved in an ejection is the reconnected magnetic flux swept by flare ribbons, as they separate during the evolution of two-ribbon flares. Using EUV dimmings as proxies and reconstructing the MC structure from one spacecraft observations, Mandrini *et al.* (2005) and Attrill *et al.* (2006) found that the magnetic flux in dimming regions was comparable to the azimuthal MC flux, while the axial MC flux was several times lower. Qui *et al.* (2007) analyzed and compared the reconnected magnetic flux to the total MC flux, finding similar results (see also Yurchyshyn *et al.* 2006; Longcope *et al.* 2007; Möstl *et al.* 2008, where MC data from two spacecraft were used).

These results led to the conclusion that the ejected flux rope is formed by successive reconnections in a sheared arcade during the eruption process, as opposed to the classical view of a previously existing flux rope being ejected. However, in extreme events that occur in not isolated magnetic configurations, it was found that the flux in dimmings did not agree with the MC flux (Mandrini *et al.* 2007). This mismatch led these authors to propose a scenario in which dimmings spread out to large distances from the initial erupt-ing region through a stepping reconnection process (in a similar process to that proposed by Attrill *et al.*, 2007, for the interpretation of EIT waves). An overview of earlier works on quantitative comparisons of solar and interplanetary global magnetohydrodynamic invariants, such as magnetic flux and helicity, can be found in Démoulin (2008).

Qualitative comparisons are also useful tools to understand the eruption process. Studying the temporal and spatial evolution of EUV dimmings, together with soft X-ray coronal observations, in conjunction with interplanetary *in situ* data of suprathermal electron fluxes, Attrill *et al.* (2006) and Crooker & Webb (2006) derive an eruption scenario in which interchange magnetic reconnection between the expanding CME loops and the open field lines of a polar coronal hole led to the opening of one leg of the erupting flux rope. Harra *et al.* (2007), combining EUV and H_α solar observations of eruptive events with *in situ* magnetic field and suprathermal electron data, were able to understand the sequence of events that produced two MCs with opposite magnetic field orientations from the same magnetic field configuration.

The simple comparison of the magnetic field orientation in the erupting configurations, which can be inferred from magnetograms, the directions of filaments, coronal arcades or loops, with the axis of the associated MCs, can give clues about the mechanism at the origin of solar eruptions. Green *et al.* (2007) analyzed in detail associations of filament eruptions and corresponding MCs, and they found that when the filament and MC axis differed by a large angle, the direction of rotation was related to the magnetic helicity sign of the erupting configuration (see also Harra *et al.* 2007). The rotation was consistent with the conversion of twist into writhe, under the ideal MHD constraint of helicity conservation, providing support for the assumption of a flux rope topology where the kink instability sets in during the eruption (see the review by Gibson *et al.* 2006).

11. Coronal mass ejections in the heliosphere (R. Harrison)

In the 1970s the *Helios* spacecraft operated from solar orbits with perihelion 0.31 AU. Zodiacal light photometers were used to detect CMEs in the inner heliosphere (see, e.g., Richter *et al.* 1982; Jackson & Leinert 1985). CME images were constructed from three photometers which scanned the sky using the spacecraft rotation. More recently, a major advance was made with the launch, in 2003, of the Solar Mass Ejection Imager (SMEI) aboard the *Coriolis* spacecraft (Eyles *et al.* 2003). This instrument maps the entire sky with three cameras each scanning 60° slices of the sky as the spacecraft moves around the Earth, and thus, it has pioneered full-sky mapping aimed specifically at the detection of CMEs propagating through the inner heliosphere (see e.g. recent papers by Kahler & Webb 2007 and Jackson *et al.* 2007).

The combination of wide-angle heliospheric mapping from out of the Sun-Earth line is now being satisfied by the Heliospheric Imagers (HI) (Harrison *et al.* 2008) aboard the NASA *STEREO* spacecraft. The development of these instruments has come very much from the SMEI heritage and, with the unique opportunities from the STEREO spacecraft locations, these instruments are able to image those CME events directed towards the Earth. Indeed, for the first time, the HI instruments provide a view of the passage of CMEs along virtually the entire Sun-Earth line and such observations represent a major milestone in investigations of the influence of solar activity on the Earth and human systems.

Each HI instrument consists of two wide-angle telescopes mounted within a baffle system enabling imaging of the heliosphere from the corona out to Earth-like distances and beyond. The low scattered light levels and sensitivity allow the detection of stars down to magnitudes of 12–13. This performance is excellent for the detection of solar ejecta and solar wind structure through the detection of Thomson scattered photospheric light off free electrons in regions of density enhancement.

The *STEREO* spacecraft were launched in October 2006 with full scientific operation of the HI instruments starting from April 2007. The spacecraft are in near Earth-like solar orbits, with one ahead and one behind the Earth in its orbit. They are drifting away at 22.5° per year (Earth-Sun-spacecraft angle). The spacecraft are labelled *STEREO A* and *STEREO B*, for ahead and behind.

The first HI observations of CMEs in the heliosphere, tracked out to Earth-like distances, were reported by Harrison *et al.* (2008). The same instruments are also reaping the benefits of wide-angle imaging of the heliosphere with observations of comets (Fulle *et al.* 2007; Vourlidas *et al.* 2007), even the imaging of co-rotating interaction regions (Sheeley *et al.* 2008a,b; Rouillard *et al.* 2008a) and impacts of CMEs at other planets (Rouillard *et al.* 2008b).

With the HI instruments we now have a real opportunity to begin to relate the coronal events that we call CMEs with their heliospheric counterparts, commonly referred to as ICMEs – Interplanetary CMEs. Most ICME studies have been performed utilizing *in situ* particle and field observations, and it is clear that heliospheric imaging can provide a thorough test of the interpretation of such *in situ* data on the topology and propagation of CMEs in the heliosphere. Indeed, the uniqueness of this opportunity is well illustrated by the fact that there are a number of extremely basic observational tests which can be made with the new facility to underline our current understanding of how CMEs travel out through the Solar System.

Crooker & Horbury (2006) have recently reviewed the propagation of ICMEs in the heliosphere, utilizing *in situ* data. They note that cartoon sketches of ICMEs commonly show magnetic field lines connected to the Sun at both ends. Furthermore, reporting on the work of Gosling *et al.* (1987), Crooker *et al.* (2002), and others, they note that it is widely accepted that counter-streaming particle beams in ICMEs are a sign that both ends of the ICME are indeed connected to the Sun. This is known as a 'closed' ICME. On the other hand, uni-directional beams may signal connection at only one end – an 'open' ICME. Logically, then, the lack of beams would appear to signal disconnection at both ends. In this case the ICME has become an isolated plasmoid.

Given this interpretation, *in situ* observations of ICMEs appear to show many events which are apparently connected to the Sun at both footpoints, and rather fewer events which appear to be connected at one end. Complete disconnection of an ICME (a plasmoid) appears to be rare. In addition, the *in situ* observations suggest that CMEs are connected to the Sun over extremely long distances; Riley *et al.* (2004) looked for the degree of 'openness' of ICMEs using observations of counter-streaming electrons from Ulysses data and could detect no trend in the openness of ICMEs with distance out to Jupiter. If ICME connectivity to the Sun is the same at 1 AU as it is at 5 AU then it can be argued that an ascending CME could still be rooted at the Sun for a week, or, indeed, much longer.

In reality, an ascending flux rope would most likely contain a mix of open and closed field lines, driven by apparently random reconnection events (Crooker & Horbury 2006; Gosling *et al.* 1995). Complete disconnection of the structure appears to be unlikely.

With the new *STEREO*-HI data we should be able to test this scenario, and this has been reported by Harrison *et al.* (2008). The HI data appear to confirm the *in situ* interpretation showing coherent structures, apparently still connected to the Sun over long distances. There is no evidence for events pinching-off. However, this in turn presents us with an anomaly. McComas (1995) has argued that the heliospheric magnetic flux does not continually build up, so flux must be shed through reconnection somehow during the ICME process. If we are rejecting the plasmoid or disconnected ICME scenario then we must find another way of limiting the flux build up over time.

In the absence of evidence for the pinching-off of CMEs, an interchange reconnection process has been suggested as the mechanism by which CMEs disconnect from the Sun (Gosling *et al.* 1995; Crooker *et al.* 2002). The basic idea is that the ascending CME can travel a considerable distance, well beyond the Earth, still connected to the Sun, and that perhaps days or even weeks after the onset, the legs of the CME, still rooted in the Sun, will interact with adjacent open field lines at low altitude in the corona; reconnection results in the formation of low-lying loops as one CME leg reconnects with the adjacent fields and an outward ascending kink-shaped structure ascends into the heliosphere from the site of one of the original CME footpoints.

This approach has a few attractive points. For example, it seems logical that the site of the greatest field density, magnetic complexity and field-line motion would be the most likely site of any reconnection in the ascending CME. However, assuming that such interchange reconnection is the 'end game' of a CME, and that this low level reconnection results in the outward propagation of a kinked field-line configuration, what might we expect to observe and, indeed, have we seen such features? Harrison *et al.* (2008) indeed point to observations of narrow V-shaped structures identified in the HI data that could be candidates for such reconnection events.

It is early days for this work using *STEREO* but the indications are that there is plenty to be gained from these studies. As the mission progresses we anticipate more opportunities where we have the chance to combine both imaging and *in situ* measurements of specific events, and their impacts, as well as to model CMEs in the heliosphere in 3D as never before. Thus, this report should be take as an early statement on the progress and direction of this work which is opening a new chapter in solar, heliospheric and space weather physics.

12. Coronal mass ejections and space weather (N. Gopalswamy)

CMEs cause adverse space weather in two ways: (*i*) when they arrive at Earth's magnetosphere, they can couple to Earth's magnetic field and cause major geomagnetic storms (Gosling *et al.* 1990); and (*ii*) they can drive fast mode MHD shocks that accelerate solar energetic particles (Reames 1999). Significant progress has been made on both these aspects over the past few years. In the case of geomagnetic storms, connecting the magnetic structure and kinematics of ICMEs observed at 1 AU to the CME source region at the Sun has received considerable attention. In the case of SEPs, assessing the contribution from flare reconnection and shock to the observed SEP intensity has been the focus. The importance of the variability in the Alfvén speed profiles in the outer corona is also under investigation because of its importance in deciding the shock formation.

12.1. *Geomagnetic storms*

High-Speed Solar Wind Streams (HSS) interacting with the slow solar wind result in co-rotating interaction regions (CIRs), which also can produce geomagnetic storms (Vršnak *et al.* 2007a), but they are generally weaker than the CME-produced storms (Zhang *et al.* 2007). Occasionally, the CIR and ICME structures combine to produce major storms (Dal Lago *et al.* 2006). Multiple CMEs are often involved producing some super-intense storms (Gopalswamy *et al.* 2007; Zhang *et al.* 2007). There are numerous effects produced by the

ICMEs in the magnetosphere and various other layers down to the ground (see Borovsky *et al.* 2006; Kataoka & Pulkkinen 2008).

The key element of ICMEs for the production of geomagnetic storms is the southward magnetic field component. While the quite heliospheric field has no out of the ecliptic field component (except for Alfvenic fluctuations in the solar wind), a CME adds this component to the interplanetary (IP) magnetic field. If an ICME has a flux rope structure, one can easily see that the azimuthal component of the flux-rope field or its axial component forms the out of the ecliptic component. In ICMEs with a flux rope structure (i.e., magnetic clouds), it is easy to locate the southward component from the structure of the cloud (Gopalswamy 2006a; Wang *et al.* 2007; Gopalswamy *et al.* 2008a). In non-cloud ICMEs, it is not easy to infer the location of the southward component. If the ICMEs are shock-driving, then the magnetosheath between the shock and the driving ICME (Kaymaz & Siscoe 2006; Lepping *et al.* 2008) can contain southward field and hence cause geomagnetic storms (Gopalswamy *et al.* 2008a). The cloud and sheath storms can be substantially different (Pulkkinen *et al.* 2007).

Once an IP structure has a southward magnetic field, the efficiency with which it causes geomagnetic storm depends on the strength of the magnetic field and the speed with which it hits the magnetosphere (Gonzalez *et al.* 2007; Gopalswamy 2008d). Statistical investigations have shown that the storm intensity (measured e.g., by the Dst index) is best correlated with the speed-magnetic field product in magnetic clouds and their sheaths. Interestingly, an equally good correlation is obtained when the magnetic cloud/sheath speed is replaced by the CME speed measured near the Sun (Gopalswamy *et al.* 2008a). This suggests that if one can estimate the magnetic field in CMEs near the Sun, the strength of the ensuing magnetic storm can be predicted. The ICME speed can be predicted based on the CME speed by quantifying the interaction between CMEs and the solar wind (Xie *et al.* 2006; Nakagawa *et al.* 2006; Jones *et al.* 2007; Vršnak & Zic, 2007). Most of the storm-causing CMEs are halo CMEs, which are subject to projection effects and hence space speeds cannot be easily measured (Kim *et al.* 2007; Gopalswamy & Xie 2008; Howard *et al.* 2007; Vršnak *et al.* 2007b). There have been several attempts to use the sky-plane speed of CMEs to obtain their space speed (Xie *et al.* 2006; Michalek *et al.* 2008; Zhao 2008) with varying extents of success. The magnetic field strength and kinetic energy of CMEs are somehow related to the free energy available in the source region. Quantifying this free energy has been a difficult task (Ugarte-Urra *et al.* 2007; Schrijver *et al.* 2008).

The solar sources of CMEs need to be close to the disk center for the CMEs to make a direct impact on Earth and they have to be fast. In fact the solar sources of magnetic clouds, storm-causing CMEs, and halo CMEs have been shown to follow the butterfly diagram suggesting that only sunspot regions have the ability to produce such energetic CMEs (Gopalswamy 2008d). The average near-Sun speed of CMEs that cause intense geomagnetic storms is $\sim 1000\,\mathrm{km\,s^{-1}}$ (Gopalswamy 2006b; Zhang *et al.* 2007), similar to the average speed of halo CMEs (Gopalswamy *et al.* 2007) because many of the storm-producing CMEs are halo CMEs. Halo CMEs are more energetic (Lara *et al.* 2006; Liu 2007; Gopalswamy *et al.* 2007, 2008a) and end up being magnetic clouds at 1 AU. Most halo CMEs ($\sim 70\%$) are geoeffective. Non-geoeffective halos are generally slower, originate far from the disk center, and originate predominantly in the eastern hemisphere of the Sun. The geoeffectiveness rate of halo CMEs has been reported to be anywhere from $\sim 40\%$ to more than 80% (Yemolaev & Yermolaev 2006), but the difference seems to be due to different definitions used for halo CMEs (some authors have included all CMEs with width $> 120°$ as halos) and the sample size (Gopalswamy *et al.* 2007). The geoeffectiveness rate of CMEs may be related to the fact that more ICMEs are observed as magnetic clouds during solar minimum than during solar maximum (Riley *et al.* 2006). It is possible that all ICMEs are magnetic clouds if viewed appropriately (Krall 2007). This suggestion is consistent with the ubiquitous nature of post eruption arcades, which seem to indicate flux rope formation in the eruption process (Kang *et al.* 2006; Qiu *et al.* 2007; Yurchyshyn 2008). While the reconnection process certainly forms a flux rope, it is not clear if the reconnection creates a new flux rope or fattens an existing one.

12.2. *SEP events*

Energetic storm particle (ESP) events are the strongest evidence for SEP acceleration by shocks, but this happens when the shocks arrive at the observing spacecraft near Earth (Cohen *et al.* 2006). This means the shocks must have been stronger near the Sun accelerating particles to much higher energies. The strongest evidence for SEPs in flares is the gamma-ray lines, which are now imaged by RHESSI (Lin 2007). All shock-producing CMEs are associated with major flares (M- or X-class in soft X-rays), so both mechanisms must operate in most SEP events. There has been an ongoing debate as to which process is dominant based on SEP properties such as the spectral and compositional variability at high energies (Tylka & Lee 2006; Cane *et al.* 2007).

The easiest way to identify shocks near the Sun are the type II radio bursts especially at frequencies below 14 MHz, which correspond to the near-Sun IP medium (Gopalswamy 2006c). Analyzing electrons and protons in SEP events, Cliver & Ling (2007) have found evidence for a dominant shock process including flatter SEP spectra, apparent widespread sources, and high association with long wavelength type II bursts. A recent statistical study finds the SEP association rate of CME steadily increases with CME speed and width especially and there is one-to-one correspondence between SEP events and CMEs from the western hemisphere with long wavelength type II bursts (Gopalswamy *et al.* 2008b). Type II burst studies have also have concluded that the variability in Alfvén speed in the outer corona decides the formation and strength of shocks (Shen *et al.* 2007; Gopalswamy *et al.* 2008c). For example, a $400 \, \mathrm{km \, s^{-1}}$ CME can drive a shock, while a $1000 \, \mathrm{km \, s^{-1}}$ CME may not drive a shock, depending on the local Alfvén speed.

James A. Klimchuk
president of the Commission

References

Abbett, W. P. 2007, *ApJ*, 665, 1469
Abramenko, V. I., Fisk, L. A., & Yurchyshyn, V. B. 2006, *ApJ (Letters)*, 641, L65
Amari, T., Aly, J. J., Mikic, Z., & Linker, J. 2007, *ApJ (Letters)* 671, L189
Antiochos, S. K., DeVore, C. R., & Klimchuk, J. A. 1999, *ApJ*, 510, 485
Aschwanden, M. J. 2008, *ApJ (Lett)*, 672, L135
Aschwanden, M. J., Winebarger, A., Tsiklauri, D. & Peter, H. 2007, *ApJ*, 659, 1673
Attrill, G. D. R., Harra, L. K., van Driel-Gesztelyi, L., & Démoulin, P. 2007, *ApJ*, 656, L101
Attrill, G. D. R., Nakwacki, M. S., Harra, L. K., *et al.* 2006, *Solar Phys.*, 238, 117
Attrill, G. D. R., van Driel-Gesztelyi, L., Démoulin, P., *et al.* 2008, *Solar Phys.*, 252, 349
Aulanier, G., Golub, L., DeLuca, E. E., Cirtain, J. W., Kano, R., *et al.* 2007, *Science*, 318, 1588
Balasubramaniam, K. S., Pevtsov, A. A., & Neidig, D. F. 2007, *ApJ*, 658, 1372
Barta, M., Karlicky, M., Vršnak, B., & Goossens, M. 2007, *Cent. Eur. Astrophys. Bull.*, 31, 165
Battaglia, M. & Benz, A. O. 2006, *A&A*, 456, 751
Bazarghan, M., Safari, H., Innes, D. E., Karami, E., & Solanki, S. K. 2008, *A&A*, 492, 13

Benz, A. O. 2004, in: A. K. Dupree & A. O. Benz (eds.), *Stars as Suns: Activity, Evolution, and Planets*, Proc. IAU Symp. No. 219 (San Francisco: ASP), p. 461

Borovsky, J. E. & Denton, M. H. 2006. *JGR*, 111, A07S08.

Bradshaw, S. J. 2008, *A&A*, 486, L5

Bradshaw, S. J. & Cargill, P. J. 2006, *A&A*, 458, 987

Burlaga, L. F. 1995, *Interplanetary Magnetohydrodynamics* (New York: Oxford University Press)

Cane, H. V., Richardson, I. G., & Rosenvinge, T. 2007, *Space Sci. Rev*, 130, 301,

Cassak, P. A., Mullan, D. J., & Shay, M. A. 2008, *ApJ* (Letters), 676, L69

Centeno, R., Socas-Navarro, H., Lites, B., Kubo, M., & Frank, Z. 2007, *ApJ* (Letters), 666, L137

Chen, P. F. 2006, *ApJ* (Letters), 641, L153

Chen, W.-Z., Liu, C., Song, H., Deng, N., Tan, C.-Y., & Wang, H. 2007, *A&A*, 7, 733

Cheung, M. C. M., Schüssler, M., Tarbell, T. D., & Title, A. M. 2008, *ApJ*, 687, 1373

Cho, K. S., Lee, J., Moon, Y. J., Dryer, M., *et al.* 2007, *A&A*, 461, 1121

Cirtain, J. W., Golub, L., Lundquist, L., van Ballegooijen, A., *et al.* 2007, *Science*, 318, 1580

Cirtain, J. W. *et al.* 2007, *ApJ*, 655, 598

Cliver, E. W. & Ling, A. G. 2007, *ApJ*, 658, 1349

Cohen, C. M. S. 2006, in: N. Gopalswamy, R. Mewaldt & J. Torsti (eds.), *Solar Eruptions and Energetic Particles*, Geophysical Monograph Series, Vol. 165, p. 275

Crooker, N. U., Gosling, J. T., & Kahler, S.W. 2002, *JGR*, 107 (A2)

Crooker, N. U., Horbury, T. S. 2006, *Sp. Sci. Rev.*, 123, 93

Crooker, N. U. & Webb, D. F. 2006, *JGR*, 111(A10), 8108

Cui, Y., Li, R., Wang, H., & He, H. 2007, *Solar Phys.*, 242, 1

Culhane, J. L., Harra, L. K., James, A. M., *et al.* 2007a, *Solar Phys.*, 243, 19

Culhane, L., Harra, L. K., Baker, D., *et al.* 2007b, *PASJ*, S751

Dahlburg, R. B., Liu, J.-H., Klimchuk, J. A., & Nigro, G. 2008, *ApJ*, submitted

Dal Lago, A. *et al.* 2006, *JGR*, 111, A07S14

Dauphin, C., Vilmer, N., & Anastasiadis, A. 2007, *A&A*, 468, 273

Dauphin, C., Vilmer, N., & Krucker, S. 2006, *A&A*, 55, 339

Delannee, C., Torok, T., Aulanier, G., & Hochedez, J.-F. 2008, *Solar Phys.*, 247, 123

Démoulin, P. 2008, *Ann. Geophys.*, 26, 3113

Deng, N., Xu, Y., Yang, G., Cao, W., Liu, C. *et al.* 2006, *ApJ*, 644, 1278

De Pontieu, B., Hansteen, V. H., Rouppe van der Voort, L., *et al.* 2007a, *ApJ*, 655, 624

De Pontieu, B., McIntosh, S. W., Carlsson, M., *et al.* 2007b, *Science*, 318, 1574

DeVore, C. R., & Antiochos, S. K. 2008, *ApJ*, 680, 740

Dolla, L. & Solomon, J. 2008, *A&A*, 483, 271

Donea, A.-C., Besliu-Ionescu, D., Cally, P. S., *et al.* 2006, *Solar Phys.*, 239, 113

Doschek, G. A. 2006, *ApJ*, 649, 515

Doschek, G. A., Mariska, J. T. & Warren, H. P. 2007, *ApJ* (Letters), 667, L109

Drake, J. F., Swisdak, M., Che, H., & Shay, M. A. 2006, *Nature*, 443, 553

Eyles, C. J., Simnett, G. M., Cooke, M. P., *et al.* 2003, *Solar Phys.*, 217, 319

Feldman, U., Landi, E., & Doschek, G. A. 2008, *ApJ*, 679, 843

Fisk, L. A. & Schwadron, N. A. 2001, *ApJ*, 560, 425

Fletcher, L., Hannah, I. G., Hudson, H. S., & Metcalf, T. R. 2007, *ApJ*, 656, 1187

Fletcher, L. & Hudson, H. S., 2008 *ApJ*, 675, 1645

Fulle, M., Leblanc, F., Harrison, R. A., *et al.* 2007, *ApJ* (Letters), 661, L93

Georgoulis, M. K. & Rust, D. M. 2007, *ApJ* (Letters), 661, L109

Gibson, S. E., Fan, Y., Török, T., & Kliem, B. 2006, *Space Science. Rev.*, 124, 131

Golub, L., Deluca, E., Austin, G., *et al.* 2007, *Solar Phys.*, 243, 63

Gontikakis, C., Anastasiadis, A., & Efthymiopoulos, C. 2007, *MNRAS*, 378, 1019

Gonzalez, W. D., Clua-Gonzalez, A. L., Echer, E., & Tsurutani, B. T. 2007, *GRL*, 34, L06101,

Gopalswamy, N. 2006a, *Space Sci. Rev*, 124, 145

Gopalswamy, N. 2006b, *J. Astrophys. Astron.*, 27, 243

Gopalswamy, N. 2006c, in: N. Gopalswamy, R. Mewaldt, & J. Torsti (eds.), *Solar Eruptions and Energetic Particles*, Geophysical Monograph Series, Vol. 165, p. 207

Gopalswamy, N. 2006d, *J. Atm. Solar Terrestrial Phys.*

Gopalswamy, N. 2006e, *Geophys. Monogr. Ser.*, 165, 207

Gopalswamy, N., Akiyama, S., Yashiro, *et al.* 2008a., *J. Atm. Solar Terrestrial Phys.*, 70, 245

Gopalswamy, N. & Xie, H. 2008, *JGR*, A10.0105

Gopalswamy, N., Yashiro, S., & Akiyama, S. 2007, *JGR*, 112, A06112

Gopalswamy, N., Yashiro, S., Akiyama, S. *et al.* 2008b, *Ann. Geophysicae*, 26, 1

Gopalswamy, N., Yashiro, S., Xie, H., *et al.* 2008c, *ApJ*, 674, 560

Gosling, J. T., Baker, D. N., Bame, S. J., *et al.* 1987, *JGR*, 92 (11), 8519

Gosling, J. T., Bame, S. J. McComas, D. J., & Phillips, J. L. 1990, *GRL*, 127, 901

Gosling, J. T., Birn, J., & Hesse, M. 1995, *GRL*, 22, 869

Green, L. M., Kliem, B., Török, T., *et al.*, *Solar Phys.* 246, 365

Grigis, P. C., & Benz, A. O. 2006, *A&A*, 458, 641

Grechnev, V. V., Uralov, A. M., Slemzin, V. A., *et al.* 2008, *Solar Phys.*, in press

Hagenaar, H. J., DeRosa, M. L., & Schrijver, C. J. 2008, *ApJ*, 678, 541

Hannah, I. G., Krucker, S., Hudson, H. S., Christe, S., & Lin, R. P. 2008, *A&A* (Letters), 481, L45

Hansteen, V. H., de Pontieu, B., & Carlsson, M. 2007, *PASJ*, 59, S699

Hara, H., Watanabe, T., Harra, L. K., *et al.* 2008, *ApJ* (Letters), 678, L67

Harra, L. K., Crooker,N. U., Mandrini,C. H., *et al.* 2007, *Solar Phys.*, 244, 95

Harra, L. K., Hara, H., Imada, S., Young, P. R., Williams, D. R., *et al.* 2007, *PASJ*, 59, S801

Harra, L. K. & Sterling, A. C. 2001, *ApJ* (Letters) 561, L215

Harrison, R. A. 2006, in: N. Gopalswamy, R. Mewaldt & J. Torsti (eds.), *Solar Eruptions and Energetic Particles*, Geophysical Monograph Series 165, p. 73

Harrison, R. A. & Bewsher, D. 2007, *A&A*, 461, 1155

Harrison, R. A., Davis, C. J., Bewsher, D., *et al.* 2008, *Adv. Space Res.*, submitted

Harrison, R. A., Davis, C. J., Eyles, C. J., *et al.* 2008, *Solar Phys.*, 247, 171

Harrison, R. A., Sawyer, E. C., Carter, M. K., *et al.* 1995, *Solar Phys.*, 162, 233

Howard, R. A., Moses, J. D., Vourlidas, A., *et al.* 2008, *SSRv*, 136, 67

Howard, T. A., D. Nandy, & A. C. Koepke 2008, *JGR*, 113, A01104

Hudson, H. S. 1991, *Solar Phys.*, 133, 357

Hudson, H. S., Acton, L. W., & Freeland, S. L. 1996, *ApJ*, 470, 629

Imada, S., Hara, H., Watanabe, T., Kamio, S., Asai, A., *et al.* 2007, *PASJ*, 59, S793

Ishikawa, R., Tsuneta, S., Ichimoto, K., *et al.* 2008, *A&A* (Letters), 481, L25

Innes, D. 2008, *A&A* (Letters), 481, L41

Isenberg, P. A. & Forbes, T. G. 2007, *ApJ*, 670, 1453

Isobe, H., Kubo, M., Minoshima, T., Ichimoto, K., Katsukawa, Y., *et al.* 2007, *PASJ*, 59, S807

Jackson, B. V., Hick, P. P., Buffington, A., Bisi, M. M., & Jensen, E. A. 2007, *Proc. SPIE*, 6689

Jackson, B. V. & Leinert, C. 1985, *JGR*, 90, 10,759

Ji, H., Huang, G., & Wang, H. 2007, *ApJ*, 660, 893

Jing, J., Chae, J., & Wang, H. 2008, *ApJ* (Letters), 672, L73

Jones, R. A., Breen, A. R., Fallows, R. A. Canals, A., Bisi, M. M., *et al.* 2007, *JGR*, 112, A08107

Judge, P. J. 2008, *ApJ*, 683, 87

Kahler, S. W. & D. F. Webb 2007, *JGR*, 112, A09103

Kamio, S., Hara, H., Watanabe, T., Matsuzaki, K., Shibata, K., *et al.* 2007, *PASJ*, 59, S757

Kang, S., Y.-J. Moon, K.-S. Cho, Y. Kim, Y. D. Park, *et al.* 2006, *JGR*, 111, A05102

Karlicky, M. & Barta, M. 2007, *A&A*, 464, 735

Karoff, C. & Kjeldsen, H. 2008, *ApJ* (Letters), 678, L73

Karpen, J. T. & Antiochos, S. K. 2008, *ApJ*, 676, 658

Kasparová, J., Kontar, E. P., & Brown, J. C. 2007, *A&A*, 466, 705

Kataoka, R. & A. Pulkkinen 2008, *JGR*, 113, A03S12

Kim, K.-H., Y.-J. Moon, & K.-S. Cho 2007, *JGR*, 112, A05104

Kliem, B. & Török, T. 2007, *Phys. Rev. L.*, 96(25), 255002

Klimchuk, J. A. 2006, *Solar Phys.*, 234, 41

Klimchuk, J. A. & Karpen, J. T. 2008, *ApJ*, in preparation

Klimchuk, J. A., Patsourakos, S., & Cargill, P. J. 2008, *ApJ*, 682, 1351

Kontar, E. P. & Brown, J. C. 2006b, *ApJ* (Letters), 653, L149

Kontar, E. P., MacKinnon, A. L., Schwartz, R. A., & Brown, J. C. 2006a, *A&A*, 446, 1157

Kosovichev, A. G. 2006, *Solar Phys.*, 238, 1

Kosovichev, A. G. 2007, *ApJ* (Letters) , 670, L65

Kosovichev, A. G. & Zharkova, V. V. 1998, *Nature*, 393, 317

Kosugi, T., Matsuzaki, K., Sakao, T., Shimizu, T., Sone, Y., *et al.* 2007, *Solar Phys.* 243, 3

Krall, J. 2007, *ApJ*, 657, 559

Krucker, S., Hurford, G. J., MacKinnon, A. L., *et al.* 2008, *ApJ* (Letters), 678, L63

Krucker, S., Kontar, E. P., Christe, S., & Lin, R. P. 2007a, *ApJ*, 663, L109

Krucker, S., White, S. M., & Lin, R. P. 2007b, *ApJ*, 669, L49

LaBonte, B. J., Georgoulis, M. K., & Rust, D. M. 2007, *ApJ*, 671, 955

Landi, R. & Feldman, U. 2008, *ApJ*, 672, 674

Lara, A., Gopalswamy, N. Xie, H., *et al.* 2006, *JGR*, 111, A06107

Lee, K. W., Buchner, J., & Elkina, N. 2008, *A&A*, 478, 889

Leka, K. D. & Barnes, G. 2007, *ApJ*, 656, 1173

Lepping, R. P., Wu, C.-C., Gopalswamy, N., & Berdichevsky, D. B. 2008, *Solar Phys.*, 248, 125

Lin, C.-H., Banerjee, D., O'Shea, E., & Doyle, J. G. 2006, *A&A*, 460, 597

Lin, J. & van Ballegooijen, A. A. 2005, *ApJ*, 629, 582

Lites, B. W. 2008, in A. Balogh (ed.), *ISSI Proc. Workshop Solar Dynamic Magnetic Field*

Lites, B. W., Kubo, M., Socas-Navarro, H., Berger, T., Frank, Z. 2008, *ApJ*, 672, 1237

Liu, C., Lee, J., Jing, J., Gary, D. E., Wang, H. 2008, *ApJ* (Letters), 672, L69

Liu, C., Lee, J., Yurchyshyn, V., Deng, N., Cho, K.-S., *et al.* 2007, *ApJ*, 669, 1372

Liu, Y. 2007, *ApJ* (Letters), 654, L171

Liu, Y. 2008, *ApJ* (Letters) 679, L151

Liu, Y. C.-M., Opher, M., Cohen, O., Liewer, P. C., & Gombosi, T. I. 2008, *ApJ*, 680, 757

Long, D. M., Gallager, P. T., McAteer, R. T. J., *et al.* 2008, *ApJ* (Letters), 680, L81

Longcope,D., Beveridge,C., Qiu,J., Ravindra,B., Barnes,G., *et al.* 2007, *Solar Phys.*, 244, 45

López Fuentes, M. C., Klimchuk, J. A., & Mandrini, C. H. 2007, *ApJ*, 657, 1127

Magdalenić, J., Vršnak, B., Pohjolainen, S., *et al.* 2008, *Solar Phys.*, 253, 305

Mancuso, S. & Avetta, D. 2008, *ApJ*, 677, 683

Mandrini, C. H., Nakwacki, M. S., Attrill, G., *et al.* 2007, *Solar Phys.*, 244, 25

Mandrini, C. H., Pohjolainen, S., Dasso, S., *et al.* 2005, *A&A*, 434, 725

McIntosh, S. W., De Pontieu, B., & Tarbell, T. D. 2008, *ApJ* (Letters), 673, L219

McTiernan, J. M. 2008, *ApJ*, submitted

Michalek, G., Gopalswamy, N., & Yashiro, S. 2008, *Solar Phys.*, 248, 113

Milligan, R. O. 2008, *ApJ* (Letters), 680, L157

Miteva, R., Mann, G., Vocks, C., & Aurass, H. 2007, *A&A*, 461, 1127

Mok, Y., Mikic, Z., Lionello, R., & Linker, J. A. 2008, *ApJ* (Letters), 679, L161

Moreno-Insertis, F., Galsgaard, K., & Ugarte-Urra, I. 2008, *ApJ* (Letters) 673, L211

Möstl, C., Miklenic, C., Farrugia, C. J., *et al.* 2008, *Annales Geophys.*, 26, 3139

Muhr, N., Temmer, M., Veronig, A., *et al.* 2008, *Cent. Eur. Astrophys. Bull.*, 32, 79

Nakagawa, T. N., Gopalswamy, N., & Yashiro, S. 2006, *JGR*, 111, A01108

Nitta, N. V., Mason, G. M., Wiedenbeck, M. E., *et al.* 2008, *ApJ* (Letters), 675, L125

Noglik, J. B. & Walsh, R. W. 2007, *ApJ*, 655, 1127

Ofman, L. 2007, *ApJ*, 655, 1134

Orozco Suárez, D., Bellot Rubio, L. R., *et al.* 2007, *ApJ* (Letters, 670, L61

Pagano, P., Reale, F., Orlando, S., & Peres, G. 2007, *A&A*, 464, 753

Parenti, S., Buchlin, E., Cargill, P. J., Galtier, S., & Vial, J.-C. 2006, *ApJ*, 651, 1219

Parenti, S. & Young, P. R. 2008, *A&A*, submitted

Pariat, E., Antiochos, S.K., & DeVore, C.R. 2008, *ApJ* in press

Patsourakos, S., Gouttebroze, P., & Vourlidas, A. 2007, *ApJ*, 774, 1214

Patsourakos, S. & Klimchuk, J. A. 2006, *ApJ*, 647, 1452

Patsourakos, S. & Klimchuk, J. A. 2007, *ApJ*, 667, 591

Patsourakos, S. & Klimchuk, J. A. 2008a, *ApJ*, 689, 1406
Patsourakos, S. & Klimchuk, J. A. 2008b, *ApJ*, submitted
Patsourakos, S., Pariat, E., Vourlidas, A., *et al.* 2008, *ApJ* (Letters), 680, L73
Pauluhn, A. & Solanki, S. K. 2007, *A&A*, 462, 311
Peter, H. 2007, *Adv. Space Res.*, 39, 1814
Peter, H., Gudiksen, B., & Nordlund, Å., 2006, *ApJ*, 638, 1086
Pohjolainen, S. 2008, *A&A*, 483, 297
Pohjolainen, S. & Lehtinen, N. J. 2006, *A&A*, 449, 359
Pulkkinen, T. I., Partamies, N., Huttunen, K. E. J., *et al.* 2007, *GRL*, 34, L02105
Qiu, J., Hu, Q., Howard, T. A., & Yurchyshyn, V. B. 2007, *ApJ*, 659, 758
Rappazzo, A. F., Velli, M., Einaudi, G., & Dahlburg, R. B. 2008, *ApJ*, 677, 1348
Reames, D. V. 1999, *Space Sci. Rev*, 90, 413
Reale, F., Parenti, S., Reeves, K. K., *et al.* 2007, *Science*, 318, 1582
Reale, F. & Orlando, S. 2008, *ApJ*, 684, 715
Reale, F. *et al.* 2008, in preparation
Regnier, S., Priest, E. R. 2007, *A&A*, 468, 701
Reiner, M. J., Krucker, S., Gary, D. E., *et al.* 2007, *ApJ*, 657, 1107
Richter, I., Leinert, C., & Planck, B. 1982, *A&A*, 110, 115
Riley, P., Gosling, J. T., & Crooker, N.U. 2004, *ApJ*, 608, 1100
Riley, P., Schatzman, C., Cane, H. V., *et al.* 2006, *ApJ*, 647, 648
Rouillard, A. P., Davies, J. A., Forsyth, R. J., *et al.* 2008a, *GRL*, 35, L10110
Rouillard, A. P., Davies, J. A., Rees, A., Zhang, T., & Forsyth, R. J. 2008b, *JGR*, in press
Rust, D. M. & LaBonte, B. J. 2005, *ApJ*, 622, 69
Sakai, J. I., Nagasugi, Y., Saito, S., Kaufmann, P. 2006, *A&A*, 457, 313
Sanchez Almeida, J., Teriaca, L., & Sütterlin, P., *et al.* 2007, *A&A*, 475, 1101
Sakamoto, Y., Tsuneta, S., & Vekstein, G. 2008, *ApJ*, 689, 1421
Sarkar, A. & Walsh, R. W. 2008, *ApJ*, 683, 516
Savcheva, A., Cirtain, J., Deluca, E. E., Lundquist, L. L., Golub, L., *et al.* 2007, *PASJ*, 59, S771
Schmelz, J. T., Kashyap, V. L., & Weber, M. A. 2007a, *ApJ* (Letters), 660, L157
Schmelz, J. T., Nasraoui, K., Del Zanna, G., *et al.* 2007b, *ApJ* (Letters), 658, L119
Schrijver, C. J. 2001, *Solar Phys.*, 198, 325
Schrijver, C. J. 2007, *ApJ* (Letters), 655, L117
Schrijver, C. J., DeRosa, M. L., Metcalf, T., Barnes, G., Lites, B., *et al.* 2008, *ApJ*, 675, 1637
Schrijver, C. J., DeRosa, M. L., Metcalf, T., *et al.* 2008, *ApJ*, 675, 1637
Schüssler, M. & Vögler, A. 2008, *A&A* (Letters), 481, L5
Shanmugaraju, A., Moon, Y.-J., Cho, K.-S., *et al.* 2006b, *Solar Phys.*, 233, 117
Shanmugaraju, A., Moon, Y.-J., Kim, Y.-H., *et al.* 2006a, *A&A*, 458, 653
Sheeley, N. R., Herbst, A. D., Palatchi, C. A., *et al.* 2008, *ApJ* (Letters), 674, L109
Sheeley, N. R., Herbst, A. D., Palatchi, C. A., *et al.* 2008, *ApJ*, 675, 853
Shen, C., Wang, Y., Ye, P., Zhao, X. P., Gui, B., & Wang S. 2007, *ApJ*, 670, 849
Siarkowski, M., Falewicz, R., Kepa, A., & Rudawy, P. 2008, *Ann. Geophys.*, 26, 2999
Silva, A. V. R., Share, G. H., Murphy, R. J., *et al.* 2007, *Solar Phys.*, 245, 311
Spadaro, D., Lanza, A. F., Karpen, J. T., & Antiochos, S. K. 2006, *ApJ*, 642, 579
Subramanian, K. R. & Ebenezer, E. 2006, *A&A*, 451, 683
Sudol, J. J. & Harvey, J. W. 2005, *ApJ*, 635, 647
Temmer, M., Veronig, A. M., Vrsnak, B., & Miklenic, C. 2007, *ApJ*, 654, 665
Temmer, M., Veronig, A. M., Vrsnak, B., *et al.* 2008, *ApJ* (Letters), 673, L95
Thompson, B. J., Cliver, E. W., Nitta, N., *et al.* 2000, *GRL*, 27, 1431
Tian, H., Marsch, E., Tu, C.-Y., *et al.* 2008b, *A&A*, 482, 267
Tian, H., Tu, C.-Y., Marsch, E., *et al.* 2008a, *A&A*, 478, 915
Titov, V. S., Démoulin, P. 1999, *A&A*, 351, 707
Tomczyk, S., McIntosh, S., Keil, S. W., *et al.* 2007, *Science*, 317, 1192
Török, T., & Kliem, B. 2007, *Astron. Nachr.*, 328, 743
Török, T., & Kliem, B. 2005, *ApJ* (Letters), 630, L97

Tripathi, D., Mason, H. E., Young, P. R., & Del Zanna, G. 2008, *A&A* (Letters), 481, L53

Tsuneta, S., Acton, L., Bruner, M., Lemen, J., Brown, W., *et al.* 1991, *Solar Phys.*, 136, 37

Tylka, A. J. & Lee, M. A. 2006, *ApJ*, 646, 1319

Ugai, M. 2008, *Phys. Plasmas*, 15, 082306

Ugarte-Urra, I., Warren, H. P., & Brooks, D. H. 2008, *ApJ*, submitted

Ugarte-Urra, I., Warren, H. P., & Winebarger, A. R. 2007, *ApJ*, 662, 1293

Uzdensky, D. A. 2007, *ApJ*, 671, 2139

van Driel-Gesztelyi, L., Attrill, G. D. R., Démoulin, P., *et al.* 2008, *Ann. Geophys.*, 26, 3077

Veronig, A. M., Karlicky, M., Vrsnak, B., *et al.* 2006, *A&A*, 446, 675

Veronig, A. M., Temmer, M., & Vršnak, B. 2008, *ApJ* (Letters) 681, L113

Veronig, A. M., Temmer, M., Vršnak, B., & Thalmann J. K. 2006, *ApJ*, 647, 1466

Vögler, A., & Schüssler, M. 2007, *A&A* (Letters), 465, 43

Vourlidas, A., Davis, C. J., Eyles, C. J., *et al.* 2007, *ApJ* (Letters) 668, L79

Vršnak, B. & Cliver E. W. 2008, *Solar Phys.*, 253, 215

Vršnak, B., Sudar, D., Rudjak, D., Zic, T. 2007b, *A&A*, 469, 339

Vršnak, B., Temmer, M., & Veronig, A. M. 2007a, *Solar Phys.*, 440, 331

Vršnak, B., Warmuth, A., Temmer, M., *et al.* 2006, *A&A*, 448, 739

Vršnak, B. & Zic, T. 2007, *A&A*, 472, 937

Wang, H. 2006, *ApJ*, 649, 490

Wang, Y., Ye, P., & Wang, S. 2007, *Solar Phys.*, 240, 373

Warmuth, A. 2007, *LNP*, 725, 107

Warmuth, A., Vršnak, B., Aurass, H., & Hanslmeier, A. 2001, *ApJ* (Letters), 560, L105

Warren, H. P., & Winebarger, A. R. 2007, *ApJ*, 666, 1245

Warren, H. P., Ugarte-Urra, I., Doschek, G. A., *et al.* 2008a, *ApJ*, submitted

Warren, H. P., Winebarger, A. R., & Mariska, J. T., 2008b, *ApJ*, 677, 1395

Webb, D. F., Lepping, R. P., Burlaga, L. F., *et al.* 2000, *JGR*, 105, 27251

White, S. M. 2007, *Asian J. Phys.*, 16, 189

Williams, D. R., Török, T., Démoulin, P., *et al.* 2005, *ApJ* (Letters), 628, L163

Wills-Davey, M. J., DeForest, C. E., & Stenflo, J. O. 2007, *ApJ*, 664, 556

Winebarger, A. R., Warren, H. P., & Falconer, D. A. 2008, *ApJ*, 676, 672

Xie, H., Gopalswamy, N., Manoharan, P. K., Lara, A., Yashiro, S., *et al.* 2006, *JGR*, 111, A01103

Xu, Y., Emslie, A. G., & Hurford, G. J. 2008, *ApJ*, 673, 576

Yashiro, S., Akiyama, S., Gopalswamy, N., & Howard, R. A. 2006, *ApJ (Lett)*, 650, L143

Yashiro, S., Michalek, G., Akiyama, S., Gopalswamy, N., & Howard, R. A. 2008, *ApJ*, 673, 1174

Yermolaev, Yu. I. & Yermolaev, M. Yu. 2006, *Adv. Space Res.*, 37 (6), 1175

Young, P. R., Del Zanna, G., & Mason, H. E. 2007, *PASJ*, 59, S727

Yurchyshyn, V. 2008, *ApJ* (Letters), 675, L49

Yurchyshyn, V. B., Liu, C., Abramenko, V., Krall, J. 2006, *Solar Phys.*, 239, 317

Zhang, J., Ma, J., & Wang, H. 2006, *ApJ*, 649, 464

Zhang, J., *et al.* 2007, *JGR* 112, A10102.

Zhang, Y., Wang, J., Attrill, G. D. R., Harra, L. K., Yang, Z., *et al.* 2007, *Solar Phys.*, 241, 329

Zhao, X. P. 2008, *JGR*, 113, A02101

Zharkova, V. V., Gordovskyy, M. 2006, *ApJ*, 651, 553

Zharkova, V. V., Zharkov, S. I. 2007, *ApJ*, 664, 573

Zhitnik, I. A., *et al.* 2006, *Solar System Research*, 40, 272

Zhukov, A. N., & Auchère, F. 2004, *A&A*, 427, 705

Žic, T., Vršnak, B., Temmer, M., & Jacobs, C. 2008, *Solar Phys.*, 253, 237

Transactions IAU, Volume XXVIIA
Reports on Astronomy 2006–2009
Karel A. van der Hucht, ed.

© 2009 International Astronomical Union
doi:10.1017/S1743921308025362

COMMISSION 12

SOLAR RADIATION AND STRUCTURE

RAYONNEMENT ET STRUCTURE SOLAIRE

PRESIDENT	Valentín Martínez Pillet
VICE-PRESIDENT	Alexander Kosovichev
SECRETARY	John T. Mariska
PAST PRESIDENT	Thomas J. Bogdan
ORGANIZING COMMITTEE	Martin Asplund, Gianna Cauzzi, Jørgen Christensen-Dalsgaard, Lawrence E. Cram, Weiqun Gan, Laurent Gizon, Petr Heinzl, Marta G. Rovira, P. Venkatakrishnan

TRIENNIAL REPORT 2006 - 2009

1. Introduction

Commission 12 encompasses investigations on the internal structure and dynamics of the Sun, mostly accessible through the techniques of local and global helioseismology, the quiet solar atmosphere, solar radiation and its variability, and the nature of relatively stable magnetic structures like sunspots, faculae and the magnetic network. A revision of the progress made in these fields is presented. For some specific topics, the review has counted with the help of experts outside the Commission Organizing Committee that are leading and/or have recently presented relevant works in the respective fields. In this cases the contributor's name is given in parenthesis.

2. Irradiance and its variability

2.1. *Measurements and models of the Total Solar Irradiance (TSI)*

Measurements of solar irradiance and its variation can only be made from space, and almost thirty years of observation have now established that the total solar irradiance (TSI) varies by only 0.1 to 0.3%, while certain portions of the solar spectrum, the ultraviolet for example, vary by orders of magnitude more. Since November 1978 a set of TSI measurements from space is available, yielding a time series of more than 25 years. The results (PMOD composite) provide reliable TSI measurements for the three solar cycles (Frohlich 2006).

Substantial work had been done to model the observed TSI variations and understand the role of magnetic and quiet Sun regions. A reconstruction of total solar irradiance (TSI) back to 1974, i.e., from the minimum of cycle 21 to the declining phase of cycle 23 based on surface magnetic fields was carried out by using data from the 512-channel Diode Array Magnetograph and the newer spectromagnetograph on Kitt Peak. Their model is based on the assumption that all irradiance changes on time-scales of a day and longer are entirely due to the variations of the surface distribution of the solar magnetic field. A good correspondence is found with the PMOD TSI composite, with no bias between the three cycles on time-scales longer than the solar rotation period, although the accuracy of the TSI reconstruction is somewhat lower when 512 channel magnetograph data are used. This suggests that the same driver of the irradiance variations, namely the evolution of the magnetic flux at the solar surface, is acting in cycles 21-23 (Wenzler 2006).

Very high-resolution filtergram and magnetogram observations of solar faculae taken at the Swedish 1-meter Solar Telescope (SST) on La Palma were used to investigate the structure of solar faculae, which provide significant contribution to TSI. These data revealed that faculae are not the interiors of small flux tubes – they are granules seen through the transparency caused by groups of magnetic elements or micropores 'in front of' the granules. Previous results which show a strong dependency of facular contrast on magnetic flux density were caused by bin-averaging of lower resolution data leading to a mixture of the signal from bright facular walls and the associated intergranular lanes and micropores. The findings are important to studies of total solar irradiance that use facular contrast as a function of disk position and magnetic field in order to model the increase in TSI with increasing sunspot activity (Berger *et al.* 2007).

The TSI variations correlate well with changes in projected area of photospheric magnetic flux tubes associated with dark sunspots and bright faculae in active regions and network. This correlation does not, however, rule out possible TSI contributions from photospheric brightness inhomogeneities located outside flux tubes and spatially correlated with them. Previous reconstructions of TSI report agreement with radiometry that seems to rule out significant 'extra-flux-tube' contributions. Using measurements with the Solar Bolometric Imager (SBI) it was shown that these reconstructions are more sensitive to the facular contrasts used than has been generally recognized. Longer term bolometric imaging will be required to determine whether the small but systematic TSI residuals are caused by remaining errors in spot and facular areas and contrasts or by extra-flux-tube brightness structures such as bright rings around sunspots or 'convective stirring' around active regions (Foukal & Bernasconi 2008).

An analysis of spatially-resolved measurements of the intensity of the photospheric continuum by the Michelson Doppler Imager (MDI) on the SOHO spacecraft indicated that, while it is possible to account for short-term (weeks to months) variation in TSI by variations in the irradiance contributions of regions with enhanced magnetic fields (larger than ten Gauss as measured by MDI), the longer-term variations are influenced significantly by variations in the brightness of the quiet Sun, defined here as regions with magnetic field magnitudes smaller than ten Gauss. The latter regions cover a substantial fraction of the solar surface, ranging from approximately 90% of the Sun near solar minimum to 70% near solar maximum. The results provide evidence that a substantial fraction, 50% or more, of the longer term ($\geqslant$ one year) variation in TSI is due to changes in the brightness of the quiet Sun (Withbroe 2006).

2.2. *Spectral solar irradiance*

There have been significant, recent advances in understanding the solar ultraviolet (UV) and X-ray spectral irradiance from several different satellite missions and from new efforts in modeling the variations of the solar spectral irradiance. The recent satellite missions with solar UV and X-ray spectral irradiance observations include the X-ray Sensor (XRS) aboard the series of NOAA *GOES* spacecraft, the *Upper Atmosphere Research Satellite* (*UARS*), the *SOHO* solar EUV Monitor (SEM), the solar XUV Photometers (SXP) on the

Student Nitric Oxide Explorer (SNOE), the solar EUV Experiment (SEE) aboard the *Thermosphere, Ionosphere, Mesosphere, Dynamics, and Energetic (TIMED)* satellite, and the *Solar Radiation and Climate Experiment (SORCE)* satellite. The combination of these measurements is providing new results on the variability of the solar ultraviolet irradiance over a wide range of time scales ranging from years to seconds. The solar UV variations of flares are especially important for space weather applications and upper atmosphere research. The new efforts in modeling these solar UV spectral irradiance variations range from simple empirical models that use solar proxies to more complicated physics-based models that use emission measure techniques. These new models provide better understanding and insight into why the solar UV irradiance varies, and they can be used at times when solar observations are not available for atmospheric studies (Woods 2008).

The International Organization for Standardization (ISO) is preparing a standard that will certify the process of developing 'reference solar spectra'. However, a few issues remain to be clarified in the current draft standard. In particular, it is not clear what methodology one should use to properly 'validate' or assess the performance of a reference composite spectrum primarily based on measured irradiance data. Excellent agreement is found between the latest composite reference spectrum and an experimental irradiance dataset limited to the UV (295–355 nm). Good agreement is also found by comparison with various single-day or average SOLSTICE-UARS spectra over a larger UV region (120 - 420 nm), even though solar activity interference in this validation attempt appears obvious. Conversely, large differences are noted when comparing the same reference spectrum (as well as other older reference spectra) to spectral data from the SORCE instruments, from 120 to 1600 nm. Furthermore, using the latest version of SORCE data, a comparison between periods of moderate activity and periods of elevated sunspot activity (with low total solar irradiance) suggests that the most part of this loss in total irradiance can be explained by small and smooth spectral changes in the 400 - 1600 nm waveband, thus confirming that sunspots are dark over all or most of the spectrum (Gueymard 2006).

The SOLAR2000 (S2K) project provides solar spectral irradiances and integrated solar irradiance proxies for space researchers as well as ground- and space-based operational users. The S2K model currently represents empirical solar irradiances and integrated irradiance proxies covering the spectral range from the X-rays through the far infrared. Variability is provided for time frames ranging from 1947 to 2052. The combination of variability through multiple time periods with spectral formats ranging from resolved emission lines through integrated irradiance proxies is a unique feature that provides researchers and operational users the same solar energy for a given day but in formats suitable for their distinctly different applications (Tobiska & Bouwer 2006).

The solar photon output from the Sun, especially true in the wavelengths shorter than 190 nm, varies considerably over time scales from seconds during solar flares to years due to the solar cycle. These variations cause significant deviations in the Earth and space environment on similar time scales, which then affects many things including satellite drag, radio communications, atmospheric densities and composition of particular atoms, molecules, and ions of Earth and other planets, as well as the accuracy in the Global Positioning System (GPS). The Flare Irradiance Spectral Model (FISM) is an empirical model that estimates the solar irradiance at wavelengths from 0.1 to 190 nm at 1 nm resolution with a time cadence of 60 s. This is a high enough temporal resolution to model variations due to solar flares, for which few accurate measurements at these wavelengths exist. This model also captures variations on the longer time scales of solar rotation (days) and solar cycle (years) (Chamberlin 2008).

NRLEUV model represents an independent approach to modeling the Sun's EUV irradiance and its variability. Instead of relying on existing irradiance observations, this model utilizes differential emission measure distributions derived from spatially and spectrally resolved solar observations, full-disk solar images, and a database of atomic physics parameters to calculate the solar EUV irradiance. Recent updates to the model include the calculation of a new quiet Sun differential emission measure distribution using data from the CDS and SUMER spectrometers on *SOHO* and the use of a more extensive database of atomic physics parameters. Although there are many areas of agreement between the modeled spectrum and the observations, there are still some major disagreements. For instance, the computed spectra cannot reproduce the observed irradiances at wavelengths below about 160Å. Also, the observed irradiances appear to overstate the magnitude of the EUV continua. Thus, more work needs to be done to develop reliable irradiance models (Warren 2006).

2.3. *Solar variability, solar forcing and climate*

Total solar irradiance changes by about 0.1% between solar activity maximum and minimum. Accurate measurements of this quantity are only available since 1978 and do not provide information on longer-term secular trends. The total solar irradiance is reconstructed from the end of the Maunder minimum to the present based on variations of the surface distribution of the solar magnetic field (Krivova *et al.* 2007). The latter is calculated from the historical record of the sunspot number using a simple but consistent physical model. The model successfully reproduces three independent data sets: total solar irradiance measurements available since 1978, total photospheric magnetic flux since 1974 and the open magnetic flux since 1868 empirically reconstructed using the geomagnetic aa-index. The model predicts an increase in the solar total irradiance since the Maunder minimum of 1.3 +0.2/-0.4 Wm^{-2}. This result is consistent with an independent analysis of Tapping *et al.* 2007.

The climate response to changes in radiative forcing depends crucially on climate feedback processes, with the consequence that solar and greenhouse gas forcing have both similar response patterns in the troposphere. This circumstance complicates significantly the attribution of the causes of climate change. Additionally, the climate system displays a high level of unforced intrinsic variability, and significant variations in the climate of many parts of the world are due to internal processes. Such internal modes contribute significantly to the variability of climate system on various time scales, and thus compete with external forcing in explaining the origin of past climate extremes. This highlights the need for independent observations of solar forcing including long-term consistent observational records of the total and spectrally resolved solar irradiance (Bengtsson 2006).

Cosmogenic isotopes are frequently used as proxy indicators of past variations in solar irradiance on centennial and millennial timescales. These isotopes are spallation products of galactic cosmic rays (GCRs) impacting Earth's atmosphere, which are deposited and stored in terrestrial reservoirs such as ice sheets, ocean sediments and tree trunks. On timescales shorter than the variations in the geomagnetic field, they are modulated by the heliosphere and thus they are, strictly speaking, an index of heliospheric variability rather than one of solar variability. Strong evidence of climate variations associated with the production (as opposed to the deposition) of these isotopes is emerging. This raises a vital question: do cosmic rays have a direct influence on climate or are they a good proxy indicator for another factor that does (such as the total or spectral solar irradiance)? The former possibility raises further questions about the possible growth of air ions generated by cosmic rays into cloud condensation nuclei and/or the modulation of the global thunderstorm electric circuit. The latter possibility requires new understanding about the required relationship between the heliospheric magnetic fields that scatter cosmic rays and the photospheric magnetic fields which modulate solar irradiance (Beer et al. 2006; Lockwood 2006).

It has been proposed that solar cycle irradiance variations may affect the whole planet's climate via the stratosphere, the Quasi-Biennial Oscillation (QBO) and Arctic Oscillation (AO). This hypothesis was tested by examining causal links between time series of sunspot number and indices of QBO, AO and ENSO activity. Various methods were employed in this study: wavelet coherence, average mutual information, and mean phase coherence to study the phase dynamics of weakly interacting oscillating systems. All methods clearly showed a cause and effect link between Southern Oscillation Index (SOI) and AO, but no link between AO and QBO or solar cycle over all scales from biannual to decadal. It is concluded that the 11-year cycle sometimes seen in climate proxy records is unlikely to be driven by solar forcing, and most likely reflects other natural cycles of the climate system such as the 14-year cycle, or a harmonic combination of multi-year cycles (Moore et al. 2006).

2.4. Measurements of solar irradiance on new space missions

LYRA is the solar UV radiometer that will embark in 2009 onboard *Proba2*, a technologically oriented ESA micro-mission. LYRA is designed and manufactured by a Belgian Swiss German consortium (ROB, PMOD/WRC, IMOMEC, CSL, MPS and BISA) with additional international collaborations. It will monitor the solar irradiance in four UV passbands. The SWAP EUV imaging telescope will operate next to LYRA on *Proba2*. Together, they will establish a high performance solar monitor for operational space weather nowcasting and research. LYRA will also demonstrate technologies important for future missions such as the ESA *Solar Orbiter* (Hochedez et al. 2006).

The solar payload of the ESA *Columbus* laboratory, which was launched to the International Space Station on 7 February 2008 with Space Shuttle Atlantis, has three solar irradiance instruments complementing each other to allow measurements of the solar spectral irradiance throughout virtually the whole electromagnetic spectrum – from 17 nm to 100 μm, in which 99% of the solar energy is emitted. *SOVIM* (SOlar Variable and Irradiance Monitor), which covers near-UV, visible and thermal regions of the spectrum (200 nm - 100 μm) is developed by PMOD/WRC (Davos, Switzerland) with one of the instrument's radiometers provided by IRM (Brussels, Belgium). *SOLSPEC* (SOLar SPECtral irradiance measurements) covers the 180 nm–3000 nm range. *SOLSPEC* is developed by CNRS (Verrires-le-Buisson, France) in partnership with IASB/BIRA (Belgium) and LSW (Germany). Very accurate irradiance measurements are expected in terms of relative standard uncertainties (RSU) ranging from 5% to 3% depending on the wavelength range (Schmidtke et al. 2006a; Schmidtke et al. 2006a).

PICARD is a new space mission dedicated to simultaneous measurements of the solar diameter, spectral, and total solar irradiance. It is presently in development for launch in 2009 on board of a microsatellite under the responsibility of Centre National d'Etudes Spatiales (France). The payload will consist of an imaging telescope, three filter radiometers with in total twelve channels, and two independent absolute radiometers (Dewitte et al. 2006).

The Atmospheric Imaging Assembly (AIA) on the *Solar Dynamics Observatory* (*SDO*) will provide revolutionary coverage of the entire visible solar hemisphere observed from photospheric to coronal temperatures at 1-arcsecond resolution with a characteristic cadence of 10 seconds for each channel The AIA comprises four dual normal-incidence telescopes that enable it to cycle through a set of EUV channels centered on strong emission lines of iron ranging from Fe IX through XXIII and helium 304 Å plus two UV channels near 1600 Å and a broad band visible channel Combined with the vector-magnetic imagery from the *SDO* Helioseismic and Magnetic Imager (HMI) the AIA observations will significantly further our understanding of the dynamics of the magnetic field in the solar atmosphere and heliosphere both in quiescent and eruptive stages The comprehensive thermal coverage of the corona will open new avenues of study for coronal energetics and seismology which will benefit from the excellent calibration against the *SDO*-EVE spectral irradiance measurements (Title *et al.* 2006).

The Extreme ultraviolet Variability Experiment (EVE) on board the NASA *Solar Dynamics Observatory* (*SDO*) will measure the solar EUV irradiance from 0.1 to 105 nm with unprecedented spectral resolution (0.1 nm), temporal cadence (10 sec), and accuracy (20%). The EVE program will provide solar EUV irradiance data for NASA's *Living With A Star* (LWS) program, including near real-time data products for use in operational atmospheric models that specify the space environment and to assist in forecasting space weather operations. The EVE program will advance understanding of the physics of the solar EUV irradiance variations on time scales from flares to the solar cycle (Woods *et al.* 2006).

3. Solar abundances: status (H. Socas-Navarro)

The solar chemical composition is under question. Recent works by Asplund *et al.* (see references in Bogdan *et al.* 2005) using a 3D hydrodynamical model of the photosphere yield abundances that are considerably lower than those obtained traditionally with 1D semiempirical models. Particularly important is the revision proposed for O, the third most abundant chemical element in the Universe after H and He and one of the largest contributors to the opacity in the solar interior. With the proposed revision (Asplund *et al.* 2004 recommend $lg\epsilon_O = 8.66 \pm 0.05$, versus 8.83 ± 0.06 of Grevesse & Sauval 1998), the excellent agreement existing between helioseismical data and theoretical predictions from stellar structure models breaks down. The far-reaching implications of a revision in the solar composition and the doubts it would cast on stellar structure and evolution models have stirred controversy in the community, to the extent that this problem is often referred to in the literature as 'the solar Oxygen crisis'. The implications of this revision for global helioseismology are discussed below.

Due to its volatility, the O abundance cannot be directly determined from meteorites and we are forced to rely solely on the photospheric abundance. Unfortunately, very few O lines exist in the visible spectrum and they are not straightforward to interpret as they are either extremely weak forbidden lines, exhibit NLTE effects and/or are strongly blended. The normal procedure to infer abundances involves using a model of the solar photosphere to synthesize the lines and then adjusting the abundance to fit the observed data. One thing that the solar crisis has bluntly put forward is how strongly model-dependent this process is and how uncertain our atmospheric models are (or, perhaps more precisely, how important the little model uncertainties are). On the one hand, a 3D model is to be preferred over a 1D model. However, when the 1D model is semiempirical, meaning that it has been derived by fitting observations (and therefore may have adjustments folded in to accurately reproduce the data) whereas the 3D model is a theoretical *ab initio* convection model, then there are advantages in using the semiempirical one. To make things more complicated, a recent work using a different 3D convection simulation produced abundances that are in the intermediate range (8.73 ± 0.07, Caffau *et al.* 2008).

At the moment, most of the argumentation in the literature has to do with the validation of the various atmospheric models employed (e.g., Ayres et al 2006; Koesterke et al 2008) since the issue of the solar abundances seems to be so critically model-dependent. An alternative approach has recently been published (Centeno & Socas-Navarro 2008) in which the authors make use of a new procedure that is nearly model-independent. This procedure is based on spectro-polarimetric observations of a sunspot and it allows for a very robust determination of the abundance ratio of O to Ni, resulting in a value that is consistent with the intermediate/high abundances (8.86 ± 0.07). It remains to be seen if this claim can be independently verified by other groups and if the new procedure is indeed robust enough. If so, the controversy would be settled in a way that would restore confidence in results obtained from seismology and stellar interior models, thus preserving the current order of things.

4. Helioseismology

4.1. *Solar interior*

Global helioseismology, based on analysis of mode frequencies, has provided accurate inferences of the solar internal structure, particularly the sound speed Christensen-Dalsgaard(2002). Standard solar models have been reasonably successful in reproducing the solar sound speed determined from helioseismology, although the remaining very significant discrepancies have raised questions, e.g., about possible mixing beneath the convection zone (e.g. Brun *et al.* 2002). However, this state of affairs was shattered by a redetermination of the solar surface composition (for a review, see Asplund 2005). The new determinations took the three-dimensional, time dependent nature of the solar atmosphere into account and in addition included departures from local thermodynamical equilibrium. This caused a very substantial reduction in the inferred abundances of oxygen, nitrogen and carbon, such that the overall abundance Z by mass of heavy elements was reduced from 0.018 to 0.012. When used in solar modelling, the result was a significant reduction in the sound speed beneath the convection zone, such that the maximum departure between the solar and model sound speed increased from 0.3 per cent to 1.5 per cent (see, e.g., Turck-Chièze *et al.* 2004, Bahcall *et al.* 2005). Also, the values of the depth of the convection zone and the envelope helium abundance in the model showed very substantial differences from the helioseismically inferred values, and substantial departures were found in the properties of the solar core, as inferred from frequencies of low-degree modes. These issues were reviewed by Basu & Antia(2008).

It is evident that this increased discrepancy represents a substantial problem for solar, and hence potentially stellar, modelling. Various attempts have been made to modify the models to bring back a more reasonable agreement between the models and the helioseismic inferences, including changes to the diffusion rates or a substantial change to the solar neon abundance; as reviewed by Guzik (2006) these have met by little success. Since the effect of the heavy-element abundances on solar structure is predominantly through their effect on the opacity, it is obvious that the models could be corrected through a change to the intrinsic opacity that compensates for the changed composition e.g., Bahcall *et al.* (2005), Christensen-Dalsgaard *et al.* (2008), although it is not clear whether the required change, up to around 30 per cent, is physically realistic. An independent verification of the revised composition would clearly be very desirable. In fact, Caffau et al. (2008) found a somewhat higher value of the oxygen abundance than did Asplund and his collaborators, using a similar technique. A very interesting possibility is to use the thermodynamic effects of the heavy-element composition on the sound speed in the convection zone as a diagnostics of the abundances, e.g., Antia & Basu (2006), Lin et al. (2007), although the effects are subtle.

Helioseismic inferences of solar internal rotation have revealed a nearly uniformly rotating radiative interior and latitudinal differential rotation in the convection zone, with a localized transition between the two regimes in a narrow so-called tachocline at the base of the convection zone for a review, see, Thompson et al. (2003). Observations over a full solar cycle have shown variations in the rotation rate extending over much of the solar convection zone, and apparently strongly correlated with the bands of activity, e.g., Howe et al. (2006), with diagnostic potential to test dynamo models of the activity cycle. An apparent variation, with a period of around 1.3 yr, in the rotation rate at the base of the convection zone was found to be active between 1996 and 2000 (Howe et al. 2000). This appears to have stopped, or at least changed its nature, around 2001 (Howe et al. 2007); it remains to be seen whether it will reappear in the corresponding part of the new solar cycle. Solar oscillation frequencies are strongly correlated with the variation in space and time of the magnetic field on the solar surface, e.g., Howe (2007), while evidence for changes in solar structure at greater depth has been elusive. However, Baldner & Basu (2008) found interesting changes between solar minimum and maximum, at a level of 0.01 per cent, in the squared sound speed at the base of the convection zone.

The acoustic modes observed in the five-minute region provide some information about the solar core, although only fairly weak constraints are available for the rotation of the core and the information remains limited about the structure of the central parts of the Sun. Far higher sensitivity would be provided by the observation of g modes in the Sun, and the detection of such modes has been a major goal for asteroseismology since its inception. Interesting recent analyses of the sensitivity of g-mode frequencies to the properties of the solar interior were presented by Mathur et al. (2007), Mathur et al. (2008) and García et al. (2008). Unfortunately, the very low amplitudes of the expected modes, in a frequency region with substantial background solar 'noise', makes detection very difficult; a recent analysis obtained an upper detection limit of around $5 \, \mathrm{mm \, s^{-1}}$ Elsworth et al. (2006). However, evidence for dipolar g modes in observations from the GOLF instrument on the SOHO spacecraft was found by García et al. (2007), on the basis of the detection of peaks in the power spectrum with the uniform period spacing expected for such modes. Interestingly, the analysis indicated that the inner core of the Sun is rotating substantially faster than the surface. However, it is probably fair to say that independent confirmation of these remarkable results will be essential, including a careful statistical analysis Appourchaux (2008).

4.2. Local helioseismic diagnostics

Methods of local helioseismology (time-distance helioseismology, ring-diagram analysis and acoustic holography) provide unique information about the structure and dynamics of large-scale convection, sunspots and active regions (Zhao 2008). In the past 3 years, most of the efforts was focused in three directions: validation and testing of local helioseismology methods in the turbulent conditions of the upper convection zone (Duvall et al. 2006; Birch et al. 2007; Braun et al. 2007; Georgobiani et al. 2007; Zhao et al. 2007) and presence of strong magnetic fields (Braun and Birch 2006; Rajaguru et al. 2006; Zhao and Kosovichev 2006; Couvidat and Rajaguru 2007; Nigam et al. 2007; Cameron et al. 2008;

Moradi and Cally 2008; Parchevsky and Kosovichev 2008), measurements and modeling of properties of supergranular scale convection (Green and Kosovichev 2006; Woodard 2006; Georgobiani *et al.* 2007; Green and Kosovichev 2007; Jackiewicz *et al.* 2007; Komm *et al.* 2007; Kosovichev 2007b; Woodard 2007; Hirzberger *et al.* 2008; Jackiewicz *et al.* 2008), and measurements of the wave-speed perturbations and mass flows associated with sunspots and emerging active regions (Couvidat *et al.* 2006; Kosovichev and Duvall 2006a; Kosovichev and Duvall 2006b; Zharkov *et al.* 2007; Braun and Birch 2008; Haber 2008; Komm *et al.* 2008; Kosovichev 2008; Kosovichev and Duvall 2008; Zhao 2008). Recent progress in realistic simulations of solar convection (Stein *et al.* 2007a; Stein *et al.* 2007b) has provided an unprecedented opportunity to evaluate the robustness of solar interior structures and dynamics obtained by methods of local helioseismology (Braun *et al.* 2007; Zhao *et al.* 2007). It has been demonstrated that in the numerical simulations properties of acoustic waves (p-modes) and surface gravity waves (f-modes) are similar to the solar conditions, and that these properties can be analyzed by the time-distance technique. The time-distance helioseismology measurements and inversions are tested by calculating acoustic travel times from a sequence of vertical velocities at the photosphere of the simulated data and inferring mean three-dimensional flow fields by performing inversion based on the ray approximation. The inverted horizontal flow fields agree very well with the simulated data in subsurface areas up to 3 Mm deep, but differ in deeper areas. These initial tests provide important validation of time-distance helioseismology measurements of supergranular-scale convection, illustrate limitations of this technique, and provide guidance for future improvements (Zhao *et al.* 2007). Among the improvements of local helioseismology techniques is the implementation of Born approximation-based travel-time sensitivity kernels that take into account finite-wavelength effects and provide more accurate results than the previously employed ray-path kernels, the inclusion of solar noise statistical properties in the inversion procedure through the noise covariance matrix, and the use of the actual variance of the noise in the temporal cross-covariances in the travel-time fitting procedure (Couvidat *et al.* 2006; Birch *et al.* 2007; Jackiewicz *et al.* 2008). Of these three improvements, the most significant is the application of the Born approximation to time-distance helioseismology. This puts the results of this discipline at the same level of confidence as those of global helioseismology based on inversion of normal-mode frequencies. It was showed that both Born and ray-path approximations return a similar two-region structure for sunspots. However, the depth of inverted structures may be offset by 1 or 2 Mm, and the spatial resolution of the results is more accurately estimated with the more realistic Born sensitivity kernels (Couvidat *et al.* 2006). Investigation of the relationship between characteristics of subsurface flows and surface magnetic flux revealed that quiet regions are characterized by weakly divergent horizontal flows and small anticyclonic vorticity (clockwise in the northern hemisphere), while locations of high magnetic activity show convergent horizontal flows combined with cyclonic vorticity (counterclockwise in the northern hemisphere). Divergence and vorticity of horizontal flows are anticorrelated (correlated) in the northern (southern) hemisphere especially at greater depth. These trends show a slight reversal at the highest levels of magnetic flux; the vorticity amplitude decreases at the highest flux levels, while the divergence changes sign at depths greater than about 10 Mm. The product of divergence and vorticity of the horizontal flows, a proxy of the vertical contribution to the kinetic helicity density, is on average negative (positive) in the northern (southern) hemisphere. The helicity proxy values are greater at locations of high magnetic activity than at quiet locations (Komm *et al.* 2007). The initial results of investigations of the magnetic flux emergence and dynamics of active regions (Hindman *et al.* 2006; Kosovichev and Duvall 2006a; Kosovichev and Duvall 2006b; Haber 2008; Komm *et al.* 2008;

Kosovichev and Duvall 2008) revel many interesting properties. In particular, it is showed that large active regions are formed by repeated magnetic flux emergence from the deep interior, and that their roots are at least 50 Mm deep (Kosovichev and Duvall 2006a). The active regions change the temperature structure and flow dynamics of the upper convection zone, forming large circulation cells of converging flows (Hindman *et al.* 2006; Kosovichev and Duvall 2006a; Haber 2008; Komm *et al.* 2008; Kosovichev and Duvall 2008). The helioseismic observations also indicate that the processes of magnetic energy release, flares and coronal mass ejections, might be associated with strong (1–2 km/s) shearing flows, 4–6 Mm below the surface (Kosovichev and Duvall 2006b).

4.3. *Imaging of the far-side of the Sun*

It is of great importance to monitor large solar active regions on the far side of the Sun for space weather forecasting, in particular, to predict their appearance before they rotate into our view from the solar east limb. Local helioseismology techniques, including helioseismic holography and time distance, have successfully imaged solar far-side active regions. The possibility of imaging and improving the image quality of solar far- side active regions by use of time-distance helioseismology was explored by (Zhao 2007). In addition to the previously used scheme with four acoustic signal skips, a five-skip scheme is also included in this newly developed technique. The combination of both four- and five-skip far-side images significantly enhances the signal-to-noise ratio in the far-side images and reduces spurious signals. The accuracy and reliability of this method are tested by using numerical simulation (Hartlep *et al.* 2008). Initial results toward a calibration of the far-side helioseismic images of active regions in terms of active region size and magnetic field strength were obtained by comparing the helioseismic maps of large active regions on the far side of the Sun, calculated from Global Oscillation Network Group (GONG) Doppler observations, with magnetic and visible-continuum images of the same active regions on the visible hemisphere before and after their far-side passage (González Hernández *et al.* 2007). The far-side seismic signature is expressed as a phase shift that a far-side active region introduces to waves from the near hemisphere as they are reflected into the solar interior on their way back to the near hemisphere. There is a significant correlation between this far-side signature and both the total area of the active region, as viewed on the near hemisphere, and the area of the sunspots contained in the active region. An approximately logarithmic increase in the seismic phase signature with increasing magnetic field strengths above a critical field of 10 G has been found. This is roughly consistent with similar helioseismic signatures measured on the near solar hemisphere concurrent with associated magnetic fields.

4.4. *Helioseismic effects of solar flares*

The helioseismic waves excited by solar flares ('sunquakes') are observed as circular, expanding waves on the Sun's surface. The first sunquake was observed for a flare on July 9, 1996, by the *Solar and Heliospheric Observatory (SOHO)* space mission. New sunquake events were detected from the SOHO/MDI of during solar flares of 2003–2006 (Besliu-Ionescu *et al.* 2006; Donea *et al.* 2006; Kosovichev 2006; Kosovichev 2007a; Kosovichev 2007c; Moradi *et al.* 2007; Zharkova and Zharkov 2007; Besliu-Ionescu *et al.* 2008; Martinez-Oliveros *et al.* 2008a; Martinez-Oliveros *et al.* 2008b; Zharkova 2008). These observations show a close association between the flare seismic waves and the hard X-ray source, indicating that high-energy particles accelerated during the flare impulsive phase produced strong compression waves in the photosphere, causing the sunquake. The results also reveal new physical properties such as strong anisotropy of the seismic waves, the amplitude of which varies significantly with the direction of propagation (Kosovichev

2006; Kosovichev 2007a). The waves travel through surrounding sunspot regions to large distances, up to 120 Mm, without significant distortion. In addition to the local observations of the sunquake waves it was found that the flares may excite global oscillations of the Sun (Karoff and Kjeldsen 2008). A new type of flare-driven waves was discovered from *Hinode* observation of the December 13, 2006, flare. These waves were detected in the sunspot umbra during the impulsive phase and probably represent a fast MHD mode (Kosovichev and Sekii 2007). The observations and analysis of helioseismic effects of solar flares open new perspectives for helioseismic diagnostics of flaring active regions on the Sun and for understanding the mechanisms of the energy release and transport in solar flares.

4.5. *Helioseismology programs on new space missions*

The great success of helioseismology investigation in recent years was due high quality data from the GONG and BISON networks and from the space mission *SOHO*. First results from initial helioseismic observations by the Solar Optical Telescope (SOT) on board the Japan *Hinode* space mission launched in 2006 (Sekii *et al.* 2007) demonstrated that intensity oscillation data from the SOT Broadband Filter Imager can be used for various helioseismic analyses. The k-omega power spectra, as well as corresponding time-distance cross-correlation function that promises high-resolution time-distance analysis below 6-Mm travelling distance, were obtained for G-band and Ca II-H data. The subsurface supergranular patterns have been observed from the initial time-distance analysis. The *Hinode* results show that the solar oscillation spectrum is extended to much higher frequencies and wavenumbers, and the time-distance diagram is extended to much shorter travel distances and times than they were observed before, thus revealing great potential for high-resolution helioseismic observations (Kosovichev and Sekii 2007; Nagashima *et al.* 2007; Sekii *et al.* 2007; Mitra-Kraev *et al.* 2008). Future helioseismology investigations will be based on high-resolution uninterrupted data from the Helioseismic and Magnetic Imager (HMI) of the NASA *Solar Dynamics Observatory (SDO)* space mission scheduled for launch in 2009. The HMI investigation encompasses three primary objectives of the *Living With a Star Program*: first, to determine how and why the Sun varies; second, to improve our understanding of how the Sun drives global change and space weather; and third, to determine to what extent predictions of space weather and global change can be made and to prototype predictive techniques. Helioseismology provides unique tools to study the basic mechanisms of the Sun's magnetic activity and variability. It plays a crucial role in all HMI investigations, which include convection-zone dynamics and the solar dynamo; origin and evolution of sunspots, active regions and complexes of activity; sources and drivers of solar activity and disturbances; links between the internal processes and dynamics of the corona and heliosphere; and precursors of solar disturbances for space-weather forecasts. The SDO mission will provide new unique opportunities for helioseismology studies in combination with data from the other instruments, Atmospheric Imaging Assembly (AIA) and Extreme-ultraviolet Variability Experiment (EVE), and also from various space and ground-based observatories (Kosovichev and Team 2007). New helioseismology data will be obtained also from the *PICARD* mission, a CNES (France) micro-satellite also scheduled for launch in 2009. Its goal is to better understand the Sun and the potential impact of its activity on earth climate by measuring simultaneously the solar total and spectral irradiance, diameter, shape and oscillations. The helioseismology program of *PICARD* aims to observe the low to medium l p-mode oscillations in intensity and search for g-mode oscillation signatures at the limb (Corbard *et al.* 2008).

5. Surface magnetism

5.1. *Sunspots*

Substantial progress in our understanding of sunspot fine structure has been made recently. Interestingly enough, it has been found that a single physical process accounts for the appearance of dark core configurations in both the umbra and the penumbra (in spite of their quite different magnetic field configurations). This magnetoconvective process is clearly responsible for the brightness observed in these sunspot zones. Dark cores, similar to those found ubiquitously in penumbral filaments by Scharmer *et al* (2002), have been predicted to occur in umbral dots (Schüssler & Vögler 2006) and actually observed with the *Hinode* satellite (Bharti 2007). Comparable dark cores are also observed in light bridges, running parallel to their elongated direction. These dark cores present conspicuous similarities for all three cases, penumbral filaments, umbral dots and light bridges (Rimmele 2008). Careful spectropolarimetric inversions obtained with *Hinode* (Riethmller 2008) have provided stratifications that show umbral dots to harbor excess temperatures of more than 500 K, fields smaller than their surroundings by 400 G and weak upflows, much in agreement with what is obtained from MHD simulations. To explain these dark cores, it is broadly accepted that magnetoconvective instabilities pile up hot material near the surface giving rise to a cusp shaped magnetic field configuration. The excess densities in these concentrations lift to higher and cooler layers the $\tau = 1$ surface and thus originate the dark core signature (in depth formulations for the penumbral case can be found in Borrero 2007, Ruiz Cobo & Bellot Rubio 2008). Where there is less agreement is in how much the field strength is reduced in this enhanced density concentrations. According to Scharmer & Spruit (2006), these high density concentrations of hot material would correspond to field free gaps radially aligned withe the penumbral direction. The field free configuration in a deep penumbra, sometimes referred as the 'gappy penumbral model, would be originated in a multitube configuration of the sunspot magnetic field below the photosphere. This model has been contrasted to the so called uncombed penumbral model (Solanki & Montavon 1993) where the hot material would flow inside a weakly magnetized tube (see the comparison of the two models in Bellot Rubio 2007). More recent MHD simulations (Heinemann *et al.* 2007, Rempel *et al.* 2008) including grey radiative transport but otherwise realistic atmospheric parameters, suggest some intermediate scenarios whereby magnetoconvective instabilities in inclined magnetic fields initiate hot upflows along reduced but not field free channels. These models are reproducing modern observations of sunspots to levels heretofore never achieved. They also provide some links between yet unsolved problems like the origin of the Evershed flow and the heat transport mechanism in the penumbra.

In this context the final fate of the Evershed flow is still a major unknown. An interesting possibility has been suggested by Vargas Domínguez (2007). They found that the moat flow that surrounds mature sunspots, and that is normally associated with some form of a supergranular cell, exits only in sunspots sides with penumbra and is aligned with its filamentary direction. That is, sunspot sides with no penumbra (as seen in irregular spots) show no moat flow and the moat flow is also absent in directions perpendicular to the filaments. This strongly suggest the possibility of a physical link between the Evershed flow and the moat flow surrounding sunspots. This hypothesis would nicely answer at once two unsolved questions such as what does it happen to the Evershed mass flow beyond the penumbral boundary and what is the exact origin of the moat flow. It is interesting to see how the connections between sunspot penumbrae and the moat flow (and fields) is indeed being found in a number of different works using diverse techniques (Balthasar & Schleicher 2008).

5.2. *Quiet Sun magnetism*

The observation of surface magnetic fields based on the Zeeman effect has always suffered an observational biased due to the quadratic dependence on the field strength of the linear polarization signals as compared to the linear dependence for the circular polarization ones. It is thus natural that past studies of solar magnetism have largely dealt with the longitudinal component of the magnetic field. The advent of high sensitive, high spatial resolution observations from the ground and space is changing this picture. One of the most outstanding results obtained by the *Hinode* -SOT telescope has been the discovery of a wealth of transverse fields over substantial fractions of the quiet solar surface, with strengths of 100–200 G and with apparent flux densities that are 5 (!) times larger than their longitudinal counterparts (Lites *et al.* 2008). One way to explain this high ratio between the transverse and the longitudinal fluxes is by assuming a small filling factor of the horizontal component, although the exact factor remains unclear. Lites *et al.* (2008) propose filling factors near 20 %. On the other hand, Orozco Suárez *et al.* (2007) find a ratio more near 2 and filling factors of 40–50 % or so and discuss that the effects of telescope diffraction must be taken into account at this high spatial resolutions (related to the distinction between stray-light and filling factor). In any case, it is clear that more detailed analysis of the superb data obtained by the *Hinode* spectropolarimeter will help consolidate a new picture of the various components of the solar magnetic fields. The same instrument has shown how dynamic these fields are at this small scale, with clear loop-like structures undergoing emergence through the photosphere in the case of quiet regions (Centeno *et al.* 2007) but also for plage areas (Ishikawa *et al.* 2007). Even before *Hinode* revealed the importance of this horizontal component in the photosphere, its presence was suggested from synoptic ground observations using, this time, the center-to-limb variation of the circularly polarized signals (of the same Fe I lines as those observed by the *Hinode*-SP). Using the SOLIS/VSM full disk instrument, another grating spectropolarimeter, Harvey *et al.* (2007) found that the quiet regions of the photosphere must be covered by patches of horizontal field lines with spatial scales of around 10 arcsec typically. The *Hinode* horizontal fields are seen to be distributed at smaller, mesogranular, scales, but a degradation of the resolution from the satellite observations to that obtained on the ground could well explain this difference. Interestingly enough, analysis of existing MHD simulations searching for the presence of the newly discovered horizontal fields has found that, indeed, such fields were present in them, with ratios as large as 5 although this ratio is seen to be strongly dependent on height in the atmosphere (Schüssler & Vögler (2008), Steiner *et al.* 2008). These simulations support the original suggestion by Lites *et al.* (2008) that the physical responsible for the large transverse fluxes is the larger spatial scale of the horizontal field lines as compared to the vertical ones.

Out of the many interesting results obtained with the *Hinode* satellite, one deserves special mentioning. For decades now, a fundamental structure thought to be present in network and plage areas were the so-called flux tubes. These structures were thought to be embedded in intergranular lanes, thus harboring strong downflows outside the field concentration that opens up the field lines with height and displays a canopy like configuration. Lines-of-sights crossing the canopy and entering the field free region would see a sharp change in physical properties, which would naturally produce asymmetric Stokes profiles (Landi Deglinnocenti & Landolfi, 1983). On the other hand, lines of sight exclusively passing through the central portion of the magnetic concentration would see a slowly varying atmosphere, which only produce symmetric profiles (or perfectly anti-symmetric ones in the case of Stokes V). The asymmetries observed at low spatial resolution (1 arcsec) have long been thought to be the average of the contributions of theses distinct lines-of-sights. Thus, a fundamental prediction of this model was that when the structures were to be resolved, and lines of sight passing through the canopy could be separated from those that go inside the magnetic tube (or sheet), different asymmetries would be observed in them. It is encouraging that *Hinode*-SP observations have provided actually this pattern of asymmetries in the observed Stokes profiles in magnetic concentrations seen at 0.3 arcsec (see Rezaei *et al.* 2007). This is a result originally predicted from theoretical grounds that has received now full observational support and that lends confidence in our understanding of the solar photospheric magnetoconvective processes.

6. Future solar telescopes: status of the ATST project (T. Rimmele)

The 4m Advanced Technology Solar Telescope (ATST) located on Haleakala will be the most powerful solar telescope and the world's leading resource for studying the dynamic solar atmosphere and its magnetic environment. As its highest priority science objective ATST shall provide high resolution and high sensitivity observations of the highly dynamic solar magnetic fields throughout the solar atmosphere, including chromosphere and corona. With its 4 m aperture and integrated adaptive optics and located at the best site identified by an extensive site survey, ATST will provide unprecedented resolution of 20 km on the solar surface and thus resolve the essential, fine-scale magnetic features and their dynamics that dictate the varying release of energy from the Sun's atmosphere. The high photon flux combined with careful polarization analysis and calibration and facility class instrumentation will allow high precision polarimetric measurements of magnetic fields as they extend into the upper atmosphere. The ATST has a factor of 64 greater collecting area than the largest existing coronagraph. The coronagraphic site and a design that is optimized for low scattered light will provide the sensitivity needed to measure the illusive weak coronal magnetic fields.

The ATST project is a collaborative effort between US and international partners. The ATST project has been recommended to the NSF Director for construction funding by the National Science Board and has progressed through the National Science Foundation MREFC process and is now ready for construction. The US Senate Appropriations Committee language contained 9.5$ million of initial funds for the Advanced Technology Solar Telescope in its proposed FY2009 Federal budget. The Environmental Impact Statement process for Haleakala is expected to be concluded at the end of 2008. The ATST construction phase is planned to begin in FY09 following a final baseline review in March of 2009. First light will be achieved by 2014. Commissioning of the initial set of four facility class instruments will take place in a phased manner during 2014–2015. During this time first science operations will be possible but have to be interlaced with significant engineering time. Full science operations will begin in 2016.

The ATST offers tremendous opportunity for the training of students and recruitment of post-docs and faculty in solar physics who will become users of the ATST and the instrument builders and theoreticians of the future. Several US and international graduate and undergraduate students have already participated in ATST development and ATST related engineering and science projects. ATST is establishing strong synergy with the education and outreach programs at the collaborating institutions. The ATST program will continue to actively involve large segments of the US and international solar physics community, helping to strengthen solar astronomy programs at universities and national centers.

7. Progress on dynamo theory during 2005–2008 (M. Dikpati)

Large-scale solar dynamo models were first built by E. Parker more than half a century ago, and since then solar dynamo modeling has been an active area of research in solar physics. Much progress has been made specifically over the past two decades, both in the area of mean-field dynamo models and full 3D MHD models.

Major progress in the mean-field dynamo models include the detailed exploration of flux-transport dynamos, which are the so-called α-Ω dynamos with meridional circulation. Flux-transport dynamos include three basic processes: (i) shearing of the poloidal magnetic fields to produce toroidal fields by the Sun's differential rotation (the Ω-effect); (ii) regeneration of poloidal fields by displacing and twisting the toroidal flux tubes by helical motions (the so-called α-effect); and (iii) advective transport of magnetic flux by meridional circulation, whereas an α-Ω dynamo involves only the first two. Meridional circulation acts as a conveyor belt in this class of models and plays an important role in determining the dynamo cycle period.

Surface meridional flow has been detected by various observational techniques. While helioseismologists are still searching for detecting the subsurface return flow, the past three years (2005–2008) have particularly been the golden era of building theoretical models for understanding the physics of meridional circulation, as well as the exploration of the predictive capability provided by this flow to flux-transport dynamos. Rempel (2005) developed a model using mean-field formalism and showed that the Coriolis force acting on convection creates Reynolds stresses that transport angular momentum toward the equator to create equatorial acceleration. Meridional circulation is then produced through the outward radial velocity created by the Coriolis force on large rotational flow at low-latitudes. Miesch *et al.* (2008) found similar physics behind the generation of meridional flow through direct 3D HD simulation. These models show primarily a large single-cell flow pattern when averaged over 4 months with a flow speed of a few m/s, which is consistent with that obtained from mass conservation.

In the case of oceanic models the Great Ocean conveyor belt carries surface forcing with a certain memory and determines the occurrence of future El Niño events. Dikpati, de Toma & Gilman (2006; see also Dikpati & Gilman 2006) demonstrated that the memory about the Sun's past magnetic field, provided by the solar meridional circulation, can be used to predict the amplitudes and timings of the future solar cycles. A $\sim 50\%$ reduction in the meridional flow speed during 1996–2003 was found by Dikpati (2005) to be the physical cause of late onset of cycle 24, and this onset-timing prediction has now been validated. Solar cycle predictions using a flux-transport dynamo have been criticized by Bushby & Tobias (2007) arguing that the solar dynamo might be operating in the highly chaotic regime, due to vigorous turbulence in the solar convection zone. But research on investigating the predictive capability of a flux-transport dynamo as well as building more physical models that can assimilate data in order to be able to predict the future cycles also continued (Cameron & Schüssler 2007; Schüssler 2007; Choudhuri, Chatterjee & Jiang 2007; Jiang, Chatterjee & Choudhuri 2007; Brandenburg & Käpylä 2007; Dikpati *et al.* 2007; Dikpati, de Toma & Gilman 2008; Dikpati, Gilman & de Toma 2008).

In view of successes of kinematic flux-transport dynamos in reproducing majority of solar cycle features, questions have arisen whether such a weak meridional flow in such models can sustain the Lorentz force back reaction due to strong (~ 100 kGauss) spot-producing fields at the base of the convection zone. Rempel (2006) developed a non-kinematic flux-transport dynamo that operates with Lorentz force back reaction on differential rotation and meridional circulation, and showed that the flux-transport dynamos are robust enough to work well for the Sun by sustaining a back reaction on the solar flow fields as long as the spot-producing fields have the peak-amplitude of ~ 30 kGauss. These models also reproduce solar torsional oscillations.

The limit of ~ 30 kGauss toroidal fields produced at the tachocline by a non-kinematic dynamo, raises another issue of how the spot-producing flux tubes of ~ 100 kGauss can be formed at the base of the solar convection zone. Are 100 kGauss flux tubes formed via some dynamical processes on ~ 30 kGauss broad toroidal fields generated by dynamo action, or are they formed not from the magnetic fields at the convection zone base but rather 3 kGauss spots are directly formed at the photospheric level by a near-surface dynamo action? While the possibility of flux tube formation from a broad toroidal sheet has been investigated in early 2000, recently Brandenburg (2005, 2006, 2007) argued in favor of the near-surface dynamo, which demonstrated efficiency in getting rid of excess small-scale magnetic helicity through coronal mass ejections.

Since meridional circulation has been identified as an important ingredient in the large-scale solar dynamo models, but details of meridional flow profiles in the bottom half of the convection zone is not observationally known yet, Bonanno, Elstner, Belvedere & Rüdiger (2005) and also Jouve & Brun (2007) studied the role of multicell meridional flow profiles in the advective transport of magnetic flux. The general conclusion was that the equatorward transport of spot-producing flux is possible via the combination of two flow cells in latitude, but multicells in latitude and radius have strong effect in cycle period as well as in the shape of butterfly diagram. Jouve *et al.* (2008) have done a very important contribution to younger generations of dynamo researchers by producing a mean-field solar dynamo benchmark that used eight different codes.

The prescription of α-effect in the mean-field, kinematic dynamos is a never-ending issue, particularly when the turbulence is anisotropic. The anisotropy leads to the generation of an electromotive force from the fluctuating flow and magnetic fields, which in turn gives rise to an α-effect tensor. Kitchatinov & Rüdiger (1992) demonstrated, with a simplified interpretation of the off-diagonal components of the α-effect tensor as diamagnetic pumping, that it could play an important role in shaping the solar butterfly diagram. Recently Käpylä, Korpi, Ossendrijver & Stix (2006) and Guerrero & de Gouveia Dal Pino (2008) have investigated the effect of turbulent pumping arising from the off-diagonal terms of α-tensor; both interpreted this pumping as a downward advection effect rather than a diffusive 'diamagnetic' effect and showed that this pumping improves the low-latitude confinement of the solar butterfly diagram; along with near-surface shear it also contributes in maintaining the dipolar parity of the large-scale solar magnetic fields.

In reality, dynamo action in the Sun occurs on many space and timescales, from the global down to granulation scales (10^{-4} of the solar radius), involving many turbulent processes. In order to capture most of these scales and processes, Browning *et al.* (2006) have built 3D global MHD models for solar differential rotation, convection and magnetic fields. These models have so far shown the sustained toroidal fields of ~ 4 kGauss strength at the tachocline region. Using a different approach of including the third dimension in a 2D mean-field kinematic model through the non-axisymmetric α-effect, Jiang & Wang (2007, 2008) have produced some non-axisymmetric dynamo modes. However, both these approaches are steps toward simulating longitude-dependent solar cycle features in addition to longitude-averaged features.

Investigation of energy cascade and energy transfer across different scales was the subject of many papers. In particular, Mininni and collaborators (Mininni *et al.* 2005; Alexakis, Mininni & Pouquet 2007; Mininni 2007) show that the turbulent velocity fluctuations in the inertial range (which are not dissipated due to eddy viscosity) is responsible for the magnetic field amplification at small scales, whereas the large-scale field is amplified mostly due to large-scale flows. MHD simulations with helical forcing indicate that the transfer of magnetic helicity is nonlocal, from the forcing scale to the global scale, implying the validity of α-effect and hence the mean-field generation.

Vögler & Schüssler (2007) showed, by performing a radiative magneto-convection simulation of dynamo action near the solar surface, that this dynamo can be a source of internetwork magnetic flux that was indicated by various observational diagnostics in past. Schüssler & Vögler (2008) have provided further evidence that the ratio of horizontal to vertical components of magnetic fields produced by their local, near-surface radiative MHD dynamo is consistent with the ubiquitous internetwork horizontal fields in the quiet photosphere (Orozco Suárez *et al.* 2007; Lites *et al.* 2008) observed by *Hinode*.

8. Quiet corona and solar wind initiation

Measurements of Doppler shifts in coronal emission lines observed in active regions with the EUV Imaging Spectrometer (EIS) on the *Hinode* satellite show large areas of persistent outflows (Harra *et al.* 2008; Doschek *et al.* 2007, 2008). These features are fainter than the core of the active region and appear to be associated with very long or possibly open magnetic field lines, suggesting that the edges of active regions may be a significant source of solar wind plasma. These results appear to confirm and extend earlier research that suggested that active regions can contribute to the heliospheric magnetic field and solar wind (e.g., Liewer *et al.* 2004).

Valentin Martínez Pillet
president of the Commission

References

Alexakis, A., Mininni, P. D., & Pouquet, A. 2006, *NJP*, 9, 298
Antia, H. M. & Basu, S. 2006, *ApJ*, 644, 1292
Appourchaux, T. 2008, *AN*, 329, 485
Asplund, M., Grevesse, N., Sauval, J., Allende Pietro, C., & Kiselman, D. 2004, *A&A*, 417, 751
Asplund, M. 2005, *Ann. Rev. Astron. Astrophys.*, 43, 481
Ayres, T. R., Plymate, C., & Keller, C.U. 2006, *ApJSS*, 165, 618
Bahcall, J. N., Basu, S., Pinsonneault, M., & Serenelli, A. M. 2005, *ApJ*, 618, 1049
Baldner, C. S. & Basu, S. 2008, *ApJ*, 686, 1349
Balthasar, H. & Schleicher, H. 2008, *A&A*, 481, 811
Basu, S. & Antia, H. M. 2008, *Phys. Rep.*, 457, 217
Beer, J., Vonmoos, M., & Muscheler, R. 2006, *Space Sci. Rev.*, 125, 67
Bellot Rubio, L. 2007, in: F. Figueras, M. Hernanz & C. Jordi (eds.), *Highlights of Spanish Astrophysics* IV (Dordrecht: Springer), p. 271
Bengtsson, L. 2006, *Space Sci. Rev.*, 125, 187
Berger, T. E., Title, A. M., Tarbell, T., *et al.* 2007, in: K. Shibata, S. Nagata & T. Sakurai (eds.), *New Solar Physics with Solar-B Mission*, ASP-CS, 369, 103
Besliu-Ionescu, D., Donea, A. C., Cally, P., & Lindsey, C. 2006, in: V. Bothmer & A. Hady (eds.), *Solar Activity and its Magnetic Origin*, Proc. IAU Symposium No. 233 (Cambridge: CUP), p. 385
Besliu-Ionescu, D., Donea, A. C., Cally, P. & Lindsey, C. 2008, in: R. Howe, R. W. Komm, K. S. Balasubramaniam & G. J. D. Petrie (eds.), *Subsurface and Atmospheric Influences on Solar Activity*, ASP-CS, 383, 297
Bharti, L., Joshi, C., & Jaaffrey, S. N. A., 2007, *ApJ* (Letters), 669, L57
Birch, A. C., Gizon, L., Hindman, B. W. & Haber, D. A. 2007, *ApJ*, 662, 730
Bonanno, A., Elstner, D., Belvedere, G., & Rüdiger, G. 2005, *AN*, 326, 170
Borrero, J. M. 2007 *A&A*, 471, 967

Brandenburg, A. 2005, *ApJ*, 625, 539

Brandenburg, A. 2006, in: H. Uitenbroek, J. Leibacher & R. F. Stein (eds.), *Solar MHD Theory and Observations*, ASP-CS 354, 121

Brandenburg, A. 2007, in: K. A. van der Hucht (ed.), *Highlights of Astronomy*, 14, 291

Brandenburg, A. & Käpylä, P. J. 2007, *NJP*, 9, 305

Braun, D. C., & Birch, A. C. 2006, *ApJ* (Letters), 647, L187

Braun, D. C., & Birch, A. C. 2008, *Solar Phys.*, 251, 267

Braun, D. C., Birch, A. C., Benson, D., Stein, R. F., & Nordlund, A. 2007, *ApJ*, 669, 1395

Browning, M., Miesch, M. S., Brun, A. S., & Toomre, J. 2006, *ApJ* (Letters), 648, L157

Brun, A. S., Antia, H. M., Chitre, S. M., & Zahn, J.-P., 2002. *A&A*, 391, 725

Bushby, P. J., & Tobias, S. M. 2007, *ApJ*, 661, 1289

Caffau, E., Ludwig, H.-G., Steffen, M., *et al.* 2008, *A&A*, 488, 1031

Cameron, R., & Schüssler, M. 2007, *ApJ*, 659, 801

Cameron, R., Gizon, L., & Duvall, T. L. 2008, *Solar Phys.*, 251, 291

Centeno, R., Socas-Navarro, H., Lites, B. W., *et al.* 2007, *ApJ* (Letters), 666, L137

Centeno, R., & Socas-Navarro, H. 2008, *ApJ*, 682, 61

Chamberlin, P. C., Woods T. N., & Eparvier, F. G. 2008, *Adv. Space Res.*, 42, 912

Choudhuri, A. R., Chatterjee, P., & Jiang, J. 2007 *Phys. Rev. Lett.*, 98, 1103

Christensen-Dalsgaard, J. 2002, *Rev. Mod. Phys.*, 74, 1073

Christensen-Dalsgaard, J., Di Mauro, M. P., Houdek, G., & Pijpers, F. 2008, *A&A*, in press

Corbard, T., Boumier, P., Appourchaux, T., & the Piccard team 2008, *AN*, 329, 508

Couvidat, S., Birch, A. C., & Kosovichev, A. G. 2006, *ApJ*, 640, 516

Couvidat, S. & Rajaguru, S. P. 2007, *ApJ*, 661, 558

Dewitte, S., Schmutz, W., & T. P. Team 2006, *Adv. Space Res.*, 38, 1792

Dikpati, M. 2005, *Adv. Space Res.*, 35, 322

Dikpati, M., de Toma, G., & Gilman, P. A. 2006, *GRL*, 33, L05102

Dikpati, M., & Gilman, P. A. 2006, *ApJ*, 649, 498

Dikpati, M., Gilman, P. A., de Toma, G., & Ghosh, S. S. 2007, *Solar Phys.*, 245, 1

Dikpati, M., Gilman, P. A., & de Toma, G. 2008, *ApJ* (Letters), 673, L99

Dikpati, M., de Toma, G., & Gilman, P. A. 2008, *ApJ*, 675, 920

Donea, A.-C., Besliu-Ionescu, D., Cally, P. S., *et al.* 2006, *Solar Phys.*, 239, 113

Doschek, G. A., Mariska, J. T., Warren, H. P., *et al.* 2007, *ApJ* (Letters), 667, L109

Doschek, G. A., Warren, H. P., Mariska, J. T., *et al.* 2008, *ApJ*, 686, 1362

Duvall, T. L., Birch, A. C., & Gizon, L. 2006, *ApJ*, 646, 553

Elsworth, Y. P., Baudin, F., Chaplin, W., *et al.* 2006, in: K. Fletcher (ed.), *Beyond the Spherical Sun*, Proc. SOHO 18 / GONG 2006 / HELAS I Conf., *ESA-SP* 624

Foukal, P., & Bernasconi, P. N. 2008, *Solar Phys.*, 248, 1

Frohlich, C. 2006, *Space Sci. Rev.*, 125, 53

García, R. A., Turck-Chièze, S., Jiménez-Reyes, S. J., *et al.* 2007, *Science*, 316, 1591.

García, R. A., Mathur, S., & Ballot, J. 2008. *Solar Phys.*, in press.

Georgobiani, D., Zhao, J., Kosovichev, A. G., *et al.* 2007, *ApJ*, 657, 1157

González Hernández, I., Hill, F., & Lindsey, C. 2007, *ApJ*, 669, 1382

Green, C. A., & Kosovichev, A. G. 2006, *ApJ* (Letters), 641, L77

Green, C. A., & Kosovichev, A. G. 2007, *ApJ* (Letters), 665, L75

Grevesse, N., Sauval, A. J. 1998, in: C. Fröhlich, M. C. E. Huber, S. K. Solanki, & R. von Steiger (eds.), ISSI Workshop (Dordrecht: Kluwer), p. 161

Gueymard, C. A. 2006, *Adv. Space Res.*, 37, 323

Guerrero, G. & de Gouveia Dal Pino, E. M. 2008, *A&A*, 485, 267

Guzik, J. A. 2006, in: K. Fletcher (ed.), *Beyond the Spherical Sun*, Proc. SOHO 18 / GONG 2006 / HELAS I Conf., *ESA-SP*, 624

Haber, D. A. 2008, in: in: R. Howe, R. W. Komm, K. S. Balasubramaniam & G. J. D. Petrie (eds.), *Subsurface and Atmospheric Influences on Solar Activity*, ASP-CS, 383, 31

Harra, L. K., Sakao, T., Mandrini, C. H., *et al.* 2008, *ApJ* (Letters), 676, L147

Hartlep, T., Zhao, J., Mansour, N. N., & Kosovichev, A. G. 2008, *ApJ*, 689, 1373

Harvey, J. W., Branston, D., Henney, C. J., & Keller, C. U. 2007, *ApJ* (Letters) 659, L177

Heinemann, T., Nordlund, A., Scharmer, G. B., & Spruit, H. C. 2007, *ApJ*, 669, 1390

Hindman, B. W., Haber D. A., & Toomre, J. 2006, *ApJ*, 653, 725

Hirzberger, J., Gizon, L., Solanki, S. K., & Duvall, T. L. 2008, *Solar Phys.*, 251, 417.

Hochedez, J.-F., Schmutz, W., Stockman, Y., *et al.* 2006, *Adv. Space Res.*, 37, 303

Howe, R. 2007, *Adv. Space Res.*, 41, 846

Howe, R., Christensen-Dalsgaard, J., Hill, F., *et al.* 2000, *Science*, 287, 2456

Howe, R., Rempel, M., Christensen-Dalsgaard, J., *et al.* 2006, *ApJ*, 649, 1155.

Howe, R., Christensen-Dalsgaard, J., Hill, F., *et al.* 2007, in: G. Maris, K. Mursula & I. Usoskin (eds.), Proc. *Second International Symposium on Space Climate, Adv. Space Res.*, 40, 915

Ishikawa, R., Tsuneta, S., Ichimoto, K., *et al.* 2007, *A&A* (Letters), 481, L25

Jackiewicz, J., Gizon, L., & Birch, A. C. 2008, *Solar Phys.*, 251, 381

Jackiewicz, J., Gizon, L., Birch, A. C., & Thompson, M. J. 2007, *AN*, 328, 234

Jiang, J. & Wang, J. 2007, *MNRAS*, 377, 711

Jiang, J., Chatterjee, P., & Choudhuri, A. R. 2007, *MNRAS*, 381, 1527

Jiang, J. & Wang, J. 2008. *Adv. Space Res.*, 41, 874

Jouve, L. & Brun, A. S. 2007, *A&A*, 474, 239

Jouve, L., Brun, A. S., Arlt, R., *et al.* 2008, *A&A*, 483, 949

Käpylä, P. J., Korpi, M. J., Ossendrijver, M., & Stix, M. 2006, *A&A*, 455, 401

Karoff, C. & Kjeldsen H., (2008). *ApJ* (Letters), 678, L73

Koesterke, L., Allende Prieto, C., & Lambert, D. L. 2008, *ApJ*, 680, 764

Komm, R., Howe, R., Hill, F., Miesch, M., Haber, D., & Hindman, B. 2007, *ApJ*, 667, 571

Komm, R., Morita, S., Howe, R., & Hill, F. 2008, *ApJ*, 672, 1254

Kosovichev, A. G. 2006, *Solar Phys.*, 238, 1

Kosovichev, A. G. 2007a, *ApJ* (Letters), 670, L65

Kosovichev, A. G. 2007b, in: F. Kubka, I. W. Roxburgh & K. L. Chan (eds.), *Convection in Astrophysics*, Proc. IAU Symposium No. 239 (Cambridge: CUP), p. 113

Kosovichev, A. G. 2007c, in: K. Shibata, S. Nagata & T. Sakurai (eds.), *New Solar Physics with Solar-B Mission*, ASP-CS, 369, 325

Kosovichev, A. G. 2008, *Adv. Space Res.*, 41, 830

Kosovichev, A. G. & Duvall, T. L. 2006a, *Space Sci. Rev.*, 124, 1

Kosovichev, A. G. & Duvall, T. L. 2006b, in: V. Bothmer & A. Hady (eds.), *Solar Activity and its Magnetic Origin*, Proc. IAU Symposium No. 233 (Cambridge: CUP), p. 365

Kosovichev, A. G. & Duvall, T. L. 2008, in: R. Howe *et al.* (eds.), *Subsurface and Atmospheric Influences on Solar Activity*, ASP-CS, 383, 59

Kosovichev, A. G. & Sekii, T. 2007, *ApJ* (Letters), 670, L147

Kosovichev, A. G. & H. S. Team 2007, *AN*, 328, 339

Krivova, N. A., Balmaceda, L., & Solanki, S. K. 2007, *A&A*, 467, 335

Landi Deglinnocenti, E. & Landolfi, M. 1983, *Solar Phys.*, 87, 221

Liewer, P. C., Neugebauer, M., & Zurbuchen, T. 2004, *Solar Phys.*, 223, 209

Lin, C.-H., Antia, H. M., & Basu, S. 2007, *ApJ*, 668, 603

Lites, B. W., Kubo, M., Socas-Navarro, H., *et al.* 2008, *ApJ*, 672, 1237

Lockwood, M. 2006, *Space Sci. Rev.*, 125, 95

Mathur, S., Turck-Chièze, S., Couvidat, S., & García, R. A. 2007, *ApJ*, 668, 594

Mathur, S., Eff-Darwich, A., García, R. A., & Turck-Chièze, S. 2008, *A&A*, 484, 517

Martinez-Oliveros, J. C., Donea, A.-C., Cally, P. S., *et al.* 2008a, *MNRAS*, 389, 1905

Martinez-Oliveros, J. C., Moradi, H., & Donea, A.-C. 2008b, *Solar Phys.*, 251, 613

Miesch, M. S., Brun, A. S., de Rosa, M. L., & Toomre, J. 2008, *ApJ*, 673, 557

Mininni, P. D. 2007, *Phys. Rev. E*, 76, 026316

Mininni, P. D., Ponty, Y., Montgomery, D. C., *et al.* 2005, *ApJ*, 626, 853

Mitra-Kraev, U., Kosovichev, A. G., & Sekii, T. 2008, *A&A* (Letters), 481, L1

Moore, J., Grinsted, A., & Jevrejeva, S. 2006, *Geophysical Research Letters*, 33, 17705

Moradi, H., & Cally, P. S. 2008, *Solar Phys.*, 251, 309

Moradi, H., Donea, A.-C., Lindsey, C., *et al.* 2007, *MNRAS*, 374, 1155

Nagashima, K., Sekii, T., Kosovichev, A. G., *et al.* 2007, *PASJ*, 59, 631

Nigam, R., Kosovichev, A. G., & Scherrer, P. H. 2007, *ApJ*, 659, 1736

Orozco Suárez, D., Bellot Rubio, L. R., *et al.* 2007, *ApJ* (Letters), 670, L61

Parchevsky, K. V. & Kosovichev, A. G. 2008, ArXiv e-prints 0806, 2897

Rajaguru, S. P., Birch, A. C., Duvall, T. L., *et al.* 2006, *ApJ*, 646, 543

Rempel, M. 2005, *ApJ*, 622, 1320

Rempel, M. 2006, *ApJ*, 647, 662

Rempel, M., Schuessler, M., & Knoelker, M. 2008, *ApJ*, in press

Rezaei, R., Steiner, O., Wedemeyer-Böhm, S., *et al.* 2007, *A&A* (Letters) 476, L33

Riethmller, T. L., Solanki, S. K., & Lagg, A. 2008, *ApJ* (Letters), 678, L157

Rimmele, T. 2008, *ApJ*, 672, 684

Ruiz Cobo, B. & Bellot Rubio, L. R. 2008 *A&A*, 488, 749

Scharmer, G. B. & Spruit, H. C. 2006, *A& A* 460, 605

Scharmer, G. B., Gudiksen, B. V., Kiselman, D., *et al.* 2002, *Nature*, 420, 151

Schmidtke, G., Brunner, R., Eberhard, D., *et al.* 2006a, *Adv.Space Res.*, 37, 273

Schmidtke, G., Frohlich C., & Thuillier, G. 2006b, *Adv. Space Res.*, 37, 255

Schüssler, M. & Vögler, A. 2006, *ApJ* (Letters), 641, L73

Schüssler, M. 2007, *AN*, 328, 1087

Schüssler, M. & Vögler, A. 2008, *A&A*, 481, L5

Sekii, T., Kosovichev, A. G., Zhao, J., *et al.* 2007, *PASJ*, 59, 637

Solanki, S. K. & Montavon, C. A. P. 1993, *A&A*, 275, 283

Stein, R. F., Benson, D., & Georgobiani, D. 2007a, in: *Unsolved Problems in Stellar Physics: A Conference in Honor of Douglas Gough*, AIP-CP, 948, 111

Stein, R. F., Benson, D., *et al.* 2007b, in: F. Kupka, I. Roxburgh & K. Chan (eds.), *Convection in Astrophysics*, Proc. IAU Symposium No. 239 (Cambridge: CUP), p. 331

Steiner, O., Rezaei, R., Schaffenberger, W., *et al.* 2008, *ApJ* (Letters), 680, L85

Tapping, K. F., Boteler, D., & Charbonneau, P. 2007, *Solar Phys.*, 246, 309

Thompson, M. J., Christensen-Dalsgaard, J., Miesch, M. S., & Toomre, J. 2003, *ARAA*, 41, 599

Title, A. M., *et al.* 2006, 36th COSPAR Scientific Assembly, meeting abstract #2600

Tobiska, W. K. & Bouwer, S. D. 2006, *Adv. Space Res.*, 37, 347

Turck-Chièze, S., Couvidat, S., Piau, L., *et al.* 2004, *Phys. Rev. Lett.*, 93, 211102-(1–4)

Vargas Domínguez, S., Bonet, J. A., Martínez Pillet, V., *et al.* 2007, *ApJ* (Letters), 660, L165

Vögler, A. & Schüssler, M. 2007, *A&A* (Letters), 465, L43

Warren, H. P. 2006, *Adv. Space Res.*, 37, 359

Wenzler, T., Solanki, S. K., Krivova, N. A., & Frohlich, C. 2006, *A&A*, 460, 583

Withbroe, G. L. 2006, *Solar Phys.*, 235, 369

Woodard, M. F. 2006, *ApJ*, 649, 1140

Woodard, M. F. 2007, *ApJ*, 668, 1189

Woods, T. N. 2008, *Adv. Space Res.*, 45, 895

Woods, T. N., Lean, J. L., & Eparvier, F. G. 2006, in: N. Gopalswamy & A. Bhattacharyya (eds.), Proc. ILWS Workshop, p. 145

Zhao, J. 2007, *ApJ* (Letters), 664, L139

Zhao, J. 2008, *Adv. Space Res.*, 41, 838

Zhao, J., Georgobiani, D., Kosovichev, A. G., *et al.* 2007, *ApJ*, 659, 848

Zhao, J. & Kosovichev, A. G. 2006, *ApJ*, 643, 1317

Zharkov, S., Nicholas, C. J., & Thompson, M. J. 2007, *AN*, 328, 240

Zharkova, V. V. 2008, *Solar Phys.*, 251, 641

Zharkova, V. V. & Zharkov, S. I. 2007, *ApJ*, 664, 573

Transactions IAU, Volume XXVIIA
Reports on Astronomy 2006–2009
Karel A. van der Hucht, ed.

© 2009 International Astronomical Union
doi:10.1017/S1743921308025374

COMMISSION 49

INTERPLANETARY PLASMA AND HELIOSPHERE

PLASMA INTERPLANÉTAIRE
ET HÉLIOSPHÈRE

PRESIDENT	Jean-Louis Bougeret
VICE-PRESIDENT	Rudolf von Steiger
PAST PRESIDENT	David F. Webb
ORGANIZING COMMITTEE	Subramanian Ananthakrishnan,
	Hilary V. Cane,
	Natchimuthuk Gopalswamy,
	Stephen W. Kahler, Rosine Lallement,
	Blai Sanahuja, Kazunari Shibata,
	Marek Vandas, Frank Verheest

TRIENNIAL REPORT 2006 - 2009

1. Introduction

Commission 49 covers research on the solar wind, shocks and particle acceleration, both transient and steady-state, e.g., corotating, structures within the heliosphere, and the termination shock and boundary of the heliosphere.

The present triennal report is particularly rich in important results and events. The crossing of the solar wind termination shock by *Voyager 2* in 2007 is a highlight and a milestone that will certainly have important consequences for astrophysical processes in general (Section 7). The fiftieth anniversary of the *International Geophysical Year* (1957–1958), which is also the fiftieth anniversary of the birth of the Space Age, was marked not only by celebrations and a strong Education and Public Outreach Program, but also by efforts in coordinating present observations and in starting new scientific programs, particularly implying developing countries (Section 8). Studies of solar energetic particles (Section 3) and the related radio bursts (Section 4) benefited from new data from a number of spacecraft. The *STEREO* mission was launched in October 2006 and has obtained new results on 3-D aspects of the inner heliosphere. Meanwhile, solar cycle 24 is expected to become active soon, following what is already the deepest solar minimum of the space age.

Heliospheric compositional signatures will be presented in Section 5, and Interplanetary Scintillation results and developments in Section 6.

One of the highlights of this present period is definitely the completion in 2008 of the Ulysses mission, after almost 18 years of scientific successes and discoveries. Ulysses marks a giant step in the exploration of the Heliosphere and the next Section will give a very brief summary of its results.

2. Ulysses

Richard G. Marsden, Ulysses Mission Manager
ESA/ESTeC (SRE-SM), P.O. Box 299, NL-2200AG Noordwijk, Netherlands
<Richard.Marsden@esa.int>

2.1. *Introduction*

The joint ESA-NASA *Ulysses* space mission is probing the most fundamental processes of our solar system from a unique, out-of-ecliptic orbit. Its principal scientific goal is to conduct as complete a survey as possible of the heliosphere within ~ 5 AU of the Sun at all solar latitudes and under a wide range of solar activity conditions. *Ulysses* was launched in October 1990, and was initially foreseen to have a 5-year lifetime. The mission has been so successful, however, that ESA and NASA have extended its operational phase a number of times, permitting three surveys of the Sun's polar regions. Nevertheless, the diminishing output from the spacecraft's power source will probably bring this historic mission to a close before the end of 2008.

2.2. *Ulysses through the solar cycle*

Ulysses orbits the Sun once every 6.2 yr in a plane that is nearly perpendicular to both the ecliptic and the solar equator. This, together with the longevity of the mission, has allowed *Ulysses* to characterize the heliosphere in 'four dimensions', i.e., three spatial dimensions and time. Solar variability, which drives many of the phenomena being investigated by *Ulysses* , occurs on a wide variety of time-scales. In the context of global heliospheric studies, two are particularly relevant: the 11-yr solar activity (sunspot) cycle, and the 22-yr magnetic (Hale) cycle. Fortuitously, the orbital period of *Ulysses* corresponds roughly to the time it takes the Sun to go from the minimum to the maximum of its activity cycle. When *Ulysses* first flew over the Sun's polar regions in 1994 and 1995, solar activity was close to minimum, providing a view of the 3-dimensional heliosphere at its most simple (Balogh *et al.* 2001, and references therein). Fast solar wind from the polar regions flowed uniformly to fill a large fraction of the heliosphere; variability was confined to a narrow region around the solar equator (McComas *et al.* 2000). When *Ulysses* returned to high latitudes in 2000 and 2001, things were very different (Balogh *et al.* 2008, and references therein). Solar activity was close to maximum and transient features were dominant. Solar wind streams from the poles appeared indistinguishable from streams at low latitudes (McComas *et al.* 2003). Amid all this apparent chaos, *Ulysses* found that the reversal of the heliospheric magnetic field polarity, which occurs every 11 yr, happens in an unexpectedly simple fashion. The main component of the field is a dipole, and this appears to simply rotate through 180 degrees to accomplish the reversal (Smith 2008). Given the complexity of the field reversal at the solar surface, this is surprising. At the time of the third polar passes, in 2006–2007, solar activity was once again close to minimum, although there were important differences when compared with the 1st high-latitude passes. In particular, the solar wind and magnetic field measured by *Ulysses* were noticeably weaker than before (McComas *et al.* 2008; Issautier *et al.* 2008; Smith & Balogh 2008).

2.3. *Scientific 'firsts' from Ulysses*

(*a*) The first direct measurements of interstellar dust and neutral helium gas: Astronomical observations suggest that the Sun is presently moving through a warm, tenuous interstellar cloud made of dust and gas, one of several that make up our local galactic neighbourhood. Using instruments on board *Ulysses* , it has been possible to make direct measurements of dust grains and neutral helium atoms from the local cloud that penetrate deep into the heliosphere for the first time (Gruen *et al.* 1993; Witte *et al.* 1993). These measurements have allowed the determination off the flow direction of the dust and gas, as well as the density and temperature of the neutral helium and the mass distribution of the dust particles.

(*b*) First high-precision measurements of rare cosmic-ray isotopes (e.g., 36Cl and 54Mn): Together with the interstellar neutral gas and dust, cosmic-ray particles are the only sample of material from outside the heliosphere that is available for direct in-situ study. *Ulysses* carries an instrument that has been able, for the first time, to make the precise measurements of rare cosmic-ray isotopes needed to test current theories of cosmic ray origin (Connell 2001).

(*c*) First measurements of so-called 'pickup' ions of both interstellar and near-Sun origin: Pickup ions are created in the heliosphere when neutral atoms become ionized by charge-exchange with solar wind ions or by photo-ionization. Measurements of pickup ions by *Ulysses* have lead to a wide range of discoveries (Gloeckler *et al.* 2001). New sources of pickup ions have been discovered. Solar wind particles appear to become embedded in dust grains near the Sun, and are subsequently released to form a pickup ion population known as the 'inner source'. Comets emit neutrals that form pickup ions, from which the composition of the comet can be determined. *Ulysses* has made detailed measurements of interstellar pickup ions, created when interstellar neutral gas becomes ionized. Interstellar neutral gas is a sample of the local interstellar medium and thus the composition of the Galaxy in the present epoch, as opposed to when the solar system was formed 4.5 billion years ago. The isotope 3He was measured in the interstellar pickup ion population by *Ulysses* , providing an important constraint on the evolution of matter in the universe.

(*d*) First observations of solar energetic particles over the solar poles: A fundamental *Ulysses* discovery is that energetic charged particles are able to move much more easily in latitude than was imagined prior to launch (Lario & Pick 2008). Large latitudinal excursions in the direction of the heliospheric magnetic field were not anticipated, and it was therefore assumed that charged particles would not be able to move easily in latitude. Surprisingly, during its first solar minimum polar pass, *Ulysses* observed particles accelerated at corotating interaction regions (CIR) well above the latitude at which the CIRs themselves occurred. Similarly at solar maximum, *Ulysses* detected large numbers of energetic particles over the solar poles, far away from the location of the solar activity that created them. Either the particles are transported across the magnetic field, or the field lines themselves undergo large excursions, enabling low-latitude sources to be connected to high latitudes. Which of these processes dominates is still a matter of debate.

2.4. *Ulysses scientific achievements 2005–2008*

Specific achievements of the *Ulysses* mission during the period covered by this report include:

- The development of a global picture of the 3-D solar wind at minimum and maximum for use in modeling the heliospheric interface with the interstellar medium.
- Showing that the magnetic flux in the heliosphere in solar cycle 23 is different from earlier cycles (weaker) (Smith & Balogh 2008).
- Showing that, from a 3-D heliospheric perspective, solar cycle 23 is in many ways different from earlier cycles.
- The discovery that the invariance of the radial magnetic field with heliolatitude,shown earlier to exist at sunspot minimum, also exists at sunspot maximum (Smith 2008).
- Acquiring observations leading to new theories for the origin of slow solar wind.

- The cataloging of abundance signatures in ICMEs.
- The discovery of the ubiquitous presence of suprathermal tails in the energy distributions of solar wind ions, leading to new models of particle acceleration (Fisk & Gloeckler 2007).
- Mapping the changing 3-D energetic particle environment between solar minimum and maximum (Lario & Pick 2008).
- The discovery of the shift in flow direction of interstellar dust in the heliosphere beginning in 2005 (Krueger *et al.* 2007).
- The discovery of in-situ magnetic reconnection in the solar wind at all heliospheric distances sampled by *Ulysses* (Gosling *et al.* 2006).

This brief summary has not been able to do justice to the full range of science to which *Ulysses* has made lasting contributions. Nevertheless, it illustrates the fundamental new insights that have been obtained through *Ulysses* into the global behaviour of the heliosphere.

References

Balogh, A., Marsden, R. G., & Smith, E. J.(eds.) 2001, *The Heliosphere Near Solar Minimum: The Ulysses Perspective* (Praxis Publishing Ltd)

Balogh, A., Lanzerotti, L. J., & Suess, S. T.(eds.) 2008, *The Heliosphere through the Solar Activity Cycle* (Springer-Praxis Ltd)

Connell, J. J. 2001, Proc. *27th Intern. Cosmic Ray Conf.*, OG, 1751

Fisk, L. A. & Gloeckler, G. 2007, *Space Sci. Rev.*, 130, 153

Gloeckler, G., Geiss J., & Fisk, L. A. 2001, in: Balogh, A., Marsden, R. G., & Smith, E. J.(eds.), *The Heliosphere Near Solar Minimum: The Ulysses Perspective* (Springer-Praxis Ltd), p. 287

Gosling, J. T., Eriksson, S., Skoug, R. M., McComas D. J., & Forsyth, R.J. 2006, *ApJ*, 644, 613

Gruen, E., Zook, H. A., Baguhl, M., *et al.* 1993, *Nature*, 362, 428

Issautier, K., Le Chat, G., Meyer-Vernet, N., *et al.* 2008, *Geophys. Res. Lett.*, 35, L19101

Krueger, H., Landgraf, M., Altobelli, N., & Gruen, E. 2007, *Space Sci. Rev.*, 130, 401

Lario, D. & Pick, M. 2008, in: A. Balogh, L. J. Lanzerotti & S. T. Suess (eds.), *The Heliosphere through the Solar Activity Cycle* (Springer-Praxis Ltd), p. 151

McComas, D. J., Barraclough, B. L., Funsten, H. O., *et al.* 2000, *J. Geophys. Res.*, 105, 10419

McComas, D. J., Elliott, H. A., Schwadron, N. A., *et al.* 2003, *Geophys. Res. Lett.*, 30, 24

McComas, D. J., Ebert, R. W., Elliott, H. A., *et al.* 2008, *Geophys. Res. Lett.*, 35, L18103

Smith, E. J. 2008, in: A. Balogh, L. J. Lanzerotti & S. T. Suess (eds.), *The Heliosphere through the Solar Activity Cycle* (Springer-Praxis Ltd), p. 79

Smith, E. J. & Balogh A. 2008, *Geophys. Res. Lett.*, 35, L22103

Witte, M., Rosenbauer, H., Banaszkiewics, M., & Fahr, H. 1993, *Adv. Space Res.*, 13, 121

3. Solar energetic particles

Hilary V. Cane

Astroparticle Physics Laboratory, NASA/GSFC, Greenbelt MD, USA

<hilary.cane@utas.edu.au>

During the period 2006 - 2008, at the end of cycle 23, the Sun was quiet except for some intense activity in December 2006. Spacecraft launched in the mid-1990s to early 2000's to make observations of the Sun, the solar wind, and solar energetic particles (SEPs), (*Wind, SOHO, ACE, RHESSI*), were still providing data and were augmented by the launch of *Hinode* and the two *STEREO* spacecraft in 2006. *STEREO* observations will provide additional information about the role of flare processes in large SEP events. It is clear from particle observations that such processes, related to magnetic reconnection, provide part of the SEP source population (Desai *et al.* 2006). However, it is not clear whether the numerous, small SEP events provide Fe-rich, ^{3}He-rich seed particles for the coronal mass ejection (CME) driven shocks (Tylka & Lee 2006) or whether the particles with enhanced abundances seen in large events come directly from the associated flare (Cane *et al.* 2006). Electron observations are also not definitive (Kahler 2007). Unfortunately in December 2006 the *STEREO* spacecraft were essentially at the same location and could not provide different viewpoints of the abundances in the SEP events that occurred at this time.

The ionic charge is an important diagnostic of SEP source regions but unfortunately measurements are not available for the majority of events and then only by indirect methods at the high energies ($>\sim 25$ MeV) that are most important for understanding processes occurring close to the Sun. A new indirect method for determining ionic charge has recently been presented by Sollitt *et al.* (2008). It uses a scaling of charge state to decay time. For the largest events, data from the *SAMPEX* spacecraft can provide charge states for high energies. The comprehensive results are not yet published. On the other hand extensive analyses have been made of the charge state measurements at energies <1 MeV/nuc returned from the *sepica* instrument on *ACE*. A review is presented by Klecker, Möbius & Popecki (2007). The important result is that the charge states are energy dependent. Modeling efforts that include stripping processes and particle propagation are continuing (e.g. Dröge *et al.* 2006, Kartavykh *et al.* 2008).

Understanding propagation processes is clearly important for interpreting observations that are primarily made at 1 AU. Interplanetary scattering must be understood if we are to make correct deductions concerning the time and location of the release of particles at the Sun (e.g., Kahler and Ragot 2006). It has also been proposed that interplanetary scattering can account for abundance ratios changing with time in individual events at a fixed energy/nuc (Mason *et al.* 2006). Particle guiding within and reflection at interplanetary structures are aspects that also need to be considered (e.g. Sáiz *et al.* 2008, Tan *et al.* 2008, Kocharov *et al.* 2008). Recent studies suggest that azimuthal spreading of particle distributions also takes place in the low corona (Wibberenz and Cane 2006, Klein *et al.* 2008) as had been proposed decades earlier.

Modeling of the shock acceleration of SEPs is ongoing and becoming more detailed with increased computing power. One aspect that is of particular interest is how quickly a shock in the low corona can accelerate high energy particles. Two studies suggest that proton energies >100 MeV may be achieved within minutes (Vainio & Laitinen 2007, Ng & Reames 2008).

Studies of small SEP events are also ongoing. Mason (2007) has recently summarised the results from the ULEIS experiment on *ACE*. The spectra of ^{3}He and ^{4}He have been modeled by assuming stochastic particle acceleration by turbulent plasma waves (Liu, Petrosian & Mason 2006). Several authors have studied the source regions of ^{3}He–rich events (Wang, Pick and Mason 2006, Nitta *et al.* 2006, 2008). These events are often associated with coronal jets that are likely to be signatures of magnetic reconnection between closed and open field lines.

The link between particles accelerated at the Sun, as evidenced by associated electromagnetic emissions, and particles detected in situ is still tenuous. However observations of the January 2005 flares from *RHESSI* have supported earlier studies that found that spectral hardening in X-rays is indicative of an event that produces interplanetary particles (Saldanha, Krucker & Lin 2008). On the other hand, comparisons between electron spectral indices deduced from X-ray observations with those measured in space do not always agree (Krucker *et al.* 2007).

References

Cane, H. V., Mewaldt, R. A., Cohen, C. M. S., & von Rosenvinge, T. T. 2006, *JGR* 111, A06S90

Desai, M. I., Mason, G. M., & Gold, R. E., *et al.* 2006, *ApJ*, 649, 470

Dröge, W., Kartavykh, Y. Y., Klecker, B., & Mason, G. M. 2006, *ApJ*, 645, 1516

Kahler, S. W. & Ragout, B. R. 2006, *ApJ*, 646, 634

Kahler, S. W. 2007, *SSR*, 129, 359

Kartavykh, Y. Y., Dröge, W., Klecker, B., *et al.* 2008, *ApJ*, 681, 1653

Klecker, B., Möbius, E., & Popecki M. A. 2007, *SSR*, 130, 273

Klein, K.-L., Krucker, S., Lointier, G., & Kerdraon, A. 2008, *A&A*, 486, 589

Kocharov, L., Pizzo, V. J., Zwickl, R. D., & Valtonen, E. 2008, *ApJ* (Letters), 680, L69

Krucker, S., Kontar, E. P., Christe, S., & Lin, R. P. 2008, *ApJ* (Letters), 663, L109

Liu, S., Petrosian, V., & Mason, G. M. 2006, *ApJ*, 636, 462

Mason, G. M., Desai, M. I., Cohen, C. M. S., *et al.* 2006, *ApJ* (Letters), 647, L65

Mason, G. M. 2007, *SSR*, 130, 231

Ng, C. K. & Reames, D. V. 2008, *ApJ* (Letters), 686, L123

Nitta, N. V., Reames, D. V., DeRosa, M. L., *et al.* 2006, *ApJ*, 650, 438

Nitta, N. V., Mason, G. M., & Wiedenbeck, M. E., *et al.* 2008, *ApJ* (Letters) 675, L125

Sáiz, A., Ruffolo, D., Bieber, J. W., Evenson, P., & Pyle, R. 2008, *ApJ*, 672, 650

Saldanha, R., Krucker, S., & Lin, R. P. 2008, *ApJ*, 673, 1169

Sollitt, L. S., Stone, E. C., Mewaldt, R. A., *et al.* 2008, *ApJ*, 679, 910

Tan, L. C., Reames, D. V., & Ng, C. K. 2008, *ApJ*, 678, 1479

Tylka, A. J. & Lee, M. A. 2006, *ApJ*, 646, 1319

Vainio, R., & Laitinen, T. 2007 *ApJ*, 658, 622

Wang, Y.-M., Pick, M., & Mason,G. M. 2006 *ApJ*, 639, 495

Wibberenz, G. & Cane, H. V. 2006, *ApJ*, 650, 1199

4. Interplanetary radio bursts

Natchimuthuk Gopalswamy
NASA/GSFC, Greenbelt MD, USA
<nat.gopalswamy@nasa.gov>

4.1. *Type II radio bursts and shocks*

The close connection among Type II radio bursts, large solar energetic particle (SEP) events, and interplanetary (IP) shocks has been examined in more detail in many recent works (Gopalswamy 2006; Cliver & Ling 2007; Cliver 2008; Gopalswamy *et al.* 2008a,b). The SEP associated rate increases with the speed and width of coronal mass ejections (CMEs) that produce type II radio bursts in the decameter-hectometric (DH) wavelength domain. In particular, presence of DH type II seems to be a necessary condition for the production of SEP events but not the type III radio bursts at low frequencies ($\sim 1\,\mathrm{MHz}$) (Cliver 2008). The three phenomena have a common cause, viz., fast and wide CMEs (speed $\geqslant 900\,\mathrm{km/s}$ and width $\geqslant 60\,\mathrm{degrees}$). Deviations from this general picture are observed as (*i*) lack of type II bursts during many fast and wide CMEs and IP shocks; (*ii*) slow CMEs associated with type II radio bursts and SEP events; and (*iii*) lack of SEP events during many type II bursts (Gopalswamy 2008). Most of the deviations can be accounted for by the large variation of the Alfvén speed in the corona, ranging from about $400\,\mathrm{km/s}$ to $1600\,\mathrm{km/s}$. A large number of fast and wide CMEs that did not produce type II radio bursts in fact came from behind the limb, which may indicate a visibility problem.

Thejappa *et al.* (2007) performed Monte Carlo simulation of the directivity of interplanetary types II and III radio bursts occurring at 120 kHz. They find that the scattering by random density fluctuations extends the visibilities of fundamental and harmonic components from 18 degrees to 90 degrees, and from 80 degrees to 150 degrees, respectively. They also reported the simultaneous observation of bursts by *Ulysses* and *Wind* spacecraft separated by more than 100 degrees. Scattering and refraction seem to play a major role in making the bursts visible over a wide range of angles.

Pulupa & Bale (2008) studied the source regions of interplanetary (IP) type II radio bursts associated with shock-driving interplanetary CMEs (ICMEs). Immediately prior to the arrival of each shock, electron beams along the interplanetary magnetic field and associated Langmuir waves are detected, implying magnetic connection to a quasi-perpendicular shock front acceleration site. The presence of a foreshock region requires nonplanar structure on the shock front. Using Wind burst mode data, the foreshock electrons are analyzed to estimate the dimensions of the curved region. Ledenev *et al.* (2007) showed that the longitudinal wave spectrum, excited in the solar wind plasma, extends with the increase of the refractive index to values > 10. This explains the broad band of emission, the constant value of the average ratio of frequency-band to radio emission frequency from interplanetary shock wave fronts. They were able to estimate the electron beam density and amplitude of Langmuir waves at the shock. The spectrum of radio emission was shown to be determined by the spectrum of Langmuir waves excited upstream of the interplanetary shock wave by heated electrons escaping from the shock wave front.

The type II burst of 2001 May 10 was tracked to very low frequencies using *Ulysses* radio data (Hoang *et al.* 2007). The associated shock was also observed in situ at *Ulysses* but the type II radio emission was observed for more than a day prior to the arrival of an interplanetary shock at *Ulysses*. By accurately subtracting the thermal noise background from the observed emission intensity they were able to deduce the type II brightness temperatures at the fundamental and harmonic near the shock crossing. The measured brightness temperature of the type II harmonic emission reached a peak a value of $\simeq 3 \times 10^{13}$ K just after the shock crossing at *Ulysses*.

Since type II bursts are good indicators of shocks, Oh *et al.* (2008) selected a set of 31 IP shocks associated with *Wind*-WAVES type II radio bursts and studied the kinematics of the shocks. They found that the mean acceleration of the IP shocks between the Sun and Earth to be about -1.02 m s^{-2}, which is smaller than the values obtained for CMEs. Using the constraints imposed by the low-frequency radio emissions generated by shocks driven by CMEs, the measured 1 AU transit times and the calculated in situ shock speeds, together with the required consistency with the white-light measurements, Reiner *et al.* (2007a) analyzed the interplanetary transport of 42 CME/shocks observed during solar cycle 23 to determine when, where, and how fast CMEs decelerate as they propagate through the corona and interplanetary medium. They found some notable correlations between the parameters that characterize the deceleration of these CMEs to 1 AU.

Kahler *et al.* (2007) found evidence for a class of shock-accelerated 'near relativistic' electron events based on particle and *Wind*-WAVES data. They compared the inferred injection times of 80 near-relativistic electron events observed by the *Wind*-3DP electron detector with 40-80 MHz solar radio observations and found no single radio signature characteristic of the inferred electron injection times. About half of the events were associated with metric or DH type II bursts, but most injections occurred before or after those bursts. Electron events with long ($\geqslant 2$ hr) beaming times at 1 AU were preferentially associated with type II bursts, which led to their conclusions on the source of the electrons.

Sakai & Karlicky (2008) performed particle-in-cell simulations of shocks to explain the band splitting of type I solar radio bursts. Near the shock front, they found some protons reflected and accelerated. The reflected protons dragged the background electrons to keep the charge neutrality, resulting in electron acceleration. The accelerated electrons excited electrostatic waves. The resulting radio emission occurred near the fundamental and second harmonic of the local plasma frequency. The band splitting of the type II burst was found to be dependent on the direction of propagation of the shock.

4.2. *Type III radio bursts*

Investigations on the type III bursts concentrated on the analysis of storm events. Reiner *et al.* (2007b) reported the detection of circular polarization ($\sim$5%) in solar type III radio storms at hectometric-to-kilometric wavelengths. The sense of the polarization is maintained for the entire duration of the type III storm (usually many days). For a given storm, the degree of circular polarization was found to peak near central meridian crossing of the associated active region. At a given time, the degree of circular polarization was found to generally vary as the logarithm of the observing frequency. These observations may provide important information on the magnitude and radial dependence of the solar magnetic field above active regions.

Using long-term observations from the *Geotail* and *Akebono* satellites, Morioka *et al.* (2007) studied the individual bursts in the type III storm ('micro-type III') radio bursts. The average power of the micro type III bursts were found to be about 6 orders of magnitude below that of normal type III bursts. They were able to identify the active regions responsible for the micro-type III bursts by examining the concurrence of their development and decay with the bursts. It was found that both micro and ordinary type III bursts can emanate from the same active region without interference, indicating the coexistence of independent electron acceleration processes. One of interesting findings was that the active regions responsible for micro-type III bursts seem to be located close to coronal holes.

4.3. *Conclusions*

The Radio and Plasma Wave Experiment (WAVES) experiment on the *Wind* spacecraft has made the frequency coverage nearly complete for investigating interplanetary radio bursts. This led to the clarification of many issues regarding the connection of radio bursts with CMEs. A similar experiment on the twin *STEREO* spacecraft (S/WAVES) will enhance our understanding of the interplanetary radio bursts by providing two view points to the radio sources (Bougeret *et al.* 2008). The S/WAVES instrument includes a suite of state-of-the-art experiments that provide comprehensive measurements of the three components of the fluctuating electric field from a fraction of a hertz up to 16 MHz. The instrument has a direction finding or goniopolarimetry capability to perform 3-D localization and tracking of radio emissions associated with streams of energetic electrons and shock waves associated with CMEs. Currently the separation between the two spacecraft is sufficient to show the directivity of radio bursts.

References

Bougeret, J. L., *et al.* 2008, *Space Science Reviews*, 136, 487
Cliver, E. W. 2008, *AIP-CP*, 1039, 190
Cliver, E. W. & Ling, A. G. 2007, *ApJ*, 658, 1349
Gopalswamy, N. 2006, *Geophysical Monograph Series*, 165, 207
Gopalswamy, N. 2008, *AIP-CP*, 1039, 196
Gopalswamy, N., Yashiro, S., Xie, H., *et al.* 2008, *ApJ*, 674, 560

Gopalswamy, N., Yashiro, S., & Akiyama, S. 2008, *Ann. Geophysicae*, 26, 1
Ledenev, V. G., Aguilar-Rodriguez, E., Tirsky, V. V., & Tomozov, V. M. 2008, *A&A*, 477, 293
Hoang, S., Lacombe, C., MacDowall, R. J., & Thejappa, G. 2007, *JGR*, 112, 9102
Kahler, S. W., Aurass, H., Mann, G., & Klassen, A. 2007, *ApJ*, 656, 567
Morioka, A., Miyoshi, Y., Masuda, S. 2007, *ApJ*, 657, 567
Oh, S. Y., Yi, Y., & Kim, Y. H. 2007, *J. Astron. & Sp. Sci*, 24, 219
Pulupa, M. & Bale, S. D. 2008, *ApJ*, 676, 1330
Reiner, M. J., Kaiser, M. L., & Bougeret, J.-L. 2007a, *ApJ*, 663, 1369
Reiner, M. J., Fainberg, J., Kaiser, M. L., & Bougeret, J.-L. 2007, *Solar Phys.*, 241, 351
Sakai, J. I. & Karlický, M. 2008, *A&A* (Letters), 478, L15
Thejappa, G., MacDowall, R. J., & Kaiser, M. L. 2007, *ApJ*, 671, 894

5. Heliospheric compositional cignature

Rudolf von Steiger
International Space Science Institute, Bern, Switzerland
<vsteiger@issibern.ch>

5.1. *Introduction*

The primary motivation for solar wind composition studies is twofold: On the one hand we seek to determine the composition of the outer convective zone of the Sun (as represented by the photosphere) in order to infer the composition of the protosolar nebula, as this represents the baseline from which the entire solar system was formed some 4.6 Gy ago (cf. von Steiger 2001, and references therein). On the other hand composition differences between different solar wind types (or other reservoirs) are indicative for the conditions and processes where these reservoirs originate. Thus composition studies naturally fall into two different types: charge state composition and elemental composition. Charge state composition, i.e., the distribution of the different charge states of a single element, probes the conditions and processes in the corona at a temperature of the order of 10^6 K, whereas elemental composition, i.e., the abundances of the elements summed over all charge states, probe the conditions and processes in the chromosphere and lower transition region at a temperature of the order of 10^4 K. Observations using composition instrumentation such as the SWICS instruments on the *Ulysses* and the *ACE* missions have revealed that both charge state composition and elemental composition are somehow related to the solar wind state, the most obvious feature being an anti-correlation of the charge state ratio O^{7+}/O^{6+} and the solar wind speed, v.

5.2. *Charge state composition*

The charge states of heavy ions observed in the solar wind are indicative of the coronal temperature at the altitude in the corona where the collision time scale equals the expansion time scale. Since the ionisation/recombination rates with hot electrons are temperature dependent each ion pair freezes in at a different altitude in the corona. In fast streams from coronal holes the charge states of each element are well represented by a single temperature. This fact was first used by Geiss *et al.* (1995) to obtain a temperature profile in the south polar coronal hole from charge states observations observed with *Ulysses*-SWICS in the southern high-speed stream. The single freezing-in temperature per element could also be used to infer that the coronal hole is thermally homogeneous to less than $\pm 100\,000$ K.

Charge state distributions obtained outside fast streams are quite different from the single-temperature ones found within. They not only have a significant excess of higher charge states, but they are also broader, indicating a mixture of sources at different, on average higher, temperatures. This has led Zurbuchen *et al.* (2002) to conclude that the slow solar wind is made up from a continuum of dynamic states. An excess of the highest charge states of each element is due to interplanetary coronal mass ejections (ICMEs); in fact, the average charge state of iron ions has been established as a very reliable ICME signature (Lepri and Zurbuchen 2004; Zurbuchen & Richardson 2006).

The direct observation of iron charge states on *Ulysses* has led to a discrepancy with *SoHO*-SUMER remote observations: Based on observations of the 1242 Å line formed by Fe XII (i.e., Fe11), Wilhelm *et al.* (1998) inferred a coronal electron temperature of just barely 1 MK at the base of a coronal hole and decreasing with altitude. This is at variance with the *Ulysses*-SWICS observation of 25% of all iron ions in the 11+ charge state, indicating a temperature of 1.23 MK at 3-4 $R_\odot$. The discrepancy has yet to be resolved (von Steiger *et al.* 2001).

5.3. *Elemental composition*

It has been known for some time known that the first ionisation potential (FIP) fractionation factor $f = (X/O)_{SW}/(X/O)_\odot$ (with X/O the abundance ration of a low-FIP element X relative to oxygen) is about 3–5 in the slow solar wind, but significantly lower than that in the fast streams from coronal holes. With *Ulysses* it was found that the slow solar wind is so variable in elemental composition (and in most other parameters as well), to the point that it becomes hardly meaningful of speak of an average FIP fractionation factor there. Daily averages of the Mg/O abundance ratio reach from only little more than the photospheric value $Mg/O_\odot = 0.074$ (Grevesse *et al.* 2007) to about 4 times that value. Analyses at higher time resolutions seem to indicate even higher FIP fractionation factors at shorter time scales, but these are difficult to ascertain since statistical variability also increases. It seems that the highest FIP factors are found at low latitudes, which helps to explain why fractionation factors found with *Ulysses* are generally smaller than the previously reported factor of $f = 3$–5. These factors were of course obtained at low latitudes, while the smaller *Ulysses* result is an average over all latitudes. The variability of the slow solar wind fits well with the Fisk field model, according to which the slow wind stems from closed loops reconnecting with open field lines as they wander along the solar streamer belt (Fisk *et al.* 1998). The natural age variability of these loops together with the fact that the FIP fractionation factor correlates with their ages (Widing & Feldman 2001) readily accounts for the observed variability.

It can be argued that the composition of the fast polar streams is as close as we can get to the solar composition with in-situ observations. This is particularly important for elements such as neon that cannot be observed by remote sensing in the photosphere for lack of transitions in the relevant energy range. The 'solar' neon abundance given in tables such as the one of Grevesse *et al.* (2007) is not really a solar value, but an approximation thereof obtained from other sources such as remote sensing of the corona or solar energetic particles. Bahcall *et al.* (2005) have used this ignorance to argue that the solar neon abundance might be higher by a factor of 2.5–3 (0.4–0.5 dex) than this estimate in order to reconcile the helioseismology results with the latest values of solar abundances, in particular of CNO, as if the neon abundance were a freely disposable parameter. But the solar wind value of neon observed with *Ulysses*-SWICS ought to be taken into account, and it makes such a high neon abundance seem quite unlikely. Von Steiger *et al.* (2000) find $Ne/O = 0.083$ in fast streams but caution that this is a difficult measurement since neon occurs in the single charge state Ne^{8+} that lies close to the most abundant of the heavy ions, O^{6+}. Nevertheless, a recent independent analysis (Gloeckler & Geiss 2007) seems to confirm this low value of Ne/O. The value is even lower than the solar estimate of Grevesse *et al.* (2007), $Ne/O = 0.15$, which makes it very difficult to believe that the real solar value could be as high as Ne/O $= 0.4$ like it would be needed for the helioseismology results to fit. The solution of this conundrum is as yet outstanding; it may well lie in the abundances of other elements than just the one of neon.

5.4. *Correlation between composition and kinetic parameters*

The anti-correlation of O^{7+}/O^{6+} and v mentioned above was studied in some detail in a pair of papers (Gloeckler *et al.* 2003; Fisk 2003). During a 166 d time period around the solar minimum in 1996 - 1997, Gloeckler *et al.* (2003) first determine a correlation $v = 144/T - 88$, where v is the solar wind speed in km/s and T is the freezing-in temperature from oxygen charge states in MK. The correlation is found to be very tight except at times when an ICME passes by. On the other hand, Fisk (2003) derives a theoretical relation between coronal electron temperature and terminal solar wind speed squared (i.e., its energy) of the form $v^2/2 = C_1/T + C_2$, where $C_{1,2}$ are constants. The theory is based on the picture of open field lines migrating across the solar surface by successively reconnecting with closed loops and thus displacing themselves by the separation of the loop's footpoints. Each of these reconnection events releases energy and mass onto the open field line, i.e., into the corona and solar wind. In turn, these two quantities determine the final energy of a solar wind parcel, or v^2. The quantities can be determined using solar observations or estimates of typical loop heights and other solar quantities, thus determining the constants $C_{1,2}$. Note that only C_1 involves quantities that are not determined in a straightforward manner, while $C_2 = -GM_\odot/r_\odot = -(437 \text{ km/s})^2$ is simply the gravitational potential at the solar surface and thus unadjustable. Fitting their data to Fisk's $V^2 \propto 1/T$ relation, Gloeckler *et al.* (2003) find an equally satisfying fit (again with the ICME periods removed) as for $V \propto 1/T$. It is noteworthy that this fit yields an intercept value very close to the unadjustable constant C_2 and has the added benefit of a physical underpinning. This can finally be used to reverse the relation and ask about the loop heights with which the migrating field lines reconnect. In the quiet Sun, loop heights were found to show a strong dependence on latitude, reaching up to $\sim 100\,000$ km at low latitudes; conversely, in polar coronal holes the lowest heights of $\sim 15\,000$–$30\,000$ km were observed with minimum fluctuation and no dependence on latitude.

References

Bahcall, J. N., Basu, S., & Serenelli A. M. 2005, *ApJ*, 631, 1281
Fisk, L. A., 2003, *JGR*, 108 (A4), 1157
Fisk, L. A., Schwadron, N., & Zurbuchen, T. H. 1998, *Space Sci. Rev.*, 86, 51
Geiss, J., Gloeckler, G., von Steiger, R., *et al.* 1995, *Science*, 268, 1033
Gloeckler, G. & Geiss, J. 2007, *Space Sci. Rev.*, 130, 139
Gloeckler, G., Zurbuchen, T. H., & Geiss, J.: 2003, *JGR*, 108(A4), 1158
Grevesse, N., Asplund, M., & Sauval, A. J. 2007, *Space Sci. Rev.*, 130, 105
Lepri, S. T. & Zurbuchen, T. H. 2004, *JGR*, 109(A18), A01112
von Steiger, R., Schwadron, N. A., Fisk, L. A., *et al.* 2000, *JGR*, 105, (27) 217
von Steiger, R., Vial, J.-C., Bochsler, P. 2001, in: R. F. Wimmer-Schweingruber (ed.), *Solar and Galactic Composition*, AIP-CP, 598, 13
Widing, K. G. & Feldman, U. 2001, *ApJ*, 555, 426
Wilhelm, K., Marsch, E., Dwivedi, B. N., *et al.* 1998, *ApJ*, 500, 1023
Zurbuchen, T. H., Fisk, L. A., Gloeckler, G., *et al.* 2002, *Geophys. Res. Lett.*, 29, 66
Zurbuchen, T. H. & Richardson, I. G. 2006, *Space Science Reviews*, 123, 31

6. Interplanetary scintillation and solar wind studies

P. K. Manoharan,
Radio Astronomy Centre, NCRA-TIFR, Udhagamandalam (Ooty), India.
<mano@wm.ncra.tifr.res.in>

6.1. *Introduction*

Interplanetary scintillation (IPS) is caused when planar wave fronts from a compact radio source (e.g., radio galaxy or quasar) pass through the solar wind. As we see in the following, IPS measurements are an essential means to probe the 3-D heliosphere in density turbulence and flow speed and to establish the connection between the solar phenomena and interplanetary consequences.

6.2. *Propagation and radial evolution of CMEs*

In a study to understand the radial evolution of 30 large CMEs, IPS images obtained from the Ooty Radio Telescope (ORT) along with white-light images from *SOHO*-LASCO have made it possible to track CMEs in the inner heliosphere. Results indicate that each CME tends to attain the speed of the ambient solar wind at 1 AU or further out of the Earth's orbit and the net acceleration imposed on a CME is determined by its initial speed and properties of the solar wind encountered on its way. Further, the radial evolution of CMEs between the Sun and 1 AU confirms the combined influence of *expansion* of the CME (i.e., the magnetic energy possessed by the CME in supporting the propagation) and *aerodynamical drag force* (i.e., interaction between CME and solar wind) at different regions of the inner heliosphere (Manoharan 2006).

In another study, IPS measurements made with the multi-antenna system operated by the Solar-Terrestrial Environment Laboratory (STEL) reveal the 3-D structure and evolution of an intense CME on October 28, 2003. This study shows that the high-density cloud associated with the above CME propagates with a speed much lower than the IP shock. Further, the IP disturbance assumes a loop-shaped distribution, which is in good agreement with the simultaneous white-light observations made with the Solar Mass Ejection Imager (SMEI). It is considered that the loop-shaped structure has resulted from the coronal ejecta confined within the magnetic flux rope, since the location and direction of the loop are consistent with the flux rope geometry inferred from the cosmic ray and *in situ* observations (Tokumaru *et al.* 2005).

The above result suggests that IPS enhancements can indicate plasma in front of the shock and also the CME ejecta behind it. While a CME observed by the coronagraph usually consists of three parts (frontal loop, cavity and core), an interplanetary CME observed *in situ* often shows a two-part structure which lacks dense material in its inner part. This discrepancy may be ascribed to limited spatial coverage of *in situ* observations. The present result is considered as observational evidence to indicate that the core material can survive and travel much farther than the field-of-view of the coronagraph (Tokumaru *et al.* 2007).

In an attempt to study the evolution of fast CMEs, speed estimates of CMEs from the following methods have been examined, radio type II bursts, white-light data, and Ooty IPS images. This study highlights the difficulties of making velocity estimates from radio observations using solar atmospheric density models, particularly under disturbed coronal conditions (Pohjolainen *et al.* 2007).

Another study to analyze events on October 28 and 29, 2003, using the IPS at STEL and other solar and interplanetary data, suggests that understanding the physical features of shock propagation is of great importance in improving the prediction efficiency (Xie *et al.* 2006). In a measurement-based MHD simulation study, the simulated interplanetary shocks are compared with the near-Earth measurement, and the well produced shock arrivals to the Earth implied the ability of the cooperation of the cone-model, the IPS-analysis and MHD simulation. It is found that the second interplanetary disturbance propagated faster in the rarefaction region of the first event, implying that the multi-event simulation is important to enhance the simulation model (Hayashi, Zhao, & Liu 2006).

The tomographic reconstruction of STEL IPS data of a CME from May 2005 has provided an excellent model fit to the EISCAT/MERLIN observation. The above sets of data show that an adjacent fast stream appears significantly deviated from the radial direction (Breen *et al.* 2008). In a follow-up study of a CME in May 2007, IPS observations from EISCAT and the heliospheric imagers on the *STEREO* spacecraft show that IPS could reveal small-scale structure within the CME front which was not resolved by the white-light imager (Dorrian *et al.* 2008). Simultaneous observations at 500 MHz, 928 MHz and 1420 MHz with baselines of up to 1200 km have been carried out. The results are found to be consistent between single- and dual-frequency correlations, allowing the range of observations possible with the EISCAT system to be expanded (Fallows *et al.* 2006).

6.3. *Solar cycle evolution*

The solar wind measurements at Ooty over the solar cycle 23 provide the large-scale changes of latitudinal features of the solar wind density turbulence and speed. The Ooty '*latitude-year*' plots show systematic changes of high speed flow from coronal hole (and its associated low-level of density turbulence). The drifting of density structures from high to low latitudes, as revealed by this study, are effected by the gradual movement of *magnetically-concentrated* coronal holes, in association with the reversal of solar magnetic field. The latitudinal results are consistent with the evolution of warping of the current sheet between the ascending and declining phases of the solar cycle. The high speed streams from these migrating coronal holes cause recurrent interaction regions, which are dominant during the year 2003 (Manoharan 2007).

6.4. *Coronal magnetic field and fast/slow solar wind flows*

It is crucial to understand the physical parameters, which determine the acceleration of solar wind from different coronal sources. In an investigation from STEL IPS made during the minimum phase of the solar cycle, combined with the extrapolated magnetic field data, it is found that the ratio between photospheric magnetic field (B) and the flux expansion factor (f) shows a considerably higher correlation with the solar wind velocity than an individual parameter of B or f (Fujiki *et al.* 2005; Kojima *et al.* 2007).

A study based on the modeling of structure functions from angular/spectral broadening observations and velocity measurements from IPS observations shows that the near-Sun solar wind is dominated by effects associated with obliquely propagating Alfvén/ion cyclotron waves. The modeling of IPS velocities reveals that the large parallel velocity spread and upward bias to the mean velocity observed near the Sun are a direct result of the density fluctuations associated with Alfvén waves along an extended line of sight (Harmon and Coles 2005).

Simultaneous observations between EISCAT and MERLIN, of baselines of up to 2000 km, suggest that two modes of fast solar wind may exist with the fastest mode above the polar regions and slower flow above equatorial extensions of the polar coronal holes (Fallows *et al.* 2007; Bisi *et al.* 2007). The mass flux measurements from *SOHO*-SWAN and mass flux estimates using LASCO-C2 density and IPS velocity show that the fast solar wind reaches its terminal velocity at $\sim 6\,R_\odot$ and expands with constant velocity beyond this distance. On the contrary, the slow solar wind attains only half of its terminal value at the above distance and is thus accelerated farther out (Quémerais *et al.* 2007).

Studies of IPS observations from STEL combined with different space missions data suggest that solar wind disappearance events (e.g., 11 May 1999 event as well as events occurred during 2002) are highly non-radial and associated with unipolar solar wind flows that originate at the boundaries of large active regions and coronal holes located at the central meridian (Janardhan *et al.* 2008; Janardhan, Tripathi & Mason 2008).

6.5. *SMEI and IPS measurements - 3-D reconstruction*

One of the goals of the Solar Mass Ejection Imager (SMEI) team is to provide the analyses for as many events as possible and for this a semi-automated system has been set up to provide the photometric results from SMEI (February 2003 - to date; refer to `<smei.ucsd.edu/>`). The reader may also refer to the review by Webb *et al.* (2006) for general observations of interplanetary CMEs by SMEI. A high-resolution analysis has allowed the 3-D reconstruction of not only the more dense regions of the CMEs observed in SMEI, but also the density enhancements behind the largest CME-driven shocks (Jackson *et al.* 2006).

The 3-D velocity reconstructions of IPS and SMEI data during 6-10 November 2004 clearly show CME structures. Ooty IPS shows very high correlations with *in situ* velocities during this complex interval of solar activity that are unparalleled at this temporal resolution (Bisi *et al.* 2008).

6.6. *Cometary scintillation*

The occultation of compact radio quasar B0019-000 by the plasma tail of comet 73P/Schwassmann-Wachmann 3-B (May 2006) has been made using the ORT, at 327 MHz. The intensity scintillation of this source shows significant increase compared to that of the control source located outside the comet tail. Further, the power spectra of intensity fluctuations show gradual changes as the target source approached the central part of the comet tail. At the point of closest distance to the comet, the spectrum reveals an enhanced level of turbulence both at large-scale (~ 500 km) and at small-scale (~ 50 km) portions of the spectrum. A density turbulence spectrum of two spatial scale sizes can explain the temporal evolution of the power spectrum during the occultation (Roy, Manoharan & Chakraborty 2007).

6.7. *New IPS arrays*

IPS studies are gaining more and more importance. For example, a new large antenna has been added to the multi-antenna IPS system at STEL, Japan. Further, new IPS arrays are being constructed at widely separated geographical longitudes, which will enable the continuous monitoring of the IP medium. The upcoming Low Frequency Array (LOFAR), in the frequency range 10 - 240 MHz, has plans to carry on IPS studies. Briefings of some of the new arrays are given below.

Pushchino Radio Telescope: A specialized antenna system for monitoring IPS at 110 MHz was installed at Pushchino Radio Astronomy Observatory, Lebedev Physical Institute, Russia, during 2006 (Shishov *et al.* 2008). It includes 16-beam array (each beam $\sim 1° \times 0.5°$), which allows to observe a few hundreds of radio sources per day at flux density 0.2 Jy and above. The preliminary observations with the above telescope indicate a quiet state of interplanetary plasma during 2006–2007.

Mexican Array Radio Telescope: This dedicated IPS array, located at Michoacan, about 350 km north-west of Mexico (19.8°N, 101.7 °W), is nearing completion. It consists of 64×64 full wavelength dipole array, operating at 140 MHz, occupying 980 m^2

(E-W 70 m × N-S 170 m) (Gonzalez-Esparza *et al.* 2006). This array supports a multiple-beam system (<www.mexart.unam.mx/>).

Murchison Widefield Array: This new radio array, under construction in Western Australia, consists of 512 antenna tiles, each of 4x4 array of crossed vertical bowtie dipoles, operating in the frequency range 80-300 MHz. The MWA plans to participate in the global IPS network. Its Faraday rotation measurement of radio sources (e.g., Jensen & Russell 2008), combined with IPS tomography, will provide the 3-D magnetic field of propagating CMEs and IP medium (<www.haystack.mit.edu/ast/arrays/MWA>).

Miyun Radio Telescope: A new IPS observing system is being setup at National Astronomical Observatories, China, to record simultaneous dual frequency data at bands of 327/611 MHz and 2.3/8.4 GHz (Zhang 2007).

References

Bisi, M. M., Fallows, R. A., Breen, A. R., Habbal, S. R., & Jones, R. A. 2007, *JGR*, 112, A06101
Bisi, M. M., Jackson, B. V., Hick, P. P., *et al.* 2008, *JGR*, 113, A00A11
Breen, A. R., Fallows, R. A., Bisi, M. M., *et al.* 2008, *ApJ* (Letters), 683, L79
Dorrian, G. D., Breen, A. R., Davies, J. A., *et al.* 2008, *GRL*, in press
Fallows, R. A., Breen, A. R., Bisi, M. M., *et al.* 2006, *GRL*, 33, L11106
Fallows, R. A., Breen, A. R., Bisi, M. M., *et al.* 2007, *Astro. Astrophys. Trans.*, 26, 489
Fujiki, K., Hirano, M., Kojima, M., *et al.* 2005, *AdSpR*, 35, 2185
Gonzalez-Esparza, J. A., Andrade, E., Carrillo, A., *et al.* 2006, *AdSpR*, 38, 1824
Harmon, J. K. & Coles, W. A. 2005, *JGR*, 110, A03101
Hayashi, A.K., Zhao,X-P., & Liu,Y. 2006, *GRL*, 33, L20103
Jackson, B. V., Buffington, A., Hick, P. P., Wang, X., & Webb, D. F. 2006, *JGR*, 111, A04S91
Janardhan, P., Fujiki, K., Sawant, H. S., *et al.* 2008, *JGR*, 113, A03102
Janardhan, P., Tripathi, D., & Mason, H. 2008, *A&A* (Letters), 488, L1
Jensen, E. A. & Russell, C. T. 2008, *Planet. Space Sci.*, 56, 1562
Kojima, M., Tokumaru, M., Fujiki, K., *et al.* 2007, in: K. Shibata, S. Nagata & T. Sakurai (eds.), *New Solar Physics with Solar-B Mission*, ASP-CS, 369, 549
Manoharan, P. K. 2006, *Solar Phys.*, 235, 345
Manoharan, P. K. 2007, in: S. S. Hasan & D. Banerjee (eds.), *Kodai School on Solar Physics*, *AIP-CP*, 919, 314
Roy, N., Manoharan, P. K., & Chakraborty, P. 2007, *ApJ*, 668, L67
Pohjolainen, S., van Driel-Gesztelyi, L., Culhane, J. L., *et al.* 2007, *Solar Phys.*, 244, 167
Quémerais, E., Lallement, R., Koutroumpa, D., & Lamy, P. 2007, *ApJ*, 667, 1229
Shishov, V. I., Tyul'Bashev, S. A., Subaev, I. A., *et al.* 2008, *Solar System Res.*, 42, 341
Tokumaru, M., Kojima, M., Fujiki, K., Yamashita, M., & Baba, D., 2005, *JGR*, 110, A01109
Tokumaru, M., Kojima, M., Fujiki, K., *et al.* 2007, *JGR*, 112, A05106
Webb, D. F., Mizuno, D. R., Buffington, A., *et al.* 2006, *JGR*, 111, A12101
Xie, Y., Wei, F., Feng, X., & Zhong, D. 2006, *Solar Phys.*, 234, 363
Zhang, X.-Z. 2007, *ChJAA*, 7, 712

7. The termination shock and heliospheric boundary

Rosine Lallement
Service d'Aéronomie du CNRS, BP 3, 91371 Verrières-le-Buisson, France
<rosine.lallement@aerov.jussieu.fr>

7.1. *Introduction*

The three years covered by the previous report had been exceptional for the heliospheric community: the events during the last three years covered by this report are even more spectacular. After the crossing of the solar wind termination shock by *Voyager 1* in

December 2004, the crossing of the shock by *Voyager 2* in August 2007 is not only a milestone in the history of space exploration, but its associated discoveries will certainly have important consequences for astrophysical plasmas in general. Following the two events is now beginning a new exciting period of exploration of the transition region between the solar wind and the interstellar medium, and hopefully to the *in situ* exploration of the ambient galactic medium. Fortuitously, fundamental and complementary information on the boundary of the heliosphere has been also provided by other spacecraft, by means of the detection of energetic neutral atoms (ENAs) from the heliosheath. All these results provide an ideal context for the launch of the *IBEX* mission, scheduled for October 2008, and entirely devoted to ENAs. All *in situ* and remote sensing observations will be combined in an unprecedented way.

7.2. *Voyager 2 historical termination shock crossing*

Thirty years after her launch, *Voyager 2* crossed the heliospheric termination shock of the solar wind several times during about 24 hours on 31 August 2007 at a distance of 83.7 AU from the Sun. The shock oscillates in and out in response to solar wind variations, which explains the multiple crossings. These crossings have been recorded by all instruments, including very fortunately the plasma analyser, whose analogon on board *Voyager 1* had stopped to function long before the *Voyager 1* crossing. Those exceptional and unprecedented data are presented by Jokipii (2008) and discussed in a special issue of Nature.

Richardson *et al.* (2008) show the very sharp transitions of the thermal plasma at the shock crossings. An enormous surprise is the low temperature (100 000 K) of the post-shock core plasma, which implies that this core plasma is still supersonic after the shock (!), and that most of the upstream solar wind energy is transferred to minor species, (see below) and not to the core. They also show from solar wind measurements that the strong heliosphere asymmetry implied by *Voyager 2* and *Voyager 1* crossings at 94 and 84 AU respectively is only partially due to temporal effects (solar wind pressure variations which move in and out the shock) and is linked to a real permanent asymmetry due to an inclined interstellar magnetic field, as suggested by neutral species measurements (Lallement *et al.* 2005) and Voyager radio and Termination Shock Particles (TSP) directional properties (Opher *et al.* 2006).

Gurnett & Kurth (2008) describe the intense plasma waves recorded at the termination shock, electron oscillations first upstream of the shock, then broadband electrostatic waves right at the shock. They compare these new data with the spectra observed at planetary bow shocks. A number of similarities are observed, but some of the differences remain to be explained. These data are mandatory for the understanding of the shock formation and temporal behaviour.

Burlaga *et al.* (2008) show the magnetic field intensity and direction along three crossings and very precise measurements of the shock foot, ramp and overshoot characteristics and thickness. These structures and the processes of quasi-perpendicular shock decrease and reformation predicted by simulations (e.g., Lembege & Savoini 1992) have been beautifully demonstrated. On the other hand, the data show that the magnetic field fluctuation spectrum is found to be transformed from a lognormal to a normal distribution as the solar wind passes through the heliosphere termination shock, which is unexpected and under study (Chen *et al.* 2008).

Decker *et al.* (2008) show the unique measurements of the low energy charged particle instrument, that cover energies between a few keV and a few MeV. These low energy accelerated particles play a central role in the new picture that emerges from the *Voyager 2* and *Voyager 1* measurements. The data (see also Decker *et al.* 2005) show that their total pressure is largely above the solar wind pressure, that they keep the same spectral index before and after the shock, despite the very strong increase in their fluxes, and that their composition is identical to the composition of the pickup ions (interstellar species after they have been ionized and convected in the solar wind, that form a suprathermal distribution). Fisk *et al.* (2006) have suggested that these particles, i.e., core pickup ions and their suprathermal tails formed in solar wind shocks and turbulence, are the main actors at the shock and follow a classical Rankine-Hugoniot compression. Most of the solar wind kinetic energy is thus used to heat these TSPs, while the core solar wind receives very little.

Stone *et al.* (2008) analyze the data of the higher energy particle telescopes, namely (*i*) the high energy tails of the previous species; (*ii*) the 10 - 100 MeV anomalous cosmic rays; and (*iii*) the ⩽ 100 MeV galactic cosmic rays. The biggest surprise is, as for *Voyager 1*, the absence at the shock of a maximum of the ACR fluxes, which contradicts standard shock acceleration as their source mechanism. Are the ACR's accelerated further in the heliosheath, as suggested by several previous works (e.g., Ferreira *et al.* 2007) or along the shock flanks (McComas & Schwadron 2006)? Galactic cosmic ray helium shows a surprisingly small gradient which suggests that the true galactic intensity may be lower than expected, or that there is a stronger gradient outside, something equally unexpected.

7.3. *The heliospheric shock entirely mediated by non-thermal ions: confirmation by heliosheath ENAs*

The new technique of ENA imaging is now providing a number of new diagnostics. The ASPERA instrument on board *Mars-Express* (ESA) is detecting energetic neutrals of both Martian and non-planetary origin. Galli *et al.* (2006) and Wurz *et al.* (2008) have measured the 0.2–2 keV spectra of the non-planetary ENAs, identified as neutrals created by charge-exchange between heliosheath protons and interstellar neutrals. More recently, the electron instruments on board the two NASA *STEREO* spacecraft have unexpectedly detected neutral particles from a large fraction of the sky, with a maximum from the front of the heliosphere. Neutral particles are identified when they do not follow the electro-magnetic fluctuations recorded in parallel. Wang *et al.* (2008) present their 3–20 keV spectra and directional fluxes, and argue based on spectral slopes and fluxes that these neutrals are formed by charge exchange between interstellar neutrals and the low energy tail of the heliosheath energetic particles detected by the *Voyagers*. A knee in the spectra at 10–12 keV is indeed clearly seen, which is very likely the counterpart in the shocked and heated heliosheath of the well known 4 keV knee in the solar wind pickup ion spectra that separates core pickup ions from their suprathermal tail. These data provide first partial maps of the heliosheath and confirm the Voyager findings, i.e., the transfer of most of the solar wind energy (more than 70%) to a minority of high energy particles, the upstream pickup ions and their suprathermal tails. The big surprise is the directionality. Instead of a broad maximum from the longitude corresponding to the front of the heliosphere, two maxima of the neutral fluxes are detected, one within 10 degrees longitude on one side of the interstellar wind direction, and one shifted by about 25 degrees longitude on the opposite side. The direction of this secondary flow corresponds to some previous measurements discussed by Collier *et al.* (2004), especially *SOHO*-CELIAS and *IMAGE*-LENA ENAs.

There is now a strong debate about the origin of the secondary maximum. Is it linked to the strong asymmetry of the heliosphere under the magnetic field influence, with shock heating and resulting energetic particles densities being function of the location? Is it solely due to temporal effects (solar wind pressure variations)? Is it due to the existence of a secondary interstellar flow? Fortunately the forthcoming *IBEX* mission (McComas

2008) will be able to answer these questions. *IBEX* will image energetic neutrals with unprecedented energy and spatial resolution. If the secondary flow is linked to the distortion of the heliosphere under the influence of the interstellar magnetic field, one can deduce from interstellar neutral hydrogen deflection and radio measurements that the secondary maximum must be located at high positive latitude. *IBEX* will also allow to determine the entire spectra of the heliosheath particles and follow their variability.

7.4. *Conclusion: lessons from the heliospheric boundary exploration*

The heliosphere boundary is a unique laboratory for the study of interstellar shocks and cosmic ray production, and answering the questions raised by the current exploration is as a consequence of fundamental importance in astrophysics. Astrospheres formed around other stellar-type stars are already shown by means of absorption spectroscopy to possess properties similar to those of our heliosphere (Wood *et al.* 2007), but there are many other structures around objects moving in interstellar plasma such as stellar winds and supernovae bubbles (confirmed production sites of galactic cosmic rays), the study of which will gain from the new findings. In particular, the role of the neutral interstellar gas fraction is found to be essential, through the neutral mass loading, pickup ion production, convection and acceleration. Their fundamental role and influence on the shock seem now to be well established. A number of major questions remain: - Where are the ACR's accelerated? - What is the origin of the double structure of the ENA's? It is impossible here to refer to the numerous works currently published and motivated by the results of the last five years and here we have focused on the recent data.

References

Burlaga, L. F., Ness, N. F., Acuña, M. H., *et al.* 2008, *Nature* 454, 75

Chen, M. Q., Chao, J. K., Lee, L. C., & Ting, N. H. 2008, *ApJ* (Letters), 680, L145

Collier, M. R., Moore, T. E., Simpson, D., *et al.* 2004, *Adv. Sp. Res.*, 34, 166

Decker, R. B., Krimigis, S. M., Roelof, E. C., *et al.* 2005, *Science*, 309, 2020

Decker, R. B., Krimigis, S. M., Roelof, E. C., *et al.* 2008, *Nature* 454, 67

Ferreira, S. E. S., Potgieter, M. S., & Scherer, K. 2007, *J.Geophys. Res. A* 112, 11101

Fisk, L. A., Gloeckler, G., & Zurbuchen, T. H. 2006, *ApJ*, 644, 631

Galli, A., Wurz, P., Barabash, S., *et al.* 2006, *ApJ* 644, 1317

Gurnett, D. A., & Kurth, W. S. 2008, *Nature* 454, 78

Jokipii, J. R. 2008, *Nature* 454, 38

Lallement, R., Quémerais, E., Bertaux, J. L., *et al.* 2005, *Science*, 307, 1447

McComas, D. J. 2008, *Sp. Sci. Rev.*, 122, in press

McComas, D. J. & Schwadron, N. A. 2006, *GRL*, 33, 4102

Opher, M., Stone, E. C., and Gombosi, T. I. 2007, *Science* 316, 875

Richardson, J. D., Kasper, J. C., Wang, C., *et al.* 2008, *Nature* 454, 63

Stone, E. C., Cummings, A. C., McDonald, F. B., *et al.* 2008, *Nature* 454, 71

Wang, L., Lin, R. P., Larson, D. E., & Luhmann, J. G. 2008, *Nature* 454, 81

Wood, B. E., Izmodenov, V. V., Linsky, J. L., & Malama, Y. G. 2007, *ApJ*, 657, 609

Wurz, P., Galli, A., Barabash, S., & Grigoriev, A. 2008, *ApJ*, 683, 248

8. The International Heliophysical Year 2007–2008

Nat Gopalswamy

NASA/GSFC, Greenbelt MD, USA

<nat.gopalswamy@nasa.gov>

8.1. *Introduction*

The *International Heliophysical Year* (IHY) commenced in February 2007, marking the fiftieth anniversary of the *International Geophysical Year* (IGY, 1957–'58). Like the IGY, the objective of the IHY is to discover the physical mechanisms that link Earth and the heliosphere to solar activity. The IHY will focus on global effects but at a much greater physical scale (from Geophysics to Heliophysics) that encompasses the entire solar system and its interaction with the local interstellar medium.

The IHY activities are centered around four key elements: Science (coordinated investigation programs or CIPs conducted as campaigns to investigate specific scientific questions), Observatory Development (an activity to deploy small instruments in developing countries), Public Outreach (to communicate the beauty, relevance and significance of the space science to the general public and students), and the IGY Gold Program (to identify and honor all those scientists who worked for the IGY program). This report concerns the science activities and the observatory development programs.

8.2. *IHY and the United Nations Basic Space Sciences Initiative*

The IHY organization has joined hands with the United Nations Office of Outer Space Affairs to promote heliophysical science activities throughout the world by deploying scientific instruments in the developing countries. Under this collaborative program known as the United Nations Basic Space Sciences (UNBSS) initiative, scientists from developed countries or those who are willing and able, donate small instruments to developing countries for studying heliophysical processes. These deployments will serve as nuclei for a sustained development of scientific activities in the host countries (Davila *et al.* 2008). The IHY/UNBSS instrument concepts can be grouped as follows:

(*i*) Solar Telescope Networks,

(*ii*) Ionospheric Networks,

(*iii*) Magnetometer Networks, and

(*iv*) Particle Detector Networks.

Extensive data on space science have been accumulated by a number of space missions. Similarly, long-term data bases are available from ground-based observations. These data can be utilized in ways different from originally intended for understanding the heliophysical processes. Most of the networks are in progress, and some of them are close to completion. The networks are designed such that observations can be made continuously. For example, the Compound Astronomical Low-cost Low-frequency Instrument for Spectroscopy and Transportable Observatory (CALLISTO) network (PI: Arnold Benz, ETH Zürich, Switzerland) covers the whole globe so the Sun can be observed continuously.

The IHY/UNBSS instrument program is supported by a series of workshops. The primary activities during the workshops can be summarized as follows:

(1) Scientists from developing and developed countries meet face-to-face to discuss collaborative projects under the UNBSS program,

(2) Scientific instrument host groups provide descriptions of the sites for instrument deployment and the facilities available for hosting the instrument,

(3) Potential providers of scientific instruments describe their instruments and the key requirements in terms of infrastructure for a successful deployment and continued operation,

(4) Progress reports after the previous workshop are presented and discussed, and

(5) Several participants provide the necessary scientific background through a series of tutorial talks.

The first IHY/UNBSS Workshop on Basic Space Science was held in Abu Dhabi and Al-Ain, United Arab Emirates during 20-23 November, 2005. Workshop participants represented 44 countries, including a significant portion of North Africa and the IHY-West Asia region. The Second IHY/UNBSS Workshop was held from 27 November - 1 December 2006 in Bangalore, India and was sponsored by UN, NASA and several institutions in India. After several sessions on background science topics, presentations by the instrument donors on the current progress and future plans for the deployment projects were made. Proceedings of the second workshop was published in the Bulletin of the Astronomical Society of India (vol. 35, December 2007). The volume contains 39 articles covering the entire range of IHY science. The third UNBSS meeting took place in Japan in June 2007, which combined both IHY and astronomy activities. Twenty two publications will soon appear in Earth, Moon and Planets. The fourth IHY/UNBSS workshop was held in Sozopol, Bulgaria in June 2008. In addition to the traditional activity, this workshop included first results from the IHY science activities. Papers presented in this workshop will be published in the on-line journal Sun and Geosphere. The final IHY/UNBSS workshop will take place in South Korea in September 2009.

8.3. *Science: Coordinated Investigation Programs*

The building blocks of the IHY science program are the Coordinated Investigation Programs (CIPs). The CIPs are essentially autonomous investigations proposed and executed by self-selected groups of scientists that span the range of IHY science. The CIPs have been grouped into seven 'Disciplines': Solar; Heliospheric/Cosmic Rays; Magnetospheres; Ionized atmospheres; Neutral atmospheres; Climate; Meteors/Meteoroids/ Interstellar Dust. One of the major achievements of IHY has been establishing more than 60 CIPs, each one involving a large number of international scientists. The CIPs will continue even after IHY ends formally in the Spring of 2009.

The Whole Heliosphere Interval (WHI) is one of the CIPs. It is an internationally coordinated observing and modeling effort to characterize the 3-dimensional interconnected solar-heliospheric-planetary system. The campaign to characterize this 'heliophysical' system was conducted during solar Carrington Rotation 2068 (20 March–16 April 2008). The previous and following rotations were also included for comparison purposes. Observers, theorists and modelers met for a workshop during 26-29 August 2008 in Boulder, CO, USA. A wealth of information on the quiet heliosphere as well as transient events has been accumulated and the analyses have been initiated during the workshop. One of the first comparisons made was between the appearance of the Sun during the solar minimum in 1996 (Whole Sun Month campaign) and the current solar minimum (2008). In August 1996, the new cycle (23) had already started, which continued to be present in 2008. Even though there are indications that cycle 24 has started in December 2007, the old cycle activity is still dominant. Comparisons were also made in the polar region using *Ulysses* data from the two minima. Many coronal mass ejections were also observed and their study assumed new significance because of the availability of data from the twin *STEREO* spacecraft.

WHI occurred during solar minimum, which optimizes our ability to characterize the 3-D heliosphere and trace the structure to the outer limits of the heliosphere. With the *Ulysses* spacecraft over the northern pole of the Sun, and the twin STEREO spacecraft in a configuration optimal for 3-D view of the Sun and inner heliosphere, there are some unique possibilities to understand the heliophysical processes. This is augmented by the observations from SOHO, Wind and ACE missions at Sun-Earth L1 point. The five THEMIS spacecraft will likewise be in an ideal configuration to perform 3-D studies of the Earth's magnetosphere.

8.4. *IAU Symposium No. 257*

The IAU Symposium No. 257 on *Universal Heliophysical Processes* was held in Ioannina, Greece, 15–19 September 2008. The symposium is cosponsored by the IHY program and the University of Ioannina. The focus of IAU Symposium No. 257 was on the universality of physical processes in the region of space directly influenced by the Sun through its mass and electromagnetic Emissions: the heliospace. The symposium also attempted to consolidate the knowledge gained in space science over the past fifty years since the birth of this discipline in 1957. The topics of discussion include: Solar sources of heliospheric variability; Origin, evolution and dissipation of magnetic structures; Planetary atmospheres, ionospheres, and magnetospheres, Plasma processes: flows, obstacles, circulation; Energetic particles in the heliosphere; Heliophysical boundaries and interfaces including shock waves; Reconnection processes; Turbulence in heliospace; Physical processes in stellar systems. The symposium had nearly 100 papers presented as keynote addresses, invited talks, contributed talks, and posters. Proceedings of the symposium will be published by Cambridge University press within six months (Gopalswamy & Webb 2009).

8.5. *The IHY Schools Program*

The IHY Schools Program was designed to provide a broad exposure of the universal processes in the heliospace to you scientists and graduate students throughout the world. The School program is designed to be synergistic with the IHY/UNBSS program. International schools have now been successfully conducted in North America (Boulder, CO, USA, August 2007), Asia-Pacific I (Kodaikanal Solar Observatory, India, December 2007), and Latin America (Sao Paulo, Brazil, February 2008). Two more schools are being planned for 2008: Asia-Pacific II in Beijing, China, 20-31 October, and Africa School in Enugu, Nigeria, 10–22 November. A book containing the lectures of the Kodaikanal IHY school is in preparation and will be published by Springer (Gopalswamy, Hasan, & Ambastha 2009). A book will also be published for the proceedings of the American IHY school in Boulder.

References

Davila, J., Gopalswamy, N., Thompson, B., *et al.* 2008, *Earth Moon and Planets*, 103, 9
Gopalswamy, N., *et al.* 2009, *Heliophysical Processes* (Springer), in preparation
Gopalswamy, N. & Webb, D. F. 2009, *Universal Heliophysical Processes*, Proc. IAU Symposium
    No. 257 (Cambridge: CUP), in preparation

9. Closing remarks

The word 'Heliophysics' was coined very recently. Heliophysics encompasses the study of the system composed of the Sun's Heliosphere and the objects that interact with it. This is precisely the scope of Commission 49; it can give us first hand knowledge on fundamental astrophysical processes, from microscopic scales to the scale of the heliosphere: turbulence, reconnection, shocks, particle acceleration, abundances, scale coupling.

The topic of dusty plasmas in the Heliosphere could not be covered in this report, despite important results. Several analyses are still ongoing at the time of this report, but preliminary results suggest that studies relating to nano-particles and their discovery in the solar wind will probably be one of the highlights of the next triennal report in this field.

Jean-Louis Bougeret
president of the Commission

Transactions IAU, Volume XXVIIA
Reports on Astronomy 2006–2009
Karel A. van der Hucht, ed.

© 2009 International Astronomical Union
doi:10.1017/S1743921308025386

DIVISION II / IAU REPRESENTATIVE REPORT
INTERNATIONAL HELIOPHYSICAL YEAR

REPRESENTATIVE   David F. Webb

TRIENNIAL REPORT 2006 - 2009

1. Introduction

The *International Heliophysical Year* (IHY) is an international program of scientific research and collaboration to understand the external drivers of the space environment and climate. Its activities were centered on the year 2008, the 50th anniversary of the *International Geophysical Year*. The IHY involves utilizing the existing assets from space and ground as a distributed Great Observatory and the deployment of new instrumentation, new observations from the ground and in space, and public and student education. The IHY officially was launched in February 2007 with an opening ceremony and workshop in Vienna. Many IHY activities, both scientific and educational, have occurred since then. In practice, these activities have taken place over the last several years, and the programs that have been established through the IHY will continue into the future as 'legacies' of the IHY.

2. IHY Activities

Within the IAU, coordination of IHY activities is within Division II *Sun and Heliosphere*, whose current president is Donald B. Melrose. David F. Webb is the IAU Representative to the IHY and Natchimuthuk Gopalswamy is the chair of the IHY subgroup within the IAU Working Group for *International Collaboration on Space Weather* (ICSW). Hans J. Haubold leads the IHY effort for the United Nations under the auspices of COPUOS and the UN Basic Space Science program. At the Prague General Assembly the IHY held a working group meeting as part of the ICSW WG and presented several talks in a sub-session of SpS5 on *Astronomy in Developing Countries*.

2.1. *IHY science*

IHY science is organized through science working groups that coordinate analysis and modeling efforts, and are responsible for planning IHY meetings, symposia and workshops through three major thrusts: scientific observing campaigns known as the Coordinated Investigation Programs (CIPs), data analysis workshops, scientific meetings and publications, and public outreach. The IHY Secretariat in Washington, D.C., USA, provides international coordination, produces newsletters, maintains the IHY website at <www.ihy2007.org/>, writes articles, coordinates media affairs, and develops outreach products. An important, recent CIP campaign is the Whole Heliosphere Interval (WHI) that occurred over one solar rotation, from 20 March - 16 April 2008. WHI is an international coordinated observing and modeling effort to characterize the 3-D interconnected solar-heliospheric-planetary system. The first WHI workshop (on Data Assessment and

Modeling) was held in Boulder, CO, USA, in August 2008. Next year the results from WHI will be presented at a science workshop as well as at other meetings and a journal special issue is planned. In addition, WHI will be the focus of JD16 at the IAU XXVII GA in Rio de Janeiro, August 2009, called *IHY Global Campaign - Whole Heliosphere Interval*. Details on WHI including data sets are at: <ihy2007.org/WHI/WHI.shtml>.

Another part of IHY science is the cooperative initiative with the UNBSS program, through which the IHY assists in deploying arrays of small instruments to make global measurements. Fifteen instrument concepts have been developed and have been or are being deployed. These include a network of radio telescopes to observe CME-related radio bursts, chains of magnetometer arrays to observed magnetic activity, and hundreds of GPS receivers to observe the ionosphere. See: <ihy2007.org/observatory/observatory .shtml>. These systems have been discussed at annual IHY-UN UNBSS workshops in the United Arab Emirates (2005), Bangalore, India (2006), Tokyo, Japan (2007) and Sozopol, Bulgaria (2008). The last workshop will be held on Jejn Island in South Korea in September 2009. The IAU is a cosponsor of all of these meetings.

2.2. *IHY science meetings and workshops*

Many scientific meetings and workshops related to IHY have been held in 2007–2008 in countries including France, India, Germany, Bulgaria, Austria, Mexico, Italy, Ethiopia and Russia. Several are highlighted here. The first major European IHY conference *The Sun, the Heliosphere, and the Earth* was held in Bad Honnef, Germany, May 2007. The World Space Week was celebrated worldwide in October 2007 with the Sputnik 50th Anniversary Celebration and Symposium. The Second IHY SCINDA Workshop and IHY-Africa Space Weather Science and Education Workshop were held in Addis Ababa, Ethiopia in November 2007. The International IHY Symposium and Sputnik 50th Anniversary Celebration was held in Zvenigorod, Moscow, Russia in November 2007. IHY themes were involved in two regional African meetings in Cairo in 2008: the first Middle East-Africa Regional IAU Meeting (MEARIM) in April and the IAGA International Symposium *Space Weather and its Effects on Spacecraft* in October. IAU Symposium No. 257, involving IHY science topics and called *Universal Heliophysical Processes*, was held 15-19 September 2008 in Ioannina, Greece. See: <iau257.uoi.gr>. The organizers were N. Gopalswamy, D. F. Webb, K. Shibata and A. Nindos.

Future IHY meetings include an AGU Chapman Conference on *Universal Processes* to be held in Savannah, GA, USA, 10–14 November 2008 and organized by Nancy Crooker and Marina Galand. At this same time a symposium on 50 years after the IGY will be held in Japan organized by M. Kono, T. Iyemori and K. Yumoto, and a European IHY meeting *Heliosphere and its Environment* will be held at ISSI in Bern, Switzerland. Finally, the IHY-Africa 2009 Workshop will be the final official IHY workshop. It is to be held in Livingstone, Zambia in June 2009. IHY special sessions have also been organized at many periodic scientific meetings, including COSPAR, AGU, EGS, IUGG, SHINE, and SORCE.

2.3. *IHY outreach*

IHY Outreach activities include spreading knowledge of space science and exploration to the public and inspiring the next generation of space scientists, and are led by Cristina Rabello-Soares. There are now outreach coordinators in 25 countries. In Thailand an IHY booth was set up by the Thai IHY group during the annual Science and Technology Fair in Bang Na 8–18 August 2007 that was attended by about 1 million students. The Center for Science Education at the University of California, Berkeley, and the

Stanford Solar Center sponsored special web-based activities to celebrate World Space Week and the 50th Anniversary of the Sputnik Launch, October 2007. They also hosted an IHY Space Weather Monitor workshop at the IHY-Africa Space Weather Science and Education Workshop in Ethiopia in November 2007. They demonstrated Sudden Ionospheric Disturbance instruments that track changes to the Earth's ionosphere caused by solar activity. These are targeted to high school students, and are being distributed world-wide as part of the IHY International Education Program. IHY also supported the Geophysical Information for Teachers (GIFT) Workshop *The International Heliophysical Year* at Addis Ababa, Ethiopia, on 10 November 2007. More recent events included Solar Week 2008 and NASA's Sun-Earth Day in March 2008 and the annual Yuri's Night Space Party on 12 April 2008. A touring art exhibit on the Sun was established at Goddard Space Flight Center in Greenbelt, MD, USA and is called Sunworks. It was first displayed in Vienna at the opening ceremony of the IHY, and is currently touring the U.S.

2.4. *IHY schools*

Part of the IHY Outreach effort is the IHY Schools Program which is assisting with five schools in 2007–2008. The purpose of these schools is to educate students about heliophysics and Universal Processes. Updated information on the schools program is at: <ihy2007.org/outreach/ihy_schools.shtml>. The first school was held in July-August 2007 and co-sponsored by NASA's *Living With a Star* program and IHY as the North America IHY School. 34 students attended the school, 14 from countries outside the U.S., with 25 lecturers and lab coordinators participating. The first Asia-Pacific School was held at the Indian Institute of Astrophysics (IIA), Bangalore, India, 10–22 December 2007. It was an intensive two-week course on heliophysics topics with about 50 students attending. The first IHY Latin America School was held 14–20 February, 2008, and organized by CRAAM and held at the Presbyterian Mackenzie University in Sao Paulo, Brazil. About 80 students attended. The fourth school is also the second Asia-Pacific school and is planned for 20–31 October 2008 in Beijing, China. The last school, for Africa and Europe, will be held at the Centre for Basic Space Science, National Space Research and Development Agency (NASRDA), the University of Nigeria, Nsukka., Nigeria on 10–22 November 2008. These schools have been very successful and have helped to educate students and spread new knowledge about heliophysics and Universal Processes throughout the world.

2.5. *IHY Gold History*

Finally, the IHY Gold History initiative has the goals of identifying and recognizing participants in the first IGY, preserving memoirs, etc., of historical significance for the IGY, making them available to historians and researchers, spreading awareness of the history of geophysics, and planning special events. The first of these was the 'IGY+50' Celebration at the IUGG meeting in Perugia, Italy, 2–13 July 2007. During the IHY session at the 2007 *Solar Extreme Events* meeting in Athens, Greece, Nat Gopalswamy presented an overview talk on IHY activities. Over 200 individuals have now been recognized as members in the IHY Gold Club.

David F. Webb
IAU representative to the International Heliophysical Year

Transactions IAU, Volume XXVIIA
Reports on Astronomy 2006–2009
Karel A. van der Hucht, ed.

© 2009 International Astronomical Union
doi:10.1017/S1743921308025398

DIVISION III   PLANETARY SYSTEMS SCIENCES
SCIENCES DES SYSTÈMES PLANETAIRES

Division III's activities focus on a broad range of astronomical research on bodies in the solar system (excluding the Sun), on extrasolar planets, and on the search for life in the Universe.

PRESIDENT	Edward L. G. Bowell
VICE-PRESIDENT	Karen J. Meech
PAST PRESIDENT	Iwan P. Williams
BOARD	Alan P. Boss, Guy J. Consolmagno,
	Régis Courtin, Julio A. Fernández,
	Bo Å. S. Gustafson, Walter F. Huebner,
	Anny-Chantal Levasseur-Regourd,
	Mikhail Ya. Marov, Michel Mayor,
	Rita M. Schulz, Pavel Spurný,
	Giovanni B. Valsecchi, Jun-ichi Watanabe,
	Iwan P. Williams, Adolf N. Witt

DIVISION III COMMISSIONS

Commission 15	Physical Study of Comets and Minor Planets
Commission 16	Physical Study of Planets and Satellites
Commission 20	Positions and Motions of Minor Planets, Comets and Satellites
Commission 21	Light of the Night Sky
Commission 22	Meteors, Meteorites and Interplanetary Dust
Commission 51	Bioastronomy
Commission 53	Extrasolar Planets

DIVISION III SERVICE

Division III WG	Minor Planet Center

DIVISION III WORKING GROUPS

Division III WG	Committee on Small Bodies Nomenclature
Division III WG	Planetary System Nomenclature

INTER-DIVISION WORKING GROUPS

Division I-III WG	Cartographic Coordinates and Rotational Elements of Planets and Satellites
Division I-III WG	Natural Satellites

TRIENNIAL REPORT 2006 - 2009

1. Division III structure

There have been some changes to the structure of Division III. This triennium, Division III is pleased to welcome the new Commission 53 on *Extrasolar Planets*, formed at the Prague General Assembly. Commission 51 has changed its name to form Bio-Astronomy to *Bioastronomy*. The Division's Working Group on Near-Earth Asteroids has now become a working group of the Executive Committee, and has changed its name to *Hazards of Near-Earth Objects*. Division III's Working Group on *Natural Satellites* is now a joint working group with Division I. Note that the report of the other Division I–III joint working group (on *Cartographic Coordinates and Rotational Elements of Planets and Satellites*) is given in Division I's report. In terms of membership, Division III, with more than 1 000 members, is the third largest of the twelve IAU Divisions.

2. Division III developments within the past triennium

Much of the Division's activity has centered on the ramifications of the definition of a planet, as adopted at the IAU XXVI General Assembly in Prague. Discussion and action have mainly involved the Division's Board, the CSBN, and the WG-PSN (see also the reports of the CSBN and WG-PSN).

Of concern soon after the Prague General Assembly was the naming of dwarf planet (136199) 2003 UB_{313} and its satellite. Following requests by the discoverers, the former was named *Eris*, and the latter *Dysnomia*. Being only the third named dwarf planet, prior discussion involved the Division Board, the CSBN, and the WGPSN. An IAU press release in September 2006 led to much media coverage.

Dwarf planet terminology led to two separate discussions. Because of the grammatical inconsistency of a dwarf planet not being a planet (whereas, for example, a dwarf star *is* a star), there was an attempt to find a single word for dwarf planet. Candidate words *nanoplanet* and *subplanet* were discussed at length, but there was no consensus to approve either. The view that the term dwarf planet is now embedded in the public's consciousness appeared to win the day. The remaining discussion concerned finding another term for 'transneptunian dwarf planet'. As the CSBN's report notes, this discussion derived from the footnote to the Prague General Assembly's Resolution 6. The CSBN, after long discussion, settled on the word *plutoid*, and this word was adopted by the IAU Executive Committee at its May 2008 meeting on the recommendation of Division III's president. Unfortunately, in part because of an email miscommunication, the WG-PSN was not involved in choosing the word *plutoid*. The WG-PSN wishes to record that it did not take part in the process of accepting the term *plutoid* for a transneptunian dwarf planet. In fact, a vote taken by the WG-PSN subsequent to the Executive Committee meeting has rejected the use of that specific term. Thus, the views of its members were seriously misrepresented at the Executive Committee meeting and in the subsequent IAU press release in June 2008.

Because of perceived urgency in the naming of the large transneptunian objects (136472) 2005 FY_9 and (136108) 2003 EL_{61}, the Executive Committee decided at its May 2008 meeting to adopt the following recommendation: *Any solar system body having (a) a semimajor axis greater than that of Neptune, and (b) absolute magnitude brighter than $H = +1$ mag shall be considered for naming purposes to be a dwarf planet and named jointly by the WG-PSN and CSBN. Name(s) proposed by the discoverer(s) will be given deference.*

Note the following: An essential phrase to bear in mind is 'for naming purposes'. It is not the intention to declare that a body having $H < +1$ mag *is* a dwarf planet. The lower diameter limit of a body having $H = +1$ mag and geometric albedo $p = 1.0$ is $D = 850$ km. For $p = 0.65$ (like Pluto-Charon), $D = 1050$ km, and for very low $p \simeq 0.03$, $D \simeq 5000$ km. While it is likely that all solar system bodies having $H < +1$ mag are dwarf planets, one cannot at this time be certain (see Section 4 for remarks on quantifying the IAU definitions of planet and dwarf planet).

The issue of merging two pairs of Commissions (Commissions 15 and 20; and Commissions 21 and 22), mooted at the Prague General Assembly, has been discussed, but no recommendations have been developed at this time (July 2008).

3. Division III science highlights

In this section, we give thumbnail sketches of the Commissions and Working Groups that comprise Division III, and we highlight the principal scientific advances since 2006.

3.1. *Commission 15 on Physical Study of Comets and Minor Planets*

The Commission's work concerns asteroids, transneptunian objects, and comets. The *Deep Impact* and *StardustNExT* missions have returned results concerning mixing in precursor materials that formed small bodies. An unprecedented international coordination resulted in many of the world's telescopes being focused on the *Deep Impact* event. The impact showed the nucleus of 9P/Tempel 1 to be extremely porous, with a density of about 350 kg/m^3. Ice is close to the surface, but below a devolatilized layer of dust. Another notable event was the disintegration of P/Schwassmann-Wachmann 3 in 2006, the huge brightness outburst of 17P/Holmes in 2007, and a close approach of comet Tuttle, which allowed Doppler radar imaging of the nucleus. The Japanese spacecraft *Hyabusa* found the surface of asteroid (25143) Itokawa to be covered in boulders without regolith.

3.2. *Commission 16 on Physical Study of Planets and Satellites*

A broad array of observational and theoretical work, from both groundbased and space-based observations, constitutes the Commission's interest. Mercury *Messenger* flew by Mercury in January 2008. The spacecraft sampled the rich plasma environment, and found that many of the species represent the first look at Mercury's surface chemistry. Mercury's magnetic field is produced by an active dynamo in the outer core. Further, the dominant tectonic landforms, lobate scarps, indicate that cooling of the core was 1/3 greater than previously believed. Groundbased explorations of new terrains on Mercury have been made, especially at the planet's north pole. A new measurement of Mercury's libration parameters has resulted from radar speckle displacement interferometry, and implies a liquid phase in the planet's core. The *Cassini-Huygens* mission to the Saturn system (currently in extended mission mode) has greatly increased our knowledge of Saturn's atmosphere, and of the surfaces and atmospheres of some of its satellites. Titan's surface morphology has been mapped in detail, and water-vapor vents on Enceladus have been discovered.

3.3. *Commission 20 on Positions and Motions of Minor Planets, Comets and Satellites*

Dynamics and ephemerides of asteroids and comets are the purview of Commission 20. The exponential increase of asteroid astrometry has led to the continued growth in number and quality of asteroid orbits, both osculating orbits and proper elements. There are now more than 189 000 numbered asteroids. Research on the Yarkovsky effect continues apace, and a particularly elegant application of the effect led to the hypothesis that collisional disruption of the parent body of (298) Baptistina 160 My ago led to a long-term enhancement in Earth's asteroid impact flux and to the source of the Cretaceous-Tertiary impactor 65 My ago.

3.4. *Commission 21 on Light of the Night Sky*

The Commission's work concerns the study of the various components making up the light of the night sky, as seen from both Earth and space. These components include airglow and tropospheric scattering in Earth's atmosphere, the zodiacal light, the Milky Way's integrated starlight, scattered light from the diffuse interstellar medium, and the cosmic microwave background. Since 2006, the most significant research has concerned continued analysis of sky maps from the *Cosmic Background Explorer* - Diffuse Infrared Background Experiment (*COBE*-sc dirbe) and cosmic microwave background data from the *Wilkinson Microwave Anisotropy Probe* (*WMAP*). Thanks to such data, our understanding of the time since the big bang, the Hubble parameter, and the space-time geometry of the universe has made unprecedented leaps.

3.5. *Commission 22 on Meteors, meteorites and Interplanetary Dust*

The very smallest bodies in the solar system, along with their interactions with planets, are the focus of the Commission's work. The most important recent scientific highlight concerned the Carancas meteorite impact in Peru in September 2007. During the present triennium, advances have been reported at several international meetings. Topics have included the formation of meteoroid streams by active or dormant comets and asteroids, the dynamics of meteoroids, meteoroids as a space hazard, the coming of age of the techniques of infrasound and radar detection of meteors, and the discovery of new meteor streams. In addition, there has been good progress made on professional-amateur cooperation in meteors, and on meteor shower nomenclature.

3.6. *Commission 51 on Bioastronomy.*

The Commission focuses on the search for planets orbiting extrasolar stars, the search for extraterrestrial technological signals, the search for biologically relevant interstellar molecules and an investigation of the chemical pathways for their formation, the investigation of detection methods for evidence of biological activity, and means of defining habitability and habitable environments within our solar system.

3.7. *Commission 53 on Extrasolar Planets.*

This newly established Commission focuses on the search for and characterization of extrasolar planets, their formation and evolution, and the study of individual extrasolar planets' internal structure and atmospheres. Recently, a number of extrasolar planets in the Earth-Neptune mass range have been found. Transit search observations, both from the ground (for example, the HAT Network and WASP) and space (*COROT*, *Spitzer*) have matured greatly. There now exist a number of accurate mean density determinations, direct measurement of a few surface temperatures, and first results on exoplanet spectra. A working definition of exoplanets has been proposed.

3.8. *Committee for Small Body Nomenclature (CSBN).*

The Committee continued its work of naming comets (701 newly named in the past three years), asteroids (2228), and asteroid satellites (4). The CSBN has also been working to refine its guidelines for naming asteroids.

3.9. *Working Group on Planetary System Nomenclature (WG-PSN).*

The Working Group approves names of natural planetary satellites and surface features on all solar system bodies except Earth. Through a *Gazetteer of Planetary Nomenclature*, it establishes and maintains a list of new names, based on established guidelines. In the current triennium, the WG-PSN has approved 22 new satellite names and 259 new surface feature names. The Working Group has also clarified the definition of divisions and gaps in planetary rings.

3.10. *Working Group on Cartographic Coordinates and Rotational Elements of Planets and Satellites.*

The Working Group is responsible for recommending models of the orientations and shapes of the Sun, planets, natural planetary satellites, asteroids, and comets. The Working Group's report is given in full under Division I.

3.11. *Working Group on Natural Satellites.*

The Working Group's main goal is to encourage and gather astrometric observations, old and current, of planets, their natural satellites, and planetary rings; and to model those observations so as to provide accurate ephemerides and, where possible, to interpret orbital evolution. This triennium, significant progress has been made in almost all aspects of the WG's work. The *Galileo* and *Cassini-Huygens* missions have proved a rich source of data.

4. Anticipated Division III activity (2008 – 2009)

Two Division III Task Groups will be established. Their goal will be to quantify two of the components of the Prague GA's definition of a planet, as given in Resolution 5A; namely, that a planet has 'cleared the neighborhood around its orbit'; and that both planets and dwarf planets are sufficiently massive that their self gravity has overcome rigid body forces so they assume near-spherical shapes (are in near-hydrostatic equilibrium). At the time of the Prague GA, there were no (known) refereed publications that centrally addressed the above two criteria. Since then, Soter (2006, *AJ*, 132, 2513) has discussed planetary orbit clearing by accretion and ejection of lesser bodies; and Tancredi & Favre (2008, *Icarus*, 195, 851) have quantified the lower diameter limits for icy and rocky dwarf planets. The two TGs need to work in concert so they can try to agree on the lower mass limit for planets and the upper mass limit for dwarf planets. It is anticipated that the TGs will present their reports to Division III at the IAU XXVII General Assembly in August 2009.

5. Division III meetings at the IAU XXVII General Assembly

We are pleased to report that Division III's interests will be served by an Invited Discourse at the IAU XXVII Genaral Asembly in Rio de Janeiro, 2009. The tentative title is 'Water on planets'.

One of six 2009 IAU GA Symposia has been proposed by Division III members: IAU Symposium No. 263 on *Icy bodies of the solar system* is supported by Commisions 7, 15, 20, 22, and 51; and is to be co-chaired by H. Campins (USA), S. Ferraz-Mello (Brazil) and R. M. Schulz (Netherlands).

One of eight 2009 GA Special Sessions was jointly proposed by Divisions III and XII and supported by Commissions 16, 51, 53, and 55. The working title is *Planetary systems as sites for other life.*

Edward L. G. Bowell
president of the Division

Transactions IAU, Volume XXVIIA
Reports on Astronomy 2006–2009
Karel A. van der Hucht, ed.

© 2009 International Astronomical Union
doi:10.1017/S1743921308025404

COMMISSION 15

PHYSICAL STUDY OF COMETS AND MINOR PLANETS

ÉTUDE PHYSIQUE DES COMÈTES
ET DES PETITES PLANÈTES

PRESIDENT	Walter F. Huebner
VICE-PRESIDENT	Alberto Cellino
PAST PRESIDENT	Edward F. Tedesco
ORGANIZING COMMITTEE	Dominique Bockelée-Morvan, Dmitrij F. Lupishko, Yuehua Ma, Harold J. Reitsema, Rita M. Schulz, Gonzalo Tancredi

COMMISSION 15 WORKING GROUPS

Div. III / Commission 15 WG	Physical Study of Comets
Div. III / Commission 15 WG	Physical Study of Minor Planets
Div. III / Commission 15 TG	Comet Magnitudes
Div. III / Commission 15 TG	Asteroid Magnitudes
Div. III / Commission 15 TG	Asteroid Polarimetric Albedo Calibration
Div. III / Commission 15 TG	Geophysical and Geological Properties of Asteroids and Comet Nuclei

TRIENNIAL REPORT 2006 - 2009

1. Introduction

The Commission 15 report was prepared primarily by the chairpersons of the two working groups: T. Yamamoto of the Comet Working Group and R. A. Gil-Hutton of the Minor Planet Working Group. In particular, the Comet section was produced by T. Yamamoto with the assistance of D. Bockelée-Morvan, H. Kawakita, and D. Prialnik, while the Minor Planet section was produced by R. A. Gil-Hutton with the assistance of A. Cellino, A. W. Harris (DLR), R. Jedicke, A.-C. Levasseur-Regourd, R. M. Schulz, T. B. Spahr, and P. Vernazza. W. F. Huebner was responsible for the Introduction. The final editing and merging of the various sections and subsections of the report was carried out by the Commission Secretary, D. C. Boice.

Scientific activity in the field has continued to grow in the past three years, as evidenced by the large number of publications in the refereed literature. The publication list cannot be accommodated in the space at our disposal. We have therefore chosen to highlight a representative subset of these publications to provide a snapshot of the current state of the field, and, as in the last several reports, without including a comprehensive bibliography. Instead, a complete list of the references used in creating this report, assembled by searching the ADS abstract service <adsabs.harvard.edu/abstract_service.html>)

to generate a list of refereed papers published between July 2005 and June 2008, inclusive, is available in the Archive section of the Division III *Physical Studies of Comets and Minor Planets* web site. This site can be reached (since it does not have a permanent home) via a link from the IAU home page.

Three task groups have been set up in Commission 15. A Task Group on *Asteroid Magnitudes* is co-led by E. F. Tedesco and R. A. Gil-Hutton, a Task Group on *Asteroid Polarimetric Albedo Calibration* is co-led by A. Cellino and R. A. Gil-Hutton, and a Task Group on *Comet Magnitudes* is co-led by G. Tancredi and T. Yamamoto. A fourth Task Group for *Geophysical and Geological Properties of Asteroids and Comet Nuclei* is being set up and co-led by K. Muinonen, R. A. Gil-Hutton and T. Yamamoto. While all task groups are of general scientific interest, they are also of great importance for the development of countermeasures against potentially hazardous objects as reported by W. F. Huebner and L. N. Johnson at the Tunguska conference in Moscow, 26-28 June, 2008.

2. Comets

Over 1100 peer-reviewed papers were published in the period of this triennial report from July 2005 through June 2008 (this number is based on a query to the Astrophysics Data System, ADS, at: `<adsabs.harvard.edu/abstract_service.html>`). It illustrates that comet science is actively pursued internationally by researchers active in the field.

2.1. *Comet nuclei*

The *Deep Impact* mission, during which a spacecraft collided with, and excavated the nucleus of Comet 9P/Tempel 1 on 4 July, 2005 (A'Hearn *et al.* 2005), was undoubtedly the major event and the main source of information on comet nuclei in the past three years. The impact showed the cometary nucleus to be very weak on scales ranging from the impactor diameter ($\sim 1\,\mathrm{m}$) to the crater diameter ($\sim 100\,\mathrm{m}$) and suggested that the strength might be low on much smaller scales as well. It also showed the cometary nucleus to be extremely porous (with a density of only $350\,\mathrm{kg\,m^{-3}}$) and the ice to be close to the surface, but below a devolatilized layer. The ambient observations showed a huge range of topography, implying ubiquitous layering on many spatial scales, frequent (more than once a week) natural outbursts, many of them correlated with the spin phase, a nucleus surface with many features that are best interpreted as impact craters, and clear chemical heterogeneity in the outgassing from the nucleus. The mission prompted a large number of studies – observational, phenomenological, and theoretical – in the endeavor to interpret and understand the large body of accumulated data. The shape, topography, temperature distribution, spin state, composition, and activity pattern of the Tempel 1 nucleus were analyzed and discussed in a long series of papers published in *Icarus* in two dedicated volumes (Vol. 187, issue 1, March 2007, pp. 1–356, and Vol. 191, issue 2, October 2007, pp. 283–674).

Another well studied comet nucleus was 67P/Churyumov-Gerasimenko, the new target of the *Rosetta* mission. Its shape, size, spin, and density were derived by Lamy *et al.* (2007) from light-curve observations and non-gravitational force modeling.

A new catalog of nuclear magnitudes, size distribution, active area fractions, and production rates of Jupiter-family comets was compiled by Tancredi *et al.* (2006). Spin rates and colors of a number of Jupiter-family comets were obtained by Snodgrass *et al.* (2006) and compared with Kuiper Belt objects, from which these comets are believed to originate.

The origin and early evolution of comet nuclei was studies in a series of papers published in April 2008 in a dedicated volume of the Space Science Reviews. Comet 81P/Wild 2 nucleus samples returned to Earth by the *Stardust* spacecraft pointed to a wide range of formation conditions, probably reflecting very different formation locations in the protoplanetary disk (Zolensky *et al.* 2006). As a result of the recent close observations of comet nuclei (Borrelly, Wild 2, and Tempel 1), the old 'rubble-pile' comet nucleus model was revised and constrained (Basilevsky & Keller 2006) and a new model for the interiors of Jupiter family comet nuclei was proposed by Belton *et al.* (2007), called the 'talps' (or 'layered pile' model), in which the interior consists of a core overlain by a pile of randomly stacked layers, primordial remnants of the early agglomeration phase.

The major nucleus outburst event during this period was that of Comet 17P/Holmes in October 2007 (Sekanina 2007), an unusual outburst both in timing and in scale: the nucleus brightened by about 9 magnitudes, at a rate of 0.5 mag/hr, at a distance of 2.4 AU on the outbound leg of the orbit. The cause of the outburst is still unknown (Moreno *et al.* 2008), but outburst observations led to the determination of the chemical composition of the nucleus (Dello Russo *et al.* 2008 and references therein; Bockelée-Morvan *et al.* 2008). Outbursts and evidence of splitting were detected in 2006 for fragments of the nucleus of Comet 73P/Schwassmann-Wachmann 3, which had split into several pieces in 1995 (Tetsuharu 2007; Bonev *et al.* 2008; Ho *et al.* 2008). Mechanisms for cometary outbursts in general were discussed by Gronkowski (2005a, b; 2007a, b) in a series of papers, and by Sekanina (2007).

Modeling of comet nuclei focused on quasi 3-D calculations and numerical approaches to non-spherical shapes (Davidsson *et al.* 2005; Skorov *et al.* 2006; Ivanova & Shulman 2006; Kossacki *et al.* 2006, DeSanctis *et al.* 2007). In addition, the first fully 3-D comet nucleus model was developed (Rosenberg & Prialnik, 2007). The topic of heat and gas diffusion in comet nuclei was reviewed, discussed and analyzed in a monograph published by ESA for the International Space Science Institute (Huebner *et al.* 2006).

2.2. *Gas coma, chemistry, plasma, and tails*

The study of compositional diversity among comets motivated several observational campaigns. In 2005 - 2008, many studies focused on the long-period Comets C/2001 Q4 (NEAT), C02002 T7 (LINEAR), C/2004 Q2 (Machholz), and C/2006 P1 (McNaught), which made spectacular apparitions, and on the short-period Comets 73P/Schwassmann-Wachmann 3 (SW3) and 8P/Tuttle, which made a close approach to Earth, and on 17P/Holmes, which underwent a mega-outburst in October 2007. Observations of the gas coma of 9P/Tempel 1 were carried out in support to the *Deep Impact* mission and are reported in a special issue of Icarus (see above).

An important result is the remarkable similarity of composition of fragments B and C of Comet SW3, in contrast to the diversity of the overall comet population, and the peculiar chemistry of this comet showing strong depletions for many species (Dello Russo et al. 2007; Kobayashi *et al.* 2007; Villanueva *et al.* 2007; Lis et al. 2008). The peculiar compositions of Comets C/2001 A2 (LINEAR), 96P/Machholz, and 8P/Tuttle provide also new evidence for chemical diversity (Biver *et al.* 2006; Langland-Shula *et al.* 2007; Magee-Sauer *et al.* 2008; Bonev *et al.* 2008). Direct measurements of the NH_3 abundances have been obtained in several comets (Biver *et al.* 2007; Magee-Sauer *et al.* 2007).

New spectral emission lines were also detected in cometary spectra, e.g., rotationally resolved H_2 transitions originating from the highly excited rovibrational levels of the $X^1\Sigma_g^+$ state using *FUSE* (Liu *et al.* 2007). Detailed modeling of the $A^1\Pi - X^1\Sigma^+$ system of CO that includes opacity effects has been performed to analyze *HST* observations of several recent comets in the UV. No evidence for distributed sources of CO was found (Lupu *et al.* 2007). Spectroscopic observations of cometary gases provided information on physical coma parameters such as gas temperature, velocity, and distribution (e.g., Tseng *et al.* 2007; Bonev *et al.* 2007; Boissier *et al.* 2007).

Several isotopic measurements were obtained: $^{16}O/^{18}O$ in water (Biver *et al.* 2006b), $^{14}N/^{15}N$ and $^{12}C/^{13}C$ in CN and HCN. The $^{14}N/^{15}N$ ratio, observed now in a dozen comets, clusters around 140 (e.g. , Hutsemékers *et al.* 2005). A similar ^{15}N enrichment with respect to the Earth value ($^{14}N/^{15}N = 272$) has been measured in Comet 17P/Holmes, which is compatible with a production of CN from HCN (Bockelée-Morvan *et al.* 2008). This high enrichment is certainly indicative of fractionation processes that occurred in the solar or presolar nebula. Other constraints on the origin of cometary material may come from the accumulating measurements of the nuclear spin temperatures of H_2O, NH_3, CH_4 and CH_3OH (e.g. , Kawakita *et al.* 2006; Bonev *et al.* 2007; Pardanaud *et al.* 2007; Woodward *et al.* 2007).

Chemical modeling of cometary atmospheres remains the topic of several papers. A complex network involving C_2H_2, C_2H_6 and C_3H_4 is proposed to explain the spatial distribution of the C_2 and C_3 radicals in Comet Hale-Bopp (Helbert *et al.* 2005). The chemistry of C, H, N, O, and S compounds corresponding to ions of masses smaller than 40 amu in the inner coma of 1P/Halley is investigated in details by Haider & Bhardwaj (2005). Noteworthy are recent hydrodynamic simulations of the gas coma which show that gas structures produced by nucleus composition inhomogeneities and nucleus shape and topography are indistinguishable (Zakharov *et al.* 2008).

The first direct imaging of the interaction a comet (2P/Encke) with a coronal mass ejection was obtained with high temporal and spatial resolution with the SECCHI Heliospheric Imager-1 (HI-1) aboard the *STEREO* mission, and strongly supports the idea that large-scale tail disconnections are magnetic in origin (Vourlidas *et al.* 2007). Frequent disruptions in the plasma tails of Comets C/2001 Q4 (NEAT) and C/2002 T7 (LINEAR), due to ubiquitous solar wind flow variations, were observed by the Earth-orbiting *Solar Mass Ejection Imager* (*SMEI*) (Buffington *et al.* 2008).

The encounter of the *Ulysses* spacecraft with the ion tail of Comet C/2006 P1 McNaught revealed the unexpected presence of magnetic turbulence and energetic ions at 1.6 AU from nucleus, and the presence of O^{3+} ions for the first time in a comet (Neugebauer *et al.* 2007). Observations of Comet McNaught with the Heliospheric Imager on *STEREO* support the presence of an iron tail in this comet (Fulle *et al.* 2007), in addition to a sodium tail (Leblanc *et al.* 2008). A number of MHD models have been developed to understand disconnection events and to analyze or prepare space investigations of solar wind interactions with cometary plasmas (e.g. Benna & Mahaffy 2006; Jia *et al.* 2007; 2008; Ekenbäck *et al.* 2008; Delamere 2006).

2.3. *Comet dust and distributed sources*

Our understanding of the complexity of cometary dust has been significantly progressing during the past triennium: The *Stardust* mission, brought for the first time dust samples from the coma of a comet (81P/Wild 2) and the *Deep Impact* mission released dust for remote observations from the subsurface of another comet (9P/Tempel 1). These missions have been complemented by ground observations, i.e., spectroscopic observations providing information about the composition of the dust and polarimetric observations providing information about the physical properties of the dust.

Numerous observations of Comet 9P/Tempel 1, in relation to the *Deep Impact* mission, have suggested the presence in the coma of low velocity, large, absorbing dust particles, which induced a high polarization and were probably covered by carbonaceous compounds (Farnham *et al.* 2007; Hadamcik *et al.* 2007). Immediately after impact, the polarization in the innermost coma decreased and a negative polarization color spectral gradient was noticed, most likely due to a drastic change in the composition of the dust ejected by the nucleus (Harrington *et al.* 2007). The event actually prompted the ejection of high speed, low albedo and high polarization dust particles from the surface, as well as of lower speed, higher albedo and slightly lower polarization particles coming from the subsurface (Sugita *et al.* 2005; Furusho *et al.* 2007). Observations near 11 μm gave clues to the presence of amorphous and crystalline olivine and pyroxene after impact (Harker *et al.* 2007).

Analyses of the samples from the *Stardust* mission have shown the presence of a very wide range of olivine and low Ca-pyroxene compositions, possibly reflecting various formation regions in the protoplanetary nebula, and of refractory organic compounds (Zolensky *et al.* 2006; Flynn, 2008). Observations of the craters on aluminum foils and tracks in aerogel indicate that both aggregates of submicrometer grains and compact particles with sizes up to a few tens of microns were present in the coma of the comet (Hörz *et al.* 2006; Burchell *et al.* 2007).

Such a structure was previously suggested from numerical and laboratory simulations of the polarimetric observations of Comet C/1995 O1 Hale-Bopp and other high-polarization comets. An excellent match was found for samples, of tens to hundreds of micrometer sized, fragile, fluffy particles corresponding to mixtures of silica-rich, magnesio-silica and ferrosilica with fluffy carbon, for an average grain size in the 0.1 μm size-range or with a few large compact grains added (Lasue & Levasseur-Regourd 2006; Hadamcik *et al.* 2007). Also, the ensemble of results obtained for interplanetary dust indicate that its light scattering and thermal properties stem from the presence of compact and fluffy particles, with compositions ranging from silicates to more absorbing materials, with a decreasing contribution of the latter with decreasing solar distance (Levasseur-Regourd *et al.* 2007).

Another important event, as far as light scattering by cometary dust is concerned, was related to polarimetric observations of the fragments of Comet 73P/Schwasmann-Wachmann 3. A high polarization, indicative of continuous fragmentation of dust particles was pointed out, together with a lower polarization in the vicinity of fragment B, with a possible inversion of the spectral dependence of the polarization (Hadamcik & Levasseur-Regourd 2007; Jones *et al.* 2008; Bonev *et al.* 2008). Infrared observations from the *Spitzer* telescope revealed silicate emission features similar to those usually noticed in the comae of active comets (Sitko *et al.* 2006).

2.4. *Cometary materials origins and laboratory experiments*

Huebner (2008) reviewed the origin of cometary materials based on recent observations and theoretical studies. Measurements of the isotopic ratios and the nuclear spin temperatures in molecules are important to investigate origin of cometary materials (see section on Gas coma, chemistry, plasma, and tails). Horner *et al.* (2007) discussed constraints on the formation region of comets from their D/H ratios. Relating to the origin of cometary materials, mechanisms of the outward transport of materials in the solar nebula are investigated theoretically by, e.g., Boss (2007, 2008), Ciesla (2007), and Mousis *et al.* (2007). Such outward transport is necessary to explain co-existence of icy materials with high-temperature processed materials like crystalline silicates and CAIs in cometary grains (see Section Comet dust and distributed sources).

Many laboratory studies were carried out. Bar-Nun *et al.* (2007a; 2007b) measured the efficiency of gas (CO, Ar, and N_2) trapping into amorphous water ice and physical properties of cometary ice analogues at low temperatures.

Measurements of transition frequencies (or wavelengths) as well as transition strengths are also important to investigate molecules in cometary comae. The pure rotational spectrum of anti-ethylamine ($CH_3CH_2NH_2$) from 10 to 270 GHz was studied in the laboratory by Apponi *et al.* (2008). Vander *et al.* (2007) reported a new set of line parameters for the transitions of C_2H_6 at $12\,\mu$m.

Photodesorption from ices of astrophysical relevance (Thrower *et al.* 2008) and desorption energies of H_2O, CH_3OH, and NH_3 for pure ices of these molecules (Brown & Bolina, 2007) were also investigated experimentally. Furthermore, many experimental studies were performed on interstellar or cometary ice analogues (made from simple molecules such as H_2O, CO_2, and CO) irradiated by high-energy particles like UV-photons, electrons, and ions. Complex molecules could be formed in these ices at low temperatures (e.g. , Baratta *et al.* 2008; Bennett & Kaiser 2007; Bennett *et al.* 2006, 2007; Brucato *et al.* 2005, 2006), while simple molecules or atoms were desorbed from the icy surfaces in some cases (e.g., Zheng *et al.* 2006; Bennett & Kaiser 2005). Charge exchange cross sections for highly ionized C, O, and Ne on H_2O, CO, and CO_2 were measured to investigate X-ray emission lines from comets by Mawhorter *et al.* (2007).

3. Asteroids, Trojans, and Centaurs

Sometimes, many detailed considerations may be nicely summarized by one simple number. In the case of this report, this number is 808. This is the number of peer-reviewed papers published between June 2005 and June 2008, having as subject the asteroids (resulting from a query to the Astrophysics Data System, ADS, at the address `<adsabs.harvard.edu/abstract_service.html>`). This number clearly demonstrates that asteroid science is experiencing a phase of active development, and the number of scientific articles produced by many teams active in the field is constantly increasing. On the other hand, this also means that it is simply impossible to produce an exhaustive list of relevant papers to be explicitly quoted in this document. For this reason, this section is mostly focused on very general facts and achievements.

The past triennium has seen one major accomplishment, namely the rendezvous of the Japanese space probe *Hayabusa* with the near-Earth asteroid (25143) Itokawa, and the subsequent experiment aimed at collecting a sample of material from the body's surface, to be sent back to Earth. While the latter part of the mission is still in progress, the detailed *in situ* exploration of this tiny object has produced many important results in terms of surface texture, regolith properties, and likely overall structure of this object.

The availability of increasingly larger telescopes and improved detectors has made possible important developments in the studies of the most distant asteroidal bodies, including Jupiter Trojans and Centaurs. In particular, important spectroscopic campaigns have been carried out by different teams. Other exciting results have been produced by high-resolution imaging techniques, which have led to the discovery of many new binary systems among asteroids and Trans-Neptunian objects, including the discoveries of the triple systems whose primary components are (45) Eugenia and (87) Sylvia.

In addition to the above topics, very important results have been obtained in a variety of studies dealing with physical and dynamical properties of the asteroid population. They will be mentioned in the following subsections.

3.1. *Photometry, shapes, and spin properties*

The last triennium has seen the developments of important techniques of inversion of lightcurves, as well as of sparse photometric data. These techniques will be extremely important in the forthcoming years, when next-generation sky surveys like PAN-Starrs and *Gaia* will start to produce huge amounts of data. In addition, the Yarkovsky - O'Keefe - Radzievskii - Paddack (YORP) effect has been widely investigated. This is a thermal radiation torque that can sensibly alter the spin state of small asteroids, and indirectly affect also the rate of the Yarkovsky-driven drift in semi-axis, with important consequences for the supply of near-Earth objects.

3.2. *RADAR, thermal IR, optical polarimetry, and light-scattering phenomena*

The period of this report has seen the publication of a large number of studies using thermal-infrared (IR) techniques to derive the sizes and albedos of various types of minor planets, including four asteroids and inactive comets (Kraemer *et al.* 2005); 25 asteroids in comet-like orbits (Fernandez *et al.* 2005); three Jupiter Trojans (Emery *et al.* 2006); the potential spacecraft target (10302) 1989 ML (Müller *et al.* 2007); 42 Centaurs and KBOs from *Spitzer Space Telescope data* reviewed by Stansberry *et al.* (2008); three Mars Trojans (Trilling *et al.*; 2007), four near-Earth asteroids (Wolters *et al.* 2008); and the Rosetta fly-by target (21) Lutetia (Müller *et al.* 2006; Carvano *et al.* 2008). A review of *Spitzer Space Telescope* observations of small bodies was given by Fernandez (2006).

The surface properties of small asteroids from thermal-IR observations were reviewed by Harris (2006). Comparison of thermal IR results with those from radar provides a valuable control on the accuracy and applicability of these techniques in different circumstances, and greater insight into the physical properties of the target objects. Objects studied in detail by means of both radar and thermal-infrared observations include the *Hayabusa* mission target (25143) Itokawa (Ostro *et al.* 2005; Müller *et al.* 2005); (1980) Betulia (Harris *et al.* 2005; Magri *et al.* 2007); binary NEO 2002 CE26 (Shepard *et al.* 2006); (33342) 1998 WT24 (Harris *et al.* 2007; Busch *et al.* 2008); and (2100) Ra-Shalom (Shepard *et al.* 2008), for which sizes, albedos, and information on surface properties were obtained.

Surface thermal inertia and conductivity are important parameters that provide information on the nature of regolith and govern the magnitude of the Yarkovsky and YORP effects. Delbo' *et al.* (2007) carried out detailed thermophysical modeling to derive the first representative value for the thermal inertia of NEOs of 200 ± 40 $\mathrm{Jm^{-2}s^{-0.5}K^{-1}}$. A larger value of 750 $\mathrm{Jm^{-2}s^{-0.5}K^{-1}}$ was reported for (25143) Itokawa by Müller *et al.* (2007), who discuss the radar and thermal IR results for Itokawa in the light of the ground-truth data from the *Hayabusa* mission.

Polarimetry is also useful for asteroidal studies, since it provides information on physical parameters such as albedo and regolith porosity. A few objects exhibiting fairly exotic polarimetric properties have been pointed out (e.g., Barbara *et al.* 2006). Surface properties of Centaurs and Kuiper Belt objects have been tentatively estimated from near back scattering photometry and polarimetry, as reviewed in Belskaya *et al.* (2008).

3.3. *Spectra, taxonomy, composition, and space weathering*

Spectroscopic observations of asteroid families have been carried out in the visible and the NIR wavelength range, measurements in the latter being a premiere in this 'family-fiel. Vernazza *et al.*'(2006a) observed members of the 5.8 Myrs old Karin family and found (*i*) little spectral variation, thus suggesting a relatively homogeneous parent body; and (*ii*) a low degree of spatial alteration for the observed surfaces in agreement with the young age of the family. Moreover, the largest member of the family, 832 Karin, was found to be homogeneous throughout its rotation (Chapman *et al.* 2007; Vernazza *et al.* 2007). In the same young family register, Willman *et al.* (2008) found that the 1 to 5 Myrs old S/K-type Iannini family members have spectral slopes compatible with those of the Karin family. Mothé-Diniz *et al.* (2008) obtained NIR spectra for 30 members of the Eos family and found a spectral diversity that may suggest a slightly differentiated parent body.

Interesting is the discovery of V-type asteroids in the middle main belt (Roig *et al.* 2008). They predict that some middle main belt V-types may originate from Vesta (up to 30%) but that most of them must have a different origin. Sunshine *et al.* (2008) discovered CAI-rich bodies (their CAIs abundance being 2-3 times that of any meteorite); they suggest that these bodies are more ancient than any known meteorite. On the basis of four S-type spectra, Hardersen *et al.* (2006) suggest that these objects experienced partial melting temperatures during their formation. Spectroscopic studies in the NIR have also looked at the spectral properties of X- and M-type asteroids (Birlan *et al.* 2007; Ockert-Bell *et al.* 2008). Sunshine *et al.* (2007) compared the spectral properties of olivine-rich asteroids and meteorites. Interesting for the Dawn mission are the following two reports: (*i*) Rivkin *et al.* (2006a) report the absence (at a 1% level) of a 3 μm water band on Vesta throughout its rotation; and (*ii*) Rivkin *et al.* (2006b) suggest the presence of carbonates and iron-rich clays on Ceres' surface. New mid-IR asteroid spectra by Lim *et al.* (2005) have shown that a SNR higher than 50-100 is necessary for unambiguously detecting emission features. Emery *et al.* (2006) performed mid-IR spectroscopy of three Trojans with the *Spitzer* telescope and suggest the presence of fine-grained silicates on their surfaces.

Space weathering (SW) processes have been intensively studied over the past three years. Laboratory experiments have simulated both solar wind implantation and dust bombardment on silicate-rich materials. It has been shown that both processes tend to redden and darken the reflectance spectra of (*i*) silicates such olivine, ortho- and clino-pyroxene (Brunetto & Strazzulla 2005; Marchi *et al.* 2005; Brunetto *et al.* 2006a; Loeffler *et al.* 2008); and (*ii*) meteorites such as OCs and HEDs (Strazzulla *et al.* 2005; Vernazza *et al.* 2006b). Recent *in-situ* measurements performed on the S-type NEA (25143) Itokawa validated these experiments (Saito *et al.* 2006; Abe *et al.* 2006; Hiroi *et al.* 2006; Ishiguro *et al.* 2007): Hiroi *et al.* (2006) found that a dark region on the small (550 meter) asteroid (25143) Itokawa is significantly more space-weathered (spectrally redder) than a nearby bright region.

Brunetto *et al.* (2006b) created a space weathering model that reproduces the spectral reddening originating form ion irradiation. This model enables the use of scattering laws (Hapke & Shkuratov models) for the interpretation of remote sensing data.

Vernazza *et al.* (2006b) exposed a eucrite meteorite, thought to originate from Vesta, to ions in the laboratory to show that its parent should indeed be substantially more weathered than it appears. They suggest that Vesta must have a magnetic field of at least 0.2 microtesla at its surface, a few hundred times smaller than Earth's field, which diverts the damaging ions.

Finally, and in relation to space weathering effects, (*i*) a general spectral slope-exposure relation for S-type main melt and near-Earth asteroids has been established (Marchi *et al.* 2006a, b); and (*ii*) a general SW model has been proposed (Lazzarin *et al.* 2006; Paolicchi *et al.* 2007).

3.4. *Space missions*

Two space mission events highlighted this triennium, the *Deep Impact* mission to Comet 9P/Tempel 1 and the first coma dust sample return from Comet 81P/Wild 2 by the Stardust mission (the comet encounter occurred in 2004). The two missions resulted in a large number of publications, which were combined, for the most part, into special issues preceded by overview and introductory papers. For *Deep Impact* see: A'Hearn *et al.* (2005), A'Hearn & Combi (2007a, b, c), and A'Hearn (2008). For *Stardust* see Brownlee *et al.* (2006).

To enhance the scientific return of the *Deep Space 1* mission a coma model of Comet 19P/Borrelly during encounter was developed (Boice & Wegmann 2007). The first comprehensive description of the *Rosetta* mission to Comet 67P/Churyumov-Gerasimenko was provided in a special issue (see Glassmeier *et al.* 2007, for the mission summary paper).

Photometric and polarimetric observations were obtained in support of the Hayabusa mission to asteroid (25143) Itokawa (Cellino *et al.*, 2005; Thomas-Osip *et al.*, 2008). Several studies were published about what advances in asteroid science should be expected from the Gaia mission; see Mignard *et al.* (2007), Mouret *et al.* (2007) and Cellino *et al.* (2007) for the latest overviews. Regular updates of the status of the Dawn mission were published by Russell *et al.* (2006, 2007a, b).

3.5. *Near Earth Objects (NEO)*

In the time period covered by this triennial report 2010 new NEO discoveries were reported to the Minor Planet Center. Of these objects 176 are larger than 1 km in diameter (assuming a reasonable albedo) and 255 are Potentially Hazardous Objects (PHO). PHOs have a maximum close approach to the Earth's orbit (not necessarily the Earth itself) of 0.05 AU and therefore pose a larger impact risk to the Earth than the remainder of the NEO population. The Catalina Sky Survey telescopes now dominate the discovery statistics reporting about 2/3 of all new objects.

In response to the 2005 George E. Brown *Near-Earth Object Survey Act*, NASA convened a study team to assess the Earth impact risk and capabilities of future Earth and space-based NEO surveys. The objectives of the act are a new survey to detect, track, catalogue, and characterize 90% of NEOs that are larger than 140 m in diameter in 15 years. The NASA report (White Paper to the US Congress) is available at <www.nasa.gov/pdf/171331main_NEO_report_march07.pdf>.

The relatively recent idea that asteroids may have internal structures more like rubble piles than solid rock was spectacularly demonstrated by images from the Japanese *Hayabusa* spacecraft of asteroid (25143) Itokawa. The 2 June 2006 issue of Science highlighted the mission's results.

Other notable results include: The first laser guide star adaptive optics imaging of an NEO (Bush *et al.* 2007), the first direct detection of asteroid spin-up rates due to the YORP effect (Lowry *et al.* 2007), and the ongoing discovery of multiple NEO systems and unusual NEO shapes using radar imaging techniques (e.g., Ostro 2006). A report about PHO countermeasures was presented by Huebner *et al.* at the Tunguska conference in Moscow, June 2008.

Walter F. Huebner
president of the Commission

Transactions IAU, Volume XXVIIA
Reports on Astronomy 2006–2009
Karel A. van der Hucht, ed.

© 2009 International Astronomical Union
doi:10.1017/S1743921308025416

COMMISSION 16

PHYSICAL STUDY OF PLANETS AND SATELLITES

ÉTUDE PHYSIQUE DES PLANÈTES ET DES SATELLITES

PRESIDENT   Régis Courtin
VICE-PRESIDENT   Melissa A. McGrath
PAST PRESIDENT   Guy J. Consolmagno
ORGANIZING COMMITTEE   Carlo Blanco, Leonid V. Ksanfomality,
Luisa M. Lara, David Morrison,
John R. Spencer, Victor G. Tejfel

TRIENNIAL REPORT 2006 - 2009

1. Introduction

This report is divided in four parts: the first part summarizes the activities of the Commission between September 2006 and June 2008; the second part reports on recent advances in the physical study of planets and satellites. However, instead of attempting to cover the large body of new knowledge gathered over the last three years, we have chosen to highlight just a few exciting results – on Mercury, the exploration of unchartered terrains with ground-based imaging and a new measurement of its libration parameters, some spectacular findings from the *Cassini* mission inside the Saturnian system, and the results of methane-band spectrophotometric monitoring of Saturn over the last 13 years; the third part summarizes future plans now being drawn by the various space agencies for the exploration of planets and satellites in the solar system; the last part tries to project the activities of the Commission over the period June 2008–August 2009, and to express a few thoughts concerning the future developments in the field, and the role of the Commission therein.

2. Activities during the past triennium (up to June 2008)

Members of the Organizing Committee met during the Annual Meeting of the Division for Planetary Sciences of the American Astronomical Society held in Orlando in October 2007. This was our first meeting since the IAU XXVI General Assembly in Prague. Besides regular organizing matters, we discussed the opportunity of proposing a Special Session at the next General Assembly (see below), as well as a recurrent concern which is the visibility of the Commission among the planetary science community, and especially the younger cohort. One idea put forward was to organize an informal event tentatively entitled *Meet the IAU/Commission 16* at the next AAS/DPS meeting to be held in Ithaca in October 2008, to present the objectives and activities of the Commission (and of the IAU in general) to those who are less informed about them. Another recurrent topic which is discussed is the idea of communicating with the membership through a regular electronic bulletin. There was a timid attempt at doing so with the first edition of *PS² News* in May 2007. However, this attempt has not been re-iterated, as the incoming of publishable material was rather low. Nevertheless, the project is still on the table and will eventually be addressed in other ways in the future.

In response to the call for Scientific Meetings to be held during the IAU XXVII General Assembly in Rio de Janeiro, the Organizing Committee, in coordination with the Organizing Committees of Commission 51 on *Bioastronomy - Searchfor Extraterrestrial Life* and Commission 53 on *Extrasolar Planets*, put together a proposal for a Special Session devoted to planetary systems seen in the light of the emergence and sustainability of life. This is a rapidly evolving field for which a review of the most recent results would be of great interest to many of our colleagues attending the General Assembly. Moreover, one of the objectives of this meeting (and not the least) is a strengthening of the relationships between the three Commissions dealing with planetary systems within Division III. The Executive Committee has responded favorably to our proposal, advising us to merge with another similar proposal emanating from Commissions 51, 53, and 55 (*Communicating Astronomy with the Public*). Undoubtedly, the individuals involved in these two proposals will work together to design an exciting session in Rio de Janeiro.

As part of his duties, the President of the Commission participated in the activities of the Working Group for *Planetary System Nomenclature* (see the report from the WG-PSN) and responded to the call for ideas emanating from the Executive Committee about the opportunity of coining a new term replacing *dwarf planet*. In May 2008, the term *plutoid* was adopted by the Executive Committee to designate transneptunian dwarf planets, as required by Resolution B6 voted by the IAU XXVI General Assembly. We deplore, however, that a breakdown in communication led to this decision without formal votes by the WG-PSN and the Board of Division III.

3. Recent advances in the study of planets and satellites

3.1. *Mercury: Ground-based surface imaging and new libration parameters*

The work of Ksanfomality & Sprague (2007) presents results from the processing of images of Mercury acquired during the 2002 spring (evening) elongation to map districts containing the largest impact craters on the surface of Mercury. These impact craters show ringed structures with and without a system of rays. Craters are found in practically all areas of the 210-290°W sector. In all versions of the synthesized images, the Northern horn near the pole shows a large shock crater of about 280 km in diameter, with an extensive rim of debris and a 90 km wide dark central spot. The latitude of this large crater is 85◦N. Closer to the North pole by 2–3° is a smaller crater of about 60 km in diameter. At the time of the observations, the North pole was off by a few degrees from the limb, and was therefore in view. Because of the proximity to the pole and some blurring on the limb, the longitudes of craters near the pole are determined with a probable error no better than 5°. Nevertheless, their positions could be compared to independent data, i.e., a detailed radar map showing a group of large impact craters, with a resolution of up to 1.5 km (Harmon *et al.* 2001). Apparently, these craters are also visible in our synthesized images. The comparison with the radar map shows that the size and position of the 85°N crater coincide almost exactly with a crater in the radar map centered at 85.5°N, 292°W, with a diameter of about 80 km. Interestingly, the rim of debris on the radar map is seen only on the Northern side of the crater, without much detail, while in the optical image, a terrace of debris surrounds the crater, and the rim is visible as a bright formation. According to Harmon *et al.* (2001), craters at the pole are unusual in that at their bottom, under a layer of regolith, are apparently large amounts of ice. The bottom of these polar craters always remains in shadow and the temperature is low enough that ice could have subsisted on a cosmogonic timescale. Another hypothesis is that elemental sulfur, which in the decimeter radar range has scattering characteristics similar to that of ice, has been deposited at the bottom of these craters (Sprague *et al.* 1995).

As mentioned in previous works, expositions at morning elongations much improve the resolution of astronomical images. Although the use of short exposures to increase the resolution of astronomical images has existed for a long time, only the advent of high quantum efficiency CCD devices made it possible to obtain well-resolved images of Mercury, by processing a large number of electronic short-exposition frames. This is illustrated by new images of the longitude sector 270–330°W, which cover significant portions of Mercury's surface not imaged by *Mariner 10*. They seem to confirm that extended relief features are asymmetrically distributed on the surface of the planet (Ksanfomality 2005). Also, during the May and November 2007 elongations, new observations of Mercury were carried out at the SAO Observatory of the Russian Academy of Science, allowing to further document areas not imaged by *Mariner 10*. A continuous sequence of video frames was acquired with real-time A/D conversion and recording. The analysis of these new data is in progress.

On the basis of the Radar Speckle Displacement Interferometry (RSDI) method suggested by Holin (1988, 2004) , Margot *et al.* (2007) used the 'Goldstone/Green Bank' interferometer to derive the inclination and amplitude of the 88 d libration of Mercury from the radar-location components of its instant spin-vector. Simultaneously, they checked for the conformity of Mercury to the Cassini state 1 (no significant deviation was found). The values obtained for the inclination and amplitude are respectively 2.11'±0.1' and 35.8"±2" (compared to a previous determination of 60"±6"). Based on these results, they concluded that the libration of the shell occurs separately from the core, which would imply the presence of a liquid phase in the core (Margot *et al.* 2007). According to Holin, however, the precision of the amplitude determination is at odds with the theoretical limit he derived earlier. Therefore, this work should be carefully verified. The work of Margot *et al.* (2007) is the first practical use of Holin's method, which could bring forth considerable progress in solar system research. There are opportunities connected to the possible creation of a new planetary radar in Euro-Asia or North Africa where, due to the set of existing radiotelescopes (mainly in Europe), the accuracy of RSDI measurements would improve considerably (Van Hoolst *et al.* 2007). One of the suggested locations for the creation of a radar-tracking complex is Northern Caucasus.

3.2. *Findings from the Cassini mission inside the Saturnian system*

The Cassini spacecraft has now completed its nominal 4-yr mission at Saturn and entered a 2-yr extended mission to last until June 2010. The last two years, especially, have been rich in spectacular findings. For instance, on Saturn, the imaging experiment ISS showed that the South polar vortex shares some properties with terrestrial hurricanes: cyclonic circulation, a warm central eye surrounded by a ring of high clouds (the eyewall), and convective clouds outside the eye (Dyudina *et al.* 2008). Thermal mapping with CIRS showed that both poles exhibit similar tropospheric and stratospheric cyclonic hotspots, though only the North pole shows a coherent hexagonal structure below 100 mbar, whereas in the South polar vortex, a circumpolar jet at 87°S coincides with the eyewall and deep-cloud clearing (Fletcher *et al.* 2008). Also from CIRS data, Fouchet *et al.* (2008) found evidence for an equatorial oscillation like those on Earth and Jupiter.

On Titan, the ionospheric composition analysis with INMS shows that the most abundant species is $HCNH^+$ (m = 28), as predicted by pre-*Cassini* models. Other confirmed predictions include $C_2H_5{}^+$, $CH_5{}^+$, $C_3H_5{}^+$, $C_7H_7{}^+$, and $C_3H_3N^+$ (Cravens *et al.* 2006).

The spectral properties of the surface in the near-infrared has been investigated with VIMS (Barnes *et al.* 2007). The gas composition has been analyzed with CIRS (Coustenis *et al.* 2007; Teanby *et al.* 2007; de Kok *et al.* 2007). The surface morphology has been partially unveiled with RADAR, which detected volcanic features (Lopes *et al.* 2007), sand dunes (Lorenz *et al.* 2006), impact craters (Lorenz *et al.* 2007), as well as vast expanses of low-reflecting material near the North pole, most probably lakes of liquid methane (Stofan *et al.* 2007). That instrument also revealed a slowdown of Titan's rotation, suggesting the presence of a subsurface ocean (Lorenz *et al.* 2008).

On Enceladus, the first close flyby led to the detection of a dynamic atmosphere by the Magnetometer (Dougherty *et al.* 2006). ISS subsequently observed jets of fine icy particles emanating from the South pole and carried aloft by water vapor probably venting from subsurface reservoirs of liquid water (Porco *et al.* 2006). CIRS confirmed hot spots associated with the vents (Spencer *et al.* 2006). Chemical analysis of the plume with INMS (Waite *et al.* 2006) leads to an internal model in which aqueous, catalytic chemistry takes place within a very hot environment (Matson *et al.* 2007).

On Iapetus, especially in the equatorial regions, dark material tends to coat the equator-facing slopes of ridges and crater walls, as well as many crater floors. This strongly suggests the warming action of the Sun in removing bright ice from these sunward-facing surfaces and leaving behind the native dark material that is normally mixed with the ice.

A further extension of the mission is now under consideration by NASA, with the aim of reaching Northern summer solstice in 2017. No doubt such a *Cassini* Solstice Mission would gather an extraordinary body of data, the analysis of which could last many years.

3.3. *Methane band spectrophotometry of Saturn in support of Cassini*

The study of the variations of the methane absorption bands on Jupiter and Saturn has a long history at the Fessenkov Astrophysical Institute (Kazakhstan), beginning in the 1960s. During the last few years, these observations served as part of the ground-based astrophysical support of the *Cassini* mission at Saturn. The most interesting data were obtained during the period 1995 - 2008, thanks to the CCD-technique. The methane absorption, as well as other optical properties of the visible atmosphere and cloud cover, shows a clear North-South asymmetry, even at the time of the 'edge-on' viewing of the rings. During this event in 1995, we found that the CH_4 absorption bands were significantly more intense at temperate latitudes in the Southern hemisphere, compared with the Northern hemisphere. A reciprocal picture was observed for the limb-darkening coefficients in the continuum. In the Northern hemisphere, the coefficients were about 0.10–0.15 lower than in the Southern hemisphere, for similar latitudes. During the following years, up until 2008, the distribution of the CH_4 absorption with latitude varied as a function of the equator tilt, and for Northern temperate latitudes, a regular increase of the methane bands intensity was observed (Tejfel 2005; Tejfel *et al.* 2007). The depth of the 7252 nm band changed from 0.53 in 1995 to 0.74 in 2007 – 2008, with a linear trend that may be described as $R(Y) = 0.530 + 0.019 \times (Y-1995)$, where Y is the year. Spectrophotometric observations of Saturn in 2007 and 2008 have detected clear differences in the behavior of weak and strong methane bands in the Northern and Southern hemispheres. The equivalent widths of the relatively strong band at 725 nm are nearly the same in both hemispheres at temperate latitudes, but for weaker bands (619 nm and others), the equivalent widths in the Northern hemisphere are significantly larger than in the Southern hemisphere (Tejfel *et al.* 2008).

4. Future plans for the in-situ exploration of planets and satellites

In the USA, NASA has two different programmes for the in-situ exploration of the planets and their satellites:

(*i*) New Frontiers missions, ranging in cost from 500 to 800 M$, and responding to strategic targets specified in the Solar System Roadmap, i.e., from Venus to giant planets, but limited in scope in terms of the complexity of operational capabilities (for instance, robotic exploration is not envisaged within the frame of these medium-class missions). Within this program, the Venus in-situ *Explorer* (VISE, launch 2020), if selected, will study the composition and surface properties of Venus. It will acquire and characterize a core sample from the surface to study the mineralogy of surface materials. On the other hand, Jupiter *Polar Orbiter* (launch in August 2011, arrival in 2016) will map the gravity field, magnetic field and atmospheric structure of Jupiter from polar orbit. Furthermore, a Jupiter *Multi-Probes* mission and a Saturn *Multi-Probes* mission are in the conceptual design phase, both with a 2020 launch timeline.

(*ii*) *Flagship* missions, ranging in cost from 800 to 1400 M$ *or* from 1400 to 2800 M$. These missions will be crucial in reaching and exploring difficult but high-priority targets: complex missions to the surface of Venus, the lower atmosphere and surface of Titan, the surface and subsurface of Europa, the deep atmosphere of Neptune and the surface of its moon Triton. Besides the *TandEM/TSSM* (to Titan and Enceladus) and *Laplace/EJSM* (to Jupiter) international missions (see below), two more are contemplated by NASA: *Enceladus Explorer* (in the planning stage and likely to be part of the *TSSM* mission), and *Europa Astrobiology Lander* (in conceptual design with a 2030 launch timeline).

Within the European Space Agency, the massive response by the scientific community to the 2004 call for themes has been synthesized into the Cosmic Vision 2015 - 2025 plan. To address the fundamental questions delineated in this programme, Announcements of Opportunities for missions were issued in March 2007, and candidate missions were selected in October 2007 for further assessment and consideration for launch in 2017/2018. For the study of planets and satellites, two missions were selected: *TandEM/TSSM* (to explore Titan and Enceladus), and *Laplace/EJSM* (to explore Jupiter and the Galilean Satellites). These are L-class missions, with a cost at completion to ESA below 650 million euros. The assessment studies are carried out jointly with NASA, and with NASA and JAXA, respectively, since both these agencies currently envisage a mission to the outer Solar System. The *TandEM/TSSM* mission aims at an exhaustive in-situ exploration of Titan and Enceladus with an orbiter (NASA responsability), a montgolfiere-type balloon, and two landers (ESA responsability). The study of Jupiter and the Galilean Satellites with *Laplace/EJSM* was conceived as a set of three satellites for which ESA, NASA and JAXA would share responsabilities: *Jupiter Europa Orbiter* (*JEO*), *Jupiter Magnetospheric Orbiter* (*JMO*) and *Jupiter Planetary Orbiter* (*JPO*). ESA is expected to make a selection between these two missions in consultation with its foreign partners in the coming years.

For its part, the Japanese space agengy JAXA is planning the exploration of Venus with the *PLANET-C* (or *Venus Climate Orbiter*) mission. The main objectives are: to elucidate the mechanism of 'super-rotation' of the atmosphere through 3-D mapping down to the surface (at UV, IR and radio wavelengths), and to search for active volcanism and lightning. The probe is scheduled to be launched in June 2010 and orbit insertion is expected in December 2010.

5. Projected activities and future developments

In the coming months, Commission 16, together with Commissions 51 and 53, will be active in preparing a Special Session for the IAU XXVII General Assembly. The scientific program has already been outlined, but the task of securing the participation of considered speakers needs to be rapidly completed, as well as the promotion of the session among the planetary science community. The latter would be an excellent motivation to implement a new attempt at editing an electronic bulletin. The Commission will also work towards a significantly stronger participation from younger researchers in its activities. This will include encouraging the new generations of planetary scientists to become members of the IAU, with a main affiliation to Commission 16.

The next decade will see the begining or the completion of a number of space missions, to Mercury, Venus, Mars, Jupiter, Saturn, and Pluto. The role of the Commission should be significant in fostering effective ground support for these planetary missions, as well as a wide participation in the interpretation of the data that will be harvested.

In the area of communicating astronomy with the public, the Commission intends to be active in encouraging and sponsoring Saturn Observing Night events, particularly during the *International Year of Astronomy* in 2009, and in the following years.

Régis Courtin
president of the Commission

References

Barnes, J. W., Brown, R. H., Soderblom, L., *et al.* 2007, *Icarus*, 186, 242
Coustenis, A., Achterberg, R. K., Conrath, B. J., *et al.* 2007, *Icarus*, 189, 35
Cravens, T. E., Robertson, I. P., Waite, J. H., *et al.* 2006, *Geophys. Res. Lett.*, 33, L07105
de Kok, R., Irwin, P. G. J., Teanby, N. A., *et al.* 2007, *Icarus*, 186, 354
Dougherty, M. K., Khurana, K. K., Neubauer, F. M., *et al.* 2006, *Science*, 311, 1406
Dyudina, U. U., Ingersoll, A. P., Ewald, S. P., *et al.* 2008, *Science*, 309, 1801
Fletcher, L. N., Irwin, P. G. J., Orton, G. S., *et al.* 2008, *Science*, 319, 79
Fouchet, T., Guerlet, S., Strobel, D. F., *et al.* 2008, *Nature*, 453, 200
Harmon, J. K., Perillat, P. J., & Slade, M. A. 2001, *Icarus*, 149, 1
Holin, I. 1988, *Radiofiz.* 31, 515. Trans.: Kholin, I. 1988, *Radiophys. Quant. Elec.*, 31, 371
Holin, I. 2004, *Sol. Syst. Res.*, 38, 449
Ksanfomality, L. 2005, *Asia Oceania Geosc. Soc. 2005 Meeting*, Abstr. #58-PS-A0974
Ksanfomality, L. & Sprague, A. 2007, *Icarus*, 188, 271
Lopes, R. M. C., Mitchell, K. L., Stofan, E. R., *et al.* 2007, *Icarus*, 186, 395
Lorenz, R. D., *et al.* 2006, *Science*, 312, 724
Lorenz, R. D., Wood, C. A., Lunine, J. I., *et al.* 2007, *Geophys. Res. Lett.*, 34, L07204
Lorenz, R. D., Stiles, B. W., Kirk, R. L., *et al.* 2008, *Science*, 319, 1649
Margot, J. L., Peale, S. J., Jurgens, R. F., *et al.* 2007, *Science*, 316, 710
Matson, D. L., Castillo, J. C., Lunine, J., & Johnson, T. V. 2007, *Icarus*, 187, 569
Porco, C. C., Helfenstein, P., Thomas, P. C., *et al.* 2006, *Science*, 311, 1393
Spencer, J. R., Pearl, J. C., Segura, M., *et al.* 2006, *Science*, 311, 1401
Sprague, A. L., Hunten, D. M., & Lodders, K. 1995, *Icarus*, 118, 211
Stofan, E. R., Elachi, C., Lunine, J. I., *et al.* 2007, *Nature*, 445, 61
Teanby, N. A., Irwin, P. G. J., de Kok, R., *et al.* 2007, *Icarus*, 186, 364
Tejfel, V. G. 2005, *Terrestr. Atmos. Oceanic Sci.*, 16, 231
Tejfel, V. G., Karimov, A. M., & Kharitonova, G. A. 2007, *Proc. Nat. Acad. Sci. Kaz.*, 4, 99
Tejfel, V. G., Vdovichenko, V. D., Karimov, A. M., *et al.* 2008, *39th LPS Conf.*, p. 1530
Van Hoolst, T., Sohl, F., Holin, I., *et al.* 2007, *Sp. Sci. Rev.*, 132, 203
Waite, J. H., Combi, M. R., Ip, W.-H., *et al.* 2006, *Science*, 311, 1419

Transactions IAU, Volume XXVIIA
Reports on Astronomy 2006–2009
Karel A. van der Hucht, ed.

© 2009 International Astronomical Union
doi:10.1017/S1743921308025428

COMMISSION 20

POSITIONS AND MOTIONS OF MINOR PLANETS, COMETS AND SATELLITES

POSITIONS ET MOUVEMENTS DES PETITES PLANÈTES, DES COMÈTES ET DES SATELLITES

PRESIDENT   Julio A. Fernández

VICE-PRESIDENT   Makoto Yoshikawa

PAST PRESIDENT   Giovanni B. Valsecchi

ORGANIZING COMMITTEE   Steven R. Chesley,
Yulia A. Chernetenko,
Alan C. Gilmore, Daniela Lazzaro,
Karri Muinonen, Petr Pravec,
Timothy B. Spahr, David J. Tholen,
Jana Tichá, Jin Zhu

TRIENNIAL REPORT 2006 - 2009

1. Naming of solar system bodies

Commission 20 has been involved in the discussion on discovery credit rules of solar system objects (mainly concerned with asteroids) in particular with the role played by dynamicists in recovering objects by linking their orbits with previous apparitions. A working group was set up to discuss the issue that was integrated by professional astronomers as well as amateurs. There was some exchange of opinions and conflicting views, but the trend of the majority was to keep the discovery credit for the discoverer whose observations led to the object's principal designation, as expressed in the MPC existing rules (see <cfa-www.harvard.edu/iau/info/HowNamed.html>), considering only exceptionally credit for dynamicists when linkage of different apparitions led to the recovery of a lost object. The precise definition of non-trivial linkage should be worked out, and for the time being, the idea is to keep the existing MPC rules and, if necessary, to improve them, rather than starting a new set of rules.

2. Support for scientific meetings

Commission 20 has given its support to several proposals of meetings. The list includes: *Mathematics and Astronomy – A Long Journey* (proposed to be held in Madrid, Spain, November 2009); *The Role of Astronomy in Society and Culture* (IAU Symposium No. 260, Paris, France, January 2009), and *Icy Bodies in the Solar System* (IAU Symposium No 263, Rio de Janeiro, Brazil, during the IAU XXVII General Assembly in August 2009). As regards the latter meeting, it was considered for us very important to have a meeting devoted to planetary sciences in the next General Assembly, in order to make it more attractive from a scientific standpoint for the community of planetary scientists. In particular, icy bodies offer clues for our understanding of how the solar system formed and evolved, and also as potential abodes of life, so this was considered to be a very appropriate and timely subject, given the new developments related to space-based missions as well as large-scale sky surveys from the Earth.

3. Commission 15 and Commission 20

A pending problem before the next IAU GA is the possible merging of Commission 15 and Commission 20 into a single commission. On the one hand, there are some advantages in the discussion of topics related to minor bodies of the solar system with a unified view covering physical as well as dynamical aspects. One example is non-gravitational forces affecting the motion of comets, or the Yarkovsky and YORP effects on asteroids. On the other hand, dynamicists and phycisists form two communities with their own specific interests, so the question of merging requires further discussion.

Julio A. Fernández
president of the Commission

Transactions IAU, Volume XXVIIA
Reports on Astronomy 2006–2009
Karel A. van der Hucht, ed.

© 2009 International Astronomical Union
doi:10.1017/S174392130802543X

COMMISSION 21

LIGHT OF THE NIGHT SKY

LUMIÈRE DU CIEL NOCTURNE

PRESIDENT   Adolf N. Witt
VICE-PRESIDENT   Jayant Murthy
PAST PRESIDENT   Bo Å. S. Gustafson
ORGANIZING COMMITTEE   W. Jack Baggaley, Eli Dwek,
                                   Anny-Chantal Levasseur-Regourd,
                                   Ingrid Mann, Kalevi Mattila,
                                   Jun-ichi Watanabe

TRIENNIAL REPORT 2006 - 2009

1. Introduction

Commission 21 consists of IAU members and consultants with expertise and interest in the study of the light of the night sky and its various diffuse components, at all accessible electromagnetic frequencies. In cosmic distance scales, the subjects of Commission 21 range from airglow and tropospheric scattering in Earth's atmosphere, through zodiacal light in the solar system, including thermal emission from interplanetary dust, integrated starlight in the Milky Way galaxy, diffuse galactic light due to dust scattering in the galactic diffuse interstellar medium, thermal emissions from interstellar dust and free free emission from ionized interstellar gas, to various diffuse extragalactic background sources, including the cosmologically important cosmic microwave background (CMB). Observations of the diffuse night sky brightness at any frequency typically include signals from several of these sources, and it has been the historic mandate of Commission 21 to foster the necessary collaboration of experts from the different astronomical sub-disciplines involved.

2. Recent significant developments

The arguably most important development since 2006 has been the ongoing analysis of the *COBE*-DIRBE diffuse sky maps and the three-year and five-year CMB data sets from *WMAP*, in which the important corrections for diffuse foreground sources relies heavily on data products generated in earlier years by astronomers affiliated with Commission 21. A detailed model of the zodiacal light is required before reliable templates for galactic dust emission can be produced. Recently, the Fabry-Perot high-resolution spectral line studies of the zodiacal light by Reynolds *et al.* (2004) have been extended (e.g., Madsen *et al.* 2005) and extensively modeled by Ipatov *et al.* (2008). Kiss *et al.* (2008) investigated the the fluctuations produced in the infrared zodiacal emission by the thermal emission from asteroids. An all-sky Hα map serves as a template for the contribution by galactic free-free emission from interstellar ionized gas, while a galactic radio survey provides information on contributions in the form of synchrotron radiation to the *WMAP* data.

All foreground sources must be known with relatively high spatial resolution so as not to introduce false-positive small-scale structure into the CMB data. Thanks to the availability of such data, unprecedented cosmological information concerning the time since the big bang, the Hubble parameter, and the space-time geometry of the universe have emerged in recent years.

In addition, the comparison of *WMAP* data with previously known foreground sources has led to the discovery of a heretofore unknown galactic emission process in the 10 - 50 GHz range, related to interstellar dust. Initially termed 'anomalous dust emission', the process is most likely related to electric dipole microwave emission from rapidly spinning interstellar grains of nanometer size (Dobler & Finkbeiner 2008a,b). The use of all-sky H-α maps as a template for galactic free-free emission may need to be reexamined in light of a finding by Mattila *et al.* (2007) that the enhanced H-α intensity seen in high-latitude galactic cirrus clouds is most likely due to dust scattering of ambient H-α photons in the galactic radiation field rather than to *in-situ* emission.

Along more traditional lines, the difficult measurements of the optical/NIR/FIR extra-galactic background light, a potentially very important source of information concerning the star formation history of the universe, continue to receive attention (Hauser & Dwek 2001; Mattila 2003; Matsumoto *et al.* 2005; Dwek *et al.* 2005; Mattila 2006; Bernstein 2007; Odegard *et al.* 2007). New measurements of the diffuse UV background have become possible thanks to *GALEX* (Henry *et al.* 2007; Sujatha *et al.* 2008). The study of extended red emission (ERE), thought to originate in UV-powered photoluminescence of interstellar PAH molecules or clusters, has now been extended to individual high-latitude clouds, which are illuminated by the galactic interstellar radiation field (Witt *et al.* 2008). The advance of detector technology has also enabled the high-resolution mapping of interstellar clouds in near-infrared light (Juvela *et al.* 2006).

3. Status and future of Commission 21

The focus and emphasis of current work on light of the night sky has undergone a strong shift toward extragalactic diffuse backgrounds and CMB data analysis. Many of the new workers in the field have backgrounds other than traditional astronomical backgrounds, and most are not members of the IAU and certainly not of Commission 21. In view of the small number of active members currently in Commission 21, it has been proposed to merge Commission 21 with Commission 22 *Meteors, Meteorites, and Interplanetary Dust* within IAU Division III on *Planetary Systems Sciences*, a move that is not supported by the majority of the active members of Commission 21, as this would effectively end the current mandate of Commission 21. Discussions concerning alternative futures of Commission 21 are currently underway.

Adolf N. Witt
president of the Commission

References

Bernstein, R. A. 2007, *ApJ*, 666, 663
Dobler, G. & Finkbeiner, D. P. 2008a, *ApJ*, 680, 1222
Dobler, G. & Finkbeiner, D. P. 2008b, *ApJ*, 680, 1235
Dwek, E., Arendt, R. G. & Krennrich, F. 2005, *ApJ*, 635, 784
Hauser, M. G. & Dwek, E 2001, *ARA&A*, 39, 249
Henry, R. C., Sujatha, N. V., Murthy, J., & Bianchi, L. 2007, *BAAS*, 210, 4406

Ipatov, S. I., Kutyrev, A. S., Madsen, G. J., Mather, J. C., Moseley, S. H., & Reynolds, R. J. 2008, Icarus, 194, 769

Juvela, M., Pelkonen, V.-M., Padoan, P., & Mattila, K. 2006, *A&A*, 457, 877

Kiss, Cs., Pal, A., Mueller, Th. G., & Abraham, P. 2008, *A&A*, 478, 605

Madsen, G. J., Reynolds, R. J., Ipatov, S. I., Kutyrev, A. S., Mather, J. C., & Moseley, S. H. 2005, *LPI Contr.* 1280, 111

Matsumoto, T., Matsuura, S., Murakami, H., Tanaka, M., Freund, M., Lim, M., Cohen, M., Kawada, M., & Noda, M. 2005, *ApJ*, 626, 31

Mattila, K. 2003, *ApJ*, 591, 119

Mattila, K. 2006, *MNRAS*, 372, 1253

Mattila, K., Juvela, M., & Lehtinen, K. 2007, *ApJ* (Letters), 654, L131

Odegard, N., Arendt, R. G., Dwek, E., Haffner, L. M., Hauser, M. G., & Reynolds, R. J. 2007, *ApJ*, 667, 11

Reynolds, R. J., Madsen, G. J., & Moseley, S. H. 2004, *ApJ*, 612, 1206

Sujatha, N. V., Murthy, J., Karnataki, A., Henry, R. C., & Bianchi, L. 2008, in press [arXiv0807.0189]

Witt, A. N., Mandel, S., Sell, P. H., Dixon, T., & Vijh, U. P. 2008, *ApJ*, 679, 497

Transactions IAU, Volume XXVIIA
Reports on Astronomy 2006–2009
Karel A. van der Hucht, ed.

© 2009 International Astronomical Union
doi:10.1017/S1743921308025441

COMMISSION 22	METEORS, METEORITES AND INTERPLANETARY DUST
	MÉTEORES, MÉTÉORITES ET POUSSIÈRE INTERPLANÉTAIRE

PRESIDENT	Pavel Spurný
VICE-PRESIDENT	Jun-ichi Watanabe
PAST PRESIDENT	Ingrid Mann
ORGANIZING COMMITTEE	Jiří Borovička (secretary), William J. Baggaley, Peter G. Brown, Guy J. Consolmagno, Peter M.M. Jenniskens, Asta K. Pellinen-Wannberg, Vladimír Porubčan, Iwan P. Williams, Hajime Yano

COMMISSION 22 WORKING GROUPS

Div. III / Commission 22 WG	Professional-Amateur Cooperation
Div. III / Commission 22 WG	in Meteors
Div. III / Commission 22 TG	Meteor Shower Nomenclature

TRIENNIAL REPORT 2006 - 2009

1. Introduction

Commission 22 is part of Division III on *Planetary System Sciences* of the International Astronomical Union. Members of Commission 22 are professional scientists studying bodies in the Solar System smaller than asteroids and comets, and their interactions with planets. The main subjects of interest are meteors, meteoroids, meteoroid streams, interplanetary dust particles, and also zodiacal cloud, meteor trains, meteorites, tektites, etc.

At present Commission 22 consists of 112 active members and 11 of them, mostly young astronomers, are new members elected during the last General Assembly of the IAU in Prague in 2006. Over the past triennium the members of the organizing committee had three business meetings, in August 2006 in Prague (Czech Republic), in June 2007 in Barcelona (Spain) and in July 2008 in Baltimore (USA). The main topics discussed there have been collected in the minutes from these meetings (see: `<meteor.asu.cas.cz /IAU/business.html>`). The official webpage of the Commission 22 was established by the commission secretary immediately after the last GA in Prague and has the address `<meteor.asu.cas.cz/IAU/>`.

2. Meetings

Several important international conferences devoted exclusively or partly to the topics covered by Commission 22 were held in the period 2006 - 2009. Specific conferences in which members of Commission 22 have played a major part are described in more detail. Commission 22 supported the proposal of Commission 20 and Commission 15 of the Division III to organize the Symposium on *Icy bodies in the solar system* during next General Assembly of the IAU in Rio de Janeiro in 2009, selected as IAU Symposium No. 263.

2.1. *Meteoroids 2007*

The *Meteoroids 2007 Conference* held at the CosmoCaixa Science Museum in Barcelona, Spain, 11-15 June 2007, was the main conference in meteor astronomy in this period and served as the official tri-annual meeting of Commission 22. It was the 6th in the series of Meteoroids conferences, which started in 1992.

The conference was very successful and was dedicated to a comprehensive review of meteor research, including topics dedicated to the physics and chemistry of meteors, meteoroid stream dynamics, activity and forecasting of meteor showers, interaction of larger meteoroids with the atmosphere, physical properties of interplanetary dust, and all techniques of meteor observations.

Researchers in meteor science and supporting fields representing more than 20 countries participated at this international conference where 126 presentations were delivered in oral and poster forms. Essential contributions were collected in the Special Issue of the *Earth, Moon and Planets* journal named *Advances in Meteoroid and Meteor Science* (EMP, Vol. 102, Nos 1-4, June 2008, eds. J. M. Trigo-Rodríguez, F. J. M. Rietmeijer, J. Llorca & D. Janches). The 69 papers included in this volume represent the work of 154 authors from about 70 different institutions across the globe (Trigo-Rogríguez *et al.* 2008). All papers underwent the rigorous refereeing process that is applied to other papers in the journal *Earth, Moon, and Planets*.

The conference provided a summarizing overview on meteoroid and meteor science in two broad-based thematic categories. The first category covered detection, observation and measurement techniques many of which were described in great detail by invited speakers. The contributed presentations in this category focused on the formation of meteoroid streams by active or dormant comets and asteroids, together with dynamical studies of meteoroids moving through the solar system. The study of meteoroids as space hazard is a topic of rapidly increasing interest due to the need of insure the safety and health of manned and unmanned space missions. Papers discussing optical techniques to observe meteor phenomena were prominent and results included the observation of enhanced activities of the 2006 Leonids and 2006 Orionids. The outcomes of years of infrasound and radar detections also showed that these methodologies are no longer stepchildren of meteor science, greatly expanding the mass range of extraterrestrial bodies which can now be studied. Radar meteor detection methodologies have evolved immensely since these instruments were first applied in the 1950s. Greater transmitted power, multi station interferometric techniques and the use of dual frequencies allow meteor radars to provide exciting new data, including the discovery of new meteoroid streams. In addition, in the past decade, the increasing use of high-power and large-aperture radars offer a new look at the meteor phenomena by allowing the routine study of the meteor head-echo, non-specular trails and a particle size range that bridge the historic gap between dust detector on board of satellites and specular meteor radars.

The second category of results included dynamical modeling exemplified by the power of reconstructing past meteor displays and accurately predicting modern meteor shower activities. Meteor observations are now providing more precise input to fine-tuned models, which is an achievement of increasing sophistication in both areas. For example, Comet Wild 2 data were preliminary explored for their relevance to cometary meteoroid properties. With the availability of this comet dust, interplanetary dust particles, micrometeorites and meteorites for laboratory studies, we are able now to take the next great step of using what we know of these samples as a starting point for experimental meteor science. Results from laboratory simulations of chemical releases during the meteor ablation process are showing that we are closer to understanding how the meteoric mass is deposited in the upper atmosphere. This particular advancement allows linking the meteoric flux with several aeronomical phenomena such as mesospheric metallic layers, noctilucent clouds and meteoric smoke particles embedded in the ionospheric plasma.

During the *Meteoroids 2007 Conference*, members of Organizing Committee of the Commission 22 decided that the next *Meteoroids 2010 Conference* will be held in Breckenridge, Colorado, USA in May 2010. The chair of the LOC will be Diego Janches. The SOC was established in late 2007 and has already started work.

2.2. *Asteroids, Comets, Meteors 2008*

The *Asteroids, Comets, Meteors (ACM)* Conference will be held in Baltimore, USA, 14-18 July 2008. This is the premier international gathering of scientists who study small bodies of our Solar System. The ACM series began in 1983 in Uppsala, Sweden, and the 2008 *ACM Conference* will be the 10th in the series and will mark the 25th anniversary of the first meeting in Uppsala. It mainly serves as a means of bringing together different groups within the asteroid, comet, and meteor communities who do not often have the opportunity to interact. The conference now takes place every three years, and it is the pre-eminent meeting for small-bodies research with great attendance.

The scope of presentations and discussion is broad, including all topics related to asteroids, comets, and meteors, including observations, theories of origin and evolution, discoveries, astrometry, dynamics, internal structure, mineralogical composition, space missions, laboratory studies, classifications, and databases. Two plenary sessions will be completely devoted to the meteor science and one session will be devoted to the interrelations among all main small bodies objects, i.e., also meteoroids.

2.3. *Bolides and Meteorites Falls 2009*

The international conference *Bolides and Meteorite Falls* will be held on the occasion of the 50th anniversary of the Pribram meteorite fall in Prague, Czech Republic, 10–15 May 2009. This conference completely covers the main interest of Commission 22 and will be held under auspices of this commission.

The Pribram meteorite fall on 7 April 1959 was the first instrumentally observed meteorite fall in history and it belongs to the most important milestones in meteor science. It initiated various observational programs and modeling efforts and our understanding of bolides and associated phenomena increased dramatically during the past 50 years. However, the conference will be devoted not only to celebrate the 50th anniversary but also to discuss numerous recent achievements in this field and future prospects in the study of larger meteoroids.

3. Commission bodies

3.1. *Working Group on Professional-Amateur Cooperation in Meteors*

The Working Group *Professional-Amateur* of Commission 22 consists of leading amateur meteor astronomers on all continents, who help provide support for international observing campaigns and facilitate contacts between professional and amateur astronomers. The working group also consists of professional meteor astronomers who recognize the importance of the amateur astronomy community for the future of our field. The members elected to this Task Group were Galina Ryabova (chair), Pavel Spurný, Chris Trayner, Juraj Tóth, Shinsuke Abe, Josep Trigo-Rodríguez, Jérémie Vaubaillon, Bob Lunsford, Hiroshi Ogawa, R. Arlt, Steve Evans (who regrettably passed away in April 2008) and Tim Cooper.

The International Meteor Organization (IMO), an association of over 200 active amateur meteor astronomers along with several professionals, plays a very important and irreplaceable role in this field. The IMO was created in 1988 in response to an ever growing need for international cooperation of amateur meteor work. As such, the IMO's main objectives are to encourage, support and coordinate meteor observing, to improve the quality of amateur observations, to disseminate observations and results to other amateurs and professionals and to make global analyses of observations received world-wide. The IMO's main instrument to achieve its goals is its bimonthly journal WGN. Annually, it contains over 220 pages of general meteor news, observing program guidelines, reports and analyses of observations, and more general articles on meteoric phenomena, some of them by professionals in the field. All important details about IMO and its work are available on the webpage `<www.imo.net/>`.

3.2. *Task Group Meteor Shower Nomenclature*

At the 2006 IAU General Assembly in Prague, a Task Group was established with the objective to formulate a descriptive list of established meteor showers that can receive official names during the next 2009 IAU General Assembly in Rio de Janeiro. This task aims to uniquely identify all existing meteor showers and establish unique names. The Commission 22 members elected to this Task Group were Peter Jenniskens (chair), Vladimír Porubčan, Pavel Spurný, William J. Baggaley, Juergen Rendtel, Shinsuke Abe, Robert Hawkes, and Tadeusz J. Jopek. The effective period of the Task Group is three years.

In order to reach this objective, a Working List of Meteor Showers ('the Working List') was adopted as a starting point and meteor shower nomenclature rules were formulated. The Working List was based on data collected in Jenniskens (2006) and published by the IAU Meteor Orbit Data Center (see: `<www.astro.sk/~ne/IAUMDC/STREAMLIST/>`). The nomenclature rules were published in IAU *Information Bulletin* 99 and announced in the scientific literature (Jenniskens 2008).

At the *Meteoroids 2007 Conference* in Barcelona, the Task Group convened and established a procedure for adding new showers to the Working List. Tadeusz J. Jopek of Poznan Observatory, Poland, was appointed to be the point of contact for reporting the discovery of new meteor showers, and for reports that established meteor showers in the Working List. Jopek established a new website for the IAU Meteor Data Center, which includes the Working List and the current IAU List of Established Showers (see: `<www.ta3.sk/IAUC22DB/MDC2007/>`) .

The first series of twelve new showers was added to the Working List based on results from the Canadian Meteor Orbit Radar of the University of Western Ontario, Canada. The new additions were announced in CBET 1142 of 17 November 2007 (Brown *et al.* 2007).

3.3. *Meteor Data Center*

The IAU Meteor Data Center (MDC) is operated by the Astronomical Institute of the Slovak Academy of Sciences under the auspices of Commission 22 and Division III. The center is responsible for the efficient collection, checking and dissemination of trajectory observations and orbits of meteors. It acts as a central depository for meteor orbits obtained by photographic, video and radar techniques. The MDC closely cooperates with the Task Group on *Meteor Shower Nomenclature* of IAU Commission 22 on the designation of all existing meteor showers. The official web page of the MDC is `<www.astro.sk/~ne/IAUMDC/Ph2003/>` (Lindblad *et al.* 2003).

4. Closing remark

The great progress made by Commission 22 members in the past triennium proves that this field of astronomy prospers. The field has an important impact on our knowledge of planetary astronomy and on space exploration. The International Astronomical Union plays an important role in organizing the community and in finding common ground on various issues.

Pavel Spurný
president of the Commission

References

Brown, P., Weryk, R. J., Wong, D. K., Jones, J., Jenniskens, P., & Jopek, T. 2007, *Central Bureau Electronic Telegrams*, 1142, 1

Jenniskens, P. 2006, *Meteor Showers and their Parent Comets* (Cambridge: CUP)

Jenniskens, P. 2008, *Earth, Moon, & Planets*, 102, 5

Lindblad, B. A., Neslušan, L., Porubčan, V., & Svoreň, J. 2003, *Earth, Moon, & Planets*, 93, 249

Trigo-Rodríguez, J. M., Rietmeijer, F. J. M., Llorca, J., & Janches, D. 2008, *Earth, Moon, & Planets* 102, 1-4, 1-561

Transactions IAU, Volume XXVIIA
Reports on Astronomy 2006–2009
Karel A. van der Hucht, ed.

© 2009 International Astronomical Union
doi:10.1017/S1743921308025453

COMMISSION 51

BIOASTRONOMY

BIOASTRONOMIE

PRESIDENT
VICE-PRESIDENT
PAST PRESIDENT
ORGANIZING COMMITTEE

Alan P. Boss
William M. Irvine
Karen J. Meech
Cristiano B. Cosmovici,
Pascale F. Ehrenfreund,
David W. Latham,
David Morrison,
Stephane Udry

TRIENNIAL REPORT 2006 - 2009

1. Introduction

Bioastronomy: Search for Extraterrestrial Life was established as Commission 51 of the IAU in 1982. The objectives of the commission included: the search for planets around other stars; the search for radio transmissions, intentional or unintentional, of extraterrestrial origin; the search for biologically relevant interstellar molecules and the study of their formation processes; detection methods for potential spectroscopic evidence of biological activity; the coordination of efforts in all these areas at the international level and the establishment of collaborative programs with other international scientific societies with related interests. In 2006, Commission 51 was renamed simply *Bioastronomy* at the IAU General Assembly in Prague, and approved for the next six years, the default extension for an IAU Commission.

2. Meetings

The primary activity of IAU Commission 51 is to organize a *Bioastronomy* meeting every three years. *Bioastronomy 2007* was held in 2007 in San Juan, Puerto Rico, and was a great success, largely due to the efforts of William Irvine, chair of the SOC for the meeting and vice-president of Commission 51, and to Karen Meech, chair of the LOC for the meeting and past president of Commission 51. A full description of the meeting can be found on the meeting's web pages at <www.ifa.hawaii.edu/UHNAI/bioast07.htm>.

While at the San Juan meeting, a group of Commission 51 members met with Antonio Lazcano, president of the International Society for the Study of the Origin of Life: The International Astrobiology Society (also known as ISSOL), several ISSOL board members, and other interested bioastronomers and astrobiologists in order to discuss plans to hold a joint meeting in the future. The motivation for having such a joint meeting is to try to ensure the continued growth of both of our fields without putting undue pressure on ourselves by having competing meetings at different times and locations. There is considerable overlap between the interests of ISSOL and Commission 51, and a common meeting seems like a reasonable means to avoid having people being able to only attend one meeting or the other due to the constraints of time and money.

We decided to plan on holding a joint meeting of ISSOL and Commission 51 in either 2010 (three years after *Bioastronomy 2007*) or 2011 (three years after the next ISSOL meeting in Florence, Italy in 2008). We anticipate having a truly joint meeting with a single, unified program of talks and events for a total of perhaps 500 participants. The dates and venue for this meeting remain to be determined. Antonio Lazcano and I have invited expressions of interest from institutions that would be willing to host the first joint ISSOL/Commission 51 meeting in either 2010 or 2011, and at least one serious indication of interest has been received. We plan to decide this issue at the ISSOL meeting in Florence in August, 2008.

3. Organizing Committee

The Commission 51 Organizing Committee (OC) has eight members, five from the USA, and three from Europe. In order to address this geographical imbalance, the OC discussed via e-mail the idea of adding new OC members, and invited a new OC member, who had to decline the invitation on the basis of not being an IAU member. In the process of deciding how to proceed, we realized that the IAU rules state that the OC should not have more than eight members, making the question of adding new members moot. In the future we intend to address this imbalance by replacing OC members from the USA by IAU members from other regions, such as Africa, Asia, and South America.

4. Symposium and Special Session sponsorship

Commission 51 has been asked to support a number of proposals to hold IAU Special Sessions and Symposia, either at the IAU XXVII General Assembly (GA) in Rio de Janeiro, 2009, or on their own. The Commission 51 OC considered four requests for solitary symposia in 2006, and found that two of the four were close enough to Commission 51's interests to merit our support. In 2007 the Commission 51 OC considered three requests for support of Special Sessions at the 2009 GA, and one request for support for an IAU Symposium at the 2009 GA. All four of these proposals were judged to be appropriate for Commission 51 support.

5. New Members

Twenty new members of Commission 51 were approved at the IAU GA in 2006, and since then another ten scientists have requested membership in Commission 51. We encourage other IAU members to join Commission 51, which can be accomplished simply by sending an e-mail to the Commission 51 President <boss@dtm.ciw.edu>.

6. Closing remarks

We look forward to holding our first joint meeting with ISSOL in 2010 or 2011. Details for this meeting will be e-mailed to the Commission 51 membership and posted on the Commission 51 web site once the decisions have been made. The web pages for Commission 51 are located at <www.dtm.ciw.edu/boss/c51index.html>.

Alan P. Boss
president of the Commission

Transactions IAU, Volume XXVIIA
Reports on Astronomy 2006–2009
Karel A. van der Hucht, ed.

© 2009 International Astronomical Union
doi:10.1017/S1743921308025465

COMMISSION 53	**EXTRASOLAR PLANETS**
	PLANÈTES EXTRA-SOLAIRE

PRESIDENT	Michel Mayor
VICE-PRESIDENT	Alan P. Boss
ORGANIZING COMMITTEE	Paul R. Butler, William B. Hubbard, Philip A. Ianna, Martin Kürster, Jack J. Lissauer, Karen J. Meech, François Mignard, Alan J. Penny, Andreas Quirrenbach, Jill C. Tarter, Alfred Vidal-Madjar

TRIENNIAL REPORT 2006-2009

1. Introduction

Commission 53 on *Extrasolar Planets* was created at the 2006 Prague General Assembly of the IAU, in recognition of the outburst of astronomical progress in the field of extrasolar planet discovery, characterization, and theoretical work that has occurred since the discovery of the pulsar planets in 1992 and the discovery of the first planet in orbit around a solar-type star in 1995. Commission 53 is the logical successor to the IAU Working Group on *Extrasolar Planets* WG-ESP, which ended its six years of existence in August 2006. The founding president of Commission 53 is Michael Mayor, in honor of his seminal contributions to this new field of astronomy. The vice-president is Alan Boss, the former chair of the WG-ESP, and the members of the Commission 53 Organizing Committee are the other former members of the WG-ESP.

2. WG-ESP

The WG-ESP was charged with acting as a focal point for research on extrasolar planets and organizing IAU activities in the field, including reviewing discovery techniques and maintaining a list of identified planets. The WG-ESP maintained a list of planetary candidates that met its criteria for acceptance as planets up until August 2006. This list also established a criterion for discovery rights, namely the date of submission for publication in a refereed journal. This list can be found on the WG-ESP web pages at <www.dtm.ciw.edu/boss/iauindex.html>.

Commission 53 has decided not to try to continue to maintain this list of planets, given the immense popularity and greater usefulness of the list maintained by Jean Schneider and his colleagues at the Extrasolar Planets Encyclopaedia web site <exoplanet.eu/>.

The WG-ESP also developed in 2001 a Working Definition of what is a 'planet', subject to change as we learn more about the population of extrasolar planets. The Working Definition was modified in 2003 to address the question of objects found by imaging surveys in regions of active star formation. The current Working Definition can be found on the WG-ESP web pages. Note that this definition does not attempt to address the

lower mass end of the range of bodies that should be considered as planets. This is
because, with the exception of certain pulsar planets, all the extrasolar planets discovered
to date are more massive than the Earth.

3. Organizing Committee

The Commission 53 Organizing Committee (OC) has thirteen members, seven from
the USA, and six from Europe, as a result of its formation from the WG-ESP. Given
that IAU rules state that the OC should not have more than eight members, we need
to allow at least five current members of the OC to rotate off. We should also attempt
to add OC members from regions that are not currently represented on the OC, such as
from Africa, Asia, and South America. This issue needs to be addressed prior to (or at)
the Commission 53 business meeting at the 2009 IAU General Assembly (GA) in Rio de
Janeiro.

4. Symposium and Special Session sponsorship

Commission 53 has been asked to support various proposals to hold IAU Special Sessions and Symposia, either at the 2009 IAU GA, or on their own. Several of these proposals were judged to be appropriate for Commission 53 support.

5. New Members

As a newly formed IAU Commission, Commission 53 is seeking astronomers who wish
to be recognized as members. We encourage interested IAU members to ask to join Commission 53, which can be accomplished simply by sending an e-mail to the Commission 53
Vice President <boss@dtm.ciw.edu>).

6. Closing remarks

We look forward to holding our first Commission 53 business meeting at the 2009 IAU
GA. At this meeting, we will need to finalize plan for changes in the OC, selecting a new
Vice President, and accepting new members of Commission 53. The web pages for Com-
mission 53 will be located at <www.dtm.ciw.edu/boss/Commission 53index.html>.

Alan P. Boss
vice-president of the Commission

Transactions IAU, Volume XXVIIA
Reports on Astronomy 2006–2009
Karel A. van der Hucht, ed.

© 2009 International Astronomical Union
doi:10.1017/S1743921308025477

DIVISION III / SERVICE
MINOR PLANET CENTER

DIRECTOR   Timothy B. Spahr
ASSOCIATE DIRECTOR   Gareth V. Williams
DIRECTOR EMERITUS   Brian G. Marsden

TRIENNIAL REPORTS 2003 - 2006, 2006 - 2009

1. Introduction

The activity of the *Minor Planet Center* continued generally to increase during the two triennia covered by this report, principally because of the continuing success of the surveys for near-earth objects. Chief among these has been the *Lincoln* (Laboratory) *Near-Earth Asteroid Research Project*, or LINEAR, which is credited with the discovery of slightly more than half of all the minor planets that have been numbered, although since 2005 the Catalina Sky Survey and Mount Lemmon Survey in Arizona and the Siding Spring Survey in New South Wales (all three of which, together with the long-lasting Spacewatch Survey, are operated from the University of Arizona) have come to dominate the field. The total number of observations of minor planets in the MPC's files more than doubled from 14.1 million in mid-2002 to 30.9 million in mid-2005, with almost another doubling, to 55.4 million, in mid-2008.

The rate of numbering new minor planets followed a similar pattern, with an increase from 43 721 in mid-2002 to 99 947 in mid-2005, almost doubling again to 189 005 in mid-2008. Since the numbering of a minor planet is the 'end product', with the assurance that such an object has been well-enough observed that it is unlikely to be lost in the foreseeable future, it is interesting to note that, while this mid-2002 figure represents only 23% of all the orbits of minor planets in the MPC collection at that time, the fraction had doubled to 46% by mid-2008. Of course, if this fractional evolution continues, there is an implication, either that the work of the MPC will soon be 'complete', or that the bulk of the recent observations are of objects recognized only on two neighboring nights. That the number of observations received actually decreased from 9.7 million during 2006 – 2007 to 8.5 million during 2007 – 2008 would seem to support the former hypothesis. This may be evidence that the current surveys are approaching the limits of what they can do. When next-generation observing programs such as Pan-STARRS and LSST start bearing fruit – i.e., routinely following up main-belt objects that are significantly fainter and smaller than now – the MPC will likely be more productive than ever.

2. Publications and archiving

The permanent archiving of data continues to be done essentially on a monthly basis, coinciding with the publication of the *Minor Planet Circulars* and the *Minor Planet Circulars Orbit Supplement*. As the traditional publication of the MPC, dating back to 1947, the former is now basically a summary of the MPC activity, the 17 236 pages published during the double triennium bringing the total up to 63 316. Through 2005 this publication was available in both printed and electronic form, although the printed edition is now essentially restricted to the pages summarizing the new numberings and providing the citations for the new namings. The *Orbit Supplement* has been issued only electronically since its inception in 2000, and the 108 182 pages published since mid-2002 make for a total now of 140 228. There is also the *Minor Planet Circulars Observation Supplement*, an entirely electronic publication started in 1997 and generally issued weekly, with the 193 368 pages published during the two triennia increasing the total to 252 006.

In addition, there are the *Minor Planet Electronic Circulars*, the 10 137 of these issued during these six years bringing to 16 054 the total since the first one was issued in 1993. The main purpose of the *MPECs*, which are in fact available on the worldwide web without subscription, is to provide immediate information concerning unusual new discoveries, principally NEOs. A 'Daily Orbit Update' *MPEC* tabulates all the orbits computed at the MPC during the previous 24 hours. This DOU issue, which is prepared entirely automatically, is consistent with the intention that the *MPECs* should be a 'temporary' (unarchived) publication, for as long as further observations are made, orbit computations will always be improved. By popular request, the DOU *MPECs* do also include continuing observations of all NEOs. Again, this publication of observations cannot be considered archival, and the automatic preparation precludes the possibility of crediting the observers in a reliable manner.

3. Near-earth objects

Prior to the preparation of an *MPEC* documenting a discovery, alerts to possible NEOs are issued on "The NEO Confirmation Page". Although this webpage has existed since 1996, recent improvements have meant that entries can appear there automatically, as the automated procedures that extract all observations reaching the MPC by e-mail have been augmented to include some estimation of the probability that an object is an NEO. If this probability is greater than 50%, under appropriate circumstances the observations, initial attempts at representing the orbit, and tabulations and plots of the orbital uncertainty appear on the NEOCP within a matter of minutes. Furthermore, the predictions for an object already on the NEOCP are automatically updated as follow-up observations arrive. A particular improvement introduced in 2004 is color-coding of the positional uncertainty plots to indicate orbital solutions that could result in near-collision with the earth.

From 2003 to the present, 3 429 separate NEOs were discovered, and of these 508 are considered potentially hazardous asteroids (PHAs). These objects have minimum orbital intersection distances with the earth less than 0.05 AU and absolute magnitude $H < 22$. While more than 98% of the discoveries are made by professional astronomers, there is room for amateur astronomers to contribute to the NEO effort by performing astrometric follow-up of NEOs and NEO candidates.

4. Comets

Some of the NEO candidates appearing on the NEOCP turn out to be comets, and in cooperation with the Central Bureau for Astronomical Telegrams (a service of Division XII/Commission 6), the MPC sometimes places suspected comets (even if they cannot be NEOs) there deliberately, in the expectation of inspiring quick follow-up observations. After the initial discovery, observations and orbit computations of a comet are routinely handled by the MPC, with temporary publication generally weekly on *MPECs* and monthly (with permanent archiving) in the *Minor Planet Circulars*. Some 260 000 observations of comets were published during 2002 - 2008, giving now a total of 412 000. Three editions of *Catalogue of Cometary Orbits* were issued, in 2003, 2005 and 2008. The 2008 edition contains 3 815 orbits for 3 708 cometary apparitions. These include 1 139 orbits for the 1 062 apparitions of the now 200 numbered comets (i.e., comets observed at multiple apparitions). There are also 1 490 near-sun comets observed only from SOHO and other space probes monitoring the sun.

5. Distant objects

Just 600 discoveries of 'Distant objects'(centaurs and transneptunian objects) were discovered during the double triennium, i.e., some 47% of the total found since 1992. Of the 600, 48% have so far been observed at only a single opposition, a number that actually suggests improvement in the recovery rate recently, given that this fraction is as high as 41% for distant objects as a whole.

Excluding the four 'plutoids', or transneptunian dwarf planets, there are 218 distant objects with orbits considered reliable enough that they have been numbered. Of these, 62 are obvious 'cubewanos', or 'classical Kuiper Belt' objects with orbital semimajor axes in the range 40-47 AU. This number increases to 71 if the range is increased to 37-52 AU. There are 25, or maybe 27 centaurs in the range of the giant planets (depending on how they are defined in terms of coupling with Jupiter), a number that increases to 41 if the semimajor axis is allowed to exceed that of Neptune. Adding what are generally considered 'scattered' (or 'scattering') disk objects (i.e., objects with perihelia somewhat beyond Neptune and generally quite eccentric orbits) augments this number to 48, while the total increases to 61 if similar objects that are considered to be 'detached' (generally with even larger perihelion distances, sometimes in excess of 45 AU) are included. There are then the objects in mean-motion resonance with Neptune, notably the 33 'plutinos', in the 2:3 resonance. The next most populated resonance may be 4:7, possibly with 13 members, although this number may be significantly overestimated, the 44 AU semimajor axis obviously inviting some confusion with cubewanos; less confusion surrounds the objects at the 3:5 resonance, at 42 AU. There is a very clear-cut group, with at least 7 members, at the 2:5 resonance (55 AU). Beyond the cubewanos there are also clearly objects at the 1:2, 3:7 and 1:3 resonances and very possibly at resonances of fifth order and even higher. Closer to the sun than the plutinos are objects at the 3:4 and 4:5 resonances, and while 'Neptune Trojans', at the 1:1 resonance, are known to exist, none has yet been numbered.

6. Outer satellites of the giant planets

Because of their potential confusion with minor planets, the MPC continues also to catalogue observations and to compute orbits for the outer satellites of the giant planets. Eleven new satellites of Jupiter discovered at the 2002 – 2003 opposition have been observed well enough to be numbered (and named), bringing the total to 49, but some listings give 62 (or more) – an unwise claim, given that these objects were observed *only* in 2003 (or in one case only in 2000). Seventeen discoveries of outer satellites of Saturn during 2003 – 2007 have also been numbered, bringing the total number of satellites of Saturn to 52; again, eight more presumed discoveries during 2004 – 2007 have not been confirmed at a second opposition. A new definite outer satellite or Uranus and five satellites of Neptune were also discovered during 2002 – 2003, the total number of satellites for these planets now being 27 and 13, respectively.

7. Personnel

B. G. Marsden retired as MPC director in 2006 after serving in that capacity for 28 years; he continues to work principally with comets, distant objects, outer-planet satellites and on the monthly production of *Minor Planet Circulars*, as well as serve as secretary of the *Committee on Small-Body Nomenclature* and as MPC-CSBN representative on the Working Group for *Planetary System Nomenclature*. Marsden was succeeded as director by T. B. Spahr, who joined the MPC staff in 2000 and has principally been involved with NEOs, for several years also serving as liaison with other organizations working on NEOs. Spahr's workload has now increased with normal (and abnormal) administrative duties. G. V. Williams continues as MPC associate director, responsible for most of the MPC's computer software (and hardware), most of the organization of observations of and orbit computations on main-belt minor planets, and the lion's share of the preparation of the permanent MPC publications. K. E. Smalley (aka S. Keys), who joined the MPC staff as a contractor in 2002, became a full member in 2005 and is also involved extensively with writing software and working with NEOs. M. Lohmiller continues to serve as administrative assistant, attending in particular to e-mail address lists and other matters related to maintaining subscriptions to the MPC publications.

Timothy B. Spahr & Brian G. Marsden
director & director emeritus of the Minor Planet Center

Transactions IAU, Volume XXVIIA
Reports on Astronomy 2006–2009
Karel A. van der Hucht, ed.

© 2009 International Astronomical Union
doi:10.1017/S1743921308025489

DIVISION III / WORKING GROUP
COMMITTEE ON SMALL BODIES NOMENCLATURE
COMITÉ DE LA NOMENCLATURE DES PETITS CORPS CÉLESTES

CHAIR	**Jana Tichá**
SECRETARY	**Brian G. Marsden**
Div. III representative	**Edward L. G. Bowell** (> 2006)
	Iwan P. Williams (until 2006)
MPC representative	**Brian G. Marsden**
CBAT representative	**Daniel W. E. Green** (for comet names only)
WG-PSN representative	**Kaare Aksnes** (until 2006)
	Rita M. Schulz (since 2006)
MEMBERS	**Michael F. A'Hearn, Julio A. Fernández,**
	Pamela Kilmartin, Yoshihide Kozai (until 2006),
	Daniela Lazzaro, Syuichi Nakano (since 2006),
	Keith S. Noll (since 2006),
	Lutz D. Schmadel (editor DMPN),
	Viktor A. Shor, Richard M. West (until 2006),
	Gareth V. Williams, Donald K. Yeomans,
	Jin Zhu

TRIENNIAL REPORT 2006-2009

1. Comets

A total of 701 comets received names between July 2005 and June 2008. Comets observed only from the *SOHO* and *STEREO* missions, as well as further comets recognized from the long-defunct *SOLWIND* satellite, accounted for 520 of these names.

2. Minor planets

Names for 2228 minor planets were approved and published in the Minor Planet Circulars in the period July 2005 – June 2008. In October 2005 the milestone of the 100 000th numbered minor planet was reached. There are now nearly 190 000 numbered minor planets. As of June 2008, the total number of named minor planets was 14 574. This marked a reduction compared to the 2766 names of the previous triennium. This was part of conscious effort to limit the naming process.

In the resolution adopted at the IAU General Assembly in July 2003 and printed on the MPC 49221 the *Committee on Small-Body Nomenclature* (CSBN) recognized the need to limit the number of minor planets named. It requested individual discoverers and observing teams to propose no more than two names every two months (corresponding to the timescale generally adopted for the publication of new names in the MPC). The CSBN also set a practical limit of 100 name proposals for consideration in each bimonthly batch. The CSBN members think that it is far better to concentrate on having a smaller number of meaningful names with broad international appeal than to name all the tens of thousands of newly numbered main-belters. The majority – but not all – of the discoverers cooperate, and this system seems to work. The CSBN continued to examine ways of limiting the naming process, particularly in the context of future years when the PanSTARRS and LSST projects become operational.

The CSBN reserves the right to accept additional names that are considered particularly meritorious, especially for traditional cases such as the those of planetary astronomers honored at the triennial *Asteroids, Comets and Meteors* meetings and of the science-fair students and mentors honored by LINEAR.

3. Most significant names

Minor planet (100000) Astronautica has been named to recognize the fiftieth anniversary of the start of the Space Age, inaugurated by the launching of the first artificial earth satellite on 4 October 1957. The name is associated with this significant number, as space is defined to begin at an altitude of 100 000 meters above the earth's surface. The name was given on the MPC 60731 in September 2007.

The first names for a recently discovered dwarf planet and its satellite were adopted in September 2006. Following near-unanimous acceptance by both the CSBN and the Working Group on *Planetary-System Nomenclature* (in consultation with the discovery team), the IAU Executive Committee approved the names Eris for (136199) 2003 UB_313 and Dysnomia for its satellite (136199) Eris I. These names were published on IAUC 8747 and MPC 57800.

Names for four smaller outer-solar system bodies and their satellites (or binary objects) were approved by the CSBN during this triennium. They are (42355) Typhon and (42355) Typhon I (Echidna), published on IAUC 8778 and MPC 57951; (65489) Ceto and (65489) Ceto I (Phorcys), published on IAUC 8778 and MPC 57952; (88611) Teharonhiawako and (88611) Teharonhiawako I (Sawiskera), published on IAUC 8840 and MPC 59388; and (66652) Borasisi and (66652) Borasisi I (Pabu), published on MPC 60731. The first two pairs are 'extended Centaurs' with mean distances beyond Neptune, and the other pairs are 'cubewanos' or 'classical Kuiper-Belt Objects'.

4. Name for transneptunian dwarf planet category

After introducing the category of dwarf planets, the IAU followed up on its footnote to Resolution 6 at the 2006 General Assembly and initiated a search for a name for transneptunian dwarf planets that are similar to Pluto. In June 2007 the CSBN received a request from the General Secretary of the IAU to propose such a name. After a long and wide discussion, at the end of August the CSBN voted on the various suggestions that had been proposed. These choices involved words related to the name of the prototype object Pluto, to the discoverer of Pluto, as well as to the transneptunian location and to the presumed physical physical nature of these objects. The name finally selected by the CSBN, *plutoid*, was approved by the IAU Executive Committee at its meeting in Oslo, Norway, in May 2008.

5. Development of guidelines

The CSBN has been working on the further development of guidelines for minor planet naming.

A naming convention has been proposed for non-resonant objects with $a > 30.07$ AU and 5.2 AU $< q < 30.07$ AU. These objects are in chaotic and short-lived orbits very similar to the orbits of the Centaurs. The proposed convention would name such objects for mythical creatures, particularly hybrids (human/animal, animal/animal) and shapeshifters. Lesser bodies in binary or multiple systems are to be named for closely related mythical characters such as siblings, spouses, parents and children, or other closely related characters. Examples are the Typhon-Echidna and Ceto-Phorcys pairs mentioned above.

Amplifying the guideline concerning commercial names, the CSBN proposed that sports-related names should not be for sports teams, but be limited to meritorious individuals in the world of sport.

6. Miscellanous

An addendum to the fifth edition of the *Dictionary of Minor Planet Names* was published by Springer-Verlag (Berlin-Heidelberg-New York) in 2006. Edited, as in the past, by L. D. Schmadel (Astronomisches Rechen-Institut, Heidelberg), it contains discovery and naming information on minor planets and their satellites that were named during 2003 - 005, together with amendments to previously published citations.

Brian G. Marsden
secretary of the Commission

Transactions IAU, Volume XXVIIA
Reports on Astronomy 2006–2009
Karel A. van der Hucht, ed.

© 2009 International Astronomical Union
doi:10.1017/S1743921308025490

DIVISION III / WORKING GROUP
PLANETARY SYSTEM NOMENCLATURE

LA NOMENCLATURE DU SYSTÈME PLANÉTAIRE

CHAIR	**Rita M. Schulz**
MEMBERS	**Kaare Aksnes, Jennifer S. Blue,**
	Jürgen Blunck († 3 July 2008),
	Edward L. G. Bowell, George A. Burba,
	Guy J. Consolmagno (since November 2008),
	Régis Courtin, Rosaly M. Lopes-Gautier,
	Mikhail Ya. Marov, Brian G. Marsden,
	Mark S. Robinson, Vladislav V. Shevchenko,
	Bradford A. Smith

TRIENNIAL REPORT 2006 - 2009

1. Introduction

The Working Group on *Planetary System Nomenclature* (WG-PSN) develops, maintains and publishes guidelines for naming natural satellites of planets and surface features on all solar system bodies except Earth. When required the WG approves lists of new nomenclature, with accompanying explanatory notes, based on the established guidelines. Approved names are immediately added into the *Gazetteer of Planetary Nomenclature*. Objections based on significant, substantive problems may be submitted within a 3-months period, and will be ruled on by Division III.

The WG-PSN is supported by six Tasks Groups dedicated to naming issues for the Moon, Mercury, Venus, Mars, Outer Solar System, and Small Bodies. To simplify the name-request process for the scientific community a Name Request Form has been made available electronically on the *Gazetteer* web page. This also ensures that a data archive of name requests can be maintained by the Working Group. A page with frequently asked questions (FAQ) was also added to the *Gazetteer*. Details on the nomenclature and the naming process can be obtained from the *Gazetteer of Planetary Nomenclature* <planetarynames.wr.usgs.gov>.

2. Activities between the IAU XXVI General Assembly in 2006 and 1 June 2008

Since the IAU General Assembly in Prague in August 2006, all discussion and approval procedures were handled via email. As of 1 June 2008, the WG-PSN has approved 259 new names for surface features on solar system bodies: 46 on Dione, 35 on Enceladus, 9 on Europa, 12 on Mercury, 2 on Io, 79 on Mars, 8 on Phobos, 33 on Tethys, 22 on Titan, and 13 on Venus. In addition 22 names for satellites of the Giant Planets were approved, 1 of Jupiter, 17 of Saturn, and 4 of Neptune, and the spelling of one Saturnian satellite was corrected. Concerning ring systems, the WG-PSN provided the following definition for divisions and gaps: 'Divisions are the separations between named rings, and gaps are the spaces within named rings. In general, divisions are large, gaps are small.' In accordance with this definition the 'Roche Division' of the Saturnian ring system was defined as the division between the outer edge of the A ring and the inner edge of the F ring, and the term 'Encke Division' was changed to 'Encke Gap'. In addition, new figures for Saturn's rings were approved and added to the Gazetteer.

A rule for the procedure of dropping names became necessary, which was defined as follows.

"Official approval by a Task Group and by the Working Group is required before a name can be dropped. Dropped names are retained in the database for reference; the code in the 'approval status' field is changed to '6' indicating the name has been dropped, the name is shown in brackets in the 'name' field, and a brief notation is made in the 'origin' field explaining why the name was dropped and when it was dropped. In general, names that have been dropped should not be re-used for other features. Dropped names could be reused in very exceptional cases; for instance, if no new names of a particular theme are available and there is strong justification, a dropped name could be considered for reuse."

Six names were dropped, 4 on Mars, 1 on Venus, and 1 on Dione, for which the descriptor terms of three names were also revised.

The existing definition of the descriptor term 'mare/maria' (as used on the Moon) was amended for use on Titan and now reads: 'Sea; large circular plain; on Titan, large expanses of dark materials thought to be liquid hydrocarbons'. The descriptor term Insula/insulae was added with the definition: 'Island (islands), an isolated land area (or group of such areas) surrounded by, or nearly surrounded by, a liquid area (sea or lake)'. A few particular themes for Mars, Phobos, Mercury and Io were expanded and four new themes were added for Titan.

3. Closing remarks

The future activities will continue to be carried out mainly via email. However, a WG-PSN meeting will be held in Ithaca, NY, USA on 10-11 October 2008, to discuss basic working group business, such as performance, need for rules and guidelines, and issues of concern. All members of the WG and the Task Groups have been invited to attend and to submit agenda items.

Rita M. Schulz
chair of the Working Group

Transactions IAU, Volume XXVIIA
Reports on Astronomy 2006–2009
Karel A. van der Hucht, ed.

© 2009 International Astronomical Union
doi:10.1017/S1743921308025507

DIVISION IV   STARS

ÉTOILES

IAU Division IV organizes astronomers studying the characteristics, interior and atmospheric structure, and evolution of stars of all masses, ages, and chemical compositions.

PRESIDENT	Monique Spite
VICE-PRESIDENT	Christopher J. Corbally
PAST PRESIDENT	Dainis Dravins
BOARD	Christine Allen, Francesca d'Antona,
	Sunetra Giridhar, John D. Landstreet,
	Mudumba Parthasarathy

DIVISION IV COMMISSIONS

Commission 26	Double and Multiple Stars
Commission 29	Stellar Spectra
Commission 35	Stellar Constitution
Commission 36	Theory of Stellar Atmospheres
Commission 45	Stellar Classification

DIVISION IV WORKING GROUPS

Division IV WG	Massive Stars
Division IV WG	Abundances in Red Giants

INTER-DIVISION WORKING GROUPS

Division IV-V WG	Active B stars
Division IV-V WG	Ap and Related Stars
Division IV-V-IX WG	Standard Stars

TRIENNIAL REPORT 2006 - 2009

1. Introduction

Stars are the main visible constituents of the galaxies, and their properties are basic to our knowledge of the Universe.

- A precise determination of the structure of the interior of the stars, and of the physical phenomena at work, is necessary to predict their evolution as a function of their masses and their chemical composition. From these predictions, the ages of globular clusters are determined, directly linking to the age of the Universe and discriminating among different populations of the distant galaxies.

- The distance scale of the Universe is based on semi-primary distance indicators as Cepheids and type-Ia supernovae , X-ray binaries and binary milliseconds pulsars, which are late stages of stellar evolution. Detailed modelling is necessary to understand the routes to their properties.

- Stellar abundances are a fundamental parameter used, for example, to check the primordial nucleosynthesis, the ages of the stars (from radioactive elements), the composition of supernovae ejecta, the star formation rates and even the initial mass function. A precise computation of these abundances in our Galaxy or in the Local Group galaxies requires very precise models of stellar atmospheres and the use of the best physics, e.g., NLTE line formation, 3D hydrodynamical model atmospheres. At the present time the uncertainties due to the interpretation of the data often exceed the uncertainties due to the measurement errors.

A role of our Division is also to promote progress in establishing the fundamental data necessary to provide the required ground truth, covering many aspects of atomic and molecular physics. All branches of astronomy benefit from progress in the understanding of stellar physics.

2. Developments within the past triennium

The scientific activity in the large field covered by Division IV has been very intense during the last triennium. It led to the publication of a very large number of papers which makes an exhaustive report impossible to produce. Here below are only some examples of recent progress:

- New wavelength regions, e.g., in the infrared, are now open to investigation with very powerful spectrographs based on mature detector technology. Large IR surveys provide extensive public data sets (UKIDSS, Spitzer Legacy Programs) and the Gould Belt Legacy survey will provide very soon samples of young stellar objects in the clouds down to the planetary mass regime, allowing time scales for different evolutionary phases and processes of clustered versus non clustered star formation be tested.

- Stellar classification once again finds itself at the forefront of stellar discovery and astrophysics, with the discovery of two new classes of cool stars (L and T) and the probable imminent discovery of yet another class (Y) of even cooler brown dwarfs.

- Recent solar observations have shown connections between the interior dynamics, surface magnetism and coronal phenomena, although it is still unclear where and how magnetic fields are generated in the Sun. Recent optical and X-ray observations of different stars with significant convective zones beneath their surfaces, reveal star spots and other surface atmospheric structures similar to solar magnetic ones. In many cases the long term evolution of these features is similar to the solar cycle. Advances in the Doppler imaging technique has allowed to obtain maps of star spot distributions and the first accurate measurements of stellar differential rotation. Moreover, the satellite *CoRoT* has detected solar-type oscillations in solar-type stars at unprecedented levels of precision.

- *CoRoT* now enables the study of photometric and spectroscopic variability of pre-main sequence stars with a rich variety of phenomena, e.g., pulsation, hot and cold spots, high modes of vibration, stellar multiplicity.

- Details of the structures of the stars are also obtained with interferometers. Stellar winds can be 'visualized'. Recently the use of ESO's Very Large Telescope Interferometer and its razor sharp eyes has helped to discover a reservoir of dust trapped into a disc that surrounds an elderly star. More extremely iron-poor stars have been discovered: three stars are now known with [Fe/H] < -4.8. The most iron-poor star known today is HE 1327−2326 with [Fe/H] = −5.4. However, all these stars are very carbon rich. Are these very peculiar stars the oldest ones in the Galaxy?

3. Activity of Division IV Commissions and Working Groups

Division IV includes five Commissions and two Working Groups (see the header). Moreover, three Working Groups are shared with Division V (Variable Stars). All the Commissions and Working Groups have been very active during the last triennium, and all desire and deserve to continue their activity during the next one.

The Division IV / Commission 45 Working Group on *Standard Stars* has become, following a request by its membership, an Inter-Division IV-V-IX Working Group.

Several IAU symposia have been (or will be) coordinated by Division IV between September 2006 and August 2009.
In 2007:
- IAU S250 on *Massive Stars as Cosmic Engines*, December 2007, Kauai, HI, USA.
In 2008:
- IAU S258 on *The Age of Stars*, October 2008, Baltimore, MD, USA.
- IAU S252 on *The Art of Modelling Stars in the 21st Century*, April 2008, Nanjing, China.
In 2009:
- IAU S265 on *Chemical Abundances in the Universe – connecting first stars to planets*, August 2009, Rio de Janeiro, Brazil.
- IAU S268 on *Light elements in the Universe*, November 2009, Geneva, Switzerland.

Moreover, several other Symposia have been supported by our Commissions or Working Groups.

The following reports from the commissions and the working groups of the Division, document the highly dynamic field of stellar astrophysics.

The author is grateful to the board of the Division for their helps in preparation of the report.

Monique Spite
president of the Division

Transactions IAU, Volume XXVIIA
Reports on Astronomy 2006–2009
Karel A. van der Hucht, ed.

© 2009 International Astronomical Union
doi:10.1017/S1743921308025519

COMMISSION 26   DOUBLE AND MULTIPLE STARS
ÉTOILES DOUBLES ET MULTIPLES

PRESIDENT	Christine Allen
VICE-PRESIDENT	Jose A. D. Docobo
PAST PRESIDENT	William I. Hartkopf
ORGANIZING COMMITTEE	Yuri I. Balega, John Davis,
	Brian D. Mason, Edouard Oblak,
	Terry D. Oswalt, Dimitri Pourbaix,
	Colin D. Scarfe

COMMISSION 26 WORKING GROUP

Div. IV / Commission 26 WG	Binary and Multiple System
	Nomenclature

TRIENNIAL REPORT 2006 - 2009

1. Introduction

Although we were happy to welcome over 20 new members at the Prague meeting, Commission 26 is still one of the smallest in the IAU. Notwithstanding its size, it continues to carry on an active and diversified program of activities. Our web site, maintained at the US Naval Observatory, contains further information on the Commission. The site includes links to other relevant sites, to databases and catalogues, an archive of our Information Circulars, a list of upcoming meetings of interest, as well as an extensive bibliography of recently published papers on double and multiple stars. The site can be accessed at <ad.usno.navy.mil/wds/dsl.html#iau>.

2. Meetings

2.1. *IAU Symposium No. 240*

The largest meeting in the history of our Commission was held during the IAU XXVI General Assembly in Prague, August 2006. IAU Symposium No. 240 on *Binary Stars as Critical Tools and Tests in Modern Astrophysics*, was jointly sponsored by Commissions 26 and 42 with support from 11 other commissions and working groups. The 3.5-day meeting was attended by some 500 astronomers from 54 countries and included talks covering all aspects of binary and multiple star research: from very long period, common proper motion pairs and other 'fragile' binaries to short-period contact binaries, binaries with degenerate components, as well as star/brown-dwarf/planet systems, with the aim of exploring interests common to all binary star researchers. Both the observational and theoretical aspects of binary and multiple star research were represented, but the main themes of the program were the new information and physical insights gleaned from the recent advances in instrumentation and techniques. The meeting also attracted those interested in the observational and theoretical aspects of modern stellar astrophysics that depend very strongly on the fundamental properties of stars found primarily from binary and multiple stars.

The format for the symposium was a mix of invited oral review presentations and more narrowly-focused topical presentations. These 40 invited talks were supplemented by over twenty short oral/poster presentations, selected by the SOC from over 180 submitted posters.

As the first major joint meeting in modern memory of the 'close' and 'wide' binary communities, it was deemed appropriate to dedicate the meeting in honor of Mirek J. Plavec†, and in memory of Charles E. Worley. Also remembered at the meeting was Wulff D. Heintz, who had died two months earlier.

Proceedings of the meeting were edited by William Hartkopf, Petr Harmanec and Ed Guinan, and published by Cambridge University Press in 2007. A companion website at Cambridge Press includes 115 complete posters from the meeting, as well as all talks and poster abstracts in the book.

2.2. *IV International Meeting on Dynamical Astronomy in Latin America*

The meeting was held in Mexico City's Historical Center and was attended by some 55 astronomers from over 14 countries. A large fraction of the papers presented dealt with various aspects of double and multiple stars. Some of the highlights included sub-milli arcsecond accuracy parallaxes applied to associations, VLBI astrometry of young multiple systems, milli-arcsecond binaries, and the effect of circumstellar and circumbinary disks in eccentric binaries. The Proceedings of the meeting are being edited by C. Allen, R. Teixeira and A. Ruelas-Mayorga, and will be published as a volume of the *Revista Mexicana de Astronomia y Astrofisica* (CS).

2.3. *IAU Symposium No. 248*

IAU Symposium No. 248 *A giant step, from milli to micro arcsecond astrometry* was held in Shanghai in October, 2007, with the support of Commission 26. It brought together results from the *Hipparchos* catalog and studies related to the upcoming *Gaia* mission, as well as ground-based projects. It was attended by a significant number of active members of our Commission.

2.4. *Workshop Multiplicity in Star Formation*

The role of multiplicity in star and planet formation was the main focus of this workshop, held in Toronto, Canada, 16-18 May 2007. Topics included multiplicity surveys of protostars, T Tauri stars and brown dwarfs, multiplicity and the initial mass function, and planet formation in multiple systems.

2.5. *ESO Workshop Multiple Stars across the HR Diagram*

This workshop was held in Garching, Germany, 12-15 July 2005, and the proceedings were published in 2008 by Springer. Editors of the volume were S. Hubrig, M. Petr-Gotzens & A. Tokovinin. Topics included observations of multiple stars from ground and space, dynamical and stellar evolution in multiple systems, environmental effects, formation and early evolution of multiples and chemically peculiar objects.

---

† Professor Plavec, who attended the meeting with his family, unfortunately passed away at his home on January 23, 2008.

3. Information Circulars

Continuing a long-time tradition, the Commission has published *Information Circulars* three times per year. The Circulars are edited by J. A. Docobo and J. Ling, of the Observatorio Astronomico R.M. Aller at the Universidad de Santiago de Compostela, Spain. The *Circulars* are distributed electronically and archived on the Commission 26 web site. During the period October 2006 to June 2008 six *Circulars* were published, announcing a total of 134 orbits, and containing also miscellaneous information, an extensive bibliography on double and multiple stars. Obituaries of Omer Nys (1931 - 2007) and Pierre Bacchus (1923 - 2007) were carried, respectively, in *Circulars* 162 and 163. We deeply regret the passing away of these two distinguished double star astronomers.

4. Catalogues and journals

The U.S. Naval Observatory double star program maintains five astrometric and photometric catalogs, each of which has seen considerable growth during the past three years. These five catalogs are updated regularly on the USNO website.

The *Washington Double Star Catalog* is the principal repository for all published astrometry of visual binary and multiple stars. As of 30 June 2008 the WDS contained 748 343 mean measures of 103 812 systems. The database continues to be improved through correction of errors, removal of duplicate discovery designations and/or measures, and improvements in coordinate, proper motion, and magnitude information. Observing lists based on the WDS are provided on demand for other observers.

As of 30 June 2008, the *Sixth Catalog of Orbits of Visual Binary Stars*, included 2 088 orbits of 1 956 systems, from a 'master file' of 6 028 orbits.

The *Fourth Catalog of Interferometric Measurements of Binary Stars* currently includes 143 903 observations of 73 275 systems. Binary measurements from speckle interferometry, adaptive optics, multi-aperture arrays, *Hipparcos*, and other high-resolution techniques are included, as are negative results from duplicity surveys.

The *Third Photometric Magnitude Difference Catalog* includes 158 946 measures of 52 752 systems. This is somewhat reduced in size from the previous version of the catalog since duplicate photometric information also in the main WDS database was excluded. The *Catalog of Rectilinear Elements* was posted to the web during the past triennium; the catalog currently includes elements for 1 170 (presumably) optical pairs, although this number is expected to increase substantially. Studies have shown that these linear fits often yield more precise differential proper motions than are available elsewhere, due to the long time base of differential astrometry measures available in the WDS database.

The second USNO Double Star CD, containing versions of all five catalogs as of 2006.5, was published just before the Prague GA and distributed to those in attendance as well as all members of the Commission. Additional copies are available upon request.

A new catalogue of 6 330 eclipsing variable stars was completed (Malkov *et al.* 2006). The catalogue was developed from the General Catalogue of Variable Stars (GCVS) including new classifications of 843 systems and correcting the GCVS data. The catalogue represents the largest list of eclipsing binaries classified from observations. Photometric and spectroscopic observations for selected *Hipparcos* eclipsing binaries were undertaken with the aim to derive absolute dimensions (Oblak *et al.* 2008).

For the near future, a catalogue of fundamental parameters of eclipsing binaries is planned (Oblak *et al.* 2009). Fundamental physical parameters will be determined for about three thousands eclipsing binaries. The transfer of the database of double stars (BDB) of the Besançon Observatory to the Astronomical Institute of the Russian Academy of Sciences in Moscow will be completed, with a mirror site of the BDB at Besançon.

The *Journal of Double Star Observations*, a quarterly publication begun in early 2005 by astronomers at the University of South Alabama, has seen considerable growth during the past triennium. This web-based refereed journal now represents a major source of published measures and other double star information by both professionals and amateurs. Other venues for publication of double star information, including the *Webb Society Circulars* and *Observations et Travaux*, also illustrate the high level (and quality) of activity in our field within the amateur observing community. The Webb Society Double Star Section continues to encourage the observation of visual double stars. During the report period two *Circulars* (Nos. 14 and 15) were issued, containing 2 643 measures of pairs with separations ranging from 0.3 to over 100 arcsec, collected by 10 observers.

5. Speckle interferometry and other single-aperture high-resolution astrometry techniques

5.1. *US Naval Observatory program*

The speckle interferometry program also continues at the USNO. Most observations have been carried out on the venerable USNO 26-inch refractor, with $\sim 4\,300$ mean measures published or in press since mid-2005 (c.f., Mason *et al.* 2006a; 2006b; 2007). Much of this observing effort has been directed toward duplicity confirmations and observations of systems not measured for many years, as well as regular monitoring of potential orbit systems. A new 'backup' speckle camera was designed and built at the USNO and installed on the 26-inch in mid-2006; this allows use of the 'primary' camera at larger telescopes with no observing time lost due to equipment in transit.

The 'primary' speckle camera was used on several larger telescopes during the past triennium, including the KPNO and CTIO 4-meters and the Mount Wilson 2.5-meter. Over 3 000 nearby G dwarf stars were examined for duplicity during this period, as were a similar number of massive (O-type, B-type, and Wolf-Rayet) stars. Results from these analyses are being prepared for publication.

Adaptive optics on the 3.6-meter AEOS telescope on Haleakala, Hawaii has been used in searches for faint companions to B-type stars (Roberts *et al.* 2007) and O-type stars (Turner *et al.* 2008). They confirm the Mason *et al.* (1998) result on the paucity of companions of O-type star runaways. Possible stellar companions to exoplanet host stars have also been searched for (Roberts *et al.* in preparation).

5.2. *The Wisconsin-Indiana-Yale-NOAO program*

E. Horch (Southern Connecticut State University) and W. van Altena (Yale University) continued to use the WIYN 3.5-m Telescope at Kitt Peak for speckle observations of double stars, particularly those discovered by the *Hipparcos* satellite. Over 90% of the observable *Hipparcos* doubles have been studied, and characterization of the motions of many pairs have been made. Horch is completing a new speckle camera, the Differential Speckle Survey Instrument, which can take speckle data in two filters simultaneously. This will make color determinations of the components of binary stars highly efficient. The program has also recently obtained an electron-multiplying CCD camera and has used it for the first time in speckle observations at WIYN in June 2008. This device has near photon-counting performance at extremely high quantum efficiency, allowing diffraction-limited imaging of binary systems as faint as at least 15th magnitude.

5.3. *The PISCO program*

The goal of this long-term program is to obtain speckle interferometry measurements of visual binary stars with the Pupil Interferometry Speckle camera and COronograph

(PISCO) instrument. It was developed in the early 1990's at the Observatoire Midi-Pyrenees, Toulouse, France, for the Pic du Midi Observatory where it was exploited on the 2 m 'Bernard Lyot' telescope from 1993 to 1998. After a period of inactivity, PISCO has been used since January 2004 on the 102-cm Zeiss telescope of the INAF - Osservatorio Astronomico di Brera, in Merate, Italy. The PISCO team participants are M. Scardia, L. Pansecchi (both at INAF-Osservatorio Astronomico di Brera, Merate, Italy), R. W. Argyle (Institute of Astronomy, Cambridge, UK), and J.-L. Prieur (Laboratoire d'Astrophysique de Toulouse-Tarbes), along with many collaborators.

From September 2006, until June 2008, 642 different binaries were observed, with a total of 724 observations. Those observations allowed the computation of twelve new or revised orbits. Results are given in the references (under Scardia and Prieur) as well as in the *Information Circulars* Nos. 160, 161, 162, 163 and 165.

Technically, both PISCO and the ICCD detector have continued to perform well. The computers and instrument control software have been updated. The reduction procedure has also been improved, allowing measurements slightly below the diffraction limit of the telescope. A new method to correct the bispectra has been developed. During the next triennial period, the PISCO program should continue to provide results at the same pace.

5.4. *The SAO program*

Speckle interferometric survey of 250 selected low-metallicity stars was performed with the 6-m telescope of the Special Astrophysical Observatory, Russia. 15 new binary and multiple systems were discovered. Taking into account already known spectroscopic pairs, the total ratio between the single : binary : triple : quadruple systems is 147 : 64 : 9 : 1. The distributions of the orbital periods and mass ratios were also determined for this sample of halo stars (Rastagaev, Balega & Malogolovets 2007).

The improved orbital parameters were derived for the massive young binary Θ^1 Ori C using the results of the 6-m speckle and VLTI observations shortly after periastron passage in 2007. The derived elements imply a short-period (P = 11.1 yr) and high-eccentricity ($e = 0.86$) orbit with a total mass-sum of $49 \pm 7 \, M_\odot$ (Kraus *et al.* 2007).

Precise speckle interferometric orbits were calculated for 10 new *Hipparcos* binaries using the results of observations with the 6-m telescope in the period 1998-2007. The accuracy of obtained dynamical masses is in the range 10-20%.

A speckle survey of 83 OB stars in the Cas-Tau association was performed at the telescope with an angular resolution of 20 mas. Twelve new binaries were discovered and the mass function was estimated for the stars in the association.

5.5. *The Calar Alto program*

First results of the optical speckle camera with the 3.5 m telescope at Calar Alto were reported by Docobo *et al.* (2008). Fifty stars with separations between 0.058 and 2.0 arcs were observed. Two new (COU 490 and A 2257) and six improved orbits and masses were calculated. Dynamical parallaxes allowed first rough distance estimates for COU 490 (145 pc) and A 2257 (210 pc). High-quality optical speckle data on binaries with separations close to the diffraction limit of the 3.5 m telescope can now routinely be obtained.

6. Long-baseline interferometry

Georgia State University's CHARA Array, located on Mt. Wilson, California, has concentrated to date on single-star science. Nevertheless, several programs aimed at detection of new binaries and orbit determinations are under way with the results to be reported in

2008 and 2009. Surveys to detect new companions to solar type stars primarily through the detection of double fringe packets corresponding to binaries with angular separations in the 5-30 milliarcsec range are being done. Such objects fall in the 'gap' separating speckle detection and spectroscopic detection.

Binaries are also occasionally being found in the course of measuring stellar angular diameters. They show up quite readily, since the distribution of visibilities cannot be modeled as the result of a resolved photosphere from a single star. In a collaboration with spectroscopists at the Center for Astrophysics and at Tennessee State University, several binaries with excellent double-lined spectroscopic orbits are being followed to determine accurate masses and orbital parallaxes. This includes a 1.0 d system with a separation of approximately 1 mas, the shortest-period system for which a 'visual' orbit has been determined. A program is also under way involving triple systems for which the long-period orbit is resolved by speckle interferometry. This then permits the 'wide' component to be used as the visibility calibrator for the short-period system.

Finally, and most dramatically, an imaging capability has been realized at the CHARA Array through a collaboration with a team from the University of Michigan who have fabricated a four-telescope beam combiner. This system has been used to image several binaries, most notable of which is the close and interacting binary Beta Lyrae, for which Roche lobe filling is apparent for one of the components. In another demonstration of the power of imaging binaries, orbital motion amounting to 15 micro-arcsec has been accurately measured for the spectroscopic binary Iota Peg from a series of observations spanning only 45 minutes in time. In contrast with visibility measurements, imaging provides the classical position angle and angular separation measurements in addition to the magnitude difference and, in some cases, the diameters of the components stars themselves.

The SUSI array produced further results. Among others, high-precision interferometric measurements of Spica and γ^2 Velorum allowed an improved determination of the fundamental parameters of these systems.

From the NPOI array new data on the binary system Θ^1 Ori C, which was spatially resolved over the period from 2006 February to 2007 March, show significant orbital motion. However, a definitive orbit solution will require more time. An ongoing survey for multiplicity among the bright stars will complement the visual orbits obtained for known close speckle and spectroscopic binaries, and detect new binary/multiple systems.

7. Other projects

7.1. *Young systems*

Zinnecker and collaborators (Correia *et al.* 2006), searching for hierarchical triples in nearby visual binaries using adaptive optics, found an enhanced fraction of young low-mass hierarchical triples (about 1/3). The *HST*-ACS survey of the outer Orion Nebula Cluster was completed by Reipurth *et al.* (2007) producing dozens of new sub-arcsec low-mass young binaries. Further, Koehler *et al.* (2006) published an adaptive optics results paper with new pre-MS binary statistics in the Orion Cluster. Koehler & Ratzka (2008) studied the orbits and masses in the prototypical T Tauri system. Multi-epoch radial velocity observations aimed at detecting massive companions of young embedded O-type stars were conducted by Apai *et al.* (2007) using VLT/ISAAC at $R = 10\,000$. They found radial velocity variations in two of 10 cases. New ideas about the origin of Trapezium type systems, and more generally, about multiplicity among massive stars, were proposed by Zinnecker at the ESO Workshop 2005 (Zinnecker 2008). Very young embedded Class I pre-main sequence binaries were studied by Connelley (2008). Surprising dissimilarities in a newly formed pair of 'identical twin' stars in the Orion nebula were recently discovered by Stassun *et al.* (2008).

7.2. *Wide binaries, trapezium systems, runaway stars*

The role of large samples of wide binaries for a great variety of galactic studies was emphasized by Chaname (2007). New faint common proper motion companions to *Hipparcos* stars were identified by Lepine & Bongiorno (2007), who also found that at least 9.5% of the nearby ($d < 100$ pc) stars have companions with separations greater than 1000 AU. The distribution of separations was found to follow Oepik's relation up to separations of 4000 AU.

Poveda *et al.* (2007) continued their studies on wide binaries, trapezium systems and runaway stars. They confirmed that wide binaries from various catalogues follow the Oepik relation for semiaxes larger than about 60 AU up to a critical separation that is a function of the age of the system. Poveda *et al.* (2008) discovered that Giclas 112-29 (NLTT 18149) is a very wide ($s = 1°.09$) companion to GJ 282 AB having a common proper motion, common parallax, common radial velocity and common age. A study of high velocity, metal-poor wide binaries led to the identification of several associated moving clusters in the galactic halo, which may be the remains of either globular clusters or captured dwarf galaxies in the process of disintegration. Chemical peculiarities allow to differentiate between both cases (Allen *et al.* 2007). The discovery (Rodriguez *et al.* 2005; Gomez *et al.* 2005; Gomez *et al.* 2008) that the runaway system BN/I/n in Orion is a multiple in the process of disintegration has confirmed the model for the acceleration of runaway stars by strong dynamical interactions in a compact few-body system (Poveda *et al.* 2008; Allen *et al.* 2008). A new technique ('diffracto-astrometry') has been developed to utilize the diffraction patterns of over-exposed *HST* images of the Orion Trapezium. With this technique recent values of the separation of components A-E have been precisely determined, (Sanchez *et al.* 2008) which, together with older determinations, have shown that the relative proper motion of A-E is 3.5 mas/yr. Further, a radial velocity study (Costero *et al.* 2008a) has shown that the systemic radial velocity of E is 8.3 km/s, larger than the average radial velocity of the ONC members and very similar to the transverse velocity relative to component A derived by Sanchez *et al.*, confirming that component E is leaving the system with a space velocity of about 11 km/s. Unexpectedly component E was found to be a double-lined spectroscopic binary with $P = 9.8952$ d, $e = 0$ (Costero *et al.* 2008b).

The discovery of a planetary-mass body ($M \sin i = 3.2\,M_{\mathrm{Jupiter}}$) orbiting the star V391 Pegasi at a distance of about 1.7 AU, with a period of 3. yr was recently reported by T. Oswalt and collaborators (Silvotti *et al.* 2007) This detection of a planet orbiting a post-red-giant star demonstrates that planets with orbital distances of less than 2 AU can survive the red-giant expansion of their parent stars.

Post-main-sequence mass loss causes orbital separation amplification in fragile binary star systems (with typical separations around 1000 AU). Such wide pairs evolve as two separate but coeval stars. Oswalt and collaborators (Oswalt *et al.* 2007; Johnston *et al.* 2007) have studied the statistical distortion of the frequency distribution of fragile binary separations caused by the mass loss. This process provides a robust test of current theories of stellar evolution and sets constraints on the dynamics of the Galactic disk.

7.3. *Stellar multiplicity*

A. Tokovinin completed several studies on multiple stars including a new catalog of multiples among bright stars (Eggleton & Tokovinin 2008) where it was found that among

4559 systems with (combined) magnitudes brighter than 6.00 the frequencies of multiplicities 1, 2, ..., 7 were 2718, 1437, 285, 86, 20, 11, and 2, respectively. Other relevant statistical results on multiple stars (triples and quadruples) appear to show that the properties of multiple stars do not correspond to the products of dynamical decay of small clusters (Tokovinin 2008). New tertiary companions to solar-type spectroscopic binaries were found using adaptive optics. The periods of tertiary companions are distributed in a wide range, from 2 to 100 000 yr, with most frequent periods of few thousand years (Tokovinin *et al.* 2008).

7.4. *Massive binaries, interacting stars*

Multicolour (*UBVRI*) polarimetric studies (Piirola *et al.* 2006) of massive interacting binaries (W Ser, SX Cas) have provided a detailed view into the dense electron scattering envelopes surrounding the mass-gaining components, and new insights into the heavy mass-transfer phenomenon taking place in the current evolutionary phase of these binaries.

The highest magnetic field ($B = 30$ MG) so far seen in an Intermediate Polar has been discovered by circular polarimetry of V405 Aur at the Nordic Optical Telescope (Piirola *et al.* 2008). This makes the system a likely candidate as a progenitor of a synchronous magnetic binary, polar. Similar studies of other probable candidates have been carried out at the ESO VLT and 3.6 m telescopes in 2006 - 2007.

A new code to calculate light curves, spectra and images of interacting binaries immersed in a moving circumstellar environment which is optically thin was developed by Jan Budaj and collaborators (Budaj, Richards & Miller 2005; Miller *et al.* 2007). It solves simple radiative transfer along the line of sight in moving media. The main applications are in the field of interacting binaries, cataclysmic variable stars, and Algol-type eclipsing binaries. The code is fully public and the latest version can be found at `<ta3.sk/ budaj/shellspec.html>`.

7.5. *Visual binaries and multiples*

H. Abt (2008a) continued his studies of fundamental parameters of multiple stars and completed the MK classifications of the components brighter than $B = 8$ mag in Aitken's Catalogue (2 403 classifications). A study of the difference between metal-poor and metal-rich binaries showed that the metal-poor stars lack short-period binaries relative to the metal-rich stars and that their period distributions are very different (Abt 2008b).

At the Royal Observatory of Belgium, P. Lampens *et al.* (2007a) reported accurate relative astrometry and differential multi-colour (*BVRI*) photometry for the components of 71 visual systems. Basic binary properties were derived for 20 physical systems. The component colors and masses were obtained for two orbital systems .

A new orbital solution based on recent high-resolution spectra and existing long-baseline optical interferometric data was determined for θ^2 Tau. Radial velocities were derived for both components applying a spectra disentangling algorithm (P. Hadrava, Ondrejov, Czech Rep.). With the improved accuracy of the orbital parameters and associated fundamental properties, the component masses and luminosities were found to be compatible with current stellar evolution models adopting the Hyades metallicity (Lampens *et al.* 2007b).

R. Argyle used the 26.5 inch Innes refractor at Johannesburg for a program of double star measurement from 16 May to 1 June 2008 inclusive. 207 mean measures were made of southern visual binaries and systems known to be in motion but not measured for 10 or 20 years.

7.6. *Spectroscopic binaries*

Roger Griffin published his 200th paper on radial velocity spectrometer orbits of binary stars in the June issue of *Observatory*. A paper by V. Trimble will appear in the August issue examining the statistics of binary star mass ratios based on those 200 papers, plus some others. Two papers published by C. Scarfe *et al.* (2007, 2008) deal with spectroscopic binaries of rather long period, whose primaries are giants, and for which secondaries were detected which are also evolved, making these systems of interest for stellar evolution models.

7.7. *Eclipsing binaries*

A new procedure for the automatic classification of eclipsing binaries was developed by Malkov, *et al.* (2007). It is based on the data from 1029 classified systems and allows for the classification of a given system based on a set of observational parameters, even if the set is incomplete. The procedure was applied to six large surveys of eclipsing variables. About 5 300 systems were classified for the first time and can be used for the determination of the astrophysical parameters of their components.

A photometric and spectroscopic observational campaign of newly discovered *Hipparcos* eclipsing binaries has been realized since 1997. The data are collected at Haute Provence (France), Cracow (Poland), Krioneri (Greece), Lvov (Ukraine) and at Ankara University Observatory (AUG) and TUBITAK National Observatory (TUG) in Turkey. Among the fist results from this campaign, out of 36 objects, 24 new double-lined eclipsing binaries were identified. Ten systems were found to be new, spectroscopic triple systems. Two triple systems show time changes of center-of-mass velocity of the eclipsing binaries due to the presence of a third body. 22 Hipparcos periods were updated. Radial velocity curves and orbital solutions were presented. Four systems were identified as good sub-solar mass candidates.

7.8. *Gaia and non-single stars*

For the past two years, Dimitri Pourbaix has been leading the Coordination Unit of the *Gaia* Data Processing and Analysis Consortium, in charge, among others, of the non single stars. He has been involved in the design and coding of methods to deal with resolved and unresolved binaries (and multiples). A review talk on binaries at IAU Symposium 248, Shanghai, 2007 (Pourbaix 2008) deals with these matters. In addition, Pourbaix still maintains the 9th catalogue of spectroscopic binary orbits on behalf of Commission 30.

7.9. *Theoretical work*

Seppo Mikkola has continued his N-body simulations with binaries. Mauri Valtonen has studied the stability of hierarchical triples with A. Myllari, V. Orlov and A. Rubinov. The books by Valtonen & H. Karttunen, *The Three-Body Problem* (Cambridge: CUP, 2006) and G. Byrd, A. Chernin & M. Valtonen, *Cosmology: Foundations and Frontiers* (Moscow: URSS Publ., 2007) deal with binaries in both Galactic and cosmological contexts. These books summarize the roles of binaries in stellar dynamics. With Jia-Qing Zheng, Valtonen has studied unequal mass binary formation in star clusters, as a way of understanding the origin of the Oort Cloud. In connection to the question of the transfer of life between different solar systems (and with Passi Nurmi as collaborator) planets around binary stars as targets for microbe-carrying asteroids were considered.

In collaboration with H. Lehto, S. Kotiranta, P. Heinamaki and A. Chernin, Valtonen has also studied the question of the 'Arrow of Time', one of the fundamental open questions in physics, and has shown that as soon as a third star is added to a binary system, the arrow of time (which is unspecified for the binary, the system being fully time-reversible) becomes specified through Kolmogorov-Sinai entropy. Papers on these topics have recently appeared (or will appear soon. Valtonen's future plans include writing a popular account of the three-body problem with J. Anosova, K. Tanikawa, V. Orlov and A. Myllari.

Acknowledgements: My sincere thanks to all of you who sent information. I am most grateful for the balanced summaries of the work of their respective groups submitted by W. Hartkopf, Y. Balega, J. Budaj, E. Horch, P. Lampens, H. McAllister, E. Oblak, D. Pourbaix, A. Poveda, M. Scardia, M. Valtonen, and H. Zinnecker. Any errors, omissions and such are, of course, solely mine.

Christine Allen
President of the Commission

References

Abt, H. A. 2008, *ApJS*, 176, 216

Abt, H. A. 2008, *AJ*, 135, 722

Allen, C., Poveda, A., & Hernandez-Alcantara, A. 2007, in: W. I. Hartkopf , P. Harmanec & E. F. Guinan (eds.), *Binary Stars as Critical Tools & Tests in Contemporary Astrophysics*, Proc. IAU Symp. 240, Prague, Czech Republic, 22-25 August 2006 (Cambridge: CUP), p. 405

Allen, C., Poveda, A. & Rodriguez, L.F. 2008, *ASP-CS*, in press

Apai, D., Arjan, B., Lex, K., Thomas, H., & Hans, Z. 2007, *ApJ*, 655, 484

Budaj, J. & Richards, M. T. 2004, *Contr. Skalnate Pleso Obs.*, 34, 167

Chaname, J. 2007, in: W. I. Hartkopf, P. Harmanec & E. F. Guinan (eds.), *Binary Stars as Critical Tools & Tests in Contemporary Astrophysics*, Proc. IAU Symp. 240, Prague, Czech Republic, 22-25 August 2006 (Cambridge: CUP), p. 316

Connelley, M. S. 2008, *AJ*, 135, 2496; 2526

Correia, S., Zinnecker, H. Ratzka, T., & Sterzik, M. F. 2006, *A&A*, 459, 909

Costero, R., Poveda, A., & Echevarria, J. 2007, in: W. I. Hartkopf , P. Harmanec & E. F. Guinan (eds.), *Binary Stars as Critical Tools & Tests in Contemporary Astrophysics*, Proc. IAU Symp. 240, Prague, Czech Republic, 22-25 August 2006 (Cambridge: CUP), p. 130

Costero, R., Allen, C., Echevarria, J., *et al.* 2008, *RevMexAA-CS*, in press

Docobo, J. A., Tamazian, V. S., Balega, Y. Y., & Melikian, N. D. 2006, *AJ*, 132, 994

Docobo, J. A., Andrade, M., Tamazian, V. S., *et al.* 2007, *RevMexAA*, 43, 141

Docobo, J. A., Tamazian, V. Balega, Y. Y., *et al.* 2008, *A&A*, 478, 187

Eggleton, P. P. & Tokovinin, A. A. 2008, *MNRAS*, 389, 869

Gomez, L., Rodriguez, L. F., Loinard, L., *et al.* 2005, *ApJ*, 635, 1166

Gomez, L., Rodriguez, L. F., Loinard, L., *et al.* 2008, *ApJ*, 685, 333

Johnston, K., Oswalt, T., & Valls-Gabaud, D. 2007, in: W. I. Hartkopf , P. Harmanec & E. F. Guinan (eds.), *Binary Stars as Critical Tools & Tests in Contemporary Astrophysics*, Proc. IAU Symp. 240, Prague, Czech Republic, 22-25 August 2006 (Cambridge: CUP), p. 429

Koehler, R., Petr-Gotzens, M. G., McCaughrean, M. J., *et al.* 2006, *A&A*, 458, 461

Koehler, R., Ratzka, T., Herbst, T. M., & Kasper, M. 2008, *A&A*, 482, 929

Kraus, S., Balega, Y. Y., Berger, J.-P., *et al.* 2007, *A&A*, 466, 649

Lampens, P., Strigachev, A., & Duval, D. 2007a, *A&A*, 464, 641

Lampens, P., Fremat, Y., De Cat, P., & Hensberge, H. 2007b, in: W. I. Hartkopf , P. Harmanec & E. F. Guinan (eds.), *Binary Stars as Critical Tools & Tests in Contemporary Astrophysics*, Proc. IAU Symp. 240, Prague, Czech Republic, 22-25 August 2006 (Cambridge: CUP), p. 213

Lepine, S. & Bongiorno, B. 2007, *AJ*, 133, 889

Malkov, O. E., Kalinichenko, L., Kazanov, M. D., & Oblak, E. 2008, *in Astronomical Data Analysis Software and Systems XVII*, *ASP-CS*, 394, 381

Malkov, O. E., Oblak, E., Avvakumova, E. A., & Torra, J. 2007, *A&A*, 465, 549

Malkov, O. E., Oblak, E., Snegivera, E. A., & Torra, J. 2006, *A&A*, 446, 785

Mason, B. D., Hartkopf, W. I., Wycoff, G. L., & Rafferty, T. J. 2006a, *AJ*, 131, 2687

Mason, B. D., Hartkopf, W. I., Wycoff, G. L., & Holdenried, E. R. 2006b, *AJ*, 132, 2219

Mason, B. D., Hartkopf, W. I., Wycoff, G. L., & Wieder, G. 2007, *AJ*, 134, 1671

Miller, B., Budaj, J., Richards, M., Koubský, P., & Peters, G. J. 2007, *ApJ*, 656, 1075

Oblak, E., Kurpinska-Winiarska, M., Carquillat, J.-M., *et al.* 2007, *A&A*, submitted

Oswalt, T. D., Johnston, K. B., Rudkin, M., Vaccaro, T., & Valls-Gabaud, D. 2007, in: W. I. Hartkopf , P. Harmanec & E. F. Guinan (eds.), *Binary Stars as Critical Tools & Tests in Contemporary Astrophysics*, Proc. IAU Symp. 240, Prague, Czech Republic, 22-25 August 2006 (Cambridge: CUP), p. 300

Piirola, V., Vornanen, T., Berdyugin, A., Coyne, S. J., G. V. 2008, *ApJ*, 684, 585

Piirola, V., Berdyugin, A., Coyne, G. V., *et al.* 2006, *A&A*, 454, 277

Pourbaix, D. 2008, in: W. J. Jin, I. Plateis & M. A. C. Perryman (eds.), *A Giant Step: form Milli to Micro Arcsecond Astrometry*, Proc. IAU Symp. 248, Shanghai, China, 15-19 October 2007 (Cambridge: CUP), p. 59

Poveda, A., Allen, C., & Hernandez-Alcantara, A. 2007, in: W. I. Hartkopf, P. Harmanec & E. F. Guinan (eds.), *Binary Stars as Critical Tools & Tests in Contemporary Astrophysics*, Proc. IAU Symp. 240, Prague, Czech Republic, 22-25 August 2006 (Cambridge: CUP), p. 417

Poveda, A., Hernandez-Alcantara, A., Costero, R., *et al.* 2008, *RevMexAA(SC)*, in press

Poveda, A., Allen, C., & Rodriguez, L. F. 2008, *RevMexAA(SC)*, in press

Prieur, J.-L., Scardia, M., Pansecchi, L., *et al.* 2008, *MNRAS*, 387, 772

Reipurth, B., Guimaraes, M. M., Connelley, M. S., & Bally, J. 2007, *AJ*, 134, 2272

Roberts, L. C., Turner, N. H., & ten Brummelaar, T. A. 2007, *AJ*, 133, 545

Rodriguez, L. F., Poveda, A., Lizano, S., & Allen, C. 2005, *ApJ (Letters)*, 627, L65

Sanchez, L. A., Ruelas-Mayorga, A., Allen, C., & Poveda, A. 2008, *RevMexAA(SC)*, in press

Scardia, M., Argyle, R. W., Prieur, J.-L., *et al.* 2007, *Astron. Nach.*, 328, 2, 146

Scardia, M., Prieur, J.-L., Pansecchi, L., *et al.* 2007, *MNRAS*, 374, 965

Scardia, M., Argyle, R. W., Prieur, J.-L., Pansecchi, L., Basso, S., Law, N. M., & Mackay, C.D. 2007, in: W. I. Hartkopf , P. Harmanec & E. F. Guinan (eds.), *Binary Stars as Critical Tools & Tests in Contemporary Astrophysics*, Proc. IAU Symp. 240, Prague, Czech Republic, 22-25 August 2006 (Cambridge: CUP), p. 558

Scardia, M., Prieur, J.-L., Pansecchi, L., *et al.* 2008, *Astron. Nach.*, 329, 1, 54

Scardia, M., Prieur, J.-L., Pansecchi, L., & Argyle, R. W. 2008, *Astron. Nach.*, 329, 4, 379

Scardia, M., Prieur, J.-L., Pansecchi, L., *et al.* 2008, *Astron. Nachr.*, 329, 54

Scarfe, C. D., Griffin, R. F., & Griffin, R. E. M. 2007, *MNRAS*, 376, 1671

Scarfe, C. D. 2008, *The Observatory*, 128, 14

Silvotti, R., Schuh, S., Janulis, R., *et al.* 2007, *Nature* 449, 189

Stassun, K. G., Mathieu, R. D., Cargile, P. A., *et al.* 2008, *Nature*, 453, 1079

Tamazian, V. S. 2007, *AJ*, 132, 2156

Tamazian, V., Docobo, J. A., Melikian, N. D., *et al.* 2007, *PASP*, 118, 814

Tokovinin, A., Thomas, S., Sterzik, M., & Udry, S. 2008, in: S. Hubrig, M. Petr-Gotzens & A. Tokovinin (eds.), *Multiple Stars across the HRD*, ESO Astrophysics Symp. (Springer-Verlag), p. 129

Tokovinin, A. 2008, *MNRAS*, 389, 925

Trimble, V. 2008, *The Observatory*, 128, 286

Turner, N. H., ten Brummelaar, T. A., Roberts, L. C., *et al.* 2008, *AJ*, 136, 554

Valtonen, M. J. 2008, *RevMexAA(SC)*, 32, 22

Zinnecker, H. 2008, in: S. Hubrig, M. Petr-Gotzens & A. Tokovinin (eds.), *Multiple Stars across the HRD*, ESO Astrophysics Symp. (Springer-Verlag), p. 265

Transactions IAU, Volume XXVIIA
Reports on Astronomy 2006–2009
Karel A. van der Hucht, ed.

© 2009 International Astronomical Union
doi:10.1017/S1743921308025520

DIVISION IV / COMMISSION 26 / WORKING GROUP
BINARY AND MULTIPLE SYSTEM NOMENCLATURE

CO-CHAIRS   **Brian D. Mason, William I. Hartkopf**
MEMBERS   **Dimitri Pourbaix, Colin D. Scarfe,**
                    Marion Schmitz, Andrei A. Tokovinin

TRIENNIAL REPORT 2006 - 2009

1. Introduction

The Working Group on *Binary and Multiple System Nomenclature* was formed within Commission 26 following Special Session 3 held during the 2003 Sydney General Assembly. Its purpose is to create the Washington Multiplicity Catalog, a comprehensive database first introduced at a multi-commission meeting at the IAU XXIV General Assembly in Manchester, 2000. Data are being compiled from the US Naval Observatory visual binary catalogs and supplemented with binary and multiple star information from other sources to include but not limited to spectroscopy, photometry, eclipsing and interacting system, as well as extra-solar planets and substellar companions. The goal being creation of a comprehensive hierarchical database and to reduce confusion from multiple nomenclature schemes used by disparate techniques.

2. Activities

Following the Manchester Multi-Commission Meeting, a sample slice of the sky was selected for implementation. This sample, of 30′ width in Right Ascension and from pole-to-pole was completed for the Sydney Special Session. Of this portion of the sky, various techniques contribute to the sample WMC in the following percentages:

95.8%	visual binaries and optical pairs
50.6%	interferometric binaries and optical pairs
1.7%	spectroscopic binaries
1.4%	cataclysmic variables or related objects
1.0%	occultation binaries
0.3%	astrometric binaries
0.2%	eclipsing binaries
0.2%	X-ray binaries
0.1%	spectrum binaries
0.1%	planets

Since the techniques are complementary, the sum is >100%. It should be noted that this breakdown is biased significantly by selection effects. For example, while visual binaries may be discovered (and cataloged) after a single observation, data on spectroscopic pairs are often not published until the full orbit has been characterized.

To identify common objects, precise (0″.1) coordinates were selected to form the sieve by which objects are brought into multiplicity arrangements. While the more modern techniques often list precise coordinates as a matter of course, the older and much larger visual database does not. To date, work has focused on completing this. To date, the completion status is as follows:

Status of WDS coordinates

0.1 arcsec level	98990	95.34%
1 arcsec level	993	0.96%
10 arcsec level	1284	1.24%
worse than 10 arcsec	1804	1.74%
rejected (bogus)	753	0.73%

Supplementary information

two proper motions	42112	40.56%
one proper motion	43501	41.90%
no proper motion	18211	17.54%
no primary magnitude	54	0.05%
no secondary magnitude	2128	2.05%
no theta	228	0.22%
no rho	36	0.03%
no observation date	28	0.03%

3. Status update

A status update will be presented at the IAU XXVII General Assembly in the Commission 26 Business Meeting in Rio de Janeiro, Brazil, August 2009.

Brian D. Mason & William I. Hartkopf
co-chairs of the Working Group

Transactions IAU, Volume XXVIIA
Reports on Astronomy 2006–2009
Karel A. van der Hucht, ed.

© 2009 International Astronomical Union
doi:10.1017/S1743921308025532

COMMISSION 29

STELLAR SPECTRA

SPECTRES STELLAIRES

PRESIDENT	Mudumba Parthasarathy
VICE-PRESIDENT	Nikolai E. Piskunov
PAST PRESIDENT	Christopher Sneden
ORGANIZING COMMITTEE	Kenneth G. Carpenter, Fiorella Castelli,
	Katia Cunha, Phillippe R. J. Eenens,
	Ivan Hubeny, Silvia Rossi,
	Masahide Takada-Hidai,
	Glenn M. Wahlgren, Werner W. Weiss

TRIENNIAL REPORT 2006 - 2009

1. Introduction

The members of IAU Commission 29 *Stellar Spectra* are actively engaged in the quantitative analysis of spectra of various types of stars. With large and medium size telescopes equipped with high resolution spectrographs LTE and Non-LTE analysis of spectra of all types stars are being carried out. Spectra of stars in our Galaxy, in globular and open clusters, stars in LMC and SMC and in nearby galaxies are being studied. Accurate chemical composition analysis of various types of stars has been carried out during the past three years. Now the analysis of stellar spectra covers the wavelength range from X-ray region to IR and sub-millimeter range. Recently stellar spectra are being analysed using time-dependent, 3D, hydrodynamical model atmospheres to derive accurate stellar abundances.

2. Activities

Spectra of massive OB stars, Wolf-Rayet stars, RGB, AGB and Horizontal Branch stars, main-sequence stars, pre-main-sequence stars, all types of variable stars, and white dwarfs were analysed. Detailed chemical composition analysis of heavy elements abundances in large number of very metal-poor stars have been carried out to understand the neutron-capture elements in the early chemical history of the Galaxy and early Universe. Detailed analysis of the chemical composition of the unevolved very metal-poor ($[Fe/H] < -5.0$) star HE1327–2326 has been made.

The number of papers on stellar spectra published during the past three years is very large and space and time does not permit me to summarize even the important results. Quantitative and detailed analysis of high resolution spectra of large number of stars with improved model atmospheres and atomic-data and LTE and Non-LTE methods is providing valuable clues to understand stellar evolution, galactic-chemical evolution, nucleosynthesis, stellar pulsations, mixing, mass-loss, circumstellar matter, etc. Spectra of stars with planets are also being studied in detail. Spectra of large number of chemically peculiar stars have been analysed during the past three years. Abundances of light elements, s-process elements and heavy elements are being derived from the analysis of spectra of large numbers of stars and they are shedding new light on early universe, s-process and r-process nucleosynthesis.

3. Meetings

During the past three years IAU Commission 29 endorsed, extended support and sponsorship to several IAU Symposia and meetings and also for several IAU Symposia, JDs and meetings that will be held during the IAU XXVII General Assembly in Rio de Janeiro, Brazil, August 2009.

Mudumba Parthasarathy
president of the Commission

Transactions IAU, Volume XXVIIA
Reports on Astronomy 2006–2009
Karel A. van der Hucht, ed.

© 2009 International Astronomical Union
doi:10.1017/S1743921308025544

COMMISSION 35   **STELLAR CONSTITUTION**

CONSTITUTION DES ÉTOILES

PRESIDENT	Francesca D'Antona
VICE-PRESIDENT	Corinne Charbonnel
PAST PRESIDENT	Wojciech Dziembowski
ORGANIZING COMMITTEE	Gilles Fontaine, Richard B. Larson,
	John Lattanzio, Jim W. Liebert,
	Ewald Müller, Achim Weiss,
	Lev R. Yungelson

COMMISSION 3 WORKING GROUPS

Div. IV / Commission 35 WG	Active B Stars
Div. IV / Commission 35 WG	Massive Stars
Div. IV / Commission 35 WG	Red Giant Abundances

TRIENNIAL REPORT 2006-2009

1. The activity

The Commission home page `<iau-c35.stsci.edu>` is maintained by Claus Leitherer and contains general information on the Commission structure and activities, including links to stellar structure resources that were made available by the owners. The resources contain evolutionary tracks and isochrones from various groups , nuclear reaction , EOS, and opacity data as well as links to main astronomical journals. As a routine activity, the Organizing Committee has commented on and ranked proposals for several IAU sponsored meetings. Our Commission acted as one of the coordinating bodies of a Symposium held at the IAU XXVI General Assembly in Prague, August 2006, (IAU Symposium No. 239 *Convection in Astrophysics*, and participated in the organization of the following Joint Discussions: JD05 *Calibrating the Top of the Stellar Mass-Luminosity Relation*, JD06 *Neutron Stars and Black Holes in Star Clusters*, JD08 *Solar and Stellar Activity Cycles*, JD11 *Pre-Solar Grains as Astrophysical Tools*; JD14 *Modelling Dense Stellar Systems*; and JD17 *Highlights of Recent Progress in the Seismology of the Sun and Sun-like Stars*.

Members of Commission 35 were involved in organization of other IAU sponsored meetings: IAU Symposium No. 241 *Stellar Populations as Building Blocks of Galaxies* in December 2006; IAU Symposium No. 246 *Dynamical Evolution of Dense Stellar Systems* in September 2007; IAU Symposium No. 250 *Massive Stars as Cosmic Engines* in December 2007; IAU Symposium No. 251 *Organic Matter in Space* in February 2008; and IAU Symposium No. 254 *The Galaxy Disk in Cosmological Context* in June 2008.

Commission 35 has been sponsoring committee for IAU Symposium No. 252 *The Art of Modelling Stars in the 21st Century*, in April 2008. Many other international meetings, in which members of the commission were involved, were held in these years.

We devote the rest of the report to examine in greater detail the scientific activity in the field of stellar constitution, considering the developments, within the past triennium, in a non-exhaustive selection of specific research topics.

2. Supernovae (E. Müller)

According to the Smithsonian/NASA ADS about 5000 publications appeared during the past three years whose title contained the word 'supernovae. This immense observational and theoretical activity concerns both thermonuclear (SNe Ia) and core collapse supernovae (see, e.g., Kotake et al. 2006; Bruenn et al. 2007; Burrows et al. 2007b; Janka et al. 2007, 2008), as well as the connection between supernovae and gamma-ray bursts (see, e.g., Della Valle 2008)

There is a consensus that SNe Ia result from the thermonuclear explosion of a carbon-oxygen white dwarf. Substantial progress in supporting this conception has been achieved during the past three years, particularly through the development of 3D SNe Ia models. However, whether the white dwarf is disrupted by a pure deflagration (Röpke et al. 2007b; Mazzali et al. 2007), a delayed detonation (Bravo & García-Senz 2008; Röpke et al. 2007a), or a gravitationally confined detonation (Plewa et al. 2004; Jordan et al. 2008) is still controversially debated. Besides the mode of propagation, the ignition conditions of the thermonuclear flame have also attracted the attention of model builders (Schmidt & Niemeyer 2006; Röpke et al. 2006, 2007a). Moreover, the widely accepted 'fact' that the mass of the exploding white dwarf is close to the Chandrasekhar mass, has been questioned by observations of the high-redshift supernova SNLS-03D3bb, which imply a super-Chandrasekhar-mass progenitor (Howell et al. 2006). Thus, their use as distance indicators in establishing the acceleration of the universe expansion (see, e.g., Astier et al. 2006; Wood-Vasey et al.2007; Bronder et al. 2008) may be problematic (Howell et al. 2006). To this end Sim et al. (2007) showed by means of 3D time-dependent radiation transport simulations that observationally significant viewing angle effects are likely to arise and may have important ramifications for the interpretation of the observed diversity of SNe Ia and their use as standard candles. Diagnosing SNe Ia by spectropolarimetry Wang et al. (2007) showed that their ejecta typically consist of a smooth, central, iron-rich core and an outer layer with chemical asymmetries. From a systematic spectral analysis of a large sample of well-observed SNe Ia Mazzali et al. (2007) conclude that all of them burned similar masses and had progenitors of the same mass. Hence, a single explosion scenario, possibly a delayed detonation, may explain most SNe Ia.

Progress in modeling core collapse supernovae was significant during the past three years. Detailed axisymmetric radiation-hydrodynamic simulations with multi-frequency, multi-angle Boltzmann neutrino transport succeeded in producing successful supernova explosions (i) for stars in the range of 8 to 10 $M_\odot$ which possess O-Ne-Mg cores at the onset of core collapse (Kitaura et al. 2006), and (ii) for stars of 11.2 and 15 $M_\odot$ which develop Fe-Ni cores (Janka et al. 2007, 2008). This progress was neither due to a sudden unexpected breakthrough nor due to the inclusion of new physical effects, but resulted from a proper treatment of the relevant physics (relativistic gravity, detailed neutrino transport, microphysics), and from performing simulations covering the activity of the standing accretion shock instability (Blondin et al. 2003). This generic hydrodynamic instability of the stagnant shock against low-mode non-radial deformation, which aids an explosion by improving the conditions for energy deposition by neutrino heating in the postshock gas, takes several hundreds of milliseconds to make an impact, thus requiring computationally extremely demanding multidimensional radiation-hydrodynamic simulations (Marek & Janka 2007). The controversially discussed excitation of gravity-waves in the surface and core of the nascent neutron star may also have supportive influence on the launch of the explosion (Burrows *et al.* 2006, 2007a; Janka *et al.* 2008; Weinberg & Quataert 2008). Studies employing simpler microphysics and/or neutrino transport, or even no transport at all, focussed on the effects of magnetic fields (see, e.g., Obergaulinger *et al.* 2006; Sawai *et al.* 2008), relativistic gravity (see, e.g., Ott *et al.* 2007; Müller *et al.* 2008), and on the gravitational wave signal produced by core collapse supernovae (see, e.g., Dimmelmeier *et al.* 2007; Kotake *et al.* 2007).

3. Asymtotic Giant Branch (J. Lattanzio & P. Ventura)

Asymptotic Giant Branch (AGB) star research remains a vital and very active area, with their role in nucleosynthesis remaining crucial to many other areas of astrophysics. Recent research has expanded the early work of Marigo (2002) in investigating the consequences of a carbon-rich envelope. The opacities for such mixtures have not been included in the past, but recent work by Cristallo *et al.* (2007) has begun to address this.

On the other hand, the most massive AGBs are currently thought to provide an essential contribution as polluters of the interstellar medium within Globular Clusters (GC), and many papers have been devoted in this triennum to modelling of the very advanced nucleosynthesis that they achieve at the bottom of their envelope, see in particular the models by Karakas & Lattanzio (2007), Ventura & D'Antona (2008a), Ventura & D'Antona (2008b). The latter models produce ejecta whose chemistry is in reasonable agreement with the abundance patterns observed in Globular Cluster stars – although work on the dynamical origin of clusters must still explain the quantitative presence of a fraction of anomalous stars reaching 50% and more. Since massive AGBs are also expected to produce great quantities of helium, the interest of these results is strengthened by the analysis of horizontal branches with an anomalous morphology, that can be naturally explained with the existence of a helium rich population Caloi & D'Antona (2007), Caloi & D'Antona (2008).

The study of Super-AGB stars (those igniting carbon in their cores, but still experiencing thermal pulses, with masses between about 8 and 12 $M_{\odot}$) has been readdressed in recent years with important papers by Siess (2006, 2007) and Poelarends *et al.* (2008), which investigate the evolution of these stars and emphasise the difficulties of such calculations. The yields from these stars may also be important for the chemical evolution of GCs. In particular, their helium yields reach the values $Y \simeq 0.37$ necessary to explain the most helium rich populations discovered in the most massive GCs.

A direct quantitative comparison between theoretical AGB models with hot bottom burning and real stars was attempted by McSaveney *et al.* (2007). Although extremely difficult because of shocks and very complicated line profiles, the comparison did verify the basic quantitative predictions of the models, although some uncertainties remain and may be due to uncertainties in the details of the properties of the convective envelope (as always).

We note again the usefulness of population synthesis for trying to understand properties of AGB stars, with recent work including that of Girardi & Marigo (2007), Marigo (2007) and Marigo & Girardi (2007), Bonačić Marinović *et al.* (2007a, 2007b, 2008) and Izzard *et al.* (2007). An area that continues to both attract and benefit from attention is the very important problem of mass-loss, with increasingly sophisticated hydrodynamical models improving our understanding (e.g., Freytag & Höfner, 2008).

4. Transport processes in stars (C.Charbonnel)

One of the main conclusions of the IAU Symposium No. 252 (April 2008, Sanya, China) was that 'The Art of Modelling Stars in the 21st Century' would strongly rely on our ability to model transport processes in stellar interiors. It is clear now that non-standard transport processes of chemicals and angular momentum have to be included in modern stellar models in order to reproduce detailed data in several parts of the HR diagram. This is a major issue in view of the wealth of new and unprecedented data expected from current and future asterosismologic instruments. Rotation appears to be a key ingredient, together with internal gravity waves and magnetic fields.

Rotation has been shown to play a determinant role at very low metallicity, bringing heavy mass loss where almost none was expected, and thus modifying drastically the evolution of the very-low metallicity stars (Meynet et al. 2006). If the stars start their evolution with a sufficiently high equatorial velocity, they reach a critical limit above which mass is lost, probably through a decretion disk (Owocki 2005). In this context, Decressin et al.(2007a) have shown that the material ejected in this slow mechanical wind is enriched in H-burning products and presents abundance patterns similar to the chemical anomalies observed in globular cluster stars. These authors (see also Prantzos et al. 2007; Decressin et al. 2007b) have thus proposed the so-called 'wind of fast rotating star' scenario to explain the origin of the abundance patterns in globular clusters.

Ekstrom et al. (2008) have studied the effects of rotation on the evolution of primordial (i.e., Pop III) stars. They find that $z=0$ models rotate with an internal profile $\Omega(r)$ close to local angular momentum conservation, due to a weak envelope-core coupling. Rotation boosts ^{14}N production that can be as much as 6 orders of magnitude higher than in the $z=0$ non-rotating models. In addition, the high rotation rate at death is expected to lead to much stronger explosion than previously thought, changing the fate of the models.

The competition between atomic diffusion and rotation-induced mixing has been studied both from the theoretical and observational point of view. Fossati et al. (2008) observed a large sample of late B-, A-, and early-F type stars belonging to open clusters of different ages, in order to study how surface chemical abundances of these objects evolve with time and correlate with stellar parameters. They found a strong correlation between the peculiarity of Am stars and $v \sin i$. This nicely confirms the predictions by Talon et al. (2006) using state-of-the art treatment for rotation-induced mixing.

The long quest for the mechanism that strongly modifies the surface abundances of $\sim 95\%$ of low-mass stars when they reach the bump on the RGB has finally been successful. Based on 3D simulations of a tip RGB star (Dearborn et al. 2006), Eggleton et al. (2006) suggested a possible cause, namely the molecular weight inversion created by the 3He(3He,2p)4He reaction in the upper part of the hydrogen-burning shell. They claimed that this mixing was due to the well-known Rayleigh-Taylor instability, which occurs in incompressible fluids when there is a density inversion. In stellar interiors, which are stratified due to their compressibility, a similar dynamical instability occurs when the Ledoux criterion for convection is satisfied, but it acts to render the temperature gradient adiabatic rather than to suppress the density inversion. Presumably it is that instability Eggleton and colleagues have observed with their 3D code, as attested by the high velocities they quote. In reality, the first instability to occur in a star, as the inverse μ-gradient gradually builds up, is the thermohaline instability, as was pointed out by Charbonnel & Zahn (2007a). This double-diffusive instability is observed in salted water in the form of elongated fingers, when the temperature is stably stratified, but salt is not, with fresh water at the bottom and salted at the top, the overall stratification being dynamically stable (Stern 1960). On Earth, this phenomenon leads to the well-known thermohaline circulation, which is the global density-driven circulation of the oceans. It was Ulrich (1972) who first noticed that the 3He(3He,2p)4He reaction would cause a μ-inversion, and he was the first to derive a prescription for the turbulent diffusivity produced by the thermohaline instability in stellar radiation zones. This prescription is based on a linear analysis, and it is certainly very crude, but it has the merit to exist. When it is applied to the μ-inversion layer in RGB stars, with the shape factor recommended by Ulrich, it yields a surface composition that reproduces the observed behavior of the carbon isotopic ratio as well as that of lithium, carbon and nitrogen in RGB stars; it simultaneously leads to the destruction of most of the 3He produced during the star's lifetime (Charbonnel & Zahn 2007a). Charbonnel & Zahn (2007b) then focussed on observations that disagree with that general scheme: a couple of evolved stars, NGC 3242 and J320, appear indeed to have eluded the thermohaline mixing because they show a high 3He abundance (Balser *et al.* 2007). Charbonnel & Zahn suggest that a fossil magnetic field suppresses the thermohaline mixing in the descendants of Ap stars, including NGC 3242 and J320. The relative number of such stars with respect to non-magnetic objects that undergo thermohaline mixing is consistent with the statistical constraint coming from observations of the carbon isotopic ratio in red giant stars. It also satisfies the Galactic requirements for the evolution of the 3He abundance.

An opposite assumption was made by Busso *et al.* (2007) who suggest that extra-mixing on the RGB originates in a stellar dynamo operated by the differential rotation below the envelope, maintaining toroidal magnetic fields near the hydrogen-burning shell. They use a phenomenological approach to the buoyancy of magnetic flux tubes, assuming that they induce matter circulation as needed by the so-called 'cool bottom processing' models. The distinction between the two opposite explanations should be made possible in the near future, thanks to the search for magnetic fields in evolved stars with spectropolarimeters.

The prediction of the spins of the compact remnants is a fundamental goal of the theory of stellar evolution. Suijs *et al.* (2008) confronted the predictions for white dwarf spins from evolutionary models including rotation with observational constraints. They calculate two sets of model sequences, with and without inclusion of magnetic fields. From the final computed models of each sequence, they deduce the angular momenta and rotational velocities of the emerging white dwarfs. They found that while models including magnetic torques predict white dwarf rotational velocities between 2 and 10 km s^{-1}, those from the nonmagnetic sequences are found to be one to two orders of magnitude larger, well above empirical upper limits. They find the situation analogous to that in the neutron star progenitor mass range, and conclude that magnetic torques may be required to understand the slow rotation of compact stellar remnants in general.

Some progress has also been made on the impact of internal gravity waves (IGW) in stellar interiors. Talon & Charbonnel (2008) showed that angular momentum transport by IGWs emitted by the convective envelope could be quite important in intermediate-mass stars on the pre-main sequence, at the end of the sub- giant branch, and during the early-AGB phase. This implies that possible differential rotation, which could be a relic of the stars main sequence history and subsequent contraction, could be strongly reduced when the star reaches the AGB-phase. This could have profound impact on the subsequent evolution. In particular, this could help explaining the observed white dwarf spins (Suijs *et al.* 2008).

5. Helio- and asteroseismology (J. Montalbán and A. Noels)

Helioseismology has proved during the last decade, with *SOHO* especially, that probing the internal structure of the Sun was indeed within our reach. As a result, researchers have made tremendous efforts in obtaining light curves for stars covering the whole HR diagram. Ground-based observations are now reaching high levels of accuracy. Space missions have been designed to even more increase the signal to noise ratios for variable stars observed during long periods of time. After the successful *MOST* mission (still running), the *CoRoT* satellite was launched on December 26, 2007. It has already achieved an initial run of ~ 60 d, short runs of ~ 30 d, and long runs of ~ 150 d. The light curves of an exceptional quality are now in the process of being analysed and frequencies have already been obtained for various types of variable stars, from solar-like to γ Dor, δ Scu, SPB, ρ Cep, and red giants. *KEPLER* and *BRITE* will soon follow, while *PLATO* is waiting for a final selection.

On the other hand, 'HELAS', the *European Helio- and Asteroseismology Network* is funded by the European Commission since 1 April 2006, as a 'Co-ordination Action' under its Sixth Framework Programme (FP6). The role of HELAS is to coordinate the activities of researchers active in the helio- and asteroseismology fields. These coordinating activities are certainly creating a dynamism among researchers mostly through colloquia and workshops (see `<www.helas-eu.org/>` for a list of HELAS meetings) and are helping to promote this new approach of the understanding of stellar interiors.

The Sun has indeed 'suffered' from the new solar abundances (Asplund *et al.* 2005). These abundances, of the order of 30% smaller for C,N,O, decrease the opacity near the internal border of the convective envelope, which destroys the good agreement between the standard solar models and helioseismology found with Grevesse & Noels (1993) abundances. A thorough discussion of all the effects of these new low abundances on the internal structure of the Sun and of the possible solutions has been made by Basu & Antia (2008). An increase of opacity in the upper radiative zone seems to be needed to reconcile the theoretical new Sun with helioseismic observations. A possible solution was proposed by Drake & Testa (2005) suggesting an increase in the Ne abundance by a factor 2.5 based on X-ray measurements. This is now totally contradicted by new observations (Robrade *et al.* 2008), which confirm the low Ne value recommended by Asplund *et al.* (2005). The 'solar problem' is still pending.

Stars are now entering the game with full force. Important questions are on the verge of being answered or put into perspective:

• What are the consequences of the new solar abundances (Asplund *et al.* 2005) on other types of stars? The lower abundances of C,N,O are due to combined NLTE effects and 3D modelling of the atmospheric layers. They should affect hot stars as well. New abundances determinations (Morel *et al.* 2006) are in favour of such a decrease in C,N,O and of a low metallicity of about $z = 0.01$ in B-type stars.

• Different abundance mixtures but also different opacities (OPAL versus OP) have been tested in SPB and β Cep stars modelling (Pamyatnykh 2007; Miglio *et al.* 2007; Dziembowski & Pamyatnykh 2008). Although the situation is improving with the new solar abundances and OP opacities, there still remain some problems with the excitation of the observed modes. There seems to be a need for an upward revision of opacities, especially in the temperature domain of the iron opacity bump, an alternative solution being an accumulation of iron due to radiative forces. Such an increase in opacity is necessary to explain the presence of pulsations in some β Cep, especially in a low-z environment such as the SMC (Diago *et al.* 2008).

• An iron enhancement in the iron opacity bump is also necessary to explain the excitation in sdB stars (Jeffery & Saio, 2006; Charpinet *et al.* 2007). Binarity seems to be rather common in sdBs (van Grootel *et al.* 2008) and the question of the origin (single or binary) of such stars is now a matter of debate.

- Modelling convection and the interaction between convection and pulsation plays a key role in establishing the red limits of the instability strips, not only for δ Scu stars but also for γ Dor stars (Dupret *et al.* 2007).

- Although progress has been made in the understanding of roAps pulsations (Théado *et al.* 2005; Théado and Cunha 2006), the coolest roAp stars are still to be explained (Théado *et al.* 2008). The presence of a magnetic field is a necessary condition to explain the lack of δ Scu-like pulsations in roAp stars. The blue to red feature in the line-profile variation can be explained by shock waves propagating through the atmospheric layers, in accordance with the oblique pulsator model (Shibahashi *et al.* 2008).

- Stochastic excitation of non radial modes in solar-like stars is the result of the presence of an outer convective zone (Belkacem *et al.* 2008). Since convective zone can also be found in hot stars due to the iron peak, solar-like oscillations could indeed be expected in B-type stars.

- New types of variables have been discovered: B-type supergiants (Lefever *et al.* 2006), SPBsg (Saio *et al.* 2006), SPBe (Cameron *et al.* 2008), hot DQV (their progenitors could be roAps), sdO (Woudt *et al.* 2006), WR (Moffat *et al.* 2008), ... and many more are to come.

6. Evolution of binary stars

In the field of formation of binaries, modelling of fragmentation and binary formation processes in 3D has shown that the latter is mainly controlled by the initial ratio of the rotational to the magnetic energy, regardless of the initial thermal energy and amplitude of the nonaxisymmetric perturbations. As the ratio of these energies decreases, formation of clouds results in formation of wide [$a \simeq (3 - 300)$ AU] binaries or close ones [$a \lesssim 0.3$ AU]. Thus, one may expect to observe a bimodal distribution of separations for young stars. Strong magnetic field suppresses fragmentation and formation of single stars is then expected Machida *et al.* (2008).

A significant attention was devoted to binaries with degenerate donors. It appeared that a larger fraction of detached double WD survive the onset of mass transfer than has been hitherto assumed, even if mass transfer is initially unstable and rises to super-Eddington levels or direct impact occurs Gokhale *et al.* (2007), Motl *et al.* (2007). Deloye & Taam (2006), found that during the early contact phase, while $\dot{M}$ is increasing, gravity wave emission continues to drive the binary to shorter $P_{\rm orb}$ for 10^3 - 10^6 yr. This may explain $\dot{P}_{\rm orb} < 0$ of RX J0806+1527 and RX J914+2456, the binaries with shortest known $P_{\rm orb}$. D'Antona *et al.* (2006) have shown that $\dot{P}_{\rm orb} < 0$ on a time scale consistent with the $P_{\rm orb}$ decrease in RX J0806+1527 is possible in a very plausible situation when the donor retained some hydrogen in the envelope and mass-transfer occurs from a not fully-degenerate envelope. The full stellar evolution of arbitrarily degenerate He-dwarf donors in AM CVn systems for the first time was computed by Deloye *et al.* (2007). Bildsten *et al.* (2007) noticed that, as the orbit if an AM CVn system widens and $\dot{M}$ drops, the mass required for the unstable ignition of accreted He increases, leading to progressively more violent flashes up to a final flash with helium shell mass ~ 0.02-0.1 $M_\odot$, which may power a faint ($M_V = -15$ to -18) and rapidly rising (few days) thermonuclear supernova nicknamed 'SN .Ia' (one-tenth as bright for one-tenth the time as a SN Ia). On the other hand, Yoon & Langer (2005) have shown that rapid rotation imposed by accretion may stabilise burning in helium shells thus facilitating the growth of WD to Chandrasekhar mass.

In the field of 'traditional' H-rich cataclysmic variables, Nomoto *et al.* (2007) addressed the physics of stable burning of accreted hydrogen by WD and confirmed that it is possible only within a narrow range of $\dot{M}$ close to Eddington limit, while Shen & Bildsten (2007) have shown that this is due to radiation pressure stabilization of burning layer. Norton et al. (2008) carried out numerical simulations of accretion flows in magnetic cataclysmic variables: disks, streams, rings, and propellers and have shown that fundamental observable determining the accretion flow, for a given mass ratio, is the spin-to-orbital-period ratio of the system. A 3D hydrodynamic simulation of the quiescent accretion with the subsequent explosive phase, showing that accumulation of mass is possible was carried out by Walder et al. (2008). Viallet & Hameury (2007) addressed numerically the problem of irradiation of secondaries during outbursts and have shown that the resulting increase in the mass transfer rate is moderate, so unlikely to be able to account for the duration of long outbursts. Matthews et al. (2006) demonstrated that combination of a weak magnetic propeller and accretion disc resonances can effectively halt accretion in short-period CVs. Epelstain et al. (2007) studied a multi-cycle nova evolution and have shown that for a low-mass WD ($0.65 M_\odot$) characteristics of the outbursts remain permanent while for more massive WD ($1 M_\odot$) they change from fast to moderate fast and then attain steady state.

One of the most popular topics remained mergers of relativistic objects, which are deemed to be associated with gamma-ray, gravity wave outbursts, chemical evolution of the galaxies. Different aspects of mergers were studied by Rantsiou et al. (2008) (black holes plus neutron stars in 3D), Surman et al. (2008) (nucleosynthesis in black hole plus neutron stars mergers), Anderson et al. (2008) (GWR from mergers of magnetized neutron stars), Shibata & Taniguchi (2008) (fully general relativistic simulation of the black hole plus neutron star binaries), Oechslin & Janka(2007) (GWR signal from merging neutron stars as function of EOS and system parameters), Setiawan et al. (2006)(3D-simulations of accretion by remnant black holes of compact object mergers aiming at possibility of production short GRB), Podsiadlowski (2007) (stability of mass transfer in black hole – neutron star mergers and conditions for formation of a hot disk around a black hole – progenitor of a short-hard GRB.

The role of binarity in producing rapidly spinning Wolf-Rayet stars – progenitors of collapsars associated with long GRB – was studied by Cantiello et al. (2007), Tutukov & Fedorova (2007). Significant efforts were invested in the studies of stellar mergers. Yoon et al. (2007) studied dynamical process of the merger of CO WD, followed by 3D SPH and found conditions in which central object becomes a CO star and a a SN Ia progenitor. Martin et al. (2006) found that if initial masses of accreting CO WD is $\sim 1.1 M_\odot$, mild carbon flashes propagating inward and converting WD into an ONe with subsequent accretion-induced collapse may be avoided, if accretion rate may be limited to ~ 0.46 Eddington. Hicken et al. (2007), based on the enormous amount of ejected ^{56}Ni $- 1.2 M_\odot$, suggested that the luminous and carbon-rich SN 2006gz might have a super-Chandrasekhar progenitor and descends from a double-degenerate. Gourgouliatos & Jeffery (2006) showed that it is necessary to lose about 50% of initial angular momentum for for formation of helium-rich giants by CO+He WD mergers.

Madhusudhan et al. (2006), Patruno et al. (2006) computed the models of ULX sources assuming a range of masses for accretors (presumably, intermediate mass black holes) and donors, while Rappaport et al. (2005) have shown that most of ULX may be reproduced assuming stellar mass black holes as accretors.

Bonačić Marinović et al. (2008) suggested a mechanism of a tidally enhanced mass loss from AGB stars, that efficiently works against the tidal circularisation of the orbit in binary systems containing a white dwarf and a less evolved companion, thus allowing to solve a long-standing problem of eccentricities of orbits of systems like Sirius and barium stars.

Francesca D'Antona
president of the Commission

References

Anderson, M., Hirschmann, E. W., Lehner, L., *et al.* 2008, *Phys. Rev. Lett.*, 100, 191101

Asplund, M., Grevesse, N., Sauval, A. J. 2005, in: T. G. Barnes & F. N. Bash (eds.), *Cosmic Abundances as Records of Stellar Evolution and Nucleosynthesis*, ASP-CS, 336, 25

Astier, P., Guy, J., Regnault, N., *et al.* 2006, *A&A*, 447, 31

Balser, D. S., Rood, R. T., & Bania, T. M., 2007, *Science*, 317, 1171

Basu, S. & Antia, H. M. 2008, *Phys. Rep.*, 457, 217

Belkacem, K., Samadi, R., Goupil, M. J., & Dupret, M. A. 2008, *A&A*, 478, 163

Bildsten, L., Shen, K. J., Weinberg, N. N., & Nelemans, G. 2007, *ApJ* (Letters), 662, L95

Blondin, J. M., Mezzacappa, A., & DeMarino, C. 2003, *ApJ*, 584, 971

Bonačić Marinović, A. A., Glebbeek, E., & Pols, O. R. 2008, *A&A*, 480, 797

Bonačić Marinović, A., Izzard, R. G., Lugaro, M., & Pols, O. R. 2007, *A&A*, 469, 1013

Bonačić Marinović, A., Lugaro, M., Reyniers, M., & van Winckel, H. 2007, *A&A*, 472, L1

Bonačić Marinović, A. A., Glebbeek, E., & Pols, O. R. 2008, *A&A*, 480, 797

Bravo, E. & García-Senz, D. 2008, *A&A*, 478, 843

Bronder, T. J., Hook, I. M., Astier, P., *et al.* 2008, *A&A*, 477, 717

Bruenn, S. W, Dirk, C. J., Mezzacappa, A., *et al.*, 2007, in press [ArXiv e-prints: 0709.0537]

Burrows, A., Livne, E., Dessart, L., Ott, C. D., Murphy, J. 2006, *ApJ*, 640, 878

Burrows, A., Livne, E., Dessart, L., Ott, C. D., Murphy, J. 2007, *ApJ*, 655, 416

Burrows, A., Dessart, L., Ott, C. D., Livne, E., 2007, *Phys. Rep.*, 442, 23.

Busso, M., Wasserburg, G. J., Nollett, K. M., Calandra, A. 2007, *ApJ*, 671, 802

Caloi, V. & D'Antona, F. 2007, *A&A*, 463, 949

Caloi, V. & D'Antona, F. 2008, *ApJ*, 673, 847

Cameron, C., Saio, H., Kuschnig, R., *et al.* 2008, *ApJ*, 685, 489

Cantiello, M., Yoon, S.-C., Langer, N., & Livio, M. 2007, *A&A*, 465, L29

Charbonnel, C. & Zahn, J.-P. 2007a, *A&A* (Letters), 467, L15

Charbonnel, C. & Zahn, J.-P. 2007b, *A&A* (Letters), 476, L29

Charpinet S., Fontaine G., Brassard P., *et al.* 2007, *CoAst*, 150, 241

Cristallo, S., Straniero, O., Lederer, M. T., Aringer, B. 2007, *ApJ*, 667, 489

D'Antona, F., Ventura, P., Burderi, L., & Teodorescu, A. 2006, *ApJ*, 653, 1429

Dearborn, D. S .P., Lattanzio, J. C., & Eggleton, P. 2006, *ApJ*, 639, 405

Decressin, T., Charbonnel, C., & Meynet, G. 2007, *A&A*, 475, 859

Decressin, T., Meynet, G., Charbonnel, C., Prantzos, N., & Ekstrom, S., 2007, *A&A*, 464, 1029

Della Valle, M. 2008, *Relativistic Astrophysics, AIP-CP*, 966, 31

Deloye, C. J. & Taam, R. E. 2006, *ApJ Letters*, 649, L99

Deloye, C. J., Taam, R. E., Winisdoerffer, C., & Chabrier, G. 2007, *MNRAS*, 381, 525

Diago, P. D., Gutiérrez-Soto, J., Fabregat, J., & Martayan, C. 2008, *A&A*, 480, 179

Dimmelmeier, H., Ott, C. D., Janka, H.-T., *et al.* 2007, *Phys. Rev. Lett.*, 98, 251101

Drake, J. & Testa, P. 2005, *Nature*, 436, 525

Dupret, M. A., Miglio, A., Grigahcène, A., & Montalbán, J. 2007, *CoAst*, 150, 98

Dziembowski, W. A., Pamyatnykh, A. A. 2008, *MNRAS*, 385, 2061

Eggleton, P., Dearborn, D. S. P., Lattanzio, J. C., *Science*, 314, 1580

Ekström, S., Meynet, G., Chiappini, C., Hirschi, R., & Maeder, A. 2008, *A&A*, 489, 685

Epelstain, N., Yaron, O., Kovetz, A., & Prialnik, D. 2007, *MNRAS*, 374, 1449

Fossati, L., Bagnulo, S., Landstreet, J., *et al.* 2008, *A&A*, 483, 891

Freytag, B. & Höfner, S. 2008, *A&A*, 483, 571

Girardi, L. & Marigo, P. 2007, *A&A*, 462, 237

Gokhale, V., Peng, X. M., & Frank, J. 2007, *ApJ*, 655, 1010

Gourgouliatos, K. N. & Jeffery, C. S. 2006, *MNRAS*, 371, 1381

Grevesse N. & Noels A. 1993, in: N. Prantzos, E. Vangioni-Flam & M. Cassé (eds.), *Origin and evolution of the elements* (Cambridge: CUP), p. 14

Hicken, M., Garnavich, P. M., Prieto, J. L., *et al.* 2007, *ApJ* (Letters), 669, L17

Howell, D. A., Sullivan, M., Nugent, P. E., *et al.* 2006, *Nature*, 443, 308

Izzard, R., Lugaro, M., Karakas, A. I., Iliadis, C., & van Raai, M. 2007, *A&A*, 466, 641

Janka, H.-T., Langanke, K., Marek, A., *et al.* 2007, *Phys. Rep.*, 442, 38

Janka, H.-T., Marek, A., Müller, B., & Scheck, L. 2008, in: *40 Years of Pulsars: Millisecond Pulsars, Magnetars and More*, AIP-CP, 983, 369

Jeffery, C. S. & Saio, H. 2006, *MNRAS*, 371, 659

Jordan, G. C., Fisher, R. T., Townsley, D. M., *et al.* 2008, in: N. V. Pogorelov, E. Audit & G. P. Zank (eds.), *Numerical Modeling of Space Plasma Flows*, ASP-CP, 385, 97

Karakas, A. I., Fenner, Y., Sills, A., Campbell, S. W., & Lattanzio, J. C. 2006, *ApJ*, 652, 1240

Karakas, A. & Lattanzio, J. C. 2007, *PASA*, 24, 103

Kitaura, F. S., Janka, H.-T., & Hillebrandt, W. 2006, *A&A*, 450, 345

Kotake, K., Sato, K., & Takahashi, K. 2006, *Rep. Prog. Phys.*, 69, 971

Kotake, K., Onishi, N., & Yamada, S. 2007, *ApJ*, 655, 406

Lefever, K., Puls, J., & Aerts, C. 2008, in: A. de Koter, L. J. Smith & L. B. F. M. Waters (eds.), *Mass Loss from Stars and the Evolution of Stellar Clusters*, ASP-CS, 388, 193

Landolt, A. 1992, *AJ*, 104, 340

Machida, M. N., Tomisaka, K., Matsumoto, T., & Inutsuka, S.-I. 2008, *ApJ*, 677, 327

Madhusudhan, N., Justham, S., Nelson, L., *et al.* 2006, *ApJ*, 640, 918

Marek, A. & Janka, H.-T. 2007, *ApJ*, in press [ArXiv e-prints 0708.3372]

Marigo, P. 2002, *A&A*, 387, 507

Marigo, P. 2007, *A&A*, 467, 1139

Marigo, P. & Girardi, L. 2007, *A&A*, 469, 239

Mazzali, P. A., Röpke, F. K., Benetti, S., Hillebrandt, W. 2007, *Science*, 315, 825

Martin, R. G., Tout, C. A., & Lesaffre, P. 2006, *MNRAS*, 373, 263

Matthews, O. M., Wheatley, P. J., Wynn, G. A., & Truss, M. R. 2006, *MNRAS*, 372, 1593

McSaveney, J. A., Wood, P. R., Scholz, M., *et al.* 2008, *MNRAS*, 378, 1089

Meynet, G., Ekstrom, S., & Maeder, A., 2006, *A&A*, 447, 623

Miglio, A., Montalbán, J., & Dupret, M.A. 2007, *MNRAS*, 375, 21

Moffat, A. F. J., Marchenko, S. V., Lefevre, L., *et al.* 2008, in: A. de Koter *et al.* (eds.), *Mass Loss from Stars and the Evolution of Stellar Clusters*, ASP-CS, 388, 29

Morel, T., Butler, K., Aerts, C., Neiner, C., & Briquet M., 2006, *A&A*, 457, 651

Motl, P. M., Frank, J., Tohline, J. E., & D'Souza, M. C. R. 2007, *ApJ*, 670, 1314

Müller, B., Dimmelmeier, H., & Müller, E. 2008, *A&A*, 489, 301

Nomoto, K., Saio, H., Kato, M., & Hachisu, I. 2007, *ApJ*, 663, 1269

Norton, A. J., Butters, O. W., Parker, T. L., & Wynn, G. A. 2008, *ApJ*, 672, 524

Oechslin, R. & Janka, H.-T. 2007, *Phys. Rev. Lett.*, 99, 121102

Obergaulinger, M., Aloy, M. A., & Müller, E. 2006, *A&A*, 457, 209

Owocki, S., 2005, in: R. Ignace & K. G. Gayley (eds.), *The Nature and Evolution of Disks Around Hot Stars*, ASP-CS, 227, 101

Ott, C. D., Dimmelmeier, H., Marek, A., *et al.* 2007, *Phys. Rev. Lett.*, 98, 261101

Pamyatnykh, A. A. 2007, *CoAst*, 150, 213

Patruno, A., Portegies Zwart, S., Dewi, J., & Hopman, C. 2006, *MNRAS* (Letters), 370, L6

Plewa, T., Calder, A. C., & Lamb, D. Q. 2004, *ApJ*(Letters), 612, L37

Podsiadlowski, P. 2007, in: N. St-Louis & A. F. J. Moffat (eds.), *Massive Stars in Interactive Binaries*, ASP-CS, 367, 541

Poelarends, A. J. T., Herwig, F., Langer, N., & Heger, A. 2008, *ApJ*, 675, 614

Prantzos, N., Charbonnel, C., & Iliadis, C. 2007, *A&A*, 470, 179

Rantsiou, E., Kobayashi, S., Laguna, P., & Rasio, F. A. 2008, *ApJ*, 680, 1326

Rappaport, S., Podsiadlowski, P., & Pfahl, E. 2005, *Interacting Binaries: Accretion, Evolution, and Outcomes*, AIP-CP, 797, 422

Robrade, J., Schmitt, J. H. M. M., & Favata, F. 2008, *A&A*, 487, 1139

Röpke, F. K., Hillebrandt, W., Niemeyer, J. C., & Woosley, S. E. 2006, *A&A*, 448, 1

Röpke, F. K., Woosley, S. E., & Hillebrandt, W., 2007, *ApJ*, 660, 1344

Röpke, F. K., Hillebrandt, W., Schmidt, W., *et al.* 2007, *ApJ*, 668, 1132

Saio, H., Kuschnig, R., Gautschy, A., *et al.* 2006, *ApJ*, 650, 1111

Sawai, H., Kotake, K., & Yamada, S. 2008, *ApJ*, 672, 465

Schmidt, W. & Niemeyer, J. C., 2006, *A&A*, 446, 627

Setiawan, S., Ruffert, M., & Janka, H.-T. 2006, *A&A*, 458, 553

Shen, K. J. & Bildsten, L. 2007, *ApJ*, 660, 1444

Shibata, M. & Taniguchi, K. 2008, *Phys. Rev. D*, 77, 084015

Shibahashi, H., Gough, D., Kurtz, D., & Kamp, E. 2008, *PASJ*, 60, 63

Siess, L. 2006, *A&A*, 448, 717

Siess, L. 2007, *A&A*, 476, 893

Sim, S. A., Sauer, D. N., Röpke, F. K., & Hillebrandt, W., 2007, *MNRAS*, 378,2.

Suijs, M. P. L., Langer, N., Poelarends, A.-J., *et al.* 2008, *A&A* (Letters), 481, L87

Surman, R., McLaughlin, G. C., Ruffert, M., *et al.* 2008, *ApJ* (Letters), 679, L117

Talon, S. & Charbonnel, C. 2008, *A&A*, 482, 597

Talon, S., Richard, O., & Michaud, G., 2006, *ApJ*, 645, 634

Théado, S. & Cunha, M. 2006, *CoAst*, 147, 101

Théado, S., Vauclair, S., & Cunha, M. 2005, *A&A*, 443, 627

Tutukov, A. V. & Fedorova, A. V. 2007, *Astr. Rep.*, 51, 291

Ulrich, R. K., 1972, *ApJ*, 172, 165

van Grootel, V., Charpinet, S., Fontaine, G., & Brassard, P. 2008, *A&A*, 483, 875

Ventura, P. & D'Antona, F. 2008a, *A&A*, 479, 805

Ventura, P. & D'Antona, F. 2008b, *MNRAS*, 385, 2034

Viallet, M. & Hameury, J.-M. 2007, *A&A*, 475, 597

Walder, R., Folini, D., & Shore, S. N. 2008, *A&A*, 484, L9

Wang, L., Baade, D., & Patat, F. 2007, *Science*, 315, 212

Weinberg, N. N. & Quataert, E. 2008, *MNRAS*, 387, L64

Wood-Vasey, W. M., Miknaitis, G., Stubbs, C. W., *et al.*, 2007, *ApJ*, 666, 694

Woudt, A., Kilkenny, D., Zietsman, E., Warner, B., *et al.* 2006, *MNRAS*, 371, 1497

Yoon, S.-C. & Langer, N. 2005, *Interacting Binaries: Accretion, Evolution, and Outcomes*, AIP-CP, 797, 651

Yoon, S.-C., Podsiadlowski, P., & Rosswog, S. 2007, *MNRAS*, 380, 933

Transactions IAU, Volume XXVIIA
Reports on Astronomy 2006–2009
Karel A. van der Hucht, ed.

© 2009 International Astronomical Union
doi:10.1017/S1743921308025556

COMMISSION 36

THEORY OF STELLAR ATMOSPHERES

THÉORIE DES ATMOSPHÈRES STELLAIRES

PRESIDENT
VICE-PRESIDENT
PAST PRESIDENT
ORGANIZING COMMITTEE

John D. Landstreet
Martin Asplund
Monique Spite
Suchitra B. Balachandran,
Svetlana V. Berdyugina,
Peter H. Hauschildt, Hans G. Ludwig,
Lyudmila I. Mashonkina,
K. N. Nagendra, Joachim Puls,
M. Sofia Randich,
Grazina Tautvaisiene

TRIENNIAL REPORT 2006 - 2009

1. Introduction

Commission 36 covers the whole field of the physics of stellar atmospheres. The scientific activity in this large subject has been very intense during the last triennium and led to the publication of a large number of papers, which makes a complete report quite impractical. We have therefore decided to keep the format of the preceding report: first a list of areas of current research, then Web links for obtaining further information.

Many conferences and workshops have been held during the period covered by this report on topics related to the interests of Commission 36. Some of them were sponsored by IAU: IAU Symposium No. 231 *Astrochemistry throughout the Universe: Recent Successes & Current Challenges*; IAU Symposium No. 233 *Solar Activity and its Magnetic Origin*; IAU Symposium No. 239 *Convection in Astrophysics*; IAU Symposium No. 240 *Binary Stars as Critical Tools and Tests in Contemporary Astrophysics*; IAU Symposium No. 243 *Star-Disk Interaction in Young Stars*; IAU Symposium No. 247 *Waves and Oscillations in the Solar Atmosphere: Heating and Magneto-Seismology*; IAU Symposium No. 250 *Massive Stars as Cosmic Engines*; and IAU Symposium No. 252 *The Art of Modelling Stars in the 21st Century*. Our members also participated in many of the Joint Discussions and Special Sessions at the IAU XXVI General Assembly in Prague, August 2006.

Meetings not organised under the auspices of the IAU also attracted interest from our members. Among others, the following meetings were attended: *Mass loss from stars and the evolution of stellar clusters*, May 2006, Lunteren, Netherlands; *The Metal Rich Universe*, June 2006, La Palma, Spain; *14th Cambridge Workshop on Cool stars, stellar systems and the Sun*, 6-10 November 2006, Pasadena, USA; *International Workshop on Corona of Stars and Accretion Disks*, 12-13 December 2006, Bonn, Germany; *International Workshop on Clumping in Hot-Star Winds*, June 2007, Potsdam, Germany; *Fifth Potsdam Thinkshop, Meridional flow, differential rotation, solar and stellar activity*, 24-29 June 2007, Potsdam, Germany; *CP#AP Workshop*, 10-14 September, Vienna, Austria; *Fifth Solar Polarisation Workshop*, September 2007, Ascona, Switzerland; and *Magnetic Fields in the Universe, II: From laboratory and stars to the primordial Universe*, 28 January - 1 February 2008, Cozumel, Mexico.

2. Primary research areas 2005 - 2008

2.1. *Physical processes*

General properties – Line-blanketed, unified NLTE models of massive star atmospheres (including winds) available to the community. Grids of synthetic fluxes and spectra. Calibrating parameterized models through physical modeling. Calibrating abundance determinations by filter photometry or low-resolution spectroscopy.

Stationary processes within stellar atmospheres – Convection (granulation) in surface layers, and its effects upon emergent spectra. Interplay between convection and nonradial pulsation. Scales of surface convection in stars in different stages of evolution. Hydrodynamic simulations of entire stellar volumes.

Transient processes – Shocks in pulsating stars. Radiative cooling of shocked gas. Emission lines as shock-wave diagnostics. Co-rotating interaction regions in radiation driven stellar winds. Particle acceleration during flares. Interaction of jets with interstellar medium. Episodic outflows and star-disk interaction.

Magnetic phenomena – Magnetic structures in single and binary stars. Dynamo generation of magnetic fields by convection. Generation of magnetic fields in massive stars. Discovery of strong magnetic fields in a few massive stars. X-ray line emission from magnetically confined winds. Interaction of magnetic fields and radiation-driven winds. Effects by magnetic fields on convective structures. Exploration of the turbulent nature of the general field of the Sun. Magnetic cycles at varying activity levels. Polarized radiation, gyrosynchrotron and X-ray emission. Deriving and interpreting Zeeman-Doppler images of stellar surfaces. Hanle effect diagnostics in stellar environments.

Radiative transfer and emergent stellar spectra – Effects on atmospheric structure by deviations from local thermodynamic equilibrium (non-LTE). Multidimensional radiative transfer. Radiative hydrodynamics. Origin and transfer of polarized light. Theory of scattering of polarized light by atoms and molecules, particularly for understanding the second solar spectrum. Numerical methods in radiative transfer. Scattering mechanisms in circumstellar disks.

Spectral lines and their formation – Line formation in convective atmospheres. Wavelength shifts as signatures of convection. Detection of strong broadening of metal lines in OB supergiants, and interpretation in terms of supersonic macroturbulence. Spectra of rapidly rotating stars viewed pole-on and equator-on. Non-LTE effects in permitted and forbidden lines. NLTE IR-diagnostics of massive hot stars, particularly those close to the Galactic Centre. Atomic and quantum processes affecting spectral lines. Databases for spectral lines. Atlases of synthetic spectra.

Forbidden lines and maser emission – Molecules in atmospheres of cool giant stars. Effects of fluorescence. Permitted and forbidden lines from shocked atmospheres of pulsating giants. Maser and laser emission from stellar envelopes.

Chemical abundances – Precise abundance measurements in BA supergiants. N abundances for massive stars as a function of rotation, which challenge present stellar evolution models. Abundance anomalies. Hydrodynamical models of metal-poor stars. Depletion of light elements through atmospheric motions. Pollution of atmospheres by interstellar dust. r- and s-process elements. Chemical stratification in stable atmospheres. Coronal versus photospheric abundances. Chemical inhomogeneities and pulsation.

Molecules – Theory of the molecular Paschen-Back effect, scattering and Hanle effect in molecular lines in the Paschen-Back regime.

2.2. *Stellar structure*

Structures across stellar disks – Doppler mapping of starspots. Radii and oblateness at different wavelengths for giant stars. Gravitational microlensing to test model atmospheres. Interaction between rotation and pulsation. Doppler tomography of stellar envelopes.

Stellar coronae – Coronal heating mechanisms (quiescent and flaring). Effects of age and chemical abundance. Multicomponent structure. Coronae in also low-mass stars and brown dwarfs. Diagnostics through X-ray spectroscopy and radio emission.

Stellar winds and mass loss – Dynamic outer atmospheres. Multi-component radiation- and dust-driven winds. Mass loss from pulsating giants. Effects of mass flows on the ionization structure. Coronal mass ejections. Instabilities in hot-star winds. X-ray emission.

Dust, grains, and shells – Formation of stellar dust shells. Grains in the atmospheres of red giants, and in T Tauri stars.

2.3. *Different classes of objects*

Stellar parameters of massive stars – significant downscaling of effective temperature scale of OB-stars, due to line-blanketing and mass-loss effects. Rotation rates for massive stars, as a function of metallicity. VLT-FLAMES survey of massive stars: 86 O-stars and 615 B-stars in 8 different clusters (Milky Way, LMC, SMC) observed and analyzed.

Stellar winds and mass-loss – Inhomogeneities in stellar winds suggest downscaling of mass-loss rates. Weak-winds from late O-/early B-dwarfs, which display signficantly lower mass-loss rates than theoretically expected. Self-consistent models for winds from Wolf-Rayet stars. Empirical mass loss-metallicity relation for O-stars. X-ray line emission from line-driven winds. Stellar winds at very low abundances (First Stars). Continuum driven winds for super-Eddington stars. Wind-diagnostics by linear polarization variability.

Pulsating stars and asteroseismology – Classically variable stars, and 'ordinary' solar-type ones. Evidence for opacity-driven gravity-mode oscillations in periodically variable B-type supergiants. Inverting observed pressure-mode frequencies into atmospheric structure. Mass-loss mechanisms in pulsating stars. Effects of rapid rotation on pulsation.

Binary stars – Atmospheric structure and magnetic dynamos in common-envelope binaries. Role of binarity on mass loss. Tidal effects. Non-LTE effects by illumination from the component. Reflection effects in close binaries. Colliding stellar winds.

New classes of very cool stars – Dust, clouds, weather, and chemistry in brown dwarfs. Cloud clearings and hot-spots. Magnetic activity. The effective temperature scale. Molecular line and continuum opacities. Transition between extrasolar giant planets and ultracool brown dwarfs.

White dwarfs and neutron stars – Radiative transfer in magnetized white-dwarf atmospheres. Stokes-parameter imaging of white dwarfs. Molecular opacities in white dwarfs. Broad-band polarization in molecular bands in white dwarfs. Atmospheres and spectra of neutron stars. Effects of vacuum polarization and accretion around magnetized neutron stars.

Special objects – Central stars of planetary nebulae. Population III stars of extremely low metallicity. Protostars. Accretion disks and coronal activity in young stars.

Interaction with exoplanets – Effects of planets on the atmospheres of evolved red giants. Characteristics of stars hosting exoplanets.

2.4. *Development of techniques*

Computational techniques – Parallel (super)computing to simulate convective surface regions, and throughout complete stars. 2- and 3-D NLTE unified models including winds. Neural networks and machine-learning algorithms. Analysis of stellar spectra using genetic algorithms. Preparing for the widely distributed network of computational tools and shared databases being developed for the forthcoming computing infrastructure GRID.

2.5. *Applications of stellar atmospheres*

Besides their study per se, stars are being used as probes for other astrophysical problems:

Exoplanets – Variable wavelength shifts in stellar spectra serve as diagnostics for radial-velocity variations induced by orbiting exoplanets. Atmospheric modeling can indicate which spectral features are suitable as such probes, and which should be avoided due to their sensitivity to intrinsic stellar activity.

Chemical evolution in the Galaxy – How accurately observations of stellar spectral features can be transformed into actual chemical abundances depends sensitively on the sophistication of the stellar model atmospheres.

Kinematics of the Galaxy – Planned space missions intend to measure radial velocities for huge numbers of stars. Model atmospheres are used to identify suitable spectral features for such measurements in different classes of stars.

Chemical evolution of Galaxies – Accurate measurements of metallicity gradients in various galaxies based on BA supergiants.

Evolution of First Stars – Effects from winds may be stronger than expected, due to fast rotation, continuum driving or self-enrichment. Strong mass-loss offers possibility to avoid Pair-Instability SNe.

Distance scales – Flux-weighted gravity-luminosity relationship as a tool to derive independent, precise extragalactic distances from A-supergiants.

Galaxies and cosmology – Stars are the main observable component of galaxies, and population synthesis for galaxies utilize model atmospheres to interpret observations. Cosmological origin of the lowest-metallicity stars.

3. Web links for further information

The following collection of links provides introductions and overviews of several significant subfields of the physics of stellar atmospheres.

3.1. *Calculating atmospheric models and spectra*

ATLAS, SYNTHE, and other model grids: `<kurucz.harvard.edu>`.
MARCS, model grids: `<www.marcs.astro.uu.se>`.
Tuebingen: Stellar atmospheres - grid of models:
   `<astro.uni-tuebingen.de/groups/stellar>`.
CCP7 - Collaborative Computational Project: `<ccp7.dur.ac.uk>`.

CLOUDY - photoionization simulations: `<www.pa.uky.edu/ gary/cloudy/>`.
MULTI - non-LTE radiative transfer: `<www.astro.uio.no/ matsc/mul22>`.
PANDORA - atmospheric models and spectra:
  `<www.cfa.harvard.edu/ avrett/pandora.lis.copy>`.
PHOENIX - stellar and planetary atmosphere code:
  `<www.hs.uni-hamburg.de/EN/For/ThA/phoenix/>`.
STARLINK - theory and modeling resources:
  `<www.astro.gla.ac.uk/users/norman/star/sc13/sc13.htx>`.
Synthetic spectra overview: `<www.am.ub.es/ carrasco/models/synthetic.html>`.
TLUSTY - model atmospheres: `<tlusty.gsfc.nasa.gov>`.
WM-Basic: unified hot star model atmospheres incl. consistent wind structure:
  `<www.usm.uni-muenchen.de/people/adi/adi.html>`.
PoWR: The Potsdam Wolf-Rayet Models, grid of synthetic spectra:
  `<www.astro.physik.uni-potsdam.de/ wrh/PoWR/powrgrid1.html>`.

3.2. *Useful links from research groups or individual researchers*

Vienna: Stellar atmospheres and pulsating stars:
  `<ams.astro.univie.ac.at/main.php>`.
Potsdam: Stellar convection:
  `<www.aip.de/groups/sternphysik/stp/convect_neu.html>`.
R. F. Stein: Convection simulations & radiation hydrodynamics:
  `<www.pa.msu.edu/ steinr/research.html#convection>`.
D. Dravins: Stellar surface structure:
  `<www.astro.lu.se/ dainis/HTML/GRANUL.html>`.
A. Collier Cameron: Starspots and magnetic fields on cool stars:
  `<star-www.st-and.ac.uk/ acc4/coolpages/imaging.html>`.
D. F. Gray: Stellar rotation, magnetic cycles, velocity fields:
  `<astro.uwo.ca/ dfgray/>`.
J. F. Donati: Magnetic fields of non degenerate stars:
  `<webast.ast.obs-mip.fr/users/donati/>`.
M. Jardine: Stellar coronal structure:
  `<capella.st-and.ac.uk/ mmj/Welcome research.html>`.
Munich: Hot stars: `<www.usm.uni-muenchen.de/people/adi/adi.html>`,
  `<www.usm.uni-muenchen.de/people/adi/wind.html>`.
S. Jeffery: Stellar model grids, hot stars: `<star.arm.ac.uk/ csj/>`.
P. Stee: Be-star atmospheres and circumstellar envelopes:
  `<www.obs-nice.fr/stee/Bemodel.html>`,
  `<www.obs-nice.fr/stee/simugb.html>`.
N. Przybilla: NLTE atmospheres of massive stars, extragalactic stellar astronomy:
  `<www.sternwarte.uni-erlangen.de/ przybilla/research.html>`.
S. Owocki: Theory of line-driven winds, hydrodynamics, rotation, magnetic fields:
  `<www.bartol.udel.edu/ owocki/>`.
R. Townsend: Magnetically-Controlled Circumstellar Environments of Hot Stars:
  `<zuserver2.star.ucl.ac.uk/ rhdt/research/magnetic/>`.
P. Crowther: Hot Lumionous Star Research Group:
  `<shef.ac.uk/physics/people/pacrowther/>`.
R. Kudritzki: Hot Stars and Winds, Extragalactic Stellar Astronomy:
  `<www.ifa.hawaii.edu/ kud/kud.html>`.
J.L.Linsky: Cool stars, stellar chromospheres and coronae:
  `<jilawww.colorado.edu/ jlinsky/>`.

G. Basri: Brown dwarfs: `<astro.berkeley.edu/ basri/bdwarfs/>`.

D. Montes *et al.*: Libraries of stellar spectra:
`<www.ucm.es/info/Astrof/spectra.html>`.

R. J. Rutten: Lecture notes: Radiative transfer in stellar atmospheres:
`<www.fys.ruu.nl/ rutten/node20.html>`.

John D. Landstreet
president of the Commission

Transactions IAU, Volume XXVIIA
Reports on Astronomy 2006–2009
Karel A. van der Hucht, ed.

© 2009 International Astronomical Union
doi:10.1017/S1743921308025568

COMMISSION 45

SPECTRAL CLASSIFICATION
CLASSIFICATION STELLAIRE

PRESIDENT Sunetra Giridhar
VICE-PRESIDENT Richard O. Gray
PAST PRESIDENT Christopher J. Corbally
ORGANIZING COMMITTEE Coryn A. L. Bailer-Jones,
Laurent Eyer, Michael J. Irwin,
J. Davy Kirkpatrick, Steven Majewski,
Dante Minniti, Birgitta Nordström

TRIENNIAL REPORT 2006 - 2009

1. Introduction

This report gives an update of developments (since the last General Assembly at Prague) in the areas that are of relevance to the commission. In addition to numerous papers, a new monograph entitled *Stellar Spectral Classification* with Richard Gray and Chris Corbally as leading authors will be published by Princeton University Press as part of their *Princeton Series in Astrophysics* in April 2009. This book is an up-to-date and encyclopedic review of stellar spectral classification across the H-R diagram, including the traditional MK system in the blue-violet, recent extensions into the ultraviolet and infrared, the newly defined L-type and T-type spectral classes, as well as spectral classification of carbon stars, S-type stars, white dwarfs, novae, supernovae and Wolf-Rayet stars.

2. Working Groups

The Working Group on *Standard Stars* (chair C. Corbally) publishes a biannual newsletter edited by Richard Gray at the `<stellar.phys.appstate.edu/ssn>`. The newsletter publishes abstracts of papers relating to all aspects of standard stars, reports of ongoing work, and contributions and discussions related to standard stars.

3. Cool Brown Dwarfs (Adam J. Burgasser)

3.1. *Overview*

As of 1 May 2008 there are 131 known T-type dwarfs archived on the DwarfArchives site `<dwarfarchives.org>`. The majority of these have been found by the 2MASS and SDSS surveys, but an increasing number are now being identified in the deeper near-infrared (NIR) successor surveys UKIDSS (Kendall *et al.* 2007; Lodieu *et al.* 2007; Warren *et al.* 2007; Chiu *et al.* 2008; CHFTLS (Delorme *et al.* 2008; Delorme *et al.* 2008).

Out of the known population, eleven are companions to nearby stars, ten are resolved binaries, and one – S Ori 70 – is a candidate member of the 3 Myr σ Orionis cluster. The known population spans distances of 3.6 pc to ~ 100 pc (~ 360 pc if S Ori 70 is a member of σ Orionis), with the vast majority lying within 20 pc of the Sun.

Apparent J-magnitudes span 12.9 to > 20. Nearly all have published low- ($R \simeq 100$) and moderate-resolution ($R \simeq 500$-2000) NIR spectroscopic data, and a fraction have published red optical and/or mid-infrared data. High-resolution ($R \simeq 20\,000$) near-infrared spectroscopic data have been reported for a few of the brightest T-type dwarfs (Zapatero Osorio *et al.* 2006; McLean *et al.* 2007; Zapatero Osorio *et al.* 2007).

3.2. *Classification*

A NIR spectral classification scheme, unifying the initial schemes has been completed and published in Burgasser *et al.* 2006. This scheme defines nine primary standards spanning subtypes T0 through T8 in integer steps, along with five alternate standards. The standards were selected to be bright and observable from both hemispheres wherever possible, and resolved binaries were excluded from the primary standard set.

Classification metrics tied to the strengths of H_2O and CH_4 absorption bands were defined in the 1-2.5 μm range for low- and moderate-resolution spectra. In addition, five classification indices were defined to enable 'quantified' classification. Cushing *et al.* 2006 have investigated spectral classification in the mid-infrared using 5.5-14.5 μm data obtained with the *Spitzer Space Telescope*. The 7.8 μm CH_4 and 10.5 μm NH_3 bands were shown to strengthen monotonically through the T sequence and could in principle be used for spectral classification. Photometric classification of T dwarfs has been demonstrated using narrow-band filters spanning the 1.6 μm CH_4 band (Tinney *et al.* 2005; Burgasser *et al.* 2006).

Several groups are now investigating the use of these filters to identify and characterize T-type dwarf candidates in deep imaging surveys.

3.3. *Secondary parameters*

As anticipated in the 2005 Commission report, a number of groups have examined secondary parameter effects in T-type dwarf spectra arising from variations in surface gravity, metallicity and vertical mixing. Key features include the 2.1 μm K-band peak modulated by pressure-sensitive, collision-induced H_2 absorption, and the 1.05 μm peak, shaped by the pressure-broadened red wing of the 0.77 μm K I doublet transition. Studies have characterized variations in these features amongst the latest-type T dwarfs, which are less affected by condensate cloud effects (see below).

Comparisons of model predictions to the spectra of empirical benchmarks (sources with astrometric distance and bolometric luminosity measurements, companions to nearby stars, resolved binary systems) have shown that these features are linked to both surface gravity and metallicity variations (Burgasser *et al.* 2006, Liu *et al.* 2006; Burgasser *et al.* 20007; Cushing *et al.* 2008; Looper *et al.* 2008). High-angular resolution imaging and spectral template analyses have verified these effects as intrinsic to the evolution of brown dwarfs across the L/T transition and not the result of age or metallicity variations (Burgasser *et al.* 2006; Liu et.al 2006; Looper *et al.*2008). Condensate cloud effects are likely responsible, although fully-consistent models are still progressing (Burrows *et al.* 2006;Helling *et al.* 2008). The L/T transition problem is of course relevant to the classification of early-type T-type dwarfs; of particular concern is the possibility that currently-defined subtypes may be biased by or entirely populated with hybrid spectra. High-resolution imaging and spectroscopic studies of early-type T-type dwarf standards to weed out unresolved pairs is ongoing.

3.4. *Beyond class T: the Y-type dwarfs*

Two discoveries made this term by UKIDSS (Warren *et al.* 2007) and CFHTLS (De-lorme *et al.* 2008) appear to be cooler and later-type than the current T8 standard 2MASS J0415–0935. Their NIR spectra also exhibit a subtle feature attributable to NH_3 absorption at $1.55 \, \mu m$. The detection of this feature at NIR wavelengths (NH_3 has already been detected in mid-infrared T-type dwarf spectra) has raised the issue as to what will define the end of the T-type dwarf class and the start of the next cooler class, already dubbed the 'Y-type dwarfs' (see Kirkpatrick 2005). Delorme *et al.* (2008) advocate that the appearance of NH_3 in the NIR – a new molecular species at these wavelengths – while Leggett *et al.* (2007) contend that only with the condensation of H_2O will spectral energy distributions change sufficiently to warrant a new class.

The current absence of a substantially cooler benchmark, comparable to the L-type dwarf (GD 165B) and T-type dwarf (Gliese 229B) prototypes, is likely to make this a con-tentious issue in the upcoming term. Continued survey work by UKIDSS and CFHTLS, and the upcoming *WISE* satellite mission (an infrared successor to both 2MASS and IRAS, with an expected launch in late 2009), will aid this endeavor by (hopefully) iden-tifying many more such cool brown dwarfs.

4. White dwarf spectral classification (James W. Liebert)

The spectral classification system currently in use is described in Sion *et al.* 1983 (Jim Liebert being one of the coauthors, theorists and observers included.) Until recently, white dwarf atmospheres came in two flavors – hydrogen-rich and helium-dominated, sometimes with traces of hydrogen. The former are classified DA, until about 5 000 K below which the hydrogen is not excited so there are no Balmer lines in absorption. Likely the helium sequence begins with the stars above 40 000 K which show He II and are classified DO, then from 12 000 to about 39 000 K they show only neutral helium (He I) and are classified DB. Below 12 000 K they can be featureless (DC), show carbon bands and/or atomic lines (DQ) or accreted heavy metals (DZ). There are hybrid classes which show traces of metals, such as DBZ. An interesting discovery of the few years is a class of cool, hydrogen atmosphere white dwarfs that show evidence of a debris disk (DAZ). Refractory heavy elements continually rain down on the atmosphere so that heavy metal lines (Ca, Mg, Fe mainly) are seen in the spectrum, and the mid-IR energy distribution shows an excess due to heating of the external debris disk by the white dwarf. The *Spitzer* satellite has contributed to the study of these.

A new class of white dwarf atmosphere has been found in the last few years, rare objects appearing among the over 9 000 white dwarfs found in the Sloan Digital Sky Survey. These are hotter DQ stars showing lines of singly-ionized carbon (C II). Now the long-established cool DQ stars, most of which show bands of diatomic carbon (C_2 Swan bands), have helium-dominated atmospheres with traces of carbon no more than one carbon for every 100 helium atoms. These have carbon dredged up from the core of the star into the helium envelope. We have recently discovered that some of the hot DQ stars with T_{eff} near 20 000 K actually have atmospheres dominated by carbon. Helium is usually not detected (see Dufour *et al.* 2007). Oxygen is also detected as a trace element in a few of these stars.

5. Carbon star classification (Thomas H. Llyod Evans)

There has been great activity in this area over the last three years. First, there have been several searches for carbon stars by classical objective prism work in galactic fields, as well as confirmation by slit spectroscopy of those discovered earlier and slit spectroscopy of candidates selected by their colours in the *IRAS* or 2MASS catalogues. Some of the important contributions are listed below. In addition, rejection of suspected members by MacConnell (2006) and detection of new RCB stars by Tisserand *et al.* (2008) has been carried out.

There have also been numerous discoveries of carbon stars in galaxies of the Local Group. A reliable survey technique, which avoids the problems with crowding and a poor limiting magnitude in objective prism surveys, uses photometry in *R*, *I* and in narrow bands centred on TiO and CN (cf. Battinelli & Demers 2006; Demers, Battinellii & Artigai 2006). Many recent surveys use infra red photometry, as many carbon stars occupy a distinctive position in colour-magnitude and two-colour diagrams.

The purity of such samples has been checked by the more specific narrow band photometry, and confusion between bluer carbon stars and M-type stars, as well as between M-type stars and hotter stars, demonstrated by Battinelli, Demers & Mannucci (2007). Much work has been done on near- and mid-infrared spectra of carbon stars in the Galaxy and nearby galaxies. Some of this work has relevance for stellar classification, and is of particular importance because it extends spectral classification to stars which are in many cases practically optically invisible, as well as extending classification to a new spectral region, which may contain new information.

6. Photometric classification (Thomas H. Llyod Evans)

The photometric classification based on Medium Band photometric systems such as Stromvil and Vilnius Systems has been carried out by Bartasiute, Straizys, Laugalys and few others and the Strömgren system had been employed. To classify stars in clusters of various metallicity by Anthony-Twarog and collaborators. Numerous broadband (Washington) as well as *UBVRI* photometry to classify stars in clusters at different Galactic locations reaching fainter limits continue to provide useful data for the study of Galactic structure. The references listed below are arranged following the classification system mentioned above.

Stromvil and Vilnius Systems 2005 - 2008
Bartasiute, S., Deveikis, V., Straizys, V., & Bogdanovicius, A. 2007, *BaltA*, 16, 199
Kazlauskas, A., Boyle, R. P., Philip, A. G. D., *et al.* 2005, *BaltA*, 14, 465
Laugalys, V., Straizys, V., Vrba, F. J., *et al.* 2006a, *BaltA*, 15, 327
Laugalys, V., Straizys, V., Vrba, F. J., *et al.* 2006b, *BaltA*, 15, 483
Laugalys, V., Straizys, V., Vrba, F. J., *et al.* 2007, *BaltA*, 16, 349
Straizys, V., Zdanavicius, J., Zdanavicius, K., *et al.* 2005, *BaltA*, 14, 555

Strömgren photometry 2005 - 2008
Anthony-Twarog, B. J., Tanner, D., Cracraft, M., & Twarog, B. A. 2006, *AJ*, 131, 461
Anthony-Twarog, B. J., Twarog, B. A., & Mayer, L. 2007, *AJ*, 133, 1585
Anthony-Twarog, B. J., *et al.* 2006, *PASP*, 118, 358
Twarog, B. A. 2006, *AJ*, 132, 299
Twarog, B. A. 2007, *AJ*, 134, 1777

Washington photometry 2005 - 2008
Clariá, J. J., Torres, M. C., & Ahumada, A. V. 2005, *BaltA*, 14, 495
Clariá, J. J., Piatti, A. E., Parisi, M. C., & Ahumada, A. V. 2007, *MNRAS*, 379, 159
Parisi, M. C., Clariá, J. J., Piatti, A.E., & Geisler, D. 2005, *MNRAS*, 363, 1247

Parisi, M. C., Clariá, J. J., Piatti, A. E., & Geisler, D. 2006, *RMxAC*, 26, 182
Piatti, A. E., Clariá, J. J., Parisi, M. C., & Ahumada, A. V. 2008, *BaltA*, 17, 67

Broadband UBVRI photometry 2005 - 2008
Busso, G., Moehler, S., Zoccali, M., Heber, U., & Yi, S. K. 2006, *BaltA*, 15, 25
Clariá, J. J., Mermilliod, J.-C., Piatti, A. E., & Parisi, M. C. 2006a, *A&A*, 453, 91
Claria, J. J., Piatti, A. E., Lapasset, E., & Parisi, M. C. 2006b, *RMxAC*, 26, 168
Liebert, J. & Givis, J. E. 2006, *PASP*, 118, 659
Pandey, A. K., Sharma, S., Upadhyay, K., *et al.* 2007, *PASJ*, 59, 547
Piatti, A. E., Clariá, J. J., & Ahumada, A. V. 2006, *NewA*, 11, 262
Piatti, A. E., Clariá, J. J., & Ahumada, A. V. 2006, *MNRAS*, 367, 599
Piatti, A. E., Clariá, J. J., Mermilliod, J.-C., *et al.* 2007, *MNRAS*, 377, 1737
Rubin, K. H. R., Williams, K. A., Bolte, M., & Koester, D. 2008, *AJ*, 135, 2163

7. Catalogues and atlases (Richard O. Gray)

7.1. *General Catalog of Stellar Spectral Classifications* (Brian A. Skiff)

Brian A. Skiff (Lowell Obs) continues to build a comprehensive catalogue of published spectral classifications. It now contains some 300 000 entries and as of mid-2008 is asymptotically complete up to about 1985 in the literature; some more recent data is included. Several important objective-prism surveys from the 1950s and 1960s are omitted at present, however. The file is updated a few times each year at the Strasbourg VizieR service as item 'B/mk'. See: <cdsarc.u-strasbg.fr/viz-bin/Cat?B/mk>.

In late 2007 Skiff also acquired most of the remaining Burrell Schmidt and some Curtis Schmidt objective-prism plates taken for the northern extension of Nancy Houk's HD reclassification project. The 1 700 plates have complete coverage north of about $+37^o$ Dec, but include substantial coverage between $+05^o$ and $+37^o$ Declination. The plate limit is about B mag 10.5. It is his intention to finish classifying the northern HD stars to the extent possible. Considerable experience in examining the plates will be necessary before being able to match the quality and consistency of Houk's work.

The classifications in this compilation include only those types determined from spectra, omitting those determined from photometry. Classifications include MK types as well as types not strictly on the MK system, such as white dwarfs, Wolf-Rayet stars, etc. This catalogue includes for the first time results from many large-scale objective-prism spectral surveys done at Case, Stockholm, Crimea, Abastumani, and elsewhere. The stars in these surveys were usually identified only on charts or by other indirect means, and have been overlooked heretofore because of the difficulty in recovering the stars. Many of these stars are not included in SIMBAD.

7.2. *Other large-scale spectral classification catalogues*

Other spectral classification catalogues, published in the past three years, include:
A Unified Near-Infrared Spectral Classification Scheme for T Dwarfs, Burgasser, A. J., Geballe, T. R., Leggett, S. K., Kirkpatrick, J. D., & Golimowski, D. A., 2006, *ApJ*, 637, 1067. This paper details a new unified near-infrared spectral classification scheme for T dwarfs, and presents spectral types for many of the known T dwarfs.

On-line spectral catalogs of T dwarfs are currently maintained by:
- A. Burgasser, (low resolution NIR: <www.browndwarfs.org/spexprism>;
red optical: <web.mit.edu/~ajb/www/tdwarf/#spectra>);
- S. Leggett (moderate resolution NIR: <www.jach.hawaii.edu/~skl/LTdata.html>),
- I. McLean (moderate resolution NIR: www.astro.ucla.edu/~mclean/BDSSarchive/>),

- J. Rayner (moderate resolution NIR: `irtfweb.ifa.hawaii.edu/~spex/WebLibrary/index.html#T>`).

Contributions to the Nearby Stars (NStars) Project: spectroscopy of stars earlier than M0 within 40 pc – The southern sample, Gray, R. O., Corbally, C. J., Garrison, R. F., McFadden, M. T., Bubar, E. J., McGahee, C. E., O'Donoghue, A. A., & Knox, E. R., 2006, *AJ* 132, 161. This paper is the second in a series of three classifying the nearby solar-type stars (stars earlier than M0 within 40 pc of the sun). Results are reported for 1676 stars south of the equator.

Visual Multiples. IX. MK Spectral Types, Abt, H. A. 2008, *ApJS* 176, 216. Classifications for 546 stars in multiple systems are given.

The Galactic O Star Catalog V.2.0, Sota, A., Maíz Apellániz, J., Walborn, N. R., & Shida, R. Y., *RMxAC* 33, 56. The Galactic O-Star Catalog continues to be updated with spectral types and other information on the known O-type stars with $V < 8$ in the Galaxy. A total of 378 stars with precise spectral classifications are featured. See `<www-int.stsci.edu/ jmaiz/research/GOS/GOSV2main.html>`.

7.3. *Stellar spectral atlases*

A number of stellar spectral atlases were published during this period, featuring stellar spectra across the electromagnetic spectrum:

X-Atlas: An On-line Archive of Chandra's Stellar High-Energy Transmission Grating Observations, Westbrook, O. W., Evans, N. R., Wolk, S. J., Kashyap, V. L., Nichols, J. S., *ApJS*, 176, 218; and *Visual Multiples. IX. MK Spectral Types*, Abt, H. A. 2008, *ApJS*, 176, 216. Classifications for 546 stars in multiple systems are given. This web-based atlas (`<cxc.harvard.edu/XATLAS/>`) features spectra of high-mass (O, B and Wolf-Rayet) and low-mass (F, G, K, M) stars as well as spectra of white dwarfs, X-ray binaries and cataclysmic binaries observed with the *Chandra* X-ray telescope.

A high resolution spectral atlas of brown dwarfs, Reiners, A., Homeier, D., Hauschildt, P. H., & Allard, F., 2007, *A&A* 473, 245.

Spectral atlas of O9.5-A1-Type supergiants, Chentsov, E.L., & Sarkisya, A.N. 2007, *Astrophysical Bulletin* 62, 257.

SpectroWeb: An Interactive Graphical Database of Digital Stellar Spectral Atlases, Lobel, A. This interactive database contains high resolution spectra of a number of standard stars. See: `<spectra.freeshell.org/spectroweb.html>`.

Spectral atlas of massive stars around He I 10830 Å, Groh, J. H., Damineli, A., & Jablonski, F. 2007, *A&A* 465, 993.

A Medium Resolution Near-Infrared Spectral Atlas of O and Early-B Stars, Hanson, M. M., Kudritzki, R.-P., Kenworthy, M. A., Puls, J., & Tokunaga, A. T. 2005, *ApJS* 161, 154. Contains spectra and spectral types of O and early-B stars in the infrared *K*-band.

7.4. *A new spectral class encoding system* (Myron A. Smith, Richard O. Gray, Christopher J. Corbally, Randall Thompson, and Inga Kamp)

A new spectral class encoding system has been developed to make archives such as in the VO and NASA's MAST more responsive to specific query needs. Although MAST has archived UV and optical wavelength spectra for well over 10 000 stars observed by NASA satellites, there has been up to now only little design work on a system that organizes these objects by spectral type in order for researchers to obtain information about targeted groups of stars.

For someone interested in identifying all stars of a chosen spectral type and luminosity class and downloading the spectra the authors have designed a 'spectral class' nomenclature system for spectra of stars across the HR Diagram into a finite number of bins. The nomenclature scheme has the form `TT.tt.LL.PPPP`, where `TT` and `tt` are numerical digits (0-9) representing spectral types and subtypes, `LL` luminosity classes, and `PPPP` represent spectral peculiarities. An entry '00' is returned for an unknown attribute. To take the example of the (incomplete) spectral classification just 'A0', the `LL` codes will be '00' because the luminosity class is unspecified. The four spectral peculiarities are arranged into two subgroups of two each. Then, `P1P2` are reserved for peculiarities that can be expected across the HR Diagram (e.g. 'e', 'p' or 'n'), and `P3P4` are codes for peculiarities common only to a smaller range of spectral types, e.g., types AF. In this way, one can represent as many as four spectral peculiarities, including the possible composite nature of the spectrum. The details of this have been published in an IVOA (International Virtual Observatory Alliance) Design Note, see: `<www.ivoa.net/Documents/latest/SpectClasses.html>`

8. Closing remarks

It is a pleasure to thank the contributors whose names appear at the head of their sections in this report. I am particularly thankful to Tom Lloyd Evans and Chris Corbally for their help and guidance over my term as vice president and later as president. I am also thankful to Richard Gray for hosting the commission page and all OC members for their support and co-operation.

Sunetra Giridhar
president of the Commission

References

Battinelli, P. & Demers, S. 2006, *A&A*, 447, 473
Battinelli, P., Demers, S., & Mannucci, F. 2007, *A&A*, 474, 35
Burrows, B., Sudarsky, D., & Hubeny, I. 2006, *ApJ*, 640, 1063
Burgasser, A. J., Geballe, T. R., Leggett, S. K., *et al.* 2006, *ApJ*, 637, 1067
Burgasser, A. J., Kirkpatrick, J. D., Cruz, K. L., *et al.* 2006, *ApJS*, 166, 585
Burgasser, A. J. 2007, *ApJ*, 659, 655
Chen, P. S. & Shan, H.-G. 2006, *ApSS*, 312, 85
Chen, P. S. & Shan, H.-G. 2008, *ApSS*, 314, 291
Chiu, K., Liu, Michael, C., & Jiang, L., *et al.* 2008, *MNRAS*, 385, 53
Cioni, M-R. L. & Habing, H. J. 2005, *A&A*, 442, 165
Cushing, M. C., Roellig, T. L., Marley, M. S., *et al.* 2006, *ApJ*, 648, 614
Cushing, M. C., Marley, M. S., Saumon, D., *et al.* 2008, *ApJ*, 678, 1372
Delorme, P., Delfosse, X., Albert, L., *et al.* 2008a, *A&A*, 482, 961
Delorme, P., Willott, C. J., Forveille, T., *et al.* 2008b, *A&A*, 484, 469
Demers, S., Battinelli, P., & Artigai, E. 2006, *A&A*, 456, 905
Dufour, P., Liebert, J., Fontaine, G., & Behara, N. 2007, *Nature*, 450, 522
Fuenmayor, F. & Stock, J. 2006, *RMAA*, 25, 107
Gigoyan, K., Mickaelian, A. M., & Mauron, N. 2006, *Astrophysics*, 49, 173
Gullieuszik, M., Rejkuba, M., Cioni, M. R., Habing, H. J., & Held, E. V. 2007, *A&A*, 475, 467
Gullieuszik, M., Greggio, L., Held, E. V., *et al.* 2008, *A&A* (Letters), 483, L5
Helling, Helling, C., Dehn, M., Woitke, P., & Hauschildt, P. H. 2008, *ApJ* (Letters) 675, L105
Jackson, D. C., Skillman, E. D., Gehrz, R. D., *et al.* 2007, *ApJ*, 656, 818
Jackson, D. C., Skillman, E. D., Gehrz, R. D., *et al.* 2007, *ApJ*, 667, 891

Kato, D., Nagashima, C., Nagayama, T., *et al.* 2007, *PASJ*, 59, 615

Kendall, T. R., Tamura, M., Tinney, C. G., *et al.* 2007, *A&A*, 466, 1059

Kirkpatrick, J. D. 2005, *ARA&A*, 43, 195

Leggett, S. K., Marley, M. S., Freedman, R., *et al.* 2007, *ApJ*, 667, 537

Liu, M. C., Leggett, S. K., Golimowski, D. A., *et al.* 2006, *ApJ*, 647, 1393

Lodieu, N., Pinfield, D. J., Leggett, S. K., *et al.* 2007, *MNRAS*, 379, 1423

Looper, D. L., Kirkpatrick, J. D., Cutri, R. M., *et al.* 2008, *ApJ*, 686, 528

MacConnell, J. 2006a, *BaltA* 15, 383

MacConnell, J. 2006b, *IBVS*, 5671,

Matsuura, M., Zijlstra, A. A., Bernard-Salas, J., *et al.* 2007, *MNRAS*, 382, 1889

Mauron, N., Kendall, T. R., & Gigoyan, K. 2005, *A&A*, 438, 867

Mauron, N., Gigoyan, K., & Kendall, T. R. 2007, *A&A*, 475, 843

Mauron, N. 2007, *A&A*, 482, 151

McLean, I.S., Prato, L., McGovern, M. R., *et al.* 2007, *ApJ*, 658, 1217

Menzies, J., Feast, M., Whitelock, P., *et al.* 2008, *MNRAS*, 385, 1045

Sion, E. M., Greenstein, J. L., Landstreet, J. D., *et al.* 1983, *ApJ*, 269, 253

Sohn, Y.-J., Kang, A., Rhee, J., Shin, M., Chun, M.-S., & Kim, H.-I. 2006, *A&A*, 445, 69

Tanaka, M., Letip, A., Nishimaki, Y., *et al.* 2007, *PASJ*, 59, 939

Tinney, C. G., Burgasser, A. J., Kirkpatrick, J. D., & McElwain, M. W. 2005, *AJ*, 130, 2326

Tisserand, P., Marquette, J. B., Wood, P. R., *et al.* 2008, *A&A*, 481, 673

Valcheva, A. T., Ivanov, V. D., Ovcharov, E. P., Nedialkov, P. L., *et al.* 2007, *A&A*, 466, 501

van Loon, J. Th., Marshall, J. R., Cohen, M., *et al.* 2007, *A&A*, 447, 971

Warren, S. J., Mortlock, D. J., Leggett, S. K., *et al.* 2007, *MNRAS*, 381, 1400

Zapatero Osorio, M. R., Martín, E. L., Bouy, H., *et al.* 2006, *ApJ*, 647, 1405

Zapatero Osorio, M. R., Martín, E. L., Béjar, V. J. S., *et al.* 2007, *ApJ*, 666, 1205

Transactions IAU, Volume XXVIIA
Reports on Astronomy 2006–2009
Karel A. van der Hucht, ed.

© 2009 International Astronomical Union
doi:10.1017/S174392130802557X

DIVISION IV / WORKING GROUP
MASSIVE STARS

CHAIR   Stanley P. Owocki
MEMBERS   Paul A. Crowther, Alexander W. Fullerton
                   Gloria Koenigsberger, Norbert Langer,
                   Claus Leitherer, Philip L. Massey
                   Georges Meynet, Joachim Puls,
                   Nicole St-Louis

TRIENNIAL REPORT 2006 - 2009

1. Background

Our Working Group studies massive, luminous stars, with historical focus on early-type (OB) stars, but extending in recent years to include massive red supergiants that evolve from hot stars. There is also emphasis on the role of massive stars in other branches of astrophysics, particularly regarding starburst galaxies, the first stars, core-collapse gamma-ray bursts, and formation of massive stars.

Before 2004, we were known as the Hot-Star Working Group, but after thorough debate, the WG Organizing Committee voted for a name change to *Massive Stars*. This name change reflects the tight interrelation of research on hot and cool stars in the upper Hertzsprung-Russell diagram.

At this time, the Organizing Committee also generated and approved a set of by-laws. These by-laws and other activities of the Working Group are posted on the web site `<www.astroscu.unam.mx/massive_stars>`. Our web master is Raphael Hirschi (University of Basel). Among other features, this web portal offers a discussion group page and an automatic Newsletter submission interface. This allows the members of the Working Group to submit their Newsletter abstracts and circulates newly received abstracts to registered members.

The *Massive Star Newsletter*, edited by Philippe Eenens (UNAM, Mexico City), continues to be the main means of communication and science dissemination in our Working Group. As of April 2008, 105 issues of the Newsletter have been published. Back issues are posted on the WG web site: `<www.astroscu.unam.mx/massive_stars/news.php>`.

2. Developments within the past triennium

Over the past three years, activities focused on co-sponsorship of IAU JD05 at the IAU XXVI GA in Prague (Czech Republic), 2006, and on the highly successful IAU Symposium No. 250 in Kauai (HI, USA). The follow gives further background on these meetings.

2.1. *IAU GA2006 JD05: Calibrating the Top of the Stellar Mass-Luminosity Relation*

The goal of this Joint Discussion, held 16 August 2006 at the IAU XXVI General Assembly in Prague (Czech Republic), was to bring together theorists and observers from the stellar and extragalactic communities to discuss the properties of the most massive stars and the implications for cosmological studies. The meeting focused on a set of themes that follow from fundamental stellar astronomy, such as mass determinations in binary stars, to recent modeling of atmospheres and evolution, to the significance of massive stars for the ecology of the host galaxy, and finally to a critical assessment of the properties of the first generation of stars in the Universe. Major topics included:

- empirical mass determinations of the most massive single stars;
- models for massive stars on and off the main sequence;
- stability near the Eddington limit with and without rotation;
- comparisons of atmospheric and evolutionary masses;
- observational efforts to detect, monitor, and analyze massive binaries;
- mass and energy return to the interstellar medium from massive stars;
- extrapolations to the first generation of stars with ultra-high masses; and
- the role of hot massive stars during the epoch of re-ionization in the early Universe.

Further information at the website: `<www.stsci.edu/science/starburst/Prague/>`.

2.2. *IAU Symposium No. 250: Massive Stars as Cosmic Engines*

This IAU Symposium held 9-14 December 2007 on Kauai (Hawaii, USA) and hosted by the Institute for Astronomy at the University of Hawaii, focused on how massive stars shape the Universe, from the nearby Universe to high-redshift galaxies and the first generation of stars.

Massive stars form in starbursts, pollute the interstellar medium (ISM), inject energy and momentum via their stellar winds and core-collapse supernovae (SNe), drive the ISM out of galaxies, polluting the intergalactic medium. Direct detection of massive stars (via their UV continua and spectral lines) and of the products of their nucleosynthesis provides some of the most stringent constraints on the physical properties of galaxies at high redshifts, whether identified via their emission at a variety of wavelengths or by the absorption they produce in quasar spectra. Within the past few years, a direct connection has been established between certain core-collapse SNe and Gamma Ray Bursts (GRBs), supporting the collapsar model in which the GRB results from the death throes of a rapidly rotating Wolf-Rayet star.

Within the past few years, great progress has been made toward our understanding of the astrophysical role played by massive stars. From an observational perspective, temperatures of OB stars have been revised downward based on the most recent observations with *FUSE*, *HST*, and ground-based facilities. The role of clumping in stellar winds has been recognized, with potentially dramatic consequences for stellar evolution, due to its influence on derived mass-loss rates. Close binaries with masses of up to $80\,M_\odot$ solar masses have been identified and studied visually and with exquisite detail using *Chandra*, *XMM*, VLA. Visibly obscured young massive clusters have been identified at our Galactic Centre, elsewhere in our own Milky Way and in external galaxies. These have been studied with *HST*, VLT, Gemini and Subaru, exploiting natural guide star Adaptive Optics (AO) techniques from the ground. Increasingly, the use of AO with laser guide stars is expected to revolutionise the study of massive star forming regions.

Quantitative spectroscopy of massive stars beyond the Local Group has been undertaken with VLT and Keck to disentangle chemical evolution of galaxies in the nearby Universe and to determine independent distances. Star formation histories have been inferred from population/spectrum synthesis of resolved/unresolved populations of nearby star forming galaxies; nearby starbursts – templates for high-redshift counterparts – have been studied with *FUSE*, *HST*, *GALEX* and *Spitzer*. Large surveys for star forming galaxies from redshifts 1 to 6, making use of colour selection techniques at optical, infrared and sub-mm wavelengths, have provided quantitative measures of their massive stellar populations over most of the age of the Universe, including their past history of star formation, the IMF, assembled stellar masses, metallicities and chemical yields. From space, *HETE* and *SWIFT* have allowed an increasing number of GRBs to be studied in detail, with rapid follow-up from ground-based facilities permitting chemical information on their host galaxies to be obtained. These are all tremendously exciting topics, at the forefront of present-day astrophysical research and providing some of the core scientific cases for the next generation of extremely large telescopes currently under development.

Theoretically, great advances have been made toward improved evolutionary and atmospheric models for massive stars allowing for rotation and magnetic fields, and toward the evolution of massive binary systems. The impact of internal waves generated at the boundary of the convective core on the transport of angular momentum and chemical species in the stellar interior. Important developments have taken place with respect to spectral synthesis of starbursts, improved spectral energy distributions of young stellar populations, hydrodynamic simulations of GRB explosions, and notably numerical simulations of star formation at the earliest epochs, including very massive Population III stars which are thought to play the dominant role in the re-ionization of the universe at redshift $z > 6$.

The key astrophysical problems for the symposium were:
- Atmospheres of massive stars;
- Physics and evolution of massive stars;
- Massive stellar populations in the nearby Universe;
- Hydrodynamics and feedback from massive stars in galaxy evolution;
- Massive stars as probes of the early Universe.

Further information at the meeting web site <www.ifa.hawaii.edu/iau250>.

3. Future outlook

Our Working Group is a sponsor of an approved 1.5-day Joint Discussion during the IAU XXVII General Assembly in Rio de Janeiro (Brazil), 2009, entitled *Eta Carinae and Interacting Massive Binaries*, for which Ted Gull and Augusto Daminelli are SOC co-chairs.

Other meetings on massive stars include:
- *Massive Stars: A Lifetime of Influence*, 13-15 October 2008, Lowell Observatory, Flagstaff (AZ, USA), in honor of the retirement of Peter Conti.
See: <www.lowell.edu/workshops/Contifest>.
- *Hot And Cool: Bridging Gaps in Massive Star Evolution*, 10-12 November 2008, CalTech, Pasadena (CA, USA).
See: <www.ipac.caltech.edu/hotandcool/CoolAndHot.html>.
- *High-Energy Phenomena in Massive Stars*, 2-5 February 2009, Jaen, Spain.
See: <www.ujaen.es/congreso/massive.stars2009/>.
- *The Interferometric View on Hot Stars*, 2-6 March 2009, Vina del Mar, Chile.
See: <www.eso.org/sci/meetings/IHOT09/>.

In the coming years exciting new results on massive stars will be obtained by interferometry and asteroseismology. These new techniques will probe both the physics of the surfaces and interiors of massive stars, allowing unprecedented constraints of key physical parameters like rotation, magnetic fields, and chemical composition.

Acknowledgements

The OC gratefully acknowledges logistic support for several teleconferences by STScI. The Institute of Astronomy of UNAM provided a new disk drive in support of the web server. UNAM and STSCI provided technical support through Liliana Hernandez and Julia Chen.

Stanley P. Owocki
chair of the Working Group

Transactions IAU, Volume XXVIIA
Reports on Astronomy 2006–2009
Karel A. van der Hucht, ed.

© 2009 International Astronomical Union
doi:10.1017/S1743921308025581

DIVISION IV / WORKING GROUP
ABUNDANCES IN RED-GIANTS

CHAIR   John C. Lattanzio
MEMBERS   Pavel A. Denissenkov, Roberto Gallino,
                    Josef Hron, Uffe Gråe Jørgensen,
                    Sun Kwok, Jacobus Th. van Loon,
                    Verne V. Smith, Christopher Tout,
                    Robert F. Wing, Ernst K. Zinner

TRIENNIAL REPORT 2006 - 2009

1. Meetings

The main activity of the WG on *Abundances in Red Giants* has been to propose a JD for the IAU GA in 2009. The increasing evidence for distinct populations within globular clusters is leading to the view that there is a continuum between globular clusters and the smallest of the galaxies. Our JD was designed to investigate this link. However, our JD was incorporated into IAU Symposium No. 266 *Star Clusters: Basic Building Blocks throughout Time and Space* for the IAU XXVII in Rio de Janeiro, 2009. We will be responsible for organising one session in the Symposium to cover the agenda put forward in our JD proposal.

On a similar topic, many researchers associated with the WG will be attending a Max-Planck Workshop on the similar topic *Chemical Evolution of Dwarf Galaxies and Stellar Clusters* in July 2008.

Members of the WG were closely involved in the organisation of a Royal Astronomical Society specialist discussion meeting on *Super-AGB stars*, in London, February 2008. The conclusion of this meeting was that the single biggest problem for these stars is mass-loss, because it is both crucial and unknown.

2. Newsletter

The AGB Newsletter, run by Jacco van Loon, a member of the WG, and Albert Zijlstra is to be more formally linked to the WG in future. It will form the main organ of communication with researchers in the field, and is distributed regularly.

3. Highlights

Recent scientific highlights of prime interest to our WG include the continuing investigation into mechanisms for driving the 'deep-mixing' known to occur in red-giants and almost certainly AGB stars as well. Such work includes investigations of thermohaline mixing and magnetic effects (Eggleton *et al.* 2006, 2008; Charbonnel & Zahn 2007; Busso *et al.* 2007; Stancliffe *et al.* 2007). Implications of these mechanisms are still being investigated but are very promising.

Years ago, Marigo (2002) emphasised the consequences of a carbon-rich envelope and the opacity effects on subsequent thermal pulsing evolution. Recent work by Cristallo *et al.* (2007) has begun to address this.

Super-AGB stars have received a lot of attention recently. These are stars that ignite C under degenerate conditions but still go on to experience thermal pulses, and have masses between about 8 and $12\,M_\odot$. Important papers are those by Siess (2006, 2007) and Poelarends *et al.* (2008), which investigate the evolution of these stars and emphasise the difficulties of such calculations. We are still awaiting a detailed calculation of yields from these stars, and these may be important for the chemical evolution of globular clusters.

McSaveney *et al.* (2007) attempted a direct quantitative test of Hot Bottom Burning by comparing models with LMC AGB stars. The basic quantitative predictions of the models were verified, although some uncertainties remain, likely due to uncertainties in the details of the properties of the convective envelope.

Recent work on very low metallicty stars has been prompted by increasingly sophisticated and complete observational studies, and includes theses by H. Lau (Cambridge) and S. Campbell (Monash) as well as Lau *et al.* (2007, 2008). Indeed, three very active cross-disciplinary areas include these and other works on interpretations of very metal-poor stars, research on the interpretation of the wealth of information present in presolar grains, and the interpretation of new measurement of neutron capture cross sections of astrophysical interest.

The field of red-giant evolution and nucleosynthesis is very active and we imagine the WG will have plenty to do in the next triennium.

John C. Lattanzio
chair of the Working Group

References

Busso, M., Wasserburg, G. J., Nollett, K. M., *et al.* 2007, *ApJ*, 671, 802

Charbonnel, C. & Zahn, J.-P. 2007, *A&A*, 476, 29

Cristallo, S., Straniero, O., Lederer, M. T., & Aringer, B. 2007, *ApJ*, 667, 489

Eggleton, P. P., Dearborn, D. S. P., & Lattanzio, J. C. 2006, *Science*, 314, 1580

Eggleton, P. P., Dearborn, D. S. P., & Lattanzio, J. C. 2007, *ApJ*, 677, 581

Lau, H. H. B., Stancliffe, R. J., & Tout, C. A. 2007, *MNRAS*, 378, 563

Lau, H. H. B., Stancliffe, R. J., & Tout, C. A. 2008, *MNRAS*, 385, 301

Marigo, P. 2002, *A&A*, 387, 507

McSaveney, J. A., Wood, P. R., Scholz, M., *et al.* 2007, *MNRAS*, 378, 1089

Poelarends, A. J. T., Herwig, F., Langer, N., & Heger, A. 2008, *ApJ*, 675, 614

Siess, L. 2006, *A&A*, 448, 717

Siess, L. 2007, *A&A*, 476, 893

Stancliffe, R. J., Glebbeek, E., Izzard, R. G., & Pols, O. R. 2007, *A&A* (Letters), 464, 57

Transactions IAU, Volume XXVIIA
Reports on Astronomy 2006–2009
Karel A. van der Hucht, ed.

© 2009 International Astronomical Union
doi:10.1017/S1743921308025593

INTER-DIVISION IV-V / WORKING GROUP
ACTIVE B-TYPE STARS

CO-CHAIRS   Juan Fabregat, Geraldine J. Peters
PAST CHAIR   Stanley P. Owocki
MEMBERS   Karen S. Bjorkman, Douglas R. Gies,
                     Hubertus F. Henrichs, David A. McDavid,
                     Coralie Neiner, Philippe Stee

TRIENNIAL REPORT 2006 - 2009

1. Introduction

The Working Group on *Active B-type Stars* (formerly known as the Working Group on *Be Stars*) was re-established under IAU Commission 29 at the IAU General Assembly in Montreal, Quebec (Canada) in 1979, and has been continuously active to the present. Its main goal is to promote and stimulate research and international collaboration on the field of the active early-type (OB) stars.

The interest of the WG were originally focused on classical Be stars research. The recent years have seen a broadening of the group interests, with an increasing contact and overlap with other research areas, particularly for closely aligned topics like pulsating OB stars.

2. Developments within the past triennium

The *Be Star Newsletter* (see <www.astro.virginia.edu/ dam3ma/benews/>) has continued to be the main source of information on new discoveries, ideas, manuscripts, and meetings on active B stars. G. Peters, D. Gies, and D. McDavid continue, respectively, as Editor-in-Chief, Technical Editor, and Web master. The issue No. 38 was published in March 2007, and contains, among other items, 18 notes and advertisements of interest for the active B stars community and 36 abstracts of papers relevant to its research subjects. The current working issue No. 39 contains 6 research notes and 18 abstracts as by June 30th.

The proceedings of the last major meeting on active B stars, *Active OB-Stars: Laboratories for Stellar and Circumstellar Physics*, held in Sapporo (Japan), were published in March 2007 (Okazaki, Owocki & Stefl 2007).

In general grounds, the field of the Be star research has been very active during the last few years, and we expect this activity to increase. The recent development of powerful observational equipment has led to significant progress in our knowledge of active B stars physics. The advent of the VLTI interferometer and the CHARA Array have allowed the spatial resolution of several Be star photospheres and disks (e.g., Domiciano de Souza *et al.* 2003; Gies *et al.* 2007), and the finding of the Keplerian rotation in the disk around α Arae (Meilland *et al.* 2007) . The application of spectropolarimetric techniques resulted in the discovery of magnetic fields in a few Be stars (Neiner *et al.* 2003), and the commissioning of new and more powerful spectropolarimeters will allow the systematic study of the prevalence of magnetic fields among Be stars and their relation with the mass ejection episodes.

Perhaps the most exciting current development is the accurate photometric monitoring of Be stars from outside our atmosphere. The Canadian space mission *MOST*, launched on 2003 June 30, has already provided photometric time series of unprecedented quality of several bright Be stars, revealing a rich spectrum of frequencies associated with radial and nonradial pulsations (Walker *et al.* 2005a, 2005b; Saio *et al.* 2007; Cameron *et al.* 2008). The French led *CoRoT* satellite was successfully launched on December 2006, and started its scientific observations on 3 February 2007. *CoRoT* is providing photometry of many B and Be stars in continuous runs lasting from 20 to 150 days. The analysis of the photometric time series and the interpretation of the results with the techniques of asteroseismology will provide important insights on the internal structure of Be stars, the role of nonradial pulsations, and its relation with the Be star outbursts (Michel *et al.* 2006).

Although opportunities to study Be stars in the FUV are currently limited, new spectra were recently obtained of Be stars in our Galaxy and the Magellanic Clouds with the *FUSE* spacecraft. This new data will provide information on the abundances of carbon, nitrogen, and the iron group elements in the photospheres of Be stars that will complement and serve as a check on the results and predictions from the planned asteroseismology investigations as well as theories for Be stars based upon the concept of critical rotation.

The rapid rotation of Be stars may be caused in some cases by past mass and angular momentum accretion in an interacting binary. The mass donor would currently appear as a hot subdwarf, stripped of its outer envelope and N-rich as the CNO-processed core is now exposed. Recently the presence of a subdwarf companion to the Be star FY CMa was detected (Peters *et al.* 2008) and thus this bright Be star becomes the second confirmed example, after ϕ Per, of a Be + sdO system.

Other important tools that will be available for Be star researchers in the near future are the ongoing and planned deep photometric galactic plane surveys. In the northern hemisphere the Hα IPHAS survey (Drew *et al.* 2005) and the near-infrared UKIDSS-GPS survey (Lawrence *et al.* 2007) are near completion, and are already releasing their data products or will start this release soon. In the southern hemisphere, the Hα VPHAS+ and near-infrared VVV ESO public surveys will start observations as soon as their associated instrumentation are available. The scientific exploitation of the survey data will produce an increase of several orders of magnitude of the number of Be stars known in our Galaxy.

All these developments configure a scenario of an exciting and rapidly evolving field of research. In this context, it seems timely for a new major international meeting of researchers on active B stars in the 2009-2011 time frame. Although it is still in the initial discussion stage, actions in this direction have already been undertaken

Juan Fabregat & Geraldine J. Peters
co-chairs of the Working Group

References

Cameron, C., Saio, H., Kuschnig, R. *et al.* 2008, *ApJ*, 685, 489
Domiciano de Souza, A., Kervella, P., Jankov, S. *et al.* 2003, *A&A* (Letters), 407, L47
Drew, J. E., Greimel, R., Irwin, M. J., *et al.* 2005, *MNRAS*, 362, 753

Gies, D. R., Bagnuolo, W. G., Baines, E. K., *et al.* 2007, *ApJ*, 654, 527

Lawrence, A., Warren, S. J., Almaini, O., *et al.* 2007, *MNRAS*, 379, 1599

Meilland, A., Millour, F., Stee, P., *et al.* 2007, *A&A*, 464, 59

Michel, E., Baglin, A., Auvergne, M., *et al.* 2006, in: M. Fridlund, A. Baglin, J. Lochard & L. Conroy (eds.), *The CoRoT Mission, ESA-SP*, 1306, 39

Neiner, C., Hubert, A. M., Frémat, Y., *et al.* 2003, *A&A*, 409, 275

Okazaki, A. T., Owoki, S. P., & Stefl, S. (eds.) 2007, *Active OB-Stars: Laboratories for Stellar and Circumstellar Physics, ASP-CS*, 361

Peters, G. J., Gies, D. R., Grundstrom, E. D., & McSwain, M. V. 2008, *ApJ*, 686, 1280

Saio, H., Cameron, C., Kuschnig, R., *et al.* 2007, *ApJ*, 654, 544

Walker, G. A. H., Kuschnig, R., Matthews, J. M., *et al.* 2005a, *ApJ* (Letters), 623, L145

Walker, G. A. H., Kuschnig, R., Matthews, J. M. *et al.* 2005b, *ApJ* (Letters), 635, L77

Transactions IAU, Volume XXVIIA
Reports on Astronomy 2006–2009
Karel A. van der Hucht, ed.

© 2009 International Astronomical Union
doi:10.1017/S174392130802560X

INTER-DIVISION IV-V / WORKING GROUP
AP AND RELATED STARS

CHAIR	Margarida S. Cunha
PAST CHAIR	Werner W. Weiss
MEMBERS	Michael M. Dworetsky, Oleg Kochukhov,
	Friedrich Kupka, Francis Leblanc,
	Richard Monier, Ernst Paunzen,
	Nikolai E. Piskunov, Hiromoto Shibahashi,
	Barry Smalley, Jozef Ziznovsky

TRIENNIAL REPORT 2006 - 2009

1. Introduction

The diversity of physical phenomena embraced by the study of Chemically Peculiar (CP) stars results in an associated research community with interests that are equally diverse. This fact became once more evident during the *CP#Ap Workshop* that took place in Vienna (Austria) in September 2007, and which gathered over 80 members of this research community. Besides the excellent scientific outcome of the meeting, during the workshop the community had the opportunity to discuss its organization and plans for the future. Following on those plans, the Working Group has submitted a proposal for a Joint Discussion during the IAU XXVII General Assembly, in Rio de Janeiro, which has meanwhile been accepted. Moreover, through an ApN newsletter forum, the Working Group has compiled requests from the community concerning atomic and related data. These requests have been put together and will be shared with Commission 14.

Besides the dedicated workshop mentioned above, a number of conferences and workshops with sessions directly related to the physics of subgroups of CP stars either took place, or will take place between September 2006 and July 2009. These include the conferences: *Magnetic Stars 2006* (SAO, Russia, September 2006); *The Future of Asteroseismology* (Vienna, Austria, September 2006); *Non-LTE Line Formation for Trace Elements in stellar atmospheres* (Nice, France, Aug 2007), *9th International Colloquium on Atomic Spectra and Oscillator Strengths for Astrophysical and Laboratory Plasmas* (Lund, Sweden, Augustus 2007); *Interpretation of Asteroseismic Data* (Wroclaw, Poland, June 2008); the symposium *Asteroseismology and Stellar Evolution* and the symposium *Star Clusters - Witnesses of Cosmic History* (both of JENAM 2008, Vienna, Austria), and IAU Symposium No. 259 *Cosmic Magnetic Fields: from Planets, to Stars and Galaxies* (Tenerife, Spain, November 2008). Members of the CP star community have been (or are) involved in all these meetings, either as organizers or as speakers.

This involvement of members of the CP star community in a vast number of scientific events highlights well the significant development of CP star research over the past few years, which was only possible due to the increasing number of excellent instruments available to the community. In particular, ground-based spectroscopic and spectropolarimetric instruments such as UVES and HARPS (ESO) and ESPaDOnS (CFHT) and space-based instruments such as that carried by the Canadian-led satellite *MOST* and the French-led satellite *CoRoT* have started, or continued, to provide data of excellent quality, feeding, simultaneously, the development of sophisticated stellar modelling. Most scientific results have been published in international journals and are available to the community in archives such as ADS or astro-ph. For those readers interested in searching the literature for articles specifically on CP stars research, we suggest a visit to the site of the ApN newsletter (`<ams.astro.univie.ac.at/apn>`). On this site, the reader will be able to find all articles that have appeared in the ADS in the recent past, which are in some way related to CP stars research. An important summary of recent and ongoing research work on CP stars is published in the proceedings of the Vienna *CP#Ap* Workshop (CoSka, vol. 38). In the following section we will summarize some recent highlights that directly concern the study of CP stars.

2. Highlights of CP stars research

In a letter *Weather in stellar atmosphere revealed by the dynamics of mercury clouds in alpha Andromedae*, Kochukhov *et al.* (2007) presented the discovery of secular evolution of mercury abundance spots in the brightest HgMn star α And. Seven-year monitoring of the surface spots in this non-magnetic chemically peculiar star revealed previously unknown non-magnetic structure formation process akin to weather in planetary atmospheres. These results have profound implications for our understanding of the radiative diffusion, mixing, and resulting heavy element enrichment of the atmospheres of non-magnetic CP stars and have consequences in the broader context of the studies of dynamical structure formation and self-organization in nature. Investigation of α And was performed using ultra-high signal-to-noise line profile observations obtained at Dominion Observatory (Canada) and Special Astrophysical Observatory (Russia). These data were analysed with the help of Doppler imaging technique.

Another very exciting result was the first interferometric determination of the angular diameter of a roAp star, specifically the brightest roAp star, α Cir Bruntt *et al.* (2008). From the latter, a new, independent, determination of the effective temperature of the star became possible. The value found $T_{\text{eff}} = 7420 \pm 170$ K is smaller that all values published in the literature, supporting the suspicion that systematic errors may affect the determinations of effective temperatures of CP stars.

Atmospheric research was also a topic of intense research. The Vienna team on Stellar Atmospheres and Pulsating Stars (SAPS) (`<ams.astro.univie.ac.at/>`) Kochukhov & Shulyak (2008), Khan & Shulyak (2007) has calculated and analysed a grid of model atmospheres with individual abundances pattern and has shown that among others, Si, Cr and Fe are the most important elements which produce most noticeable effects in the model atmosphere structures of CP stars. On the more theoretical side, diffusion in the presence of magnetic fields has been considered by Alecian & Stift (2007) under the assumption of equilibrium stratification in the atmosphere, and shown to be consistent with a number of abundance profiles and surface maps derived from Doppler imaging of Ap stars.

Concerning pulsations, the combination of space photometry from MOST (Canada) and ground-based high-resolution spectroscopy from UVES (ESO) has provided a unique opportunity to study the pulsational characteristics of several roAp stars Gruberbauer *et al.* (2008), Ryabchikova *et al.* (2007), Huber *et al.* (2008), Sachkov *et al.* (2008). The pulsation analysis in the stellar atmospheres was supplemented by the NLTE line formation study of the rare-earth elements Nd and Pr, which spectral lines usually show the highest pulsation radial velocity amplitudes. All together, these studies provide unique new material for the further development of theoretical models that may explain the details of pulsations in this most interesting class of pulsators.

On the theoretical side, the long debated issue of mode reflection in the atmospheres of roAp stars was addressed by Sousa & Cunha (2008). The authors have shown that the magnetic field provides a natural mechanism for the reflection of a fraction of the mode energy. Although part of the mode energy is indeed lost every pulsation cycle through running acoustic waves in the atmosphere and through running magnetic waves in the interior, the energy kept in the mode might be sufficient for the perturbations to be over stable, so far as enough energy is input in each cycle through the opacity mechanism.

Another important step that is being undertaken in this field of research is the observation of statistically meaningful samples of stars, to characterize their observational properties and to learn about aspects of the underlying physics. Kudryavtsev *et al.* (2006) published the results of spectropolarimetric observation of 96 chemically peculiar stars acquired at a 6 m telescope (SAO RAS, Russia), among which magnetic fields have been detected in 72 stars. The authors demonstrate that selecting candidate magnetic stars by considering their photometric indices, in particular strong anomalies of the continuum flux depression at 5200 Å, considerably increases the detection rate. Adelman & Woodrow (2007) summarized the published variability studies of mCP star performed by Adelman and his collaborators at the 0.75 m Four College APT (FCAPT) at the Fairborn Observatory. Sixty-eight of 70 mCP stars studied were found to be variable. This means that all mCP stars can be considered to be variable and by inference that all class members are variable. Moreover, Adelman(2008) noted that about 12% of the mCP stars with published data from the FCAPT have variable seasonal light curves in the Strömgren system, indicating that their rotational axis must be precessing about the magnetic axis. Finally, an important step towards the understanding of the origin and evolution of stellar magnetic fields in being undertaken, based on a survey of magnetic fields in Ap/Bp stars and their progenitors e.g.]landstreet08,alecian08 with the spectropolarimeter ESPaDOnS (CFHT).

3. A look into the future

Most of the exciting results from high spectral- and time-resolution spectroscopy and spectropolarimetry of various subgroups of CP stars are very recent. Several new instruments, have just became or are expected to become available in the near future, opening new windows in studies of CP stars.

The NARVAL spectropolarimeter at TBL (Pic du Midi Observatory (`<www.ast.obs-mip.fr/projets/narval/v1/>`) was commissioned in November 2006. It is the world's first astronomical facility worldwide fully dedicated to stellar spectropolarimetry, and in particular to the study of stellar magnetic fields. It will provide large amounts of observing time to the community and will allow for long-term monitoring and for surveys of various types of CP stars.

The project *Magnetism in Massive Stars* (MiMes, PI G. Wade (`<www.physics.queensu.ca/~wade/mimes>`) was selected as a CFHT Large Program and was allocated large amount of observing time with the ESPaDOnS spectropolarimeter for the period 2008-2012. This observing project, which includes 42 co-investigators, aims at improving our knowledge of the basic statistical properties and the structure of massive star magnetic fields, promising a new insight into the understanding of stellar magnetic field origin and evolution.

The 0.5 m ASTRA Spectrophotometric Telescope is nearing completion at the Fairborn Observatory (USA) <astra.citadel.edu>] (Adelman *et al.* 2007) with scientific observations expected to begin before the 2009 IAU GA. These data, with a resolution of 14 Å in first order and Å in second order, along with Balmer line profiles for each star should greatly help in the determination of effective temperature and surface gravity of CP stars. Improvements in the understanding of the photometric variability of mCP stars and binarity of Am and HgMn stars are also expected.

The Uppsala Astronomical Observatory in cooperation with Utrecht University, Rice University, and Space Telescope Science Institute is developing a polarimetric module for the ultra-stable spectrometer HARPS at the 3.6 m ESO telescope in La Silla. The polarimeter design was approved by ESO and assembly of the instrument will soon commence. Commissioning at the telescope is expected in 2009. The polarimetric upgrade of HARPS will enable high-resolution Stokes parameter observations from southern hemisphere and thus will be of great interest for the studies of magnetic fields in CP stars.

Looking ahead, beyond 2009, we may expect that ambitious projects such as *Gaia* (ESA: <www.rssd.esa.int/Gaia>) and WSO-UV (Russia: <wso.inasan.ru/>) will have a major impact in the understanding of CP stars' observational properties. That, in turn, will provide us with further opportunities to progress in the understanding of different physical processes, such as rotation, magnetic fields, diffusion and pulsation, that contribute to the diversity of observational properties that are found among these fascinating objects.

The diversity of theoretical and observational challenges faced in the study of CP stars has potential to attract a vast community, including researchers whose central interests are spread over different astrophysical contexts. The Working Group on *Ap and Related Stars* provides a privileged way to link together this vast community, promoting the constant interaction among its members.

Margarida S. Cunha
chair of the Working Group

References

Adelman, S. J. 2008, *PASP*, 120, 595
Adelman, S. J. & Woodrow, S. L. 2007, *PASP*, 119, 1256
Adelman, S. J., Gulliver, A. F., Smalley, B., *et al.* 2007, C. Sterken (ed.), *The Future of Photometric, Spectrophotometric and Polarimetric Standardization*, ASP-CS, 364, 255
Alecian, E., Wade, G. A., Catala, C., *et al.* 2008, *CoSka*, 38, 235
Alecian, G. & Stift, M. J. 2007, *A&A*, 475, 659
Bruntt, H., North, J. R., Cunha, M., *et al.* 2008, *MNRAS*, 386, 2039
Gruberbauer, M., Saio, H., Huber, D., *et al.* 2008, *A&A*, 480, 223
Huber, D., Saio, H., Gruberbauer, M., *et al.* 2008, *A&A*, 483, 239
Khan, S. A. & Shulyak, D. V. 2007, *A&A*, 469, 1083
Kochukhov, O. & Shulyak, D. 2008, *CoSka*, 38, 419
Kochukhov, O., Adelman, S. J., Gulliver, A. F., & Piskunov, N. 2007, *NatPh*, 3, 526
Kudryavtsev, D. O., Romanyuk, I. I., Elkin, V. G., & Paunzen, E. 2006, *MNRAS*, 372, 1804
Landstreet, J. D., Silaj, J., Andretta, V., *et al.* 2008, *A&A*, 481, 465
Ryabchikova, T., Sachkov, M., Weiss, W. W., *et al.* 2007, *A&A*, 462, 1103
Sachkov, M., Kochukhov, O., Ryabchikova, T., *et al.* 2008, *MNRAS*, 389, 903
Sousa, S. G. & Cunha, M. S. 2008, *MNRAS*, 386, 531

Transactions IAU, Volume XXVIIA
Reports on Astronomy 2006–2009
Karel A. van der Hucht, ed.

© 2009 International Astronomical Union
doi:10.1017/S1743921308025611

INTER-DIVISION IV-V-IX / WORKING GROUP STANDARD STARS

CHAIR   **Christopher J. Corbally**
NEWSLETTER EDITOR   **Richard O. Gray**

TRIENNIAL REPORT 2006 - 2009

1. Introduction

The Working Group, created in 1982 during the IAU General Assembly in Patras (Greece), exists to examine, organize, and coordinate the various sets of standard stars for the fields in which they are used: radial velocities, spectral classification, photometry, astrometry, and others. It also provides a forum for discussion and education. All those working with and upon standard stars are considered members of the WG.

During 2007 the WG, with the support of Monique Spite, the Division IV president of its hosting Commission 45 on *Stellar Classification*, was changed to an Inter-Division Working Group. This was done since it was realized that some of the original Commissions from which the WG was formed now belonged to different Divisions. These are Commissions 25 on *Stellar Photometry and Polarimetry*, 29 on *Stellar Spectra*, and 30 on *(Radial Velocities*, to all of which standard star issues are essential. Hence the WG now spans Divisions IV, V, and IX.

2. Activities

As its main activity the WG publishes *The Standard Star Newsletter* in April and October of each year. This has contained research notes, abstracts of papers, websites of interest, and reports and announcements of meetings, all with relevance to standard stars and their use in astronomy. These newsletters can be downloaded from <stellar.phys.appstate.edu/ssn/>.

Other activities in this period have included: work on a spectral-type coding system, spanning the range of standard and peculiar stars, which will be incorporated into MAST's (Multi-Mission Archive at Space Telescope) search engine; advice to the U.S. Naval Observatory on incorporation of spectrophotometric standards into the Astronomical Almanac; and discussion of a 'Standard Field' in which different surveys could make direct comparisons of their calibrations for stellar data.

Papers from the meeting on *Future of Photometric, Spectrophotometric and Polarimetric Standardization* have been published in the ASP Conference Series 2007 under the editorship of Christiaan Sterken.

With sadness I report the passing of a founding member of the WG and the first editor of *The Standard Star Newsletter*, Laura E. Pasinetti, on 12 September 2006.

Christopher J. Corbally
chair of the Working Group

Transactions IAU, Volume XXVIIA
Reports on Astronomy 2006–2009
Karel A. van der Hucht, ed.

© 2009 International Astronomical Union
doi:10.1017/S1743921308025623

DIVISION V   VARIABLE STARS

ÉTOILES VARIABLES

Division V deals with all aspects of stellar variability, either intrinsic or due to eclipses by its companion in a binary system. In the case of intrinsic stellar variability the analysis of pulsating stars, surface inhomogeneities, stellar activity and oscillations are considered. For close binaries, classical detached eclipsing binaries are studied as well as more interacting systems, like contact and semi-detached binaries, or those with compact components, like cataclysmic variables and X-ray binaries, including the physics of accretion processes.

PRESIDENT   Alvaro Giménez
VICE-PRESIDENT   Steven D. Kawaler
SECRETARY   Conny Aerts
PAST PRESIDENT   Jørgen Christensen-Dalsgaard
BOARD   Michael Breger, Edward F. Guinan,
                     Donald W. Kurtz, Slavek M. Rucinski

DIVISION V COMMISSIONS

Commission 27   **Variable Stars**
Commission 42   **Close Binary Stars**

DIVISION V WORKING GROUP

Division V WG   **Spectroscopic Data Archiving**

INTER-DIVISION WORKING GROUPS

Division IV-V WG   **Active B-type Stars**
Division IV-V WG   **Ap and Related Stars**

TRIENNIAL REPORT 2006 - 2009

1. Introduction

Division V on *Variable Stars* consists of Commission 27, also called *Variable Stars*, and Commission 42 *Close Binaries*. Thus the former deals with stars whose variations are intrinsic, whereas in the latter the variations are caused by the interactions between the components in the binary. It is evident that the definition of the Division is predominantly observational, and there may be cases where the assignment of an object to one of the two commissions may be in doubt. For example, the observation of pulsating stars in eclipsing binaries within nearby galaxies, or the relation between some types of oscillation modes and membership to binary systems, have been widely discussed.

The report of the Division for the triennium is obviously documented in the reports of each of the two Commissions, as well as the associated Working Groups, and reference to them has to be made here. The preparation of the report of the only Working Group of Division V proper, on Spectroscopic Data Archiving, has not been produced in time due to ongoing discussions about its orientation and future, which will be the subject of decisions to be taken in the coming IAU XXVII General Assembly.

2. Variability

The progress of studies on variable stars has been reviewed in a series of international meetings in almost all the domains of interest for Division V. Particular attention has been paid to the launch and operation of the French-led space observatory *CoRoT*, especially designed for the study of asteroseismology as well as the search for extra-solar planets. Of course, in addition to the core programme targets, a good number of pulsating stars and eclipsing binaries of different types have been discovered and are now the subject of in-depth analysis. The literature of close binary stars, like that of other variables, suffers from a multiplicity of names for the objects of study. Discussions on nomenclature and different possible classification schemes have triggered a lot of interest, no doubt at least partly due to the wealth of data provided by surveys of different type, like the Sloan Digital Sky Survey but also MACHO, OGLE or ASAS. They are providing excellent new data for astronomers working in the field of variable stars, both intrinsic variables or binary stars, requiring adequate tags to help data mining.

Among the most interesting activities, the study of solar-like oscillations, now including sub-giant and giant, has found important new applications in the analysis of stars hosting planets. Main sequence pulsators are showing a growing number of 'hybrid' cases blurring the usually adopted classifications, while new interferometry results complement traditional sources of information. In particular, new data cast light on the possible mechanisms for energy deposition on magnetic waves in roAp stars. The study of the Cepheids *Period-Luminosity* relation has also focus the attention of researchers. Stellar activity continues to develop thanks to space missions like *MOST* and large X-ray observatories. Equally, during the triennium there was a continued growth in the study of pulsating subdwarf B-type stars.

3. Close binaries

Concerning close binaries, there has also been important new advances. Most short-period double-lined eclipsing binaries with M-type components continue to indicate that the sizes of M-type dwarfs are larger than predicted by theory, and their temperatures are lower. The discrepancies with models are no longer confined to low-mass systems. Both heavy spot coverage and reduced convective efficiency due to strong magnetic fields may play a role. For W UMa binaries we know now that almost certainly they have companions. The list of cataclysmic white dwarf pulsators has grown and are being probed with asteroseismological techniques to address the extent to which accretion affects the white dwarf parameters. A likely classical nova shell has been detected around the prototype dwarf nova. This would be one of the first direct links between classical and dwarf novae. The large ratio of short- to long-period CVs predicted by standard CV evolution theory with observed data can only be reconciled if the rate of angular momentum loss below the period gap is increased. A mechanism of angular momentum loss in addition to gravitational radiation is required.

Important analyses of Algol systems has continued. Evolutionary models suggest that a circumbinary disk could extract angular momentum from the binary, thereby causing the orbit to shrink. This disk may explain the low mass ratio systems undergoing rapid mass transfer. Synthetic spectra of cool Algol secondaries can now be calculated. In active binaries, Doppler maps show low latitude spots with a temperature contrast and some weak polar features. Weak solar-type differential rotation is derived. Observational evidence of interacting coronae of the two components of a young binary system has been reported giving the first evidence that even if the flare origin is magnetic reconnection due to inter-binary collision, both stars independently emit in the radio range with structures of their own.

Alvaro Giménez
President of the Division

Transactions IAU, Volume XXVIIA
Reports on Astronomy 2006–2009   © 2009 International Astronomical Union
Karel A. van der Hucht, ed.   doi:10.1017/S1743921308025635

COMMISSION 27   **VARIABLE STARS**

ÉTOILES VARIABLES

PRESIDENT   Steven D. Kawaler
VICE-PRESIDENT   Gerald Handler
PAST PRESIDENT   Conny Aerts
ORGANIZING COMMITTEE   Timothy R. Bedding, Márcio Catelán,
                                  Margarida Cunha, Laurent Eyer,
                                  C. Simon Jeffery, Peter Martinez,
                                  Katalin Olah, Karen Pollard,
                                  Seetha Somasundaram

TRIENNIAL REPORT 2006 - 2009

1. Introduction

The Organizing Committee of Commission 27 has decided to again provide a somewhat abbreviated bibliography as part of this triennial report, as astronomy-centered search engines and on-line publications continue to blossom. We focus on selected highlights in variable star research over the past three years. Further results can be found in numerous proceedings of conferences held in the time frame covered by this report.

Following the IAU XXVI General Assembly, 2006, a number of international meetings considered intrinsically variable stars as a significant element of their program. In September 2006, the *Vienna Workshop on the Future of Asteroseismology* (Handler & Houdek 2007) celebrated Michel Breger's 65th birthday. The November 2006 joint *CoRoT*/ESTA workshop on *Solar/Stellar Models and Seismic Analysis Tools*, has proceedings edited by Straka, Lebreton & Monteiro. The community gathered in July 2007 to honor Prof. Douglas Gough with a conference on *Unsolved Problems in Stellar Physics*, the proceedings of which were edited by Stancliffe *et al.* (2007). The long series of pulsating star meetings continued in July 2007 with *Stellar Pulsation & Cycles of Discovery*. Also in July 2007, the *3rd Meeting on Hot Subdwarf Stars and Related Objects* was held in Bamberg (Heber *et al.* 2008). In April 2008, IAU Symposium No. 252 on *The Art of Modeling Stars in the 21st Century* was held in Sanya, China (Deng *et al.* 2008).

2. Solar-like variables

General reviews of asteroseismology that concentrate on solar-like oscillations include Bazot *et al.* (2008) and Bedding & Kjeldsen (2007), which discuss how we are now exploring interesting stars using their solar-like oscillations to extend our understanding of solar phenomena to other realms. Carrier, Eggenberger and colleagues measured oscillations in the visual binary 70 Oph (Carrier & Eggenberger 2006) and used these for an asteroseismic analysis (Eggenberger *et al.* 2008). Vauclair *et al.* (2008) used the oscillation frequencies in the star ι Hor, already known to host an exoplanet, to determine that this star is an evaporated member of the primordial Hyades cluster.

Subgiant stars also show solar-like oscillations; examples include the metal-poor subgiant ν Indi (Bedding *et al.* 2006; Carrier *et al.* 2007), β Hydri (Bedding *et al.* 2007; North *et al.* 2007), and μ Her (Bonanno *et al.* 2008). Climbing to and up the giant branch, solar-like oscillations were examined in a number of targets. While radial velocity measurements were used to uncover oscillations from ϵ Oph (De Ridder *et al.* 2006). Notably, space-based photometry with *SMEI*, *MOST*, and *WIRE* has now revealed solar-like oscillations in K-type giants (Tarrant *et al.* 2007; Barban *et al.* 2007; Stello *et al.* 2008).

3. Main Sequence Stars

Pigulski & Pojmański (2008) significantly update earlier compilations of β Cep stars, confirm the basic properties of these variables, and more than doubles the number of known variables. The authors also identify interesting new members such as pulsators with eclipses. There are a growing number of 'hybrid' pulsators that show p and g modes among the main sequence pulsators, blurring the lines between these variable types. A new example of a hybrid β Cep/SPB star (Handler *et al.* 2006) was modeled by Dziembowski & Pamyatnykh (2008). In the δ Scuti/ γ Dor stars realm, two hybrid examples were found using *MOST* (King *et al.* 2006; Rowe *et al.* 2006).

New interferometry results complement and reinforce asteroseismic inference (Cunha *et al.* 2007). The first direct determination of the angular diameter of a roAp star (α Cir; Bruntt *et al.* 2008), gave an independent, and lower, determination of the star's effective temperature. This confirmed earlier suspicions that effective temperatures of cool Ap stars determined by traditional methods might be biased towards hotter values. With the new effective temperature of α Cir, the seismic data acquired with *WIRE* now agrees with that derived from theoretical models.

High-resolution spectroscopic data on roAp stars probes the dynamics associated with pulsations in their atmospheres. Such data provide evidence for the co-existence of running and standing waves (e.g., Ryabchikova *et al.* 2007), and reveal the effect of shock waves driven by pulsations (Shibahashi *et al.* 2008).

The physical process responsible for reflecting the high frequency oscillations at the surface of roAp stars has been debated for many years. Sousa & Cunha (2008) show that mode conversion in these stars can deposit a significant amount of the mode energy on magnetic waves in the outer magnetically dominated layers. The energy in the magnetic waves will be kept in the oscillation, providing a natural explanation for the (partial) reflection of the oscillations. This new understanding should allow the development of a new generation of theoretical models with an appropriate outer boundary condition, which is essential to study pulsations in these stars.

4. Large amplitude radial pulsators

A recent analysis of the Cepheid *P-L* relation by Fouqué *et al.* (2007) that includes traditional metrics as well as newer interferometric studies (e.g., Kervella *et al.* 2008) and parallaxes (Benedict *et al.* 2007; van Leeuwen *et al.* 2007) reveals a universality to the slope of the period-luminosity (PL) relation with negligible dependence on color. The parallax studies provide a zero point favoring a distance modulus for the LMC of $(m - M)_0 = 18.40 \pm 0.05$ mag. *Spitzer* data also show that the slope of the PL relation is constant through the mid-IR (Ngeow & Kanbur 2008; Freedman *et al.* 2008).

Among RV Tauri and other evolved stars, recent evidence (optical and infrared photometry [Gielen *et al.* 2007], interferometry [Deroo *et al.* 2007] and polarimetry) suggests the existence of long-lived and highly-processed dust disks. There is mounting evidence that those variables which show long-term (500 - 2500 d) photometric periods are in binary systems and possess unusual abundances in their atmospheres (Reyniers & Van Winckel 2007). The observed abundances can result from accretion of material from a circumbinary dust disk which has been depleted of refractory elements.

The Sloan Digital Sky Survey (SDSS) led to important developments through identification of RR Lyrae stars in several of the dwarf spheroidal satellites of the Milky Way. The RR Lyrae content in these galaxies can tell us about the small building blocks that we now think were assembled into big galaxies like the Milky Way. This assembly happened nearly 10 Gyr ago, which is precisely the age of the RR Lyrae stars. In a very real sense, any RR Lyrae stars in existence today were eyewitnesses to the formation history of the Milky Way. Several recent studies compared the properties of the RR Lyrae stars in the Milky Way and in the newly discovered dwarf spheroidal galaxies, to see if the former could have been assembled from protogalactic fragments resembling the early counterparts of the latter. Such papers include Siegel (2006) and Dall'Ora *et al.* (2006) for the Bootes dSph, Kuehn *et al.* (2008) for the CVn I dSph, and Greco *et al.* (2008) for the CVn II dSph. Similar such studies are being carried out in other dSph galaxies, e.g., Greco *et al.* 2007, for a globular cluster in the Fornax dSph.

These results reveal that the ancient RR Lyrae population of the dwarf spheroidal satellite system of the Milky Way differs in some fundamental respects from the bona-fide RR Lyrae population in the Galactic halo. This suggests that the Milky Way cannot have been formed from fragments resembling the dSph galaxies, even as they looked more than 10 Gyr ago. In particular, it turns out that the famous 'Oosterhoff dichotomy', which is a marked property of the Milky Way, is not present in the oldest populations of the dwarf satellites of the Milky Way. Had the Galaxy formed from accretion of smaller structures like these dwarf galaxies, the Oosterhoff dichotomy would simply not exist.

Another notable development is the fact that PL and period-color relations have recently become available for RR Lyrae stars in the near-infrared bandpasses of the Johnson-Cousins-Glass system and in the bandpasses of the Strömgren and SDSS systems. With these new calibrations, Catelan & Cortés (2008) use the trigonometric parallax of RR Lyrae itself to show that it is intrinsically brighter than previously thought, by 0.064 ± 0.013 mag. They provide a recalibrated relationship between the absolute magnitudes of RR Lyrae stars and the metallicity, and apply it to the LMC to obtain a revised distance modulus of $(m - M)_0 = 18.44 \pm 0.11$ mag.

Martin *et al.* (2007) used ultraviolet observations with *GALEX* to discover a spectacular comet-like tail extending 2 degrees behind Mira, the prototype long-period variable. The bow shock and tail appear to arise from the interaction between Mira's wind and the surrounding interstellar medium (Wareing *et al.* 2007).

Meanwhile, MACHO and OGLE observations of pulsating red giants continue to produce interesting new results on their PL relations that have motivated theoretical studies (Xiong & Deng 2007), but we still await a definitive solution to the origin of the mysterious long secondary periods (Derekas *et al.* 2006; Soszyński 2007).

5. Stellar activity

Satellite observations have benefited the study of stellar activity. The high quality and continuity of *MOST* data can reveal differential rotation coefficients directly through spot modeling. Equatorial velocities and spot patterns on the stellar surfaces are given for ϵ Eri by Croll *et al.* (2006) and for κ^1 Ceti by Walker *et al.* (2007). Using ground- and space-based instruments together offers new possibilities. For example, the study of the coronal structure of the young, single cool star AB Doradus by Hussain *et al.* (2007) combined ground-based circularly polarized spectra from the AAT, and X-ray light curves and spectra from *Chandra*. They showed that the X-ray corona must be concentrated close to the stellar surface, with a height $H \simeq 0.3\text{-}0.4\,R_*$, determined by the high coronal density and complex multipolar magnetic field from the surface maps. The theoretical results of Morin *et al.* (2008) are very important: they find a stable magnetic field for the fully convective star V374 Peg from phase-resolved spectropolarimetric observations. They find that the contrast and fractional coverage of spots are much lower than those of non-fully convective active stars with similar rotation, and that the large-scale magnetic topology is remarkably stable on the time-scale of one year.

6. Evolved and compact pulsating stars

This period saw continued growth in the study of pulsating subdwarf B-type (sdB) stars. Reviews include Charpinet *et al.* (2007) and Kilkenny (2007). Further developments were made in mode identification using multicolor photometry (Baran *et al.* 2008), and time-series spectroscopy (Tillich *et al.* 2007; Telting & Ostensen 2006; Vučković *et al.* 2007). Large multisite campaigns also explored the photometric nature of the pulsations (i.e., Vučković *et al.* 2006). The long period sdB stars (PG 1716 variables) were explored in a number of investigations. Hybrid sdB stars that show both long and short periods were found and continue to be investigated (Baran *et al.* 2008; Schuh *et al.* 2006).

Models of pulsating sdB stars continue to teach us about diffusion in these stars in terms of the excitation and driving mechanism (Jeffery *et al.* 2006a,b). Comparison between the observed pulsation frequencies and model frequencies also continues to be an active area of exploration (i.e., Charpinet *et al.* 2006). A certain highlight during this interval involved the star HS 2201 (V391 Pegasi), a pulsating sdB star. Silvotti *et al.* (2007) announced the discovery of a giant planet orbiting the star that used timing of the pulsations to determine the star's reflex orbital motion.

White dwarf pulsation studies have been energized by the SDSS, which has uncovered a huge number of white dwarfs, resulting in discovery of a large number of well characterized (statistically) pulsators. In addition to nearly doubling the number of known pulsators (i.e., Gianninas *et al.* 2006), we can now explore the details of the boundary of the white dwarf instability strips (i.e., Castanheira *et al.* 2007). A number of new members of the class of ZZ Ceti-like pulsating white dwarfs that live within cataclysmic variable systems were also found (i.e., Nilsson *et al.* 2006).

The past three years has seen an increase in the sophistication of models of the internal structure of PG 1159 (GW Vir) stars based on asteroseismic constraints (Althaus *et al.* 2008a,b; Corsico & Althaus 2006). These investigations are in response to newly published observations of GW Vir stars (Costa *et al.* 2008; Fu *et al.* 2007).

7. Closing remarks – the space age

The future of variable star research is very bright. A newly developing capability for space-based observations has the potential to transform the field in ways that are (more than) hinted at with the success of *MOST* and *CoRoT*. Future missions of interest include *BRITE* (i.e Zwintz & Kaiser 2008), *Kepler*, and *Gaia*.

We anxiously await the launch of the *Kepler* mission in February 2009. One of the science components is the Kepler Asteroseismology Investigation (Christensen-Dalsgaard 2007), which oversees asteroseismic analysis of Kepler targets. With the potential of well over 100,000 stars observed continuously for up to 3.5 yr, the data flood should provide a dramatic expansion in the use of asteroseismology for studying a wide variety of stars.

ESA's *Gaia* satellite will survey one billion stars down to magnitude 20 over 5 years and detect about 50-100 million variable stars. It will provide distances, proper motions, and provide multi-epoch radial velocities, photometric (G-band), blue (BP) and red (RP) photometric data for a large fraction of these stars. The consortium to reduce the data has been formed; one international group is taking care of the Variability Analysis aiming at a systematic automated description and classification of the variable objects. Launch of *Gaia* is planned for the end of 2011. For more information, see <www.rssd.esa.int/Gaia>.

Steven D. Kawaler
president of the Commission

References

Althaus, L. G., Corsico, A. H., Miller Bertolami, M. M., *et al.* 2008, *ApJ* (Letters), 677, L35

Althaus, L. G., Corsico, A. H., Kepler, S. O., & Miller Bertolami, M. M. 2008, *A&A*, 478, 175

Baran, A., Pigulski, A., & O'Toole, S. J. 2008, *MNRAS*, 385, 255

Barban, C., Matthews, J. M., de Ridder, J., *et al.* 2007, *A&A*,468, 1033

Bazot, M., Monteiro, M. J. P. F. G, & Straka, C. 2008, in: L. Gizon *et al.* (eds.), *Helioseismology, Asteroseismology and MHD Connections*, JPhCS, 118, 012008.

Bedding, T. R., Butler, R. P., Carrier, F., *et al.* 2006, *ApJ*, 647, 558

Bedding, T. R., Kjeldsen, H., Arentoft, T., *et al.* 2007, *ApJ*, 663, 1315

Bedding, T. R. & Kjeldsen, H. 2007, in: R. Stancliffe, J. Dewi, G. Houdek & Martin, R. (eds), *Unsolved Problems in Stellar Physics*, AIP-CP, 948, 117

Benedict, G. F., McArthur, B. E., Feast, M. W., *et al.* 2007, *AJ*, 133, 1810

Bonanno, A., Benatti, S., Claudi, R., *et al.* 2007, *ApJ*, 676, 1248

Bruntt, H., North, J. R., Cunha, M., *et al.* 2008, *MNRAS*, 386, 2039

Carrier, F. & Eggenberger, P. 2008, *A&A*, 450, 695

Carrier, F., Kjeldsen, H. E., Bedding, T. R., *et al.* 2007, *A&A*, 470, 1059

Castanheira, B., Kepler, S. O., Costa, A. F. M., *et al.* 2007, *A&A*, 462, 989

Catelan, M. & Cortés, C. 2008, *ApJ* (Letters), 676, L135

Charpinet, S., Silvotti, R., Bonanno, A., *et al.* 2006, *A&A*, 459, 565

Charpinet, S., Fontaine, G., Brassard, P., *et al.* 2007, *CoAst*, 150, 241

Christensen-Dalsgaard, J. 2007, *CoAst*, 150, 350

Corsico, A. H. & Althaus, L. G. 2006, *A&A*, 454, 863

Costa, J. E. S., Kepler, S. O., Winget, D. E., *et al.* 2008, *A&A*, 477, 627

Croll, B., Walker, G. A. H., Kuschnig, R., *et al.* 2006, *ApJ*, 648, 607

Cunha, M., Aerts, C., Christensen-Dalsgaard, J., *et al.* 2007, *A&AR*, 14, 217

Dall'Ora, M., Clementini, G., Kinemuchi, K., *et al.* 2006, *ApJ* (Letters), 653, L109

De Ridder, J., Barban, C., Carrier, F., *et al.* 2006, *A&A*, 448, 689

Deng, L. & Chan, K. L. (eds.) 2008, *The Art of Modelling Stars in the 21st Century*, Proc. IAU Symposium No. 252 (Cambridge: CUP)

Derekas, A., Kiss, L., Bedding, T. R., *et al.* 2006, *ApJ* (Letters), 650, L55

Deroo, P., Acke, B., Verhoelst, T., Dominik, *et al.* 2007, *A&A* (Letters), 474, L45

Dziembowski, W. & Pamyatnykh, A. 2008, *MNRAS*, 385, 2061

Eggenberger, P., Miglio, A., Carrier, F., Fernandes, J., & Santos, N. C. 2008, *A&A*, 482, 631

Freedman, W., Madore, B., Rigby, J., Persson, S., & Sturch, L. 2008, *ApJ*, 679, 71

Fouqué, P., Arriagada, P., Storm, J., *et al.* 2007, *A&A*, 476, 73

Fu, J.-N., Vauclair, G., Solheim, J.-E., *et al.* 2007, *A&A*, 467, 237

Gielen, C., Van Winckel, H., Waters, L. B. F. M., Min, M., & Dominik, C. 2007, *A&A*, 475, 629

Gianninas, A., Bergeron, P., & Fontaine, G. 2006, *AJ*, 132, 831

Greco, C., Clementini, G., Catelan, M., *et al.* 2007, *ApJ*, 670, 332

Greco, C., Dall'Ora, M., Clementini, G., *et al.* 2008, *ApJ* (Letters), 675, L73

Handler, G., Jerzykiewicz, M., Rodríguez, E., *et al.* 2006, *MNRAS*, 365, 327

Handler, G. & Houdek, G. (eds.) 2008, *Vienna Workshop on the Future of Asteroseismology*, *CoAst*, 150

Heber, U., Jeffery, C. S., & Napowotzki, R. (eds.) 2008, *3rd Meeting on Hot Subdwarf Stars and Related Objects*, ASP-CS, 392

Hussain, G. A. J., Jardine, M., Donati, J.-F., *et al.* 2007, *MNRAS*, 377, 1488

Jeffery, C. S. & Saio, H. 2006a, *MNRAS*, 371, 659

Jeffery, C. S. & Saio, H. 2006b, *MNRAS*, 372, 48

Kervella, P., Mérand, A., Szabados, L., *et al.* 2008, *A&A*, 480, 167

Kilkenny, D. 2007, *CoAst*, 150, 234

King, H., Matthews, J. M., Rowe, J. F., *et al.* 2006, *CoAst*, 148, 28

Kuehn, C., Kinemuchi, K., Ripepi, V., *et al.* 2008, *ApJ* (Letters), 674, L81

Martin, D. C., Seibert, M., Neill, J. D., *et al.* 2007, *Nature*, 448, 780

Morin, J., Donati, J.-F., Forveille, T., *et al.* 2008, *MNRAS*, 384, 77

Ngeow, C. & Kanbur, S. 2008, *ApJ*, 679, 76

Nilsson, R., Uthas, H., Ytre-Eide, M., Solheim, J.-E., & Warner, B. 2006, *MNRAS*, 370, L56

North, J. R., Davis, J., Bedding, T. R., *et al.* 2007, *MNRAS*, 380, L80

Pigulski, A. & Pojmański, G. 2008, *A&A*, 477, 917

Reyniers, M. & Van Winckel, H. 2007, *A&A* (Letters), 463, L1

Rowe, J. F., Matthews, J. M., Cameron, C., *et al.* 2006, *CoAst*, 148, 34

Ryabchikova, T., Sachkov, M., Kochukhov, O., & Lyashko, D. 2007, *A&A*, 473, 907

Schuh, S., Huber, J., Dreizler, S., *et al.* 2006, *A&A* (Letters), 445, L31

Shibahashi, H., Gough, D., Kurtz, D., & Kamb, E. 2008, *PASJ*, 60, 63

Siegel, M. H. 2006, *ApJ* (Letters), 649, L83

Silvotti, R., Schuh, S., Janulis, R., *et al.* 2007, *Nature*, 449, 189

Soszyński, I. 2007, *ApJ*, 660, 1486

Sousa, J. & Cunha, M. 2008, *MNRAS*, 386, 531

Stancliffe, R., Dewi, J., Houdek, G., Martin, R., & Tout, C. (eds.) 2007, *Unsolved Problems in Stellar Physics*, AIP-CP, 948

Stello, D., Bruntt, H., Preston, H., & Buzasi, D. 2008, *ApJ* (Letters), 674, L53

Straka, C., Lebreton, Y., & Monteiro, M. J. P. F. G. (eds.) 2007, *Solar/Stellar Models and Seismic Analysis Tools*, EAS-PS, 26

Tarrant, N., Chaplin, W., Elsworth, Y., *et al.* 2007, *MNRAS* (Letters), 382, L48

Telting, R. & Ostensen, R. 2006, *A&A*, 450, 1149

Tillich, A., Heber, U., O'Toole, S. J., Ostensen, R., & Schuh, S. 2007, *A&A*, 473, 219

Vauclair, S., Laymand, M., Bouchy, F., *et al.* 2008, *A&A* (Letters), 482, L5

van Leeuwen, F., Feast, M. W., Whitelock, P. A., & Laney, C. D. 2007, *MNRAS*, 379, 723

Vučković, M., Kawaler, S. D., O'Toole, S., *et al.* (the WET collaboration) 2006, *ApJ*, 646, 1230

Vučković, M., Aerts, C., Østensen, R., *et al.* 2007, *A&A*, 471, 605

Walker, G. A. H., Croll, B., Kuschnig, R., *et al.* 2007, *ApJ*, 659, 1611

Wareing, C. J., Zijlstra, A. A., O'Brien, T. J., & Seibert, M. 2007, *ApJ* (Letters), 670, L125

Xiong, D. R. & Deng, L. 2007, *MNRAS*, 378, 1270

Zwintz, K. & Kaiser, A. (eds.) 2008, Proc. *First BRITE Workshop*, *CoAst*, 152

Transactions IAU, Volume XXVIIA
Reports on Astronomy 2006–2009
Karel A. van der Hucht, ed.

© 2009 International Astronomical Union
doi:10.1017/S1743921308025647

COMMISSION 42

CLOSE BINARY STARS
ÉTOILES DOUBLES SERRÉES

PRESIDENT	Slavek M. Rucinski
VICE-PRESIDENT	Ignasi Ribas
PAST PRESIDENT	Alvaro Giménez
ORGANIZING COMMITTEE	Petr Harmanec, Ronald W. Hilditch, Janusz Kaluzny, Panayiotis Niarchos, Birgitta Nordström, Katalin Oláh, Mercedes T. Richards, Colin D. Scarfe, Edward M. Sion, Guillermo Torres, Sonja Vrielmann

TRIENNIAL REPORT 2006 - 2009

1. Introduction

Two meetings of interest to close binaries took place during the reporting period: A full day session on short-period binary stars – mostly CV's – (Milone *et al.* 2008) during the 2006 AAS Spring meeting in Calgary and the very broadly designed IAU Symposium No. 240 on *Binary Stars as Critical Tools and Tests in Contemporary Astrophysics* in Prague, 2006, with many papers on close binaries Hartkopf *et al.* 2007. In addition, the book by Eggleton (2006), which is a comprehensive summary of evolutionary processes in binary and multiple stars, was published.

The report that follows consists of individual contributions of the Commission 42 Organizing Committee members. Its goal has been to give very personal views of a few individuals who are active in the field, so the report does not aim at covering the whole field of close binaries.

2. Close binary research from the perspective of BCB (C. D. Scarfe)

Some light can be thrown on the state of research on close binaries by considering the papers listed in the semi-annual Bibliography of Close Binaries (BCB). Since no similar summary appeared in the last few triennial reports of the Commission, we review here the patterns seen in the last 14 issues, since the most recent change of editorship in 2001, but with emphasis on the most recent six issues.

The total output of work on close binaries, appears to have decreased slightly during this decade. The two largest recent issues were those of December 2001 and June 2003, and the two smallest were those of December 2005 and December 2006. Part of this decline can be attributed to the editors' efforts to list only once, in the 'Collections of Data' section, papers giving the same kind of results for numerous binary systems. This avoids repetition of the listing for each system in the 'Individual Stars' section.

There is a slow trend toward larger numbers of authors per paper. The mean number of authors per paper has grown from 3.80 for the 420 entries in #72 to 4.88 for the 380 entries in #85. These means are affected by the few papers with very large numbers of authors – over 100 in rare cases – but the medians show the same trend. There is also a consistent systematic difference between 'Individual Star' (mostly observational) papers and those in the 'General' (mainly theoretical) category, the latter having about two fewer authors than the former, on average.

The literature of close binary stars, like that of other variables, suffers from a multiplicity of names for the objects of study. We have tried to follow the normal variable-star nomenclature of letters or numbers followed by constellation names, or their numbers in widely used catalogues such as the HD catalogue. Failing that, we prefer a coordinate-based nomenclature, and do our best to avoid other names. We encourage authors of papers to use standard nomenclature, making use of the SIMBAD identifier lists.

We turn now to the individual objects that are of abiding interest, or have attracted great temporary interest, based on the number of papers about them in the most recent 14 issues of BCB. Consistent with the preceding paragraph, we refer to each object by the name preferred for BCB, usually the standard variable-star designation. Papers have been listed in every issue for only two objects: V1357 Cyg (77 papers) and V1487 Aql (62 papers). Three other objects have more than 30 papers listed: KV UMa with 37 in 12 issues, V1343 Aql with 34 in 11, and HZ Her, with 33 in 12. However, PSR J0737−3039 was vigorously studied between 2004 and 2006, resulting in 29 papers in only 6 issues, including a record number of 12 in a single issue (#80). The remaining objects with at least 20 papers are V381 Nor (25 in 11), V615 Cas (24 in 9), V821 Ara (24 in 12), WZ Sge (23 in 10), V1033 Sco (23 in 9), V1521 Cyg and BR Cir (both 22 in 12), V818 Sco (21 in 10) and V4580 Sgr (20 in 11). Clearly, X-ray binaries dominate this list, but objects as the recurrent nova RS Oph, CH Cyg, AE Aqr and η Car are close behind (averaging > 1 paper per issue). All but RS Oph are of continuing interest, whereas that object's recent outburst has led to more papers in the latest three issues than for any other object.

3. Low-mass binaries and model discrepancies (G. Torres)

The properties of short-period double-lined eclipsing binaries (EBs) with M-type components have long been known to present disagreements compared to predictions from stellar evolution models (e.g., Popper 1997), particularly in their radii and effective temperatures. Since mid-2006 there have been considerable efforts to find additional low-mass systems in order to test theory, such as the Monitor Project that focuses on young open clusters (Aigrain et al. 2007). A number of other discoveries of potentially useful systems have occurred serendipitously or as a result of searching existing photometric databases (e.g., Bayless & Orosz 2006; Young et al. 2006; López-Morales & Shaw 2007; Vaccaro et al. 2007; Becker et al. 2008). Most of these systems continue to indicate that the sizes of M dwarfs are larger than predicted by theory, and their temperatures are lower. A few single-lined F+M systems resulting from searches for transiting planets have also provided useful comparisons (e.g., Beatty et al. 2007). The discrepancies with models are not confined to low-mass systems, however, and have been shown to extend up to masses almost as large as the Sun (Torres et al. 2006).

Chromospheric activity that is common in late-type and tidally locked binaries has long been suspected to be the culprit, as pointed out by many of the above investigators, and observational evidence for this has been mounting recently (see López-Morales 2007; Reiners et al. 2007, Morales et al. 2008). Both heavy spot coverage and reduced convective efficiency due to strong magnetic fields may play a role. The problem has been reviewed, e.g., by Ribas *et al.* (2007). Although much remains to be done both observationally and on the theoretical side, recent modelling studies of the evolution of low-mass stars under the influence of chromospheric activity are very encouraging. They appear to be able to explain the observed effects quite well with reasonable magnetic field strengths and typical spot filling factors (Chabrier *et al.* 2007). Similar progress has been made in our understanding of the origin and strength of magnetic fields in fully convective stars that lack a tachocline (Browning 2008).

4. W UMa-type binaries (S. M. Rucinski)

Although numbers of publications remain high, the field of contact binaries does not enjoy any obvious progress. The literature is still abundant in single-object photometric solutions of dubious usefulness (is anybody going ever to combine them?), with unrealistically small parameter uncertainties; they contribute little and just inflate citation lists. While very few new theoretical studies have been done – with some notable exceptions, e.g., the model of AW UMa (Paczyński *et al.* 2007), a model of evolution including the mass reversal (Stępień 2006a) – most efforts concentrate on observations and their interpretation.

New, shallow but extensive surveys of variability such as ASAS (Paczyński *et al.* 2006) have resulted in large numbers of contact binaries to about $V \simeq 12$ mag. W UMa type binaries are not very common, with one per about 500 late-type dwarfs ($\simeq 1.0 \times 10^{-5}$ pc^{-3}; Rucinski 2006), but they are easy to discover, hence their strong representation in variable star catalogues. The catalogues are biased in terms of physical properties: an elementary correction for the sample volume shifts the peak of the period distribution from 0.35 d to 0.27 d (Rucinski 2007). This is not far from the sharp cut-off at 0.22 d, where CC Comae is accompanied by the new shortest-period record holder discovered by ASAS, GSC 01387–00475 (Rucinski & Pribulla 2008). Stępień (2006b) explains the cut-off by the strong dependence of the angular-momentum-loss efficiency on mass, $-dH/dt \propto M^3$, so that low-mass objects simply have had no time to evolve beyond it.

We know now that W UMa binaries almost certainly have companions (Rucinski *et al.* 2007; and citations therein), in full confirmation of the work of Tokovinin *et al.* (2006) of the increasing multiplicity at short binary periods.

Only AW UMa is worth mentioning as an individual object: This flagship contact binary, when analyzed spectroscopically in detail, appears not to be in contact at all, but is a complex and hard to interpret semi-detached binary (Pribulla & Rucinski 2008). This teaches us a lesson that light curve fits even with most elaborate synthesis codes may yield an entirely incorrect picture. There is also a major question if the contact model applies to W UMa binaries at all ... .

Spectroscopy of bright contact binaries is essential for progress. Faint, accidentally discovered systems continue to be observed photometrically because this is easy, but usefulness of this is questionable. Note that hundreds W UMa's still remain to be detected to $V = 10$ mag, but most of them have small amplitudes because of low inclinations and/or dilution by the light of companions.

5. Cataclysmic Variables (E. M. Sion)

Since the discovery of the first CV containing a pulsating white dwarf with ZZ Ceti-like non-radial g-mode oscillations, the list of such cataclysmic white dwarf pulsators has grown to 12 (Gaensicke *et al.* 2006; Mukadam *et al.* 2007). These objects are being probed with asteroseismological techniques to address the extent to which accretion affects the white dwarf mass, temperature, and composition and how efficiently angular momentum is transferred into the core.

Classical novae and dwarf novae are powered by two entirely different outburst mechanisms and have never been seen closely spaced in time. However, Sokoloski *et al.* (2006) discovered the first example (Z And) of an accretion disk instability depositing enough material that a 2 magnitude flare was triggered, followed by a thermonuclear explosion on an accreting white dwarf which brightened the system to $10^4 \, L_\odot$, accompanied by a mass ejection. The two closely spaced events occurred in a symbiotic variable, but this so-called 'combination nova' holds potential implications for CVs as well. A likely classical nova shell has been detected around the prototype dwarf nova Z Cam by Shara *et al.* (2007). This would be one of the first direct links between classical novae and dwarf novae.

Darnley *et al.* (2006) estimate that if recent X-ray surveys of the galactic plane are correct in predicting more than $> 2 \times 10^4$ CVs with X-ray luminosity $< 2 \times 10^{33}$ erg/s, then each must undergo a nova explosion once per millennium to keep up the expected nova rate of 34 per year. Shara & Hurley (2006) reported a new class of CVs, found exclusively in globular clusters, whose members never went through a common envelope binary phase. They predict that their distribution of orbital periods should not have the standard CV period gap at 2-3 hr. Pretorius & Knigge (2008) presented an independent new sample of CVs, selected by $H\alpha$ emission. They cannot reconcile the large ratio of short- to long-period CVs predicted by standard CV evolution theory with their sample unless the rate of angular momentum loss below the period gap is increased by a factor of at least 3. Schmidtobreick & Tappert (2006) reported that there are now 11 known CVs in the period gap, but a shortage of expected dwarf novae that have evolved past the turn-around in orbital period. This again suggests that a mechanism of angular momentum loss in addition to gravitational radiation is required. Recent studies of the temperature distribution of CV white dwarfs versus orbital period below the period gap also suggest additional angular momentum loss.

Knigge (2007) has introduced a valuable method for determining the distances to cataclysmic variables using 2MASS *JHK* photometry and a semi-empirical relationship between donor absolute magnitude in the *K*-band and orbital period. The long sought-after and controversial mass of the white dwarf in WZ Sge was determined by Steeghs *et al.* (2007) in 2004. The WD mass is $0.85 \pm 0.03 \, M_\odot$. Finally, Kromer *et al.* (2007) have succeeded in constructing model accretion disks with emission lines formed through irradiation of the inner disk by the hot white dwarf. This is a significant breakthrough since these accretion disk models enable emission line modelling *ab initio*.

6. Algol-type binaries (M. T. Richards)

The important fundamental light curve and spectroscopic analyses of several systems has continued (e.g., Angione & Sievers 2006; Soydugan *et al.* 2007) We also need additional studies of infrared light curves like those of Lázaro *et al.* (2006). Analyses that include a third body (e.g., Li 2006; Hoffman *et al.* 2006) suggest that many Algols have a companion in a long-period orbit and that the third body has a non-negligible effect on the binary through long-term variations in the binary orbital period. Both light curves and spectra have been combined by Van Hamme & Wilson (2007) to refine the derivation of third-body parameters and to challenge the existence of a third body in VV Ori. The six largest coronal X-ray flares ever detected by *Chandra* were studied by Nordon &

Behar (2007) to derive the flare properties. Flare size and position have been constrained by studying eclipsed X-ray flares on Algol and VW Cep (Sanz-Forcada *et al.* 2007).

The angular momentum evolution of 74 detached binaries and 61 semidetached binaries was studied by Ibanoğlu *et al.* (2006). Evolutionary models suggest that a circumbinary disk could extract angular momentum from the binary, thereby causing the orbit to shrink. This disk may explain the low mass ratio systems undergoing rapid mass transfer (Chen *et al.* 2006). Conservative Roche Lobe overflow explains the observed range of orbital periods in systems with B-type primaries but not for those with high mass ratios (van Rensbergen *et al.* 2006). However, hot spots and the spin-up of the primary may lead to significant mass loss.

Synthetic spectra of cool Algol secondaries can now be calculated with the LinBrod code (Bitner & Robinson 2006). In addition, synthetic spectra of the accretion disk and gas stream were calculated for TT Hya using the SHELLSPEC code (Miller *et al.* 2007). The physical properties of the accretion disk and stream were derived from the direct comparison between the observed and synthetic spectra of the system and also by using Doppler tomography to demonstrate visually that the accretion structures have been properly modelled.

K-band direct imaging of SS Lep with VINCI/VLTI and photometry from the UV to far-IR has been used by Verhoelst *et al.* (2007) to reveal the stars and a dust shell or disk. Indirect images of W Cru based on the eclipse mapping method identified a clumpy disk structure (Pavlovski *et al.* 2006). Finally, new 3D Doppler tomograms of the U CrB binary demonstrate that the alternating accretion disk and stream structures have significant flow velocities beyond the central 2D velocity plane and there is evidence of a gas jet at the star-stream impact site Agafonov *et al.* (2006).

7. The oEA stars (P. G. Niarchos)

Following the introduction of the name 'oEA' (oscillating EA) for (B)A-F spectral type, mass-accreting, main-sequence, pulsating stars in semi-detached Algol-type EBs by Mkrtichian *et al.* (2004), several studies having been published in the recent years. The oEA stars are the former secondaries of evolved, semi-detached EBs which are (still) undergoing mass transfer and form a class of pulsators close to the main-sequence.

Soydugan *et al.* (2006a), considering a sample of 20 EBs with δ Sct primaries, discovered that there is a possible relation among the pulsational periods of the primaries and the orbital periods of the systems. An important contribution was made by Soydugan *et al.* (2006b), who presented a catalogue of close binaries (25 known and 197 candidate binaries with pulsating components) located in the δ Sct region of the instability strip. A status report on the search for pulsations in primary components of Algol-type systems was presented by Mkrtichian *et al.* (2006), while a revised list of presently known oEA stars and a discussion on the pulsation mode visibility and strategies of mode identification was published by Mkrtichian *et al.* (2007).

Studies of individual systems: The following oEA systems were studied: AB Per, QU Sge, RZ Cas, IV Cas, Y Cam, and CT Her. Moreover, Pigulski & Michalska (2007) and Michalska & Pigulski (2007) detected pulsating components in 25 EBs in the ASAS-3 database.

8. Effects of binarity on stellar activity (K. Oláh)

Kővári *et al.* (2007) present a detailed spot modeling analysis and Doppler images for the primary stars of the RS CVn-type binary ζ And. The photometric light modulation originates in the distorted geometry of the primary, and additionally, comes from spots, which preferably appear on the stellar surface towards the companion star and opposite to it. Doppler maps also show low latitude spots with a temperature contrast of about 1000 K, and some weak polar features. Weak solar-type differential rotation was derived from the cross-correlation of the consecutive Doppler maps.

Massi *et al.* (2008) reported observational evidence of interacting coronae of the two components of the young binary system V773 Tau A. The VLBA and Effelsberg radio images show two distinctive structures, which are associated with each star. In one image the two features are extended up to $18 R_\star$ each and are nearly parallel revealing the presence of two interacting helmet streamers, observed for the first time in stars other than the Sun. During the stellar rotation, these helmet streamers come into collision producing periodical flares. This is the first evidence that even if the flare origin is magnetic reconnection due to inter-binary collision, both stars independently emit in the radio range with structures of their own. The helmet streamers appear to interact throughout the whole orbit, although the radio flares become stronger when the stars approach. Around periastron the stellar separation is only $30 R_\star$, where the two streamers overlap producing the observed giant flares.

Dunstone *et al.* (2008) present the first measurements of surface differential rotation of the young, pre-main sequence binary system HD 155555. Both components are found to have high rates of differential rotation, similar to those of the same spectral type Main Sequence single stars. The results for HD 155555 are therefore in contrast to those found in other, more evolved binary systems where negligible or weak differential rotation has been discovered. The rotation of both stars of HD 155555 is synchronous and the system is tidally locked. The authors found that more likely the convection zone depth is the cause of the low differential rotation rates of evolved giants, rather than the effects of tidal forces. The strong differential rotation provides extra stresses on the fields, and the reconnection of these long binary field loops significantly contribute to the X-ray luminosity of the system and to the elevated frequency of large flares.

Slavek M. Rucinski & Ignasi Ribas
president & vice-president of the Commission

References

Agafonov, M., Richards, M., & Sharova, O. 2006, *ApJ*, 652, 1547

Aigrain, S., Irwin, J., Hebb, L., *et al.* 2007, *The Messenger*, 130, 36

Angione, R. J. & Sievers, J. R. 2006, *AJ*, 131, 2209

Bayless, A. J. & Orosz, J. A. 2006, *ApJ*, 651, 1155

Beatty, T. G., Fernández, J. M., Latham, D. W., *et al.* 2007, *ApJ*, 663, 573

Becker, A. C., Agol, E., Silvestri, N. M., *et al.* 2008, *MNRAS*, 386, 416

Bitner, M. A. & Robinson, E. L. 2006, *AJ*, 131, 1712

Browning, M. K. 2008, *ApJ*, 676, 1262

Chabrier, G., Gallardo, J., & Baraffe, I. 2007, *A&A* (Letters), 472, L17

Chen, W.-C., Li, X.-D., & Qian, S.-B. 2006, *ApJ*, 649, 973

Darnley, M. J., Bode, M. F., Kerins, E., *et al.* 2006, *MNRAS*, 369, 257

Dunstone, N. J., Hussain, G. A. J., Cameron, A. C., *et al.* 2008, *MNRAS*, 387, 481; 387, 1525

Eggleton, P. P., 2006, *Evolutionary Processes in Binary and Multiple Stars*, Cambridge Ap. Ser., 40 (Cambridge: CUP)

Gänsicke, B., Rodríguez-Gil, P., Marsh, T. R., *et al.* 2006, *MNRAS*, 365, 939

Hartkopf, W. I., Guinan, E. F., & Harmanec, P. 2007, *Binary Stars as Critical Tools and Tests in Contemporary Astrophysics*, Proc. IAU Symposium No. 240 (Cambridge: CUP)

Hoffman, D. I., Harrison, T. E., McNamara, B. J., *et al.* 2006, *AJ*, 132, 2260

İbanoğlu, C., Soydugan, F., Soydugan, E., & Dervişoğlu, A. 2006, *MNRAS*, 373, 435

Knigge, C. 2007, *MNRAS*, 382, 1982

Kővári, Zs., Bartus, J., Strassmeier, K. G., *et al.* 2007, *A&A*, 463, 1071

Kromer, M., Nagel, T., & Werner, K. 2007, *A&A*, 475, 301

Lázaro, C., Arévalo, M. J., & Antonopoulou, E. 2006, *MNRAS*, 368, 959

Li, L.-S. 2006, *AJ*, 131, 994

López-Morales, M. 2007, *ApJ*, 660, 732

López-Morales, M. & Shaw, J. S. 2007, in: Y. W. Kang, H.-W. Lee, K.-C. Leung & K.-S. Cheng (eds.), Proc. *Seventh Pacific Rim Conference on Stellar Astrophysics*, ASP-CS, 362, 26

Massi, M., Ros, E., Menten, K. M., *et al.* 2008, *A&A*, 480, 489

Michalska, G. & Pigulski, A. 2007, *CoAst*, 150, 71

Miller, B., Budaj, J., Richards, M., Koubský, P., & Peters, G. J. 2007, *ApJ*, 656, 1075

Milone, E. F., Leahy, D. A., & Hobill, D. W. (eds.) 2008, *Short Period Binary Stars: Observations, Analysis, and Results*, Ap & Space Sci. Lib., Vol. 352 (Berlin: Springer)

Mkrtichian, D., Kusakin, A. V., Rodriguez, E., *et al.* 2004, *A&A*, 419, 1015

Mkrtichian, D., Kim, S.-L., Kusakin, A. V., *et al.* 2006, *Ap&SS*, 304, 169

Mkrtichian, D., Kim, S.-L.; Rodríguez, E., *et al.* 2007, in: O. Demircan, S. O. Selam & B. Albayrak (eds.), *Solar and Stellar Physics Through Eclipses*, ASP-CS, 370, 194

Morales, J. C., Ribas, I., & Jordi, C. 2008, *A&A*, 478, 507

Mukadam, A. S., Szkody, P., Fraser, O. J., *et al.* 2007, in: R. Napiwotzki & M. R. Burleigh (eds.), Proc. *15th European Workshop on White Dwarfs*, ASP-CS, 372, 603

Nordon, R. & Behar, E. 2007, *A&A*, 464, 309

Paczyński, B., Sienkiewicz, R., & Szczygieł, D. M. 2007, *MNRAS*, 378, 961

Paczyński, B., Szczygieł, D., Pilecki, B., Pojmański, G., 2006, *MNRAS*, 368, 1311

Pavlovski, K., Burki, G., & Mimica, P. 2006, *A&A*, 454, 855

Pigulski, A. & Michalska, G. 2007, *AcA*, 57, 61

Popper, D. M. 1997, *AJ*, 114, 1195

Pretorius, M. & Knigge, C. 2008, *MNRAS*, 385, 1485

Pribulla, T. & Rucinski, S. M. 2008, *MNRAS*, 386, 377

Reiners, A., Seifahrt, A., Stassun, K. G., *et al.* 2007, *ApJ* (Letters), 671, L149

Ribas, I., Morales, J., Jordi, C., *et al.* 2007, *MmSAI*, 79, 562

Rucinski, S. M. 2006, *MNRAS*, 368, 1319

Rucinski, S. M. 2007, *MNRAS*, 382, 393

Rucinski, S. M. & Pribulla, T. 2008, *MNRAS*, 388, 1831

Rucinski, S. M., Pribulla, T., & van Kerkwijk, M. H. 2007, *AJ*, 134, 2353

Sanz-Forcada, J., Favata, F., & Micela, G. 2007, *A&A*, 466, 309

Schmidtobreick, L. & Tappert, C. 2006, *A&A*, 455, 255

Shara, M. M. & Hurley, J. 2006, *ApJ*, 646, 464

Shara, M. M., Martin, C. D.; Seibert, M., *et al.* 2007, *Nature*, 446, 15

Sokoloski, J., Kenyon, S. J., Espey, B. R., *et al.* 2006, *ApJ*, 636, 1002

Soydugan, E., İbanoğlu, C., Soydugan, F., *et al.* 2006a, *MNRAS*, 366, 1289

Soydugan, E., Soydugan, F., Demircan, O., & İbanoğlu, C. 2006b, *MNRAS*, 370, 2013

Soydugan, F., Frasca, A., Soydugan, E., *et al.* 2007, *MNRAS*, 379, 1533

Steeghs, D., Howell, S. B., Knigge, C., *et al.* 2007, *ApJ*, 667, 442

Stępień, K. 2006a, *AcA*, 56, 199

Stępień, K. 2006b, *AcA*, 56, 347

Tokovinin, A., Thomas, S., Sterzik, M., & Udry, S. 2006, *A&A*, 450, 681

Torres, G., Lacy, C. H., Marschall, L. A., Sheets, H. A., & Mader, J. A. 2006, *ApJ*, 640, 1018

Vaccaro, T. R., Rudkin, M., Kawka, A., *et al.* 2007, *ApJ*, 661, 1112

Van Hamme, W., & Wilson, R. E. 2007, *ApJ*, 661, 1129

van Rensbergen, W., de Loore, C., & Jansen, K. 2006, *A&A*, 446, 1071

Verhoelst, T., van Aarle, E., & Acke, B. 2007, *A&A* (Letters), 470, L21

Young, T. B., Hidas, M. G., Webb, J. K., *et al.* 2006, *MNRAS*, 370, 1529

Transactions IAU, Volume XXVIIA
Reports on Astronomy 2006–2009
Karel A. van der Hucht, ed.

© 2009 International Astronomical Union
doi:10.1017/S1743921308025659

DIVISION VI   INTERSTELLAR MATTER

MATIÈRE INTERSTELLAIRE

Division VI provides a focus for astronomers studying a wide range of problems related to the physical and chemical properties of interstellar matter in the Milky Way and other galaxies.

PRESIDENT   Thomas J. Millar
VICE-PRESIDENT   You Hua Chu
PAST PRESIDENT   John E. Dyson
BOARD   Dieter Breitschwerdt, Michael G. Burton,
                   Sylvie Cabrit, Paola Caselli,
                   Elisabete M. de Gouveia Dal Pino,
                   Gary J. Ferland, Mika J. Juvela,
                   Bon-Chul Koo, Sun Kwok,
                   Susana Lizano, Michal Rozyczka,
                   L. Viktor Tóth, Masato Tsuboi, Ji Yang

DIVISION VI COMMISSIONS

Commission 34   Interstellar Matter

DIVISION VI WORKING GROUPS

Division VI WG   Astrochemistry
Division VI WG   Star Formation
Division VI WG   Planetary Nebulae

TRIENNIAL REPORT 2006 - 2009

1. Introduction

Division VI, which is itself a single Commission (Commission 34), has around 900 members whose research interests cover the wide spectrum of activities, theoretical, experimental and observational, associated with the study of the interstellar medium in the Universe. As such, the Division has close links with Divisions VIII, IX and X.

Our report on activity since 2006 is divided into four sections, covering important meetings and conferences, relevant proceedings published in the period under review, a list of important review articles, and Working Group reports.

2. Meetings and conferences

The reporting period has been one of great activity with many conferences devoted in whole or in part to the scientific interests of Division members. Below, we list some of the most significant meetings:

- *Planetary Nebulae in Our Galaxy and Beyond*, IAU Symposium No. 234, 3-7 April 2006, HI, USA
- *Triggered Star Formation in a Turbulent ISM*, IAU Symposium No. 237, 14-18 August 2006, Prague, Czech Republic
- *Science with ALMA, A New Era for Astrophysics*, 13-17 November 2006, Madrid, Spain
- *Molecular Databases for Herschel, ALMA, and SOFIA*, 6-8 December 2006, Leiden, Netherlands
- *4th Korea-Mexico Joint Workshop on Astrophysics: Interstellar Medium*, 5-7 May 2007, Daejeon, Korea
- *Molecules in Space and Laboratory*, 14-18 May 2007, Paris, France
- *Asymmetrical Planetary Nebulae. IV*, 18-22 June 2007, La Palma, Spain
- *Star Formation Through Cosmic Time*, 13-17 August 2007, Santa Barbara, CA, USA
- *Massive Star Formation: Observations Confront Theory*, 10-14 September 2007, Heidelberg, Germany
- *From Protostellar Cores to Disk Galaxies*, September 2007, Zurich, Switzerland
- *FIR Workshop 2007, Far-Infrared and Submillimeter Emission of the Interstellar Medium: Models meet Extragalactic and Galactic Observations*, 5-7 November 2007, Bad Honnef, Germany
- *The Evolving Interstellar Medium in the Milky Way and Nearby Galaxies*, 2-5 December 2007, Pasadena, CA, USA
- *Magnetic Fields in the Universe. II. From Laboratory and Stars to the Primordial Structures*, 28 January - 1 February 2008, Cozumel, Mexico
- *Organic Matter in Space*, IAU Symposium No. 251, 18-22 February 2008, Hong Kong, China
- *The Local Bubble and Beyond. II*, 21-24 April 2008, Philadelphia, PA, USA
- *The Molecular Universe: an International Meeting on the Physics and Chemistry of the Interstellar Medium*, 5-8 May 2008, Arcachon, France
- *Protostellar Jets in Context*, 7-12 July 2008, Rhodes, Greece
- *10th Asian-Pacific Regional IAU Meeting*, 3-6 August 2008, Kunming, China
- *The Role of Disk-Halo Interaction in Galaxy Evolution: Outflow vs. Infall?*, 18-22 August 2008, Espinho, Portugal
- *Praha2008, the 20th International Conference on High Resolution Molecular Spectroscopy*, 2-6 September 2008, Prague, Czech Republic
- *Cosmic Dust Near and Far*, 8-12 September 2008, Heidelberg, Germany
- *JENAM 2008, Symposium No. 7 on Grand Challenges in Computational Astrophysics*, 8-11 September 2008, Vienna, Austria
- *Cosmic Dust and Radiative Transfer*, 15-17 September 2008, Heidelberg, Germany
- *Future Directions in Ultraviolet Spectroscopy*, 20-22 October 2008, Annapolis, MD, USA
- *New Light on Young Stars: Spitzer's View of Circumstellar Disks*, 26-30 October 2008, Pasadena, CA, USA
- *Cosmic Magnetic Fields: from Planets, to Stars and Galaxies*, 3-7 November 2008, Tenerife, Spain
- *When the Universe Formed Stars*, 17-21 November 2008, La Martinique Island, France

The Division is supporting a number of Symposia at the IAU XXVII General Assembly in Rio de Janeiro, August 2009, and is the coordinating Division for three Joint Discussions and two Special Sessions during the GA.

- *AKARI international conference*, 16-22 February 2009, Tokyo, Japan (Exact title and the date are to be determined.)
- *Gas and Dust Grains: From the ISM to the Solar System* (Preliminary), 17-19 June 2009, Seoul, Korea
- *Astrophysical Outflows and Associated Accretion Phenomena*, IAU XXVII GA Joint Discussion (JD7), 6-7 August 2009, Rio de Janeiro, Brazil (Coordinating Division)
- *FIR2009: the ISM of Galaxies in the Far-Infrared and Sub-Millimetre*, IAU XXVII GA Joint Discussion (JD14), 12-14 August 2009, Rio de Janeiro, Brazil (Coordinating Division)
- *Magnetic Fields in Diffuse Media*, IAU XXVII GA Joint Discussion (JD15), 12-14 August 2009 (Coordinating Division)
- *Young Stars, Brown Dwarfs, and Protoplanetary Disks*, IAU XXVII GA Special Session (SpS7), 11-14 August 2009, Rio de Janeiro, Brazil (Coordinating Division)
- *The Galactic Plane - In Depth and Across the Spectrum*, IAU XXVII GA Special Session (SpS8), 11-14 August 2009, Rio de Janeiro, Brazil (Coordinating Division)

3. Proceedings and monographs

3.1. *Conference proceedings*

- *Planetary Nebulae in our Galaxy and Beyond*, Proc. IAU Symposium No. 234, Eds. M. J. Barlow & R. H. Méndez (Cambridge: CUP), 2006.
- *Far-Infrared and Submillimeter Emission of the Interstellar Medium*, Eds. C. Kramer, S. Aalto & R Simon, EAS-PS, 2007
- *Triggered Star Formation in a Turbulent Interstellar Medium*, Proc. IAU Symposium No. 237, Eds. B. G. Elmegreen & J Palous (Cambridge: CUP), 2007
- *Astrophysical Masers and Their Environments*, Proc. IAU Symposium No. 242, Eds. J. M. Chapman & W. A. Baan (Cambridge: CUP), 2008
- *Massive Stars as Cosmic Engines*, Proc. IAU Symposium No. 250, Eds. F, Bresolin, P. A. Crowther & J. Puls (Cambridge: CUP), 2008
- *Magnetic Fields in the Universe. II. From Laboratory and Stars to the Primordial Structures*, Eds. A. Esquivel, *et al.*, *RMAA-CP*, 2008, in press
- *Asymmetrical Planetary Nebulae. IV*, Eds. R. L. M. Corradi, A. Manchado & N. Soker. See <www.iac.es/proyecto/apn4/pages/proceedings.php>.

3.2. *Research monographs*

- *Physics and Chemistry of the Interstellar Medium*, Sun Kwok, University Science Books, 2006
- *The Elemental Abundances in Bare Planetary Nebula Central Stars and the Shell Burning in AGB Stars*, Klaus Werner & Falk Herwig, *PASP*, 118, 183, 2006
- *Wonderful Mira*, Christopher J. Wareing, 2008, *Phil. Trans. R. Soc. A*, 366. Triennial Issue *Astronomy*: Ed. J. M. T. Thompson,

4. Review articles

Recent articles on interstellar matter published in the Annual Reviews of Astronomy and Astrophysics have included:

- Snow, T. P., & McCall, B. J. 2006, *ARA&A*, 44, 367, *Diffuse Atomic and Molecular Clouds*
- Harris, D. E., & Krawczynski, H. 2006, *ARA&A*, 44, 463, *X-Ray Emission from Extragalactic Jets*

- Bergin, E.A., & Tafalla, M. 2007, *ARA&A*, 45, 339, *Cold Dark Clouds: The Initial Conditions for Star Formation*
- Zinnecker, H., & Yorke, H. W. 2007, *ARA&A*, 45, 481, Toward Understanding Massive Star Formation
- McKee, C. F., & Ostriker, E. C. 2007, *ARA&A*, 45, 565, *Theory of Star Formation*
- Reynolds, S. P. 2008, *ARA&A*, 46, 89, *Supernova Remnants at High Energy*
- Hester, J. J. 2008, *ARA&A*, 46, 127, *The Crab Nebula: An Astrophysical Chimera*
- Tielens, A. G. G. M. 2008, *ARA&A*, 46, 339, *Interstellar Polycyclic Aromatic Hydrocarbon Molecules*
- Wyatt, M. C. 2008, *ARA&A*, 46, 339, *Evolution of Debris Disks*
- Wang, L., & Wheeler, J. C. 2008, *ARA&A*, 46, 433, *Spectropolarimetry of Supernovae*

5. Working Group reports

5.1. *Working Group Astrochemistry*

MEMBERS: Ewine F. van Dishoeck (chair, Netherlands), Eric Herbst (secretary, USA), Louis J. Allamandola (USA), John H. Black (Sweden), Geoffrey A. Blake (USA), Paola Caselli (Italy, now UK), Pascale F. Ehrenfreund (Netherlands), Guido Garay (Chile), Micel Guelin (France), Christian Henkel (Germany), Uffe G. Jørgensen (Denmark), John P. Maier (Switzerland), Karl M. Menten (Germany), Thomas J. Millar (UK), Young Chol Minh (South Korea), Masatoshi Ohishi (Japan), Alejandro C. Raga (Mexico), Jonathan Rawlings (UK), Bertrand R. Rowe (France), and Ji Yang (China).

The Working Group on *Astrochemistry* planned and organized the highly successful symposium on *Astrochemistry – Recent Successes and Current Challenges*, IAU Symposium No. 231, held at Asilomar, California, USA, in August 2005. The volume of the proceedings, edited by Darek Lis, Geoffrey Blake & Eric Herbst, which appeared in early 2006, is a valuable addition to the field of astrochemistry.

Currently the Working Group is starting to prepare for the next symposium in the astrochemistry series, planned to be held in southern Europe in either 2010 or 2011. Chair Ewine van Dishoeck and secretary Eric Herbst have been stimulating other interim meetings in the field of astrochemistry and have also been searching for gifted young speakers for the next IAU Symposium, including related fields in chemical physics.

The astrochemistry working group website at `<www.strw.leidenuniv.nl/~iau134>` is regularly maintained and updated, especially on links with molecular databases.

Within the next several years, both the *Herschel Space Observatory* and the Atacama Large Millimeter Array (ALMA) will help to revolutionize the field of astrochemistry by dramatically improving the spatial resolution of small objects such as protoplanetary disks, vastly increasing the intensity of molecular spectral signals, and greatly enlarging the region of the electromagnetic spectrum available to study molecules. We trust that our next IAU Symposium will be bursting with new information and analysis.

5.2. *Working Group Star Formation*

MEMBERS: Francesco Palla (chair, Italy), Rafael Bachiller (Spain), Michael Burton (Australia), Lawrence E. Cram (Australia), Yasuo Fukui (Japan), Garay Garay (Chile), Thomas Henning (Germany), Charles J. Lada (USA), Maria T. V. T. Lago (Portugal), Susana Lizano (Mexico), Jan Palouš (Czech Republic), Bo Reipurth (USA), Annelia I. Sargent (USA), and Stephen E. Strom (USA).

The Working Group on *Star Formation* has been involved in the organization of a number of IAU conferences on various aspects related to the field of star formation. In recent years, there has been a great interest and development both on the large scale aspects of star formation and feedback effects on the interstellar medium, and on the small scale implication on protoplanetary disk formation and evolution.

IAU Symposium No. 237 on *Triggered Star Formation in a Turbulent Environment* was held at the IAU XXVII General Assembly in Prague, August 2006. The proceedings of the conference edited by Bruce Elmegreen & Jan Palous appeared in 2007 and represents a valuable reference book on the topic.

IAU Symposium No. 242 on *Astrophysical Masers and Their Environments* was organized in Alice Springs, Australia, in May 2007 and followed a previous IAU conference on masers held in Brazil in 2001. The volume of the Symposium was edited by J. M. Chapman & W. A. Baan, appeared in 2008, and provides a thorough summary of the impact of maser research on star formation studies.

Finally, IAU Symposium No. 243 on *Star-Disk Interaction in Young Stars* was held in Grenoble in 2007. The proceedings have been edited by J. Bouvier & I. Appenzeller and appeared in 2007.

The WG has been active at promoting and supporting requests for the organization of Symposia, Joint Discussions and Special Sessions during the IAU XXVII General Assembly in Rio de Janeiro, Brazil, 2009. As a result, we look forward to participating in IAU Symposium No. 266 on *Star Clusters – Basic Building Blocks throughout Time and Space*, in JD7 on *Astrophysical Outflows and Accretion Phenomena*, and in SpS7 on *Young Stars, Brown Dwarfs, and Protoplanetary Disks*. The WG web page <www.arcetri.astro.it/sfwg> is maintained and updated with links to forthcoming meetings on star formation.

5.3. *Working Group Planetary Nebulae*

MEMBERS: Arturo T. Manchado (chair, Spain), Michael J. Barlow (UK), Romano L. M. Corradi (Spain), You-Hua Chu (USA), Shuji Deguchi (Japan), Adam Frank (USA), George H. Jacoby (USA), Sun Kwok (Hong Kong, China), Alexander López (Mexico), Walter J. Maciel (Brazil), Roberto H. Méndez (USA), Quentin A. Parker (Australia), Detlef Schoenberner (Germany), Letizia Stanghellini (USA), and Albert Zijlstra (UK).

The Working Group on *Planetary Nebulae* planned and organized IAU Symposium No. 234 on *Planetary nebulae in our Galaxy and Beyond*, held in Waikoloa Beach (HI, USA), April 2006. The attendance was 150 astronomers from 18 countries. This meeting followed the time-honored tradition of having an IAU Symposium on planetary nebulae every five years. The symposium proceedings were edited by M. J. Barlow & R. H. Mendez, and published timely in 2006. This volume offers contributions on a wide variety of subjects: Galactic and extragalactic PN surveys; the relationships between AGB stars, post-AGB objects, PN central stars and white dwarfs; nucleosynthesis and the properties of the central stars, including the role of binarity; PN properties across the full electromagnetic spectrum; atomic processes and PN abundances; mechanisms for the formation of PN structures; and properties and applications of extragalactic PNs, currently detected out to Coma cluster distances.

The WG is starting to plan its proposal for the next IAU Symposium in the planetary nebulae series, to take place on Tenerife, Spain, in summer 2011.

Some of the WG members have been organizing additional meetings in the field of planetary nebulae. Arturo Manchado (WG chair) & Romano Corradi organized the conference *Asymmetrical Planetary Nebulae. IV*, held in June 2007 on La Palma, Spain.

Albert Zijlstra, Romano Corradi & Alberto Lopez, are organizing the follow-up conference *Asymmetrical Planetary Nebulae. V* , to be held in Bowness-on-Windermere, UK,
in June 2010. See: <www.astrophysics.manchester.ac.uk/apn5.html>.

The WG web page <www.iac.es/proyecto/PNgroup/wg/index.html> is maintained
and updated with links to past and forthcoming meetings on planetary nebulae.

Thomas J. Millar
president of the Division

Transactions IAU, Volume XXVIIA
Reports on Astronomy 2006–2009
Karel A. van der Hucht, ed.

© 2009 International Astronomical Union
doi:10.1017/S1743921308025660

DIVISION VII   THE GALACTIC SYSTEM

SYSTÈME GALACTIQUE

Division VII gathers astronomers studying the diffuse matter in space between stars, ranging from primordial intergalactic clouds, via dust and neutral and ionized gas in galaxies, to the densest molecular clouds and the processes by which stars are formed.

PRESIDENT	Ortwin Gerhard
VICE-PRESIDENT	Despina Hatzidimitriou
PAST PRESIDENT	Patricia A. Whitelock
BOARD	Charles J. Lada, Ata Sarajedini,
	Rosemary F. Wyse, Joseph Lazio

DIVISION VII COMMISSIONS

Commission 33	Structure and Dynamics of the Galactic System
Commission 37	Star Clusters and Associations

DIVISION VII WORKING GROUP

Division VII WG	Galactic Center

TRIENNIAL REPORT 2006 - 2009

1. Introduction

Division VII provides a forum for astronomers studying the Milky Way Galaxy and its constituents. The tremendous detail in which the distribution of Milky Way stars and gas can be observed makes our Galactic System an important cornerstone for understanding galaxies and their formation in general.

2. Activities

At the Business Meeting during the IAU XXVI General Assembly in Prague, 2006, it was decided that Division VII would form an IAU Working Group on the Galactic Centre. The first chair would be Joseph Lazio who would select an organizing committee. The organizing committee, besides the usual considerations of representation, should take into account those at the meeting who expressed an interest and ensure that all the major groups working in the field were represented. This is still pending. The decision to form the Working Group followed a very successful scientific meeting on the Galactic Centre, which is summarized in the triennial report of the Working Group. The Working Group should be discussed further at the IAU XXVII General Assembly in Rio de Janeiro.

3. Symposia

Sponsoring IAU Symposia is one of the main activities of the Division. Since the General Assembly in Prague in 2006, Division VII has supported the following IAU meetings:

- IAU Symposium No. 246, *Dynamical Evolution of Dense Stellar Systems*, Capri, Italy, September 2007
- IAU Symposium No. 254, *Galactic Disk in Cosmological Context*, Copenhagen, Denmark, June 2008
- IAU Symposium No. 256, *The Magellanic System: Stars, Gas, and Galaxies*, Keele, UK, June 2008
- IAU Symposium No. 258, *The Ages of Stars*, Baltimore, MD, USA, October 2008
- IAU Symposium No. 259, *Cosmic Magnetic Fields: from Planets, to Stars, and Galaxies*, Puerto Santiago, Spain, November 2008
- IAU Symposium No. 265, *Chemical Abundances in the Universe – Connecting First Stars to Planets*, Rio de Janeiro, Brazil, August 2009
- IAU Symposium No. 266, *Star Clusters – Basic Building Blocks throughout Time and Space*, Rio de Janeiro, Brazil, August 2009

Many other conferences of interest to members of the division were held in the past triennium. In a number of these the Milky Way Galactic System played a prominent rôle as a template for galaxy evolution and formation.

4. Closing remarks

Further information relevant to the Division can be found on the Division web page `<www.mpe.mpg.de/IAU_DivVII/>` and the web pages of Division VII Commissions 33 and 37.

Ortwin Gerhard
president of the Division

Transactions IAU, Volume XXVIIA
Reports on Astronomy 2006–2009
Karel A. van der Hucht, ed.

© 2009 International Astronomical Union
doi:10.1017/S1743921308025672

COMMISSION 33

STRUCTURE AND DYNAMICS OF THE GALACTIC SYSTEM

STRUCTURE ET DYNAMIQUE DU SYSTÈME GALACTIQUE

PRESIDENT   Ortwin Gerhard
VICE-PRESIDENT   Rosemary F. Wyse
PAST PRESIDENT   Patricia A. Whitelock
ORGANIZING COMMITTEE Yuri N. Efremov, Wyn Evans,
                         Chris Flynn, Jonathan E. Grindlay,
                         Birgitta Nordström, Chi Yuan

TRIENNIAL REPORT 2006 - 2009

1. Introduction

Commission 33 provides a forum for astronomers studying the structure and dynamics of the Milky Way Galaxy, a unique laboratory for exploring the stellar and gaseous components of galaxies and the processes by which they form and evolve.

2. Symposia and reviews

Sponsoring IAU Symposia is one of the main activities of the Commission. Since the General Assembly in Prague in 2006, Commission 33 has supported the following IAU meetings:

- IAU Symposium No. 241, *Stellar Populations as Building Blocks of Galaxies*, La Palma, Spain, December 2006
- IAU Symposium No. 245, *Formation and Evolution of Galaxy Bulges*, Oxford, UK, July 2007
- IAU Symposium No. 246, *Dynamical Evolution of Dense Stellar Systems*, Capri, Italy, September 2007
- IAU Symposium No. 254, *The Galactic Disk in Cosmological Context*, Copenhagen, Denmark, June 2008
- IAU Symposium No. 258, *The Ages of Stars*, Baltimore, MD, USA, October 2008
- IAU Symposium No. 261, *Relativity in Fundamental Astronomy - Dynamics, Reference Frames an Data Analysis*, Virginia Beach, USA, April 2009
- IAU Symposium No. 262, *Stellar Populations – Planning for the Next Decade*, Rio de Janeiro, Brazil, August 2009
- IAU Symposium No. 266, *Star Clusters – Basic Building Blocks throughout Time and Space*, Rio de Janeiro, Brazil, August 2009

In addition to the reviews presented in the proceedings of these conferences, Helmi's review on the Galactic Halo (2008, *A&ARv* 15, 145) is worth noting.

3. Closing remarks

In addition to the Symposia, Commission 33 has been supporting a Joint Discussion on *Modeling the Milky Way in the Era of Gaia*, and a Special Session on *The Galactic Plane - in Depth and Across the Spectrum*, both to be held during the IAU XXVII General Assembly in Rio de Janeiro, Brazil, 2009. The topics of these two events show the enormous progress we may expect in Milky Way studies over the next years, due to on-going or planned new surveys of unprecedented depth and sky coverage across the entire wavelength spectrum, culminating in the planned *Gaia* mission, and their modeling and theoretical analysis. Further information relevant to the Commission can be found on the Commission web page <www.mpe.mpg.de/IAU_Comm33/>.

Ortwin Gerhard
president of the Commission

Transactions IAU, Volume XXVIIA
Reports on Astronomy 2006–2009
Karel A. van der Hucht, ed.

© 2009 International Astronomical Union
doi:10.1017/S1743921308025684

COMMISSION 37

STAR CLUSTERS AND ASSOCIATIONS

AMAS STELLAIRES ET ASSOCIATIONS

PRESIDENT   Despina Hatzidimitriou
VICE-PRESIDENT   Charles J. Lada
PAST PRESIDENT   Ata Sarajedini
ORGANIZING COMMITTEE   Russell D. Cannon,
                     Kyle McC. Cudworth,
                     Gary S. Da Costa, LiCai Deng,
                     Young-Wook Lee, Ata Sarajedini,
                     Monica Tosi

TRIENNIAL REPORT 2006 - 2009

1. Introduction

Star clusters are valuable tools for theoretical and observational astronomy across a wide range of disciplines from cosmology to stellar spectroscopy. For example, properties of globular clusters are used to constrain stellar evolutionary models, nucleosynthesis and chemical evolution, as well as the star formation and assembly histories of galaxies and the distribution of dark matter in present-day galaxies. Open clusters are widely used as stellar laboratories for the study of specific stellar phenomena (e.g., various emission-line stars, pulsating pre-MS stars, magnetic massive stars, binarity, stellar rotation, etc.). They also provide observational constraints on models of massive star evolution and of Galactic disk formation and chemical evolution.

2. Publications

Star clusters have continued to receive much attention within the scientific community during the past three years. In the period from January 2006 until September 2008, about 420 papers related to globular clusters have been published in refereed journals (measuring already more than 3100 citations), 270 on open clusters (with 665 citations) and 30 on associations (140 citations). The papers cover a multiplicity of subjects, including issues related to the formation and dynamical evolution of star clusters, stellar evolution and ages, star clusters as tracers of stellar populations, studies of specific types of objects within clusters, nuclear clusters and extragalactic cluster systems, while the authors utilize observations covering an increasing portion of the electromagnetic spectrum, ranging from X-rays to the far-infrared, as well as advanced N-body simulations.

3. Symposia, colloquia and reviews

During the reporting period several international conferences, workshops, meetings and schools, related directly or indirectly to star clusters, have taken place:

- MPA/ESO/USM/MPE 2008 Joint Astronomy Conference *Chemical Evolution of Dwarf Galaxies and Stellar Clusters*, 21-25 July 2008, Garching, Germany
- IAU Symposium No. 255 *Low-Metallicity Star Formation: From the First Stars to Dwarf Galaxies*, 16-20 June 2008, Rapallo, Liguria, Italy
- *Nuclear Star Clusters across the Hubble Sequence*, 25-27 February 2008, MPI for Astronomy, Heidelberg, Germany
- *Modeling Dense Stellar Systems (MODEST-8)*, 5-8 December 2007, Bonn/Bad Honnef, Germany
- IAU Symposium No. 246 *Dynamical Evolution of Dense Stellar Systems*, 5-9 September 2007, Capri, Italy
- ESO workshop: *12 Questions on Star and Massive Star Cluster Formation*, 3-6 July 2007, Garching, Germany
- *Milky Way Halo Conference*, 29 May - 2 June 2007, Bonn, Germany
- *Structure formation in the Universe*, 27 May - 1 June, 2007, Chamonix, France
- *Galactic & Stellar Dynamics in the Era of High-Resolution Surveys*, 16-18 March 2007, Strasbourg, France
- *The Dynamics of Star Clusters and Star Cluster Systems*, 6-8 November 2006, Sheffield, UK
- IAU XXVI General Assembly Joint Discussion 14, *MODEST-7*, 17 - 23 August 2006, Prague,Czech Republic
- IAU Symposium No. 235 *Evolution of Galaxies across the Hubble Time*, 14-17 August 2006, Prague, Czech Republic
- IAU Symposium No. 237 *Triggered Star Formation in a turbulent ISM*, 14-18 August, 2006, Prague, Czech Republic
- *International School on Galactic and Cosmological N-Body Simulations*, 23 July - 5 August 2006, Tonantzintla, Puebla, Mexico.
- *Cambridge N-body School*, 30 July - 11 August 2006, Cambridge, UK
- *Mass Loss from Stars and the Evolution of Stellar Clusters Workshop*, 29 May - 1 June 2006, Lunteren, Netherlands

Commission 37 has endorsed three IAU Symposia to be held during the IAU XXVII General Assembly in Rio de Janeiro, Brazil, 3-14 August 2009:

- IAU Symposium No. 262 *Stellar Populations - Planning for the Next Decade*
- IAU Symposium No. 265 *"Chemical Abundances in the Universe - from Stars to Planets*
- IAU Symposium No. 266 *"Star Clusters: Basic Galactic Building Blocks throughout Time and Space*

An excellent review on the use of globular cluster systems as tracers of galaxy formation and assembly was given by J. P. Brodie & J. Strader (2006, *ARAA* 44, 193). One of the primary aims of the authors was "to emphasize the current and potential links with results from galaxy surveys at high redshift and interpretations from stellar population synthesis, numerical simulations, and semi-analytical modeling".

Additional reviews regarding a variety of aspects of star cluster research have also appeared in the proceedings of the conferences and meetings listed above.

4. Databases

Data on Open Clusters in the Milky Way and the Magellanic Clouds can be found in the WEBDA site `<www.univie.ac.at/webda/>`, which was originally developed by Jean-Claude Mermilliod from the Laboratory of Astrophysics of the EPFL (Switzerland)

and is now maintained and updated by Ernst Paunzen from the Institute of Astronomy of the University of Vienna, Austria.

Data on Galactic Globular Clusters can be found in the *Catalog of parameters for Milky Way globular clusters* by W. E. Harris <`www.physics.mcmaster.ca/Globular`>), as well as in the *Galactic Globular Clusters Database* at the Astronomical Observatory of Rome (INAF-OAR: <`venus.mporzio.astro.it/ marco/gc/`>.

A *Catalogue of Variable Stars in Globular Clusters* developed and maintained by Christine Clement can be found at <`www.astro.utoronto.ca/ cclement/`>.

5. Discussion and closing remarks

Recent advances in instrumentation (wide-field imaging cameras and multiplexing spectrographs on several 8-10 meter class telescopes; enhanced sensitivity and spatial resolution both at short wavelengths - *Chandra / XMM* and *GALEX* - and in the mid and far infrared – *Spitzer Space Telescope* – in conjunction with advanced numerical and semi-analytic simulations and models are expected to bring up a revolution in our understanding of star clusters allowing us to follow their spatial, kinematic, chemical, and structural evolution both in the Milky Way and in external galaxies.

Despina Hatzidimitriou
president of the Commission

Transactions IAU, Volume XXVIIA
Reports on Astronomy 2006–2009
Karel A. van der Hucht, ed.

© 2009 International Astronomical Union
doi:10.1017/S1743921308025696

DIVISION VII / WORKING GROUP
GALACTIC CENTER

CHAIR Joseph Lazio

TRIENNIAL REPORT 2006 - 2009

1. Introduction

The Working Group *Galactic Center* was created following a Business Meeting meeting of Division VII *Galactic Systems*, held with the concurrence of Division VIII *Galaxies and the Universe*, at the IAU XXVI General Assembly in Prague, 2006. The objective of the meeting was to highlight both recent progress on understanding the Galactic Center as well as to illustrate the way in which the center of the Milky Way Galaxy serves as a laboratory for understanding other galactic nuclei.

The scientific meeting itself featured talks across the wavelength range, from radio to γ-ray, on two general topics. The first topic was the Galactic Center itself, particularly with respect to probing Sgr A*, the central supermassive black hole. Among the papers presented were a description of the possibility of imaging in the radio the shadow of Sgr A* against its accretion disk, the theory and observations of IR and high-energy flares from Sgr A*, and the Galactic Center at very-high-energy (VHE) γ-rays. The second topic was the environment of the Galactic Center and its relation to other galactic nuclei. Among the papers presented were theoretical considerations on star formation in the strong gravitational potential of the Galactic Center, censuses of the stars in the inner parsecs, and how the Milky Way Galaxy's nucleus compares to that of other disk-dominated galaxies.

Additionally, part of the motivation for the formation of the WG was the series of successful *Galactic Center Workshops* (Falcke *et al.* 1999; Cotera *et al.* 2002; Schödel *et al.* 2006), focused workshops, that bring the Galactic Center research community together to discuss recent progress as well as plan for future observations.

2. Developments within the past triennium

Primary activity of the WG is to publish GCNews <www.aoc.nrao.edu/~gcnews/> the newsletter for Galactic Center research. GCNews has more than 300 subscribers and serves as a forum for keeping the community informed about recent events relating to Galactic center research. GCNews consists of two components, a newsletter and news-flashes. Newsflashes are distributed by email as new abstracts are received.

Newsletters are Web publications, with the following major sections. An Abstract section compiles abstracts received since the last publication of a newsletter. An Announcement section keeps the community informed about upcoming conferences, symposia, and workshops; as they become available, brief reports from previous conferences, symposia, and workshops are also presented. Finally, original feature articles headline the newsletter. The feature articles are designed to highlight recent results, explain aspects of Galactic center research in slightly more detail, or illustrate currently unresolved questions. In the past triennium, feature articles have included a re-evaluation of the extent to which non-thermal emission is present in the star forming region Sgr B (Lang, Palmer & Goss 2008); a discussion of the mass function of black holes in the local Universe *vis-á-vis* the extent to which Sgr A* is 'typical' (Greene & Ho 2007); and a discussion of stellar orbits in the central parsec and the implication for star formation in the Galactic center (Lu *et al.* 2006).

A secondary focus for the WG has been planning for the next Galactic Center workshop. The previous three workshops have been held in either the United States or Germany, and there was broad agreement that it would be desirable to expand the sites considered for as hosts for GC'09. Given that the Galactic Center is overhead in the Southern Hemisphere, there was initial thought given to a Southern Hemisphere site. More recently, however, there has been an expression of interest from Asia, and it seems likely that the next Galactic Center workshop will be held there. Allowing for the prime observing season for the Galactic center being the (Northern Hemisphere) summer, the original target date was late in 2008. However, it was then decided that this might nonetheless be difficult to schedule with the Olympics. The current target date is in 2009, though careful attention is also being given to the dates of the IAU XXVII General Assembly in Rio de Janeiro, Brazil, so that GC'09 does not conflict with it.

3. Closing remarks

This triennium has seen a number of remarkable scientific discoveries and advances related to the Galactic center. High-energy observations have become routine and are producing an entirely new probe of the region. A number of hypervelocity stars, on escape trajectories from the Galaxy, have been identified and used to probe the conditions in the Galactic Center that gave rise to their extreme velocities. Infrared and X-ray flares continue to provide strong constraints on the accretion environment around Sgr A*. And recent sub-millimeter very long baseline interferometric (VLBI) observations of Sgr A* continues to indicate the promise of future imaging of the accretion shadow of the supermassive black hole itself. With new instruments and techniques continuing to be developed, the next triennium looks quite exciting for continued understanding of the Galactic Center.

Acknowledgements

It is a pleasure to thank the other individuals responsible for the smooth running of GCNews during the past triennium, including Loránt O. Sjouwerman, Cornelia C. Lang, Rainer Schödel, Masaaki Sakano, Feng Yuan, and Robeson M. Herrnstein, as well as Angela S. Cotera and Heino D. Falcke, who started GCNews. Basic research in radio astronomy at the Naval Research Laboratory is supported by 6.1 Base funding.

Joseph Lazio
chair of the Working Group

References

Cotera, A., Markoff, S., Geballe, T. R., & Falcke, H. (eds.) 2002, *The Central 300 Parsecs of the Milky Way*, Proc. Galactic Center Workshop 2002, *Astron. Nach.*, Suppl. 1

Falcke, H., Cotera, A., Duschl, W. J., Melia, F., & Rieke, M. J. (eds.) 1999, *The Central Parsecs of the Galaxy*, Proc. Galactic Center Workshop 1998, *ASP-CS* 186

Greene, J. & Ho, L. 2008, *GCNews*, 26

Lang, C., Palmer, P., & Goss, M. 2008, *GCNews*, 27

Lu, J. R., Ghez, A. M., Hornstein, S. D., *et al.* 2006, *GCNews*, 25

Schödel, R., Bower, G. C., Muno, M. P., Nayakshin, S., & Ott, T. (eds.) 2006, Proc. *Galactic Center Workshop 2006*, *JPh-CS*, 54

Transactions IAU, Volume XXVIIA
Reports on Astronomy 2006–2009
Karel A. van der Hucht, ed.

© 2009 International Astronomical Union
doi:10.1017/S1743921308025702

DIVISION VIII GALAXIES AND THE UNIVERSE

LES GALAXIES ET l'UNIVERS

Division VIII provides a focus for astronomers studying a wide range of problems related to galaxies and cosmology. Objects of the study include individual galaxies, groups and clusters of galaxies, large scale structure, comic microwave background radiation and the universe itself. Approaches are diverse from observational one to theoretical one including computer simulations.

PRESIDENT   Sadanori Okamura
VICE-PRESIDENT   Elaine Sadler
PAST PRESIDENT   Francesco Bertola
BOARD   Mark Birkinshaw, Françoise Combes,
                             Roger L. Davies, Thanu Padmanabhan,
                             Rachel L. Webster

DIVISION VIII COMMISSIONS

Commission 28	Galaxies
Commission 47	Cosmology

DIVISION VIII WORKING GROUPS

Division VIII WG	Supernovae

TRIENNIAL REPORT 2006 - 2009

1. Introduction

As documented by the report of the president for the triennium 2003 - 2006, there have been dramatic developments in the fields of extragalactic research and cosmology recently. During the period 2006 - 2009 even more rapid progress has been made.

2. Observations

On the observational side, this progress has largely come about because of large and/or deep surveys of galaxies and galaxy systems, and because of new observing windows opened by such satellites as *Spitzer* <www.spitzer.caltech.edu/>), *AKARI* <www.ir.isas.jaxa.jp/ASTRO-F/>), and *GALEX* <www.galex.caltech.edu/>). *HST* <www.stsci.edu/hst/>) continues to produce valuable data, and we look forward to its further operation after Servicing Mission 4, planned for October 2008.

Over the past three years, cosmological parameters have been more strongly constrained through a number of approaches. Chief among these have been studies of the CMB: their importance for cosmology was recognized by a Nobel Prize in 2006. The *WMAP* satellite has now accumulated five years of data, which provide an accurate determination of important cosmological parameters (Hinshaw, G., *et al.* 2008, in press;

Komatsu *et al.* 2008, in press). Further pressure on the cosmological model comes from supernova work and from studies of structure formation (e.g., Zhen *et al.* 2008, *AJ*, 135, 1766; Wright 2007, *ApJ*, 664, 633), and from projects investigating the locations of baryons and the epoch of re-ionization through studies of quasar/AGN absorption lines (e.g., Danforth & Shull 2008, *ApJ*, 679, 194), the population of early galaxies, and the problem of the causes and effects of feedback.

3. Theory

On the theoretical side, several groups have performed sophisticated cosmological simulations to study the evolution of galaxies and large scale structure in the Universe, including the effects that supermassive black holes can have on their host galaxies. Such simulations include

- the Millennium Simulation Project `<www.mpa-garching.mpg.de/galform/virgo/millennium/>`,
  - the Cosmic Data ArXiv `<t8web.lanl.gov/people/heitmann/test3.html>`
  - the Marenostrum Numerical Cosmology Project `<astro.ft.uam.es/marenostrum/>`
  - the HORIZON project `<www.projet-horizon.fr/rubrique3.html>`

4. Surveys

Shown here is an (inevitably incomplete) list of major recent non-targeted surveys that are intimately related to the activities of our Division.

- SDSS: Sloan Digital Sky Survey `<www.sdss.org/>`
- COSMOS: Cosmic Evolution Survey `<cosmos.astro.caltech.edu/index.html>`
- DEEP2: Deep Extragalactic Evolutionary Probe 2 `<deep.berkeley.edu/>`
- UKIDSS: The UKIRT Infrared Deep Sky Survey `<www.ukidss.org/>`
- VIMOS-VLT Deep Survey `<www.oamp.fr/virmos/index.html>`
- Subaru Deep Surveys
  - SXDS: Subaru-XMM Newton Deep Survey (Furusawa *et al.* 2008, *ApJS*, 176, 1)
  - SDF: Subaru Deep Field (Kashikawa *et al.* 2006, *ApJ*, 648, 7)
  - MOIRCS Deep Survey (Ichikawa *et al.* 2008, *PASJ*, 59, 1081)
- Hubble Ultra Deep Field `<www.stsci.edu/hst/udf>`
- 2dFGRS: 2df Galaxy Redshift Survey `<www2.aao.gov.au/2dFGRS/>`
- 2SLAQ: 2dF-SDSS LRG and QSO survey `<www.2slaq.info>`
- 6dF Galaxy Survey `<www.aao.gov.au/local/www/6df/>`
- CFHT Legacy Survey `<www.cfht.hawaii.edu/Science/CFHLS/>`
- 2MASS: The Two Micron All Sky Survey `<www.ipac.caltech.edu/2mass/>`
- GOODS `<www.stsci.edu/science/goods/>`; `<www.eso.org/science/goods/>`
- GEMS `<www.mpia.de/GEMS/gems.htm>`
- GRAPES `<www.stsci.edu/~san/Grapes/>`
- COMBO-17 `<www.mpia.de/COMBO/>`
- EdisCS `<www.mpa-garching.mpg.de/galform/ediscs/>`
- SHADES `<www.roe.ac.uk/ifa/shades/>`
- The Phoenix Deep Survey `<www.physics.usyd.edu.au/~ahopkins/phoenix/>`
- MUSYC `<www.astro.yale.edu/MUSYC/>`
- The Hα Galaxy Survey. VI (James, P. A., *et al.* 2008, *A&A* 486, 131)
- The Gemini Deep Deep Survey `<lcirs.ociw.edu/gdds.html>`
- SAURON Project. XI (Peletier, R.F., *et al.* 2007, *MNRAS*, 379, 445)

- Chandra Deep Field <`www.astro.psu.edu/~niel/hdf/hdf-chandra.html`>; <`www.mpe.mpg.de/~mainieri/cdfs~pub/`>
- The Arecibo Galaxy Environment Survey <`www.naic.edu/~ages/`>
- NVSS: NRAO VLA Sky Survey <`www.cv.nrao.edu/nvss/`>
- VLA FIRST survey <`sundog.stsci.edu/`>
- SUMSS <`www.astrop.physics.usyd.edu.au/SUMSS/`>
- HIPASS: HI Parkes All-Sky Survey <`www.atnf.csiro.au/research/multibeam/`>
- WiggleZ <`wigglez.swin.edu.au/Welcome.html`>
- GAMA: Galaxies And Mass Assembly <`www.eso.org/~jliske/gama/`>

The availability of large amounts of data, which cover wide areas of the sky and/or the majority of cosmic history from several hundred million years after the Big Bang to the present, and the many predictions that can be made from numerical simulations, have reduced the heights of the barriers between theoreticians and observers, and between galaxies and cosmology.

Division VIII, the largest division (with 1544 members), will continue to contribute to our understanding of the properties of galaxies, the formation and evolution of galaxies and large scale structure, and the content and fate of the Universe.

Sadanori Okamura
president of the Division

Transactions IAU, Volume XXVIIA
Reports on Astronomy 2006–2009
Karel A. van der Hucht, ed.

© 2009 International Astronomical Union
doi:10.1017/S1743921308025714

COMMISSION 28

GALAXIES

GALAXIES

PRESIDENT
VICE-PRESIDENT
PAST PRESIDENT
ORGANIZING COMMITTEE

Françoise Combes
Roger L. Davies
Elaine M. Sadler
Avishai Dekel, Marijn Franx,
John S. Gallagher,
Valentina Karachentseva,
Gillian R. Knapp,
Renée C. Kraan-Korteweg,
Bruno Leibundgut, Naomasa Nakai,
Jayant V. Narlikar, Monica Rubio

TRIENNIAL REPORT 2006 - 2009

1. Highlights in extragalactic research 2005 - 2008

This short report describes some highlights in extragalactic research over the past three years, and lists the main symposia and meetings in the domain.

1.1. *Star formation in galaxies with Spitzer and GALEX*

The combination of results from the two satellites, infrared and ultraviolet, has given a new and more complete view of star formation in nearby galaxies. In particular, the *Spitzer* Infrared Nearby Galaxies Survey (SINGS) has obtained a comprehensive set of imaging and spectroscopy for 75 nearby galaxies. By comparison with optical and ultraviolet SFR tracers, robust extinction-corrected star formation rate indices have been developed, They have been applied to measure the form of the spatially-resolved SFR vs gas density Schmidt law (Kennicutt *et al.* 2007). Among the various observed tracers, the flux at $24\,\mu$m has been found a good tracer of star formation, while $8\,\mu$m depends more on dust and metallicity (Calzetti *et al.* 2007).

Combined with *HST* optical and *XMM* X-ray results, infrared and UV trace star formation as a function of environment. This gives insight into galaxy interactions and ram pressure triggering mechanisms. Star formation histories have been obtained. While the average SFR increases as z increases, the fluctuations in SFR is not large, and most galaxies spend their time within a factor 2 of the average SFR since $z = 1$ (Noeske *et al.* 2007). The dominant mode of the evolution of SF since $z \simeq 1$ is more a gradual decline of the average SFR than a decreasing frequency of starburst episodes, or a decreasing starburst efficiency. The gradual decline is suggesting gas exhaustion, as a slow quenching of star formation. Poggianti *et al.* (2006) have determined the influence of environment on star formation history. At high z, the star formation in clusters is decreasing with the cluster mass. Galaxy clusters had more star forming galaxies in the past then now, and the quenching occurs earlier for higher masses.

1.2. *Extended UV (XUV) disks*

GALEX has unveiled the existence in UV emission of extended disks of spiral galaxies. While the Hα emission is truncated at the optical disk radius, the UV, which is a tracer of older star formation, extends almost as far as the H I 21 cm emission, and correlates with it. Gil de Paz *et al.* (2007) show through abundance measurements, that the outer disks are not made of pristine material, but are enriched like young spiral disks 1 Gyr after these first stars would have formed. The amount of gas in the XUV disks allows maintaining the current level of star formation for at least a few Gyr.

A large survey of UV disks has shown that extended disks are frequent, beyond the threshold of traditional star formation and beyond the limits of old stellar disks (Thilker *et al.* 2007). The star formation in these outer disks could be episodic; these disks are gas rich, and the phenomenon could be due to significant, continued gas accretion from the intergalactic medium or neighboring objects.

1.3. *Bimodality in galaxy colour and classification*

Bimodality has been found in large galaxy surveys, in colour-magnitude diagrams, and many properties follow this feature, like the Sersic index, or the surface brightness (Driver *et al.* 2006). It has been recently found that the number of red galaxies has almost doubled since $z = 1$, while the blue sequence was stationary. Galaxies appear to enter the red sequence via different quenching modes, which vary with time. The red sequence builds up in different ways (merging, external accretion, environment harassment), and the concept of a single process to explain 'downsizing' is not likely (Faber *et al.* 2007).

The theory of downsizing however has gained some further support (Cimatti *et al.* 2008). Old, passive galaxies at $z > 1.4$ have been followed in the GOODS fields, and it was shown that major star formation and assembly processes for these galaxies occurred at $z > 2$.

A population of heavily-obscured and distant AGN has been found via mid-infrared excess. About 30% of $z \simeq 2$ galaxies detected at 24 μm with *Spitzer* show an excess mid-IR emission, with respect to the SED expected from star formation. When their X-ray spectra are stacked, X-ray emission is detected, revealing Compton-thick AGN. The density of these objects on the sky is twice that of X-ray detected AGN, and their fraction increases with galaxy mass, up to 60% at 10^{11} M$_\odot$ (Daddi *et al.* 2007). These AGN were expected from the observed mass ratio of black holes to stars in local spheroids, and confirms the concurrent growth of both in distant massive galaxies.

An interpretation of the bimodality has been proposed in terms of halo mass categories, and external gas accretion. The latter is cold, under a certain mass threshold, and hot above, since accretion shocks are induced by the depth of the potential well. In halos below a critical shock-heating mass of $\sim 10^{12}$ M$_\odot$, discs are built by cold streams, not heated by a virial shock, yielding efficient early star formation (Dekel & Birnboim 2006).

1.4. *Evolution and high-z galaxies*

Large scale structures and galaxy evolution have been studied with deep and wide surveys, such as COSMOS at $z < 1.1$ (Scoville *et al.* 2007), or GOODS (Vanzella *et al.* 2008). The comparison with the large-scale CDM simulations reveal higher overdensities in the observed distribution. The downsizing phenomenon is evident, and galaxies in dense environments tend to be older.

The molecular clouds have been mapped with interferometers (Tacconi *et al.* 2006), and it has been possible to obtain, the first gas rotation curve at high redshifts

(Förster-Schreiber *et al.* 2006). Galaxies reveal disk rotation quite similar to the analogs at low redshifts.

Very deep near-infrared and optical imaging have searched for galaxies at $z \simeq 7\text{-}8$, about 700 Myr after the Big Bang (Boowens & Illingworth 2006). Only one candidate galaxy at $z \simeq 7\text{-}8$ was found, where ten would be expected if there were no evolution in the galaxy population between $z \simeq 7\text{-}8$ and $z \simeq 6$. This demonstrates that very luminous galaxies are quite rare 700 Myr after the Big Bang, which can be explained since the Universe is just too young to have built up many luminous galaxies at $z \simeq 7\text{-}8$ by the hierarchical merging of small galaxies.

1.5. *Scaling laws and time evolution*

The Tully-Fisher relation has been observed not to evolve strongly up to $z \simeq 1$, except for a brightening of B-magnitudes, due to star formation. This result is obtained by only taking into account the third of disks that are rotating, and excluding the perturbed ones (Flores *et al.* 2006).

Sheth *et al.* (2008) determine in the COSMOS field that the bar fraction decreases with redshift, refining the previous GEMS survey (Jogee *et al.* 2004).

The SAURON project has studied the kinematics of a large sample of early-type galaxies, and deduced scaling relations and their M/L ratio. The comparison of the dynamical M/L with that inferred from the stellar population, indicates a median dark matter fraction in early-type galaxies of 30% of the total mass inside the characteristic radius Re. There appears to be a variation in the total to stellar mass ratio, linked to the galaxy dynamics: fast-rotating galaxies have lower dark matter fractions than the slow-rotating and generally more-massive ones (Cappellari *et al.* 2006).

1.6. *Numerical simulations*

There has been remarkable advance in numerical simulations, in particular in the baryonic physics, to probe galaxy formation scenarios. The standard ΛCDM theory encounters difficulties at the galaxy scale, to form disks without bulges, or sufficiently large spiral galaxies, with enough angular momentum. The complex baryon processes, including gas cooling, star formation, the effects of a uniform ultraviolet background and feedback from supernovae or supermassive black holes have been studied (e.g., Bower *et al.* 2006; Cox *et al.* 2006; Governato *et al.* 2007).

The black hole formation has been studied through empirical recipes, and their concurrent formation with bulges has been found compatible with was is expected from the $M(BH)\text{-}\sigma$, or $M(BH)\text{-}M(\text{stars})$ relation. The BH accretion rate density peaks at lower redshift and evolves more strongly at high redshift than the star formation rate density, while the ratio of black hole to stellar mass density shows only a moderate evolution at low redshifts (Di Matteo *et al.* 2008).

Attempts have been made to solve the missing satellite problem, in adapting the cooling and feedback during the hierarchical scenario of structure formation. Suppression of cooling into lower sigma peaks suppresses efficiently satellite galaxies. Re-ionization at $z = 12$ is a way to wash out the formation of these small systems (Moore *et al.* 2006).

The 'Via Lactea' N-body simulation, one of the biggest dark matter simulation for a single spiral galaxy, shows that the mass function of CDM subhalos is steeply rising below 10^8 M$_\odot$. But the dark to stellar mass ratio is not likely to be constant over the satellite distributions. To reproduce the observations, the brightest satellites must correspond to the earliest forming subhalos (Strigari *et al.* 2007).

2. Conferences

2.1. *IAU Symposia and Joint Discussions*

The following lists IAU meetings where Commission 28 was widely involved:

- IAU Symposium No. 230 *Populations of High Energy Sources in Galaxies*, 15-19 2005, Dublin, Ireland
- *11th Latin-American Regional IAU Meeting* (LARIM-2005), 12-16 December 2005, Pucon, Chile
- IAU Symposium No. 235 *Galaxy Evolution across the Hubble Time*, 14-17 August 2006, Prague, Czech Republic
- IAU XXVI General Assembly Joint Discussion 07 *The Universe at $z > 6$*, 17-18 August 2006, Prague, Czech Republic
- IAU Symposium No. 238 *Black Holes – from Stars to Galaxies*, 21-25 August 2006, Prague, Czech Republic
- IAU Symposium No. 241 *Stellar Populations as Building Blocks of Galaxies*, 10-16 December 2006, La Palma, Canary Islands, Spain
- IAU Symposium No. 244: *Dark Galaxies and Lost Baryons*, 25-29 June 2007, Cardiff, UK
- IAU Symposium No. 245 *Formation and Evolution of Galaxy Bulges*, 16-20 July 2007, Oxford, UK
- IAU Symposium No. 254 *The Galaxy Disk in Cosmological Context*, 9-13 June 2008, Copenhagen, Denmark
- IAU Symposium No. 255 *Low-Metallicity Star Formation: From the First Stars to Dwarf Galaxies*, 16-20 June 2008, Rapallo, Liguria, Italy

2.2. *Other meetings*

There were a long list of international meetings, during the last three years, and we have selected those related to the research domain of Commission 28, from July 2005 to June 2008. The list is not complete, but gives an idea of the wealth of the domain. The list compiled by Liz Bryson at CADC has been extremely useful to build such a summary. See `<www1.cadc-ccda.hia-iha.nrc-cnrc.gc.ca/meetings/>`.

- *Island Universes: the Structure and Evolution of Disk Galaxies*, 3-8 July 2005, Terschelling, Netherlands
- *Mass Profiles and Shapes of Cosmological Structures*, 4-9 July 2005, Paris, France
- *Stellar Populations, a Rosetta Stone for Galaxy Formation*, 4-8 July 2005, Ringberg, Germany
- *Mass and Mystery in the Local Group*, 17-22 July 2005, Cambridge, UK
- *Nearly Normal Galaxies 2005*, 7-13 August 2005, Santa Cruz, CA, USA
- *Extreme Starbursts: Near and Far*, 14-20 August 2005, Lijiang, China
- *Stellar Evolution at Low Metallicity: Mass Loss, Explosions, Cosmology*, 15-19 August 2005, Tartu, Estonia
- *QSO Host Galaxies : Evolution and Environment*, 22-26 August 2005, Leiden, Netherlands
- *Outer Edges of Disk Galaxies: A Truncated Perspective*, 4-8 October 2005, Leiden, Netherlands
- *Galactic and Extragalactic ISM Modelling in an ALMA Perspective*, 13-15 October 2005, Onsala, Sweden
- *Origin of Matter and Evolution of Galaxies* (OMEG05), 8-11 November 2005, Tokyo, Japan

- *The Spitzer Science Center 2005 Conference: Infrared Diagnostics of Galaxy Evolution*, 14-16 November 2005, Pasadena, CA, USA
- *Groups of Galaxies in the Nearby Universe*, 5-9 December 2005, Santiago, Chile
- *From Z-Machines to ALMA: (Sub)millimeter Spectroscopy of Galaxies*, 13-14 January 2006, Charlottesville, VA, USA
- *Galactic Bulges, 25-26 January 2006*, Padova, Italy
- *Aspen Center for Physics: Local Group Cosmology*, 5-11 February 2006, Aspen, CO, USA
- *Globular Clusters - Guides to Galaxies*, 6-10 March 2006, Concepcion, Chile
- *Dwarf Galaxies as Astrophysical and Cosmological Probes*, 12-17 March 2006, Ringberg, Germany
- XXVI Astrophysics Rencontre de Moriond: *From Dark Halos to Light*, 12-18 March 2006, La Thuile, Italy
- *Galaxies and Structures Through Cosmic Times*, 26-31 March 2006, Venezia, Italy
- 36th Advanced Course Swiss Society for Astrophysics and Astronomy, *First Light in the Universe*, 3-8 April 2006, Les Diablerets, Switzerland
- *Cosmology, Galaxy Formation and Astro-particle Physics on the Pathway to the SKA*, 10-12 April 2006, Oxford, UK
- Galactic Center Workshop 2006 *From the Center of the Milky Way to Nearby Low-Luminosity Galactic Nuclei*, 18-22 April 2006, Bad Honnef, Germany
- *Galaxies in the Cosmic Web*, 15-19 May 2006, New Mexico, USA
- Fourth Harvard-Smithsonian Conference on Theoretical Astrophysics *The History of Nuclear Black Holes in Galaxies*, 15-18 May 2006, Cambridge, MA, USA
- *Making the most of the Great Observatories*, 22-24 May 2006, Pasadena, CA, USA
- *Studying Galaxy Evolution with Spitzer and Herschel*, 28 May - 2 June 2006, Agios Nikolaos, Crete, Creece
- *Galaxy Evolution from Large Surveys*, 29 May - 18 June 2006, Aspen, CO, USA
- *Deconstructing the Local Group: Dissecting Galaxy Formation in our own Backyard*, 12 June - 2 July 2006, Aspen, CO, USA
- *The Metal Rich Universe*, Monday, 12 June 2006 - Friday, 16 June 2006, Los Cancajos, La Palma, Canary Islands, Spain
- *Blazars: Disk-jet Connection. Observations and theories*, 12-17 June 2006, Nauchny, Crimea, Ukraine
- *The First Stars and Evolution of the Early Universe*, 19-21 June 2006, Seattle, WA, USA
- *Fine-Tuning Stellar Population Models*, 26-30 June 2006, Leiden, Netherlands
- *Mapping the Galaxy and Nearby Galaxies*, 26-30 June 2006, Ishigaki, Okinawa, Japan
- *Physics and Astrophysics of Supermassive Black Holes*, 9-14 July 2006, Santa Fe, NM, USA
- CRAL Conference Series. I. *Chemodynamics: from First Stars to Local Galaxies*, 10-14 July 2006, Lyon, France
- *Galaxy Redshift Surveys of the Future*, 20 July 2006, Portsmouth, UK
- *Gravitational Lensing*, 31 July - 5 August 2006, Leiden, Netherlands
- *Heating vs. Cooling in Galaxies and Clusters of Galaxies*, 6-11 August 2006, Garching, Germany
- *The Role of Black Holes in Galaxy Formation and Evolution*, 10-13 September 2006, Potsdam, Germany
- EARA Workshop 2006 *Supernovae and their host galaxies*, 2-3 October 2006, Paris, France

- *Applications Of Gravitational Lensing: Unique Insights Into Galaxy Formation And Evolution*, 3-6 October 2006, Santa Barbara, CA, USA
- *Galaxy Mergers: From the Local Universe to the Red Sequence*, 4-6 October 2006, Baltimore, MD, USA
- *Radiation Backgrounds from the First Stars, Galaxies and Black Holes*, 9-11 October 2006, College Park, MD, USA
- *The Central Engine of Active Galactic Nuclei*, 16-21 October 2006, Xi'an, China
- *From Stars to Galaxies: Building the Pieces to Build up the Universe*, 16-20 October 2006, Venice, Italy
- *Massive Galaxies over Cosmic Time. II*, 1-3 November 2006, Tucson, AZ, USA
- *Extragalactic Surveys:A Chandra Science Workshop*, 6-8 November 2006, Cambridge, MA, USA
- 37th Saas-Fee Advanced Course *The Origin of the Galaxy and Local Group*, 4-10 March 2007, Murren, Switzerland
- *Extragalactic surveys with LOFAR*, 6-8 March 2007, Leiden, Netherlands
- *The Origin of Galaxies: Exploring Galaxy Evolution with the New Generation of Infrared-Millimetre Facilities*, 24-29 March 2007, Obergurgl, Austria
- *A New Zeal for Old Galaxies*, 25-30 March 2007, Rotorua, New Zealand
- *ANGLES School on Gravitational Lens Modelling*, 27-30 March 2007, Valencia, Spain
- New Quests in Stellar Astrophysics. II. *The Ultraviolet Properties of Evolved Stellar Populations*, 16-20 April 2007, Puerto Vallarta, Mexico
- *The Nuclear Region, Host Galaxy and Environment of Active Galaxies*, 18-20 April 2007, Huatulco, Oax., Mexico
- *Black Holes*, 23-26 April 2007, Baltimore, MD, USA
- *Pathways Through an Eclectic Universe*, 23-27 April 2007, Playa de La Arena, Tenerife, Spain
- *Black Holes VI*, 12-15 May 2007, White Point, Nova Scotia, Canada
- *Structure Formation in the Universe: Galaxies, Stars, Planets*, 27 May - 1 June 2007, Chamonix, France
- *Obscured AGN across Cosmic Time*, 5-7 June 2007, Bavaria, Germany
- *HI Survival through Cosmic Times*, 11-15 June 2007, Sarteano (Siena), Italy
- *Searching for Strong Lenses in Large Imaging Surveys*, 14-15 June 2007, Batavia, IL, USA
- *Tracing Cosmic Evolution with Clusters of Galaxies: Six Years Later*, 25-29 June 2007, Sesto Pusteria, Italy
- *Galaxy Interactions and Mergers*, 4-5 July 2007, Nottingham, UK
- *Galaxies in the Local Volume*, 8-13 July 2007, Sydney, Australia
- *From IRAS to HERSCHEL/PLANCK: cosmology with infrared and submillimetre surveys*, 9-11 July 2007, London, UK
- *Galaxy Growth in a Dark Universe*, 16-20 July 2007, Heidelberg, Germany
- *Dynamics of Galaxies*, 6-10 August 2007, Saint Petersburg, Russia
- *Star Formation, Then and Now*, 13-17 August 2007, Santa Barbara, CA, USA
- *The Globular Clusters - Dwarf Galaxies Connection*, 27-29 August 2007, Ann Arbor, MI, USA
- *X-rays from Nearby Galaxies*, 5-7 September 2007, Madrid, Spain
- MPA/ESO/MPE/USM 2007 Joint Astronomy Conference *Gas Accretion and Star Formation in Galaxies*, 10-14 September 2007, Garching, Germany
- *Next Generation of Computational Models of Baryonic Physics in Galaxy Formation: From Protostellar Cores to Disk Galaxy Formation*, 17-21 September 2007, Zurich, Switzerland

- *Formation and Evolution of Galaxy Disks*, 1-5 October 2007, Rome, Italy
- *Spectroscopy in Cosmology and Galaxy Evolution 2005 - 2015*, 3-5 October 2007, Granada, Spain
- *Indian Conference on Cosmology and Galaxy Formation*, 3-5 November 2007, Allahabad, India
- *ELSA school on the Science of Gaia*, 19-28 November 2007, Leiden, Netherlands
- *Galaxy and Black Hole Evolution: Towards a Unified View*, 28-30 November 2007, Tucson, AZ, USA
- *The Evolving Insterstellar Medium in the Milky Way and Nearby Galaxies*, 2-5 December 2007, Pasadena, CA, USA
- 1st Subaru International Conference *Panoramic Views of Galaxy Formation and Evolution*, 11-15 December 2007, Hayama, Japan
- *Surveys and Simulations of Large-Scale Structure: A Celebration of Marc Davis' 60th Birthday*, 16-18 January 2008, Berkeley, CA, USA
- *Galaxy Evolution from Mass-Selected Samples*, 28 January - 1 February 2008, Leiden, Netherlands
- *The Evolution of Galaxies through the Neutral Hydrogen Window*, 1-3 February 2008, Arecibo, Puerto Rico
- *The Dark Side of the Universe through Extragalactic Gravitational Lensing*, 4-8 February 2008, Leiden, Netherlands
- *Observational Evidence for Black Holes in the Universe*, 10-15 February 2008, Kolkata, India
- *The First Two Billion Years of Galaxy Formation: The Reionization Epoch and Beyond*, 11-15 February 2008, Aspen, CO, USA
- *EARA Workshop: Herschel Promises for Galaxy Evolution Studies*, Monday, 18 February 2008 - Tuesday, 19 February 2008, Paris, France
- *Galactic & Stellar Dynamics in the Era of High Resolution Surveys*, 16-20 March 2008, Strasbourg, France
- *Galactic Structure and the Structure of Galaxies*, 17-21 March 2008, Ensenada, Baja California, Mexico
- *Deep Surveys of the Radio Universe with SKA Pathfinders*, 31 March - 4 April 2008, Perth, Australia
- *An XXL Extragalactic Survey: Prospects for the XMM Next Decade*, 14-16 April 2008, Paris, France
- *Galaxy Formation and Evolution as Revealed by Cosmic Gas*, 17-19 April 2008, Irvine, CA, USA
- *Blazars Variability across the Electromagnetic Spectrum*, 22-25 April 2008, Paris, France
- *Cosmic Web: Galaxies and the Large-Scale Structure*, 16-17 May 2008, Socorro, NM, USA
- OXFORD-COSMOCT Workshop on *The Interface between Galaxy Formation and AGN*, 19-22 May 2008, Island of Vulcano, Messina, Italy
- *The Central Kiloparsec: Active Galactic Nuclei and Their Hosts*, 4-6 June 2008, Ierapetra, Crete, Greece
- *Gas and Stars in Galaxies:A Multi-Wavelength 3D Perspective*, 10-13 June 2008, Garching, Germany
- *Tidal Dwarf Galaxies: Ghosts from Structure Formation*, 15-20 June 2008, Bad Honnef, Germany
- *Merging Black Holes in Galaxies: Galaxy Evolution, AGN and Gravitational Waves*, 15-20 June 2008, Sydney, Australia

3. Review articles

- Athanassoula, E. 2008, in: M. Bureau, E. Athanassoula & B. Barbuy (eds.), *Formation and Evolution of Galactic Bulges*, Proc. IAU Symposium No. 245 (Cambridge: CUP), p. 93, *Boxy/Peanut and Discy Bulges: Formation, Evolution and Properties*
- Bregman, J. 2007, *ARAA*, 45, 221, *The Search for the Missing Baryons at Low Redshift*
- Brodie, J. P., & Strader, J. 2006, *ARAA*, 44, 193, *Extragalactic Globular Clusters and Galaxy Formation*
- Combes, F. 2008, in: M. Bureau, E. Athanassoula & B. Barbuy (eds.), *Formation and Evolution of Galactic Bulges*, Proc. IAU Symposium No. 245 (Cambridge: CUP), p. 151, *Gaseous Flows in Galaxies*
- Fabbiano, G. 2006, *ARAA* 44, 323, *Populations of X-Ray Sources in Galaxies*
- Gadotti, D. A. 2008, in: G. Contopoulos, & P. A. Patsis (eds.) *Chaos in Astronomy*, in press [arXiv:0802.0495], *Barred Galaxies: An Observer's Perspective*
- Li, A. 2007, in: L. C. Ho & J.-M. Wang (eds.), *The Central Engine of Active Galactic Nuclei*, ASP-CS 373, 561, *Dust in Active Galactic Nuclei*
- Mayer, L., Governato, F., & Kaufmann, T. 2008, *Advanced Science Letters*, in press [arXiv:0801.3845], *The Formation of Disk Galaxies in Computer Simulations*
- McNamara, B. R., & Nulsen, P. E. J. 2007, *ARAA*, 45, 117, *Heating Hot Atmospheres with Active Galactic Nuclei*
- Omont, A. 2007, *Rep. Prog. Phys.*, 70, 1099, *Molecules in Galaxies*
- Renzini, A. 2006, *ARAA*, 44, 141, *Stellar Population Diagnostics of Elliptical Galaxy Formation*

4. Closing remarks

Commission 28 is more lively than ever. The era of large galaxy surveys, such as the SDSS, and the search for statistical samples at higher and higher redshifts, have triggered wide modelisation efforts and numerical simulations, to tackle galaxy formation and evolution. At the IAU XXVI General Assembly in Prague, 2006, the Symposia organised by Commission 28 had by far the widest attendance. Out of the 40 IAU Commissions, Commission 28 has about 850 members, i.e. nearly 10% of all members.

With the prospective of future space or ground instruments, such as ALMA, ELTs or SKA, the progress in the domain is expected to grow at even higher rate, ensuring an exciting future.

Françoise Combes
president of the Commission

References

Bouwens, R. & Illingworth, G. D. 2006, *Nature 443*, 189
Bower, R. G., Benson, A. J., Malbon, R., *et al.* 2006, *MNRAS*, 370, 645
Calzetti D., Kennicutt, R. C., Engelbracht, C. W., *et al.* 2007, *ApJ*, 666, 870
Cappellari, M., Bacon, R., Bureau, M., *et al.* 2006, *MNRAS*, 366, 1126
Cimatti, A., Cassata, P., Pozzetti, L., *et al.* 2008, *A&A*, 482, 21
Cox, T. J., Jonsson, P., Primack, J. R., & Somerville, R. S. 2006, *MNRAS*, 373, 1013
Daddi, E., Alexander, D. M., Dickinson, M., *et al.* 2007, *ApJ*, 670, 173
Dekel A. & Birnboim Y. 2006, *MNRAS*, 368, 2
Di Matteo, T., Colberg, J., Springel, V., *et al.* 2008, *ApJ*, 676, 33

Driver, S. P., Allen, P. D., Graham, A. W., *et al.* 2006, *MNRAS*, 368, 414

Faber, S. M., Willmer, C. N. A., Wolf, C., *et al.* 2007, *ApJ*, 665, 265

Flores, H., Hammer, F., Puech, M., *et al.* 2006, *A&A*, 455, 107

Förster-Schreiber, N. M., Genzel, R., Lehnert, M. D., *et al.* 2006, *ApJ*, 645, 1062

Gil de Paz, A., Madore, B. F., Boissier, S., *et al.* 2007, *ApJ*, 661, 115

Governato, F., Willman, B., Mayer, L., *et al.* 2007, *MNRAS*, 374, 1479

Jogee, S., Barazza, F. D., Rix, H.-W., *et al.* 2004, *ApJ* (Letters), 615, L105

Kennicutt, R., Calzetti, D., Walter, F., *et al.* 2007, *ApJ*, 671, 333

Moore, B., Diemand, J., Madau, P., *et al.* 2006, *MNRAS*, 368, 563

Noeske K. G., Weiner B. J., Faber S. M., *et al.* 2007, *ApJ* (Letters), 660, L43

Poggianti, B. M., von der Linden, A., De Lucia, G., *et al.* 2006, *ApJ*, 642, 188

Scoville, N., Aussel, H., Benson, A., *et al.* 2007, *ApJS* 172, 150

Sheth, K., Elmegreen, D. M., Elmegreen, B. G., *et al.* 2008, *ApJ*, 675, 1141

Strigari, L. E., Bullock, J. S., & Kaplinghat, M. 2007, *ApJ*, 669, 676

Tacconi, L. J., Neri, R., Chapman, S. C., *et al.* 2006, *ApJ*, 640, 228

Thilker D.A., Bianchi L., Meurer G., *et al.* 2007 *ApJS, 173*, 538

Vanzella, E., Cristiani, S., Dickinson, M., *et al.* 2008, *A&A*, 478, 83

Transactions IAU, Volume XXVIIA
Reports on Astronomy 2006–2009
Karel A. van der Hucht, ed.

© 2009 International Astronomical Union
doi:10.1017/S1743921308025726

DIVISION VIII / WORKING GROUP
SUPERNOVA

CO-CHAIRS	Wolfgang Hillebrandt and Brian P. Schmidt
MEMBERS	Edward Baron, Stefano Benetti, Sergey I. Blinnikov,
	Sergey I. Blinnikov, David R. Branch,
	Enrico Cappellaro, Alexei V. Filippenko,
	Claes Fransson, Peter M. Garnavich,
	Daniel W. E. Green, Ariel M. Goobar, Mario Hamuy,
	Peter H. Hauschildt, Robert P. Kirshner,
	Bruno Leibundgut, Daniel J. Lennon, Eric J. Lentz,
	Peter Lundqvist, Robert Mc Graw, Paolo A. Mazzali,
	W. Peter S. Meikle, Anthony Mezzacappa,
	Jens C. Niemeyer, Ken'ichi Nomoto, Reynald Pain,
	Nino Panagia, Ferdinando Patat, Mark M. Phillips,
	Elena Pian, Guiliano Pignata, Philipp Podsiadlowski,
	María Pilar Ruiz-Lapuente, Elaine M. Sadler,
	Brian P. Schmidt, Peter O. Shull, Jason Spyromilio,
	Nicholas B. Suntzeff, Friedrich-Karl Thielemann,
	Christopher Tout, Virginia L. Trimble,
	James W. Truran, Dmitry Yu. Tsvetkov,
	Massimo Turatto, Massimo della Valle,
	Schuyler Van Dyk, Wolfgang H. Voges,
	Nicholas A. Walton, Lifan Wang, J. Craig Wheeler,
	Kurt W. Weiler, Patricia A. Whitelock,
	Stanford E. Woosley, Hitoshi Yamaoka, Gang Zhao

TRIENNIAL REPORT 2006 - 2009

1. Introduction

The Supernova Working Group was re-established at the IAU XXV General Assembly in Sydney, 21 July 2003, sponsored by Commissions 28 (*Galaxies*) and 47 (*Cosmology*). Here we report on some of its activities since 2005.

A first WG on supernovae was founded by Fritz Zwicky to coordinate and provide uniform standards for photographic supernova searches in the era of the Palomar 48″ and other Schmidt telescopes. It died with him in 1974. Commission 28 (*Galaxies*) re-established a WG-SN (with Virginia Trimble as chair) at the 1982 Patras IAU General Assembly, largely because of concerns that the few events then being discovered (mostly serendipitously) were not receiving enough follow-up observations to be very information. Among other things, the WG had some small part in preparing the astronomical community to take full advantage of SN 1987A. This event made astronomy so supernova-conscious that the WG hardly seemed needed, and it died a natural death at the 1991 IAU General Assembly.

The Supernova Working Group was re-established at the IAU XXV General Assembly in Sydney, 21 July 2003, sponsored by Commissions 28 (*Galaxies*) and 47 (*Cosmology*). Here we report on some of its activities since 2005.

The wish to request permission of the IAU Executive Committee to re-establish a Supernova Working Group arose from the large number of discoveries (up to several 100 per year), some at very large distances, being made by more than half a dozen teams in the past few years. This was beginning to lead to competition for telescope time for follow-up work which would be unnecessary if the scientists involved had a platform for information exchange. LOTOSS was taking up speed. The Nearby Supernova Factory, the Carnegie Supernova Project, ESSENCE, and other projects to discover and observe SNe Ia at low and intermediate redshifts had left the planning stage in Europe, Japan, and the USA and were thought also to benefit from better coordination. Also, special space probes (*SNAP*) were being discussed aiming at the discovery of hundreds of supernovae at all redshifts alone. Finally, considerable progress in modeling the various types of events made it desirable also to have theoretical results, including explosion models, synthetic light curves, and predicted spectra, easily accessible.

Therefore, the aim of the new WG-SN was (*a*) to take care of coordinating observational activities, including searches and surveys (e.g., SCP, High-z Team, GOODS, ESSENCE, LOTOSS, CSP, RTN/ESC, SDSS, Pan-STARRS, etc.); (*b*) to advocate suitable follow-ups; (*c*) to establish ways to archive SN data and to make them available to the community; (*d*) to edit an electronic newsletter complementary to the IAU Circulars (the GCN network for γ-ray bursters being a possible model); and (*e*) to be the interface with the coming virtual observatory initiatives (AVO, NVO, ASTRO-Grid, GAVO...) for data storage and handling.

2. Tasks of the Supernova Working Group

At an inaugural meeting (which was part of the Division VIII/Commission 28 Business Meeting in Sydney) the objectives of the WG were discussed and approved. In particular it was agreed

• to coordinate the nearby searches and the follow-up observations;

• to launch a community effort for a Treasury or Large *HST* programme to observe type Ia supernovae in the UV;

• to investigate the possibility of having a robotic (2m-class) telescope devoted to supernova observations; and

• to find ways to save supernova data in some form of database, freely accessible to the community.

In addition, it was suggested that the WG should take an active role in the discovery and observation reports of supernovae, presently done via IAU Circulars, which involves some unwanted delay. Here the problem is, however, that rapid communication in an 'unfiltered' way increases the risk of 'false alarms'. A second problem is going to be how to cope with 100's of supernovae per year. A possible solution could be that IAU Circulars could continue to be the place for 'confirmed' discoveries, preceded by rapid communications in a web-based platform.

Finally, Wolfgang Hillebrandt (Garching, Germany) and Brian Schmidt (Mt. Stromlo, Australia) were elected as co-chairs of the new *Supernova* Working Group.

3. Activities of the Supernova Working Group since 2005

Here we summarize briefly some of the WG-SN activities in the past three years.

3.1. *Coordinated observations of nearby type Ia supernovae*

A new attempt was made to observe a sample of (nearby) type Ia supernovae with *HST* in the UV in Cycle 17 after the service mission SM4 scheduled for October 2008. Two proposals were submitted with the goal to improve substantially our physical knowledge of SNe Ia and their calibration as cosmological distance probes. One of these proposals was approved (PI: R. Ellis), together with with about 15 other proposals dealing with various aspects of both, thermonuclear and core-collapse supernovae and their progenitors and environment.

A new on-line supernova spectrum archive (SUSPECT) at the University of Oklahoma and is organized and maintained by David Branch and Eddie Baron. It is steadily extending and at present, it contains photometric and/or spectroscopic data of 175 supernovae of various types as well as links to other archives. Similar efforts are underway at LBNL which in the future should include model data also.

3.2. *Conferences and meetings*

In the past years the WG-SN took an active role in organizing several supernova conferences and workshops.

A Joint Discussion on *Supernovae: One Millennium after 1006* at the IAU XXVI General Assembly in Prague to celebrate the SN 1006 millennium by reviewing recent progress in understanding supernovae, their remnants, and their application to cosmology. In a stimulating day and a half there were 25 (mostly invited) oral papers, as well as some two dozen posters touching on many observational and theoretical aspects of supernova research. The oral papers focused primarily on type Ia supernovae and their remnants, including observations of SN1006 itself and its more recent cousins, models for SN Ia progenitors, explosion mechanisms, and how they interact with the interstellar and circumstellar medium on the way to becoming remnants, through the application of SN Ia to cosmology.

There was as SNAP workshop at Berkeley in September 2005, with the main emphasis on the use of type Ia supernovae as probes for cosmology. A conference at Cefalu on Sicily in 2006 dealt with the *Multicolour Landscape of Compact Objects and their Explosive Origins* and another supernova conference at the same site is coming up in the fall of 2008. There was an extended (type Ia) supernova program at the KITP, Santa Barbara, USA, from January through May, 2007, and several conferences and workshops devoted to SN 1987A (*20 years after ...*).

4. Future activities: Supernova data bases and related issues

Setting-up and maintaining supernova data bases will continue to be in the focus of the WG-SN. In the future it will become even more important to combine observational data with detailed model predictions, and virtual observatory initiatives (AVO, NVO, ASTRO-Grid, GAVO ...) can provide the platform for data storage and handling needed for this purpose. Whether or not the WG-SN can play a significant role in these efforts will have to be seen.

Wolfgang Hillebrandt & Brian P. Schmidt
co-chairs of the Working Group

Transactions IAU, Volume XXVIIA
Reports on Astronomy 2006–2009
Karel A. van der Hucht, ed.

© 2009 International Astronomical Union
doi:10.1017/S1743921308025738

DIVISION IX   OPTICAL AND INFRARED TECHNIQUES

TECHNIQUES OPTIQUES ET INFRAROUGES

Division IX provides a forum for astronomers engaged in the planning, development, construction, and calibration of optical and infrared telescopes and instrumentation, as well as observational procedures including data processing.

PRESIDENT	**Rolf-Peter Kudritzki**
VICE-PRESIDENT	**Andreas Quirrenbach**
PAST PRESIDENT	**Christiaan L. Sterken**
BOARD	**Michael G. Burton, Xiangqun Cui,**
	Martin Cullum, Michel Dennefeld,
	Peter Martinez, Guy S. Perrin,
	Andrei A. Tokovinin, Guillermo Torres,
	Stephane Udry

DIVISION IX COMMISSIONS

Commission 25	**Stellar Photometry and Polarimetry**
Commission 30	**Radial Velocities**
Commission 54	**Optical and Infrared Interferometry**

DIVISION IX WORKING GROUPS

Division IX WG	**Detectors**
Division IX WG	**Site Testing Instruments**
Division IX WG	**Sky Surveys**

INTER-DIVISION WORKING GROUP

Division IX-X WG	**Encouraging the International Development of Antarctic Astronomy**

TRIENNIAL REPORT 2006 - 2009

1. Introduction

Division IX is currently in a state of transition regarding its structure and constituency, reflecting larger shifts in the way most observational astronomers carry out their work. While traditionally most optical astronomers used local telescopes and built their own instruments suited for their scientific interests, today large national and international facilities as well as space observatories dominate many fields. In this context questions revolving around the planning for future facilities and their instrumentation, access to existing observatories, and optimization of their use are becoming important topics for Division IX.

A related question concerns the education of instrumentation specialists, and the standing of instrumentalists within the astronomical community. All too often astronomy departments and observatories report that they cannot find qualified candidates for open positions that require strong technical skills, yet gifted instrument builders complain that they get short-changed in promotions and tenure review processes in a culture that uses publication and citation indices as the main criteria for academic achievement. Division IX can clearly play an important role here to maintain excellence in astronomical instrumentation and techniques, thus serving the whole astronomical community.

In response to these general considerations, Division IX has started to change its Commission and Working Group structure, following discussions in the Board, and deliberations at the 2006 IAU XXVI General Assembly in Prague. Commission 9, with a name identical to that of the whole Division was dissolved, and the successful Working Group on *Optical and Infrared Interferometry* elevated to Commission status (now Commission 54). At the same time the process of establishing new Working Groups was started by creating one on site testing instruments; discussions for several additional working groups are underway that could be established in the 2009-2012 time frame; in the long run these could become Commissions of Division IX if they prove to be successful.

2. Developments within the past triennium

2.1. *Commission 25 Photometry and Polarimetry*

Commission 25 currently has 230 members from 40 countries. The Commission's membership represents 2.4% of the total IAU membership. During the triennium 2006-2008 the Commission devoted a considerable amount of effort to updating its membership records. This was done by contacting members individually and requesting them to confirm or update their personal details. Many of the members' details were outdated or incorrect. The updated lists will be sent to the IAU Secretariat in Paris.

In 2007, the Commission's web site was moved from its previous host site at the Vrije Universiteit Brussel to the South African Astronomical Observatory. The URL of the new Commission 25 web site is `<iauc_25.saao.ac.za>`. We thank Christiaan Sterken for having established the Commission's web site and for having maintained it for a number of years. The Commission plans to change the web site to bring it more in line with the 'look and feel' of the official IAU web site.

The Commission has started a long-term project to develop an IAU standard star database. This is motivated by the various large-scale surveys that are generating a flood of new standard stars and standard star observations. The Commission believes that some form of coordination among the various initiatives would be helpful to the astronomical community at large. The idea is to develop a web site that would be a one-stop location for all standard star data access. It would be a data depository and a gateway to the various recommended data servers via internet access. Such a database would be quality-controlled by a group of experts, so that observers could easily choose suitable and reliable standard stars that they need to use for their observations. The envisaged database should also be compatible with the Virtual Observatory (VO) data standards in order to take full advantage of the suite of data manipulating tools developed by the International VO Alliance.

During the triennium, there were a number of important developments and conferences in the fields of polarimetry and polarimetry. Some of the highlights are described in the report of Commission 25 following this chapter.

2.2. *Commission 30 Radial Velocities*

Radial velocity observations have seen tremendous developments during the past few years driven by the need of reaching higher and higher precision to detect planets of always smaller masses, or to be applied to stars of special properties. Five years of measurements with the HARPS spectrograph on the ESO 3.6 m telescope have demonstrated that a long-term stability below 1 m/s is indeed possible for solar-type stars. These high-precision measurements are now revealing the existence of a large population of Neptune-mass planets and super-Earths around solar-type stars in our neighborhood. Important progress has also been made for earlier type dwarfs (A-F), for which the high rotation rate and the smaller number of available spectral lines were considered for a long time as a hard limit for the radial-velocity precision. New dedicated techniques to extract the Doppler information are now providing good velocities for those stars, as small as a few tens of m/s for rotation rates at the 100 km/s level, on high S/N high-resolution spectra.

Several large scale consultations have been initiated by space agencies (mainly NASA and ESA) or by the community itself to plan in an efficient way their efforts for the coming decade in the field of exoplanets (e.g., the NASA *Exoplanet Task Force*; the ESA EP-RAT; the *Exoplanet Forum*; the *Blue Dot Team*). Published reports make all a strong point on intensifying ground-based radial-velocity programs, in preparation for (good target list), or complementary (e.g., for transit candidates) to future space missions. The volume of activities related to radial velocities in the domain of extrasolar planets will thus still increase in the coming years, accompanied by new instrumental developments on large facilities.

Administrative activities of Commission 30 have been rather quiet during the past three years, mainly dedicated to support conferences related to the interests of the commission (exoplanet meetings, dynamics in the Galaxy, etc.). The Working Groups in the Commission (*Stellar Radial Velocity Bibliography*; *Spectroscopic Binary Catalog SB9*; *Radial-Velocity Standards*) are progressing at their own pace, engaged in long-term efforts.

2.3. *Commission 54 Optical and Infrared Interferometry*

The Commission was created at the IAU XXVI General Assembly held in Prague, August 2006. Its goal is to coordinate international collaborations on scientific and technical matters relating to long-baseline optical and infrared interferometry. The Commission continues the work begun through the Working Group on *Optical/IR Interferometry*. As a Commission within Division IX its focus is to establish scientific and technical standards that facilitate the future growth of the field.

The work of the Commission takes place primarily within the Commission's working groups. The current active working groups are:
- *Interferometry Data Format*
- *Imaging Algorithms*
- *Calibrator Stars*
- *Advances in Astronomy with Interferometry*
- *Intensity Interferometry*

The first two Working Groups were created prior to the Commission. The Working Group *Interferometry Data Format* has established the OPTICAL INTERFEROMETRY DATA EXCHANGE FORMAT (published in 2005) and its supporting software. Some work remains to update this format consistently with the needs of future instruments and software both in Europe (VLTI second generation) and in the USA (CHARA, MROI, NPOI,

KeckI). The Working Group *Imaging Algorithms* expands the original aim of optical/IR interferometry imaging contests (held in June 2004 and May 2006) towards the goal of developing and disseminating these software tools in the community. A new contest was successfully organized in June 2008. A lot of progress has been made in this field and this Working Group activity fosters exchanges and comparisons between groups.

The Working Group *Calibrator Stars* is now active. A web page has been created and discussions are taking place among its members. Methods to obtain suitable calibrators and a list of bad stars for calibrations are being put together.

The Working Group *Advances in Interferometry* has undertaken several actions among which the writing of science briefs to advertise for results obtained with interferometers (28 are available at `<olbin.jpl.nasa.gov/iau/briefs/>`). Other actions focus on active lobbying to get interferometry talks in relevant conferences or including stellar diameter measurements in the list of measurements of stars provided to astronomers by SIMBAD.

The Working Group *Intensity Interferometry* has been created recently. This is a newly re-born field since the end of the Intensity Interferometer in Narrabri in the 1970s. This restart is motivated by the advent of fast detectors with an increased bandwidth used by ground-based γ-ray observatories. The first goal of this Working Group is to write a white paper on the technique and discuss how it can complement amplitude interferometry.

Two other Working Groups are being discussed but have not been started yet:
- *Future Large Arrays*
- *Vademecum of Interferometry*

The Commission promotes the use and scientific impact of interferometry through collaborations with individual science Commissions within the IAU, most significantly with Commission 8 *Astrometry*, Commission 26 *Double and Multiple Stars*, Commission 27 *Variable Stars*, and Commission 36 *Theory of Stellar Atmospheres*.

The web site of Commission 54 `<olbin.jpl.nasa.gov/iau/index.html>` is hosted at the Optical Long Baseline Interferometry News (OLBIN). This web site and its associated Email Forum exist to further the interests of the optical interferometry community and goals of Commission 54.

The Commission currently has 14 members. However, a total of 68 IAU members have indicated their intent to become members. In addition 13 non-IAU members are applying for membership to join the Commission. After the next General Assembly, Commission 54 should be 80 members strong.

2.4. *Division IX Working Group Detectors*

Upon dissolution of Commission 9, the Working Group on *detectors* changed its status from a Commission 9 Working Group to a Division IX WG. It provides information on detectors (mainly CCDs) to the interested community. However, as optical and infrared detector technologies have matured substantially over the past years, many parallel information channels are now available. The future role of the Working Group *Detectors* will therefore have to be re-defined over the coming months.

2.5. *Division IX Working Group Site-Testing Instruments*

The Working Group has been organized following the decision of the IAU Division IX taken in 2006. A web site has been implemented where information on various techniques and their correct usage is assembled; see `<www.ctio.noao.edu/science/iauSite/>`. This is a distributed effort, with different techniques being covered by different groups or individuals world-wide. However, certain techniques are not yet covered by the abovementioned web resource. A list of registered participants contains 38 WG members (not all of them are IAU members). Two major conferences on the subject took place in Kona and San Pedro Martir in 2007, and a workshop in Sardinia as well as a SPIE meeting in 2008.

2.6. *Division IX Working Group Sky Surveys*

The Working Group *Sky Surveys* was also a Commission 9 WG before that Commission was disbanded. Now directly under Division IX, this working group maintains information on sky surveys, catalogs, astronomical data bases, and on conferences on these topics. With the advent of very large survey projects, the role of this Working Group in the coming years will have to be re-defined to reflect advances in the ways astronomers use the resulting very large data bases.

2.7. *Inter-Division IX-X Working Group Encouraging the International Development of Antarctic Astronomy*

The Inter-Division IX-X Working Group *Encouraging the International Development of Antarctic Astronomy* monitors technical activities characterizing various Antarctic sites for astronomical use, including the South Pole, Dome A, Dome C, and Dome F. In addition, it maintains contacts between the astronomical community and other agencies and scientific bodies involved in Antarctic research. In 2008, the IAU became a member of SCAR, the Scientific Committee for Antarctic Research.

3. Closing remarks

Although the restructuring and rejuvenation of Division IX is showing first results, much remains to be done in the coming years. With a new generation of extremely large telescopes on the horizon, space observatories serving large communities, and massive data bases created by digital sky surveys, the environment in which optical and infrared astronomy is conducted keeps changing profoundly. Division IX will have to put a structure into place that supports these changes and helps astronomers world wide to meet the many technological challenges.

Andreas Quirrenbach
vice-president of the Division

Transactions IAU, Volume XXVIIA
Reports on Astronomy 2006–2009
Karel A. van der Hucht, ed.

© 2009 International Astronomical Union
doi:10.1017/S174392130802574X

COMMISSION 25

STELLAR PHOTOMETRY AND POLARIMETRY

PHOTOMÉTRIE ET POLARIMÉTRIE STELLAIRE

PRESIDENT	Peter Martinez
VICE-PRESIDENT	Eugene F. Milone
PAST PRESIDENT	Arlo U. Landolt
ORGANIZING COMMITTEE	Carme Jordi, Aleksey V. Mironov, Shengbang Qian, Edward G. Schmidt, Christiaan L. Sterken

COMMISSION 25 WORKING GROUP

Div. IX / Commission 25 WG	Infrared Working Group

TRIENNIAL REPORT 2006 - 2009

1. Introduction

According to the IAU membership database, Commission 25 currently has 230 members from 40 countries. The Commission's membership represents 2.4% of the total IAU membership of 9658.

This report presents a small selection of the work of many members of the Commission during the triennium 2006 - 2008. During the current triennium considerable effort was devoted to updating the Commission's membership records. This was done by contacting members individually and requesting them to confirm or update their personal details. Many of the members' details were outdated or incorrect. The updated lists will be sent to the IAU Secretariat in Paris.

In 2007, the Commission's website was moved from its previous host site at the Vrije Universiteit Brussel to the South African Astronomical Observatory. The URL of the new Commission 25 website is `<iauc_25.saao.ac.za>`. We thank Chris Sterken for having established the Commission's website and for having maintained it for a number of years. The Commission plans to change the website to bring it more in line with the 'look and feel' of the official IAU website.

In January 2008 the community of photometrists was saddened by the news that Theo Walraven had passed away. He was a staff member, lecturer and professor at the Leiden Observatory from the 1940s until 1980. He began his career with Anton Pannekoek in Amsterdam working on variable stars and contributed in a fundamental way to our insight into variable stars (together with Paul Ledoux, with whom he wrote the famous article in the *Handbuch der Physik*). He also helped understand the synchrotron radiation in the Crab nebula (with Jan Oort). Highly talented in instrumentation, he developed the 'Walraven photometer' with its corresponding 'Walraven photometric system' on the 0.75 m 'Light collector' at the Leiden Southern Station, near Hartebespoortdam in South

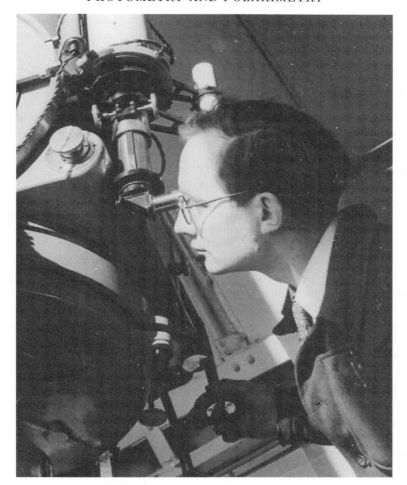

Figure 1. Theodore (Fjeda) Walraven at the 50 cm Zunderman Telescope of the Leiden Observatory, sometime in the 1950s, where he did his extensive work on RR Lyrae stars, published in the *BAN*. Image courtesy of Leiden Observatory.

Africa. From the 1960s onward he and his wife Johanna, who was his close collaborator, lived with their family for most of the time at the Leiden Southern Station, with short visits to Leiden for teaching and developing new instrumentation. The instrument and telescope were moved to ESO in Chile in 1978/79, and the photometric system was in use until some quite some time after the decommissioning of the 'Walraven photometer' in 1991.

2. Photometry

2.1. *Photometric standards and calibration*

Work on photometric standards and calibration, which underpins all photometric applications, proceeded during the triennium under review. Many issues of standardization were discussed in the conference reviewed in more detail in Section 5 below. That conference also provided updates on the Strömvil system and the Walraven system.

Landolt (2007) published $UBVRI$ photoelectric observations of 109 stars around the sky, centered more or less at $-50°$ declination. The majority of the stars fall in the magnitude range $10.4 < V < 15.5$ and in the colour index range $-0.33 < (B-V) < +1.66$.

Rodgers *et al.* (2006) commenced a project to expand the DDO standard star list by integrating Landolt's equatorial standard star fields. The new primary DDO standard stars include the principal $UBVR_CI_C$ stars in the equatorial SA fields, with fainter stars becoming secondary standards for that field. They published preliminary results for the primary standard stars in SA fields 92 through 106, with magnitudes between $9.0 < M_{48} < 16.0$.

Smith *et al.* (2007) reported at the 2007 AAS meeting their progress on an extension of the $u'g'r'i'z'$ standard star network developed to support calibration of the Sloan Digital Sky Survey (SDSS). There were some limitations in the original system due to tailoring the network to support the survey goals, namely a lack of fainter and redder stars. Their latest work includes both fainter and redder stars. Clem *et al.* (2007) presented a new set of secondary standard stars for the $u'g'r'i'z'$ photometric system established in selected open and globular star clusters. These standards are calibrated to the original standard system defined by Smith *et al.* with an accuracy of 1% or better, extend as faint as $r' \sim 20$, and are concentrated in small regions of the sky. As a result, they can serve as viable calibrators of photometry obtained on large-aperture telescopes.

Leggett *et al.* (2006) presented near-infrared JHK photometry of 115 stars. Of these, 79 are faint standard stars drawn from the UKIRT Faint Standards and the Las Campanas standards (but see Milone & Young (2007) and references therein for comments about such 'standardization'). The average brightness of the sample in all three bandpasses is 11.5 mag, with a range between 10 and 15.

Kilkenny *et al.* (2007) presented homogeneous and standardized $UBV(RI)_C JHK$ photometry for over 100 M-type stars selected on the basis of apparent photometric constancy. Most of the stars have a substantial number of $UBV(RI)_C$ observations and could prove useful as red supplementary standards.

2.2. *Photometric studies of the near-Earth space environment*

Photometric techniques have also been used to study space debris, an issue of growing concern to users of near-Earth space. Several papers were presented at the *Advanced Maui Optical and Space Surveillance Technologies Conferences,* held in Wailea, Maui (HI, USA), 10-14 September 2006 and 12-15 September 2007.

A satellite's spectrum has two components: reflected sunlight and blackbody radiation. Hamilton (2006) discussed the use of the extended Strömgren system, $uvby$H$_\beta$Ca, as a self-calibrating method of subtracting out reflected sunlight. While the blackbody radiation contribution is weak, it does not have the H$_\beta$ and Ca II H and K absorption features. Hamilton argues that the signal-to-noise ratio of the blackbody radiation to the solar radiation may be detectable, and that this information may be used for satellite characterization. If so, this would provide a complement to radar-based studies of satellites in the near-Earth orbital environment.

Seitzer *et al.* (2007) presented results of a 14-night study at CTIO to detect and follow-up space debris in geostationary orbit. In this project one telescope was dedicated to survey operations, while a second telescope was used for follow-up observations for orbits and colors. The goal was to obtain orbital and photometric information on every faint object found with the survey telescope. Their survey concentrated on objects fainter than $R = 15$ mag, reaching a limiting R magnitude of 18.0 for a S/N of 10.

2.3. *New space-based and ground-based instruments*

A number of new instruments, both ground-based and space-based, were commissioned during the triennium under review. The list of instruments mentioned here is indicative of the current developments, and is not meant to be a comprehensive review.

Canada's *MOST* (Microvariability and Oscillations of STars) satellite is currently the smallest dedicated space-based photometer in operation. It has been collecting photometric time-series data for asteroseismology of main sequence stars and planetary transit observations since 2003. Huber *et al.* (2008) present *MOST* time series photometry of the rapid oscillations in the roAp star 10 Aql showing the presence of three pulsation frequencies in this star. Miller-Ricci *et al.* (2008) present transit observations of the exoplanet system HD 189733. Another space photometry mission that entered operation during the triennium is *CoRoT*, launched in December 2006. Barge *et al.* (2008) report the detection of the first planet discovered by *CoRoT*. The planet orbits a mildly metal-poor G0V star of magnitude $V = 13.6$ in 1.5 d.

EPOXI is a NASA Discovery Program Mission of Opportunity using the *Deep Impact* fly-by spacecraft. The EPOCh (Extrasolar Planet Observation and Characterization) is a 30-cm visible imager being used to gather photometric time series of known transiting exoplanet systems from January through August 2008. Christiansen *et al.* (2008) present an overview of the project and some preliminary results from observations of six systems.

Earth-bound telescopes are constrained by the diurnal cycle to observe targets for only a fraction of each day, except at the poles, where it is possible to acquire long and continuous photometric time series. Strassmeier *et al.* (2008) present stellar time-series optical photometry from Dome C in Antarctica and analyze approximately 13 000 CCD frames taken in July 2007. They conclude that Dome C in Antarctica can be used successfully for uninterrupted time-domain studies of long duration.

New ground-based instruments under development include the ASTRA spectrophotometer (Adelman *et al.* 2007). The soon-to-be-completed ASTRA spectrophotometer uses an automated 0.50-m telescope with a spectrograph that has a CCD as a detector and produces data with bandpasses of 7 or 14 Å. Once it becomes operational this instrument is expected to produce high-quality fluxes through at least the $\lambda\lambda$ 3300-9000 Å region for application to a wide variety of astrophysical problems.

McGraw *et al.* (2006) are working towards the development of the CCD/Transit Instrument with Innovative Instrumentation (CTI-II). This is a 1.8 m stationary, meridian-pointing telescope capable of millimagnitude photometry and milliarcsecond astrometry. It will also introduce new capabilities in faint object detection and characterization for low Earth orbit and geosynchronous transfer orbit satellites.

Another large ground-based instrument under development is the Advanced Liquid-mirror Probe of Astrophysics, Cosmology and Asteroids (ALPACA), an 8 m optical telescope destined for Cerro Tololo and designed to scan a strip of sky passing overhead and extending over 1000 square degrees (Crotts 2006). The imaging survey will be conducted in five photometric bands covering the optical region. It will allow photometric discrimination of many source types, including supernovae and asteroids, as well as photometric redshift determinations for galaxies and supernovae. The ALPACA is intended to operate over at least three years and is expected reach a cumulative point-source detection limit of about 28 mag at 10 sigma. ALPACA will deliver nightly photometry for many classes of variable and moving objects.

2.4. *Large-scale surveys*

A number of large-scale surveys announced major data releases during the period under review. The Sloan Digital Sky Survey announced its 5th data release (Adelman-McCarthy *et al.* 2007). It includes five-band photometric data for 217 million objects selected over 8000 square degrees. The COSMOS survey announced its 1st data release (Capak *et al.* 2007). They have published imaging data and photometry for the COSMOS survey in 15 photometric bands between 0.3 and 2.4 μm. These include data taken on the Subaru 8.3 m telescope, the KPNO and CTIO 4 m telescopes, and the CFHT 3.6 m telescope. Special techniques are used to ensure that the relative photometric calibration is better than 1% across the field of view. The absolute photometric accuracy from standard-star measurements is found to be 6%.

3. Infrared Working Group

A separate report for the WG-IR can be found in a subsequent chapter of this volume, so the report of the working group here will be brief. The work of the WG-IR during the current triennium was to promote further the benefits of the WG-IR filters as described by Young, Milone, & Stagg in 1994, and demonstrated in Milone & Young (2005). The increased signal-to-noise characteristics and lower extinction coefficients of the corresponding IRWG passbands were further elaborated in presentations to the meeting on standardization held at Blankenberge, Belgium, in May, 2006, and to a meeting of three amateur astronomy organizations (RASC, AAVSO, and ALPO) at Calgary, Canada in July, 2007. The written forms of these presentations can be found in Milone & Young (2007, 2008). The most important tasks at present are to expand the list of standards for the near-infrared part of this system (*iz*, *iJ*, *iH*, and *iK*) to more and fainter stars. This is a priority because of increasing interest by amateur astronomers in the infrared, a desire to obtain the increased variable-star precision that the IR offers, and because these (and perhaps the *iN* and *in* passbands in the relatively clean longer-wavelength 10 μ-m window) can permit meaningful data to be obtained at lower elevation sites.

To date, the filters needed to test the *iM*, *iN*, *in*, and *iQ* passbands have yet to be manufactured because of increasing filter manufacturing costs. Moreover, the *iL* and *iL'* passbands have not yet been tested because of technical problems at the one site where we have had sufficient time to test these filters, the Rothney Astrophysical Observatory. Photometrists with a keen interest in these developments, and who have access to facilities where these tests can be carried out, are urged to contact the chair of the working group.

4. Polarimetry

4.1. *Polarimetric studies - some highlights*

Significant progress has been made measuring magnetic fields in Herbig Ae and Be and other types of stars during the last four years. Polarimetry is used as an important tool for studying debris disks and disks around young stars.

Over the past few years polarimetric observations (e.g., Andersson & Potter 2007) have shown the importance of radiative alignment of dust grains in the interstellar medium. Consequently, more theoretical work has been carried out in this area (e.g., Hoang & Lazarian 2008; Lazarian & Hoang 2008).

Polarization of comets, asteroids and Trans-Neptunian objects as a function of phase angle reveals many properties of surfaces and dust grains. Berdyugina *et al.* (2008)

reported the first detection of polarized scattered light from an exoplanetary atmosphere for the short-period transiting planet orbiting HD 189733b.

Polarimetry, and in particular spectropolarimetry, provides key information about the structure of the inner regions in Seyfert and other active galaxies. Finally, CMB polarimetry, with recent *WMAP* results, the upcoming *Planck* mission and other planned future missions should provide a wealth of information on the early-Universe physics.

Landi Degl'Innocenti *et al.* (2007) presented a review of the current situation regarding polarimetric standardisation. Fossati *et al.* (2007) presented a consolidated analysis of the observations of standard stars for linear polarization obtained from 1999 to 2005 within the context of the calibration plan of the FORS1 instrument of the ESO VLT.

4.2. *New polarimetric instruments*

The South African Astronomical Observatory reports the completion of a new two-channel, High-speed Photo-POlarimeter (HIPPO) for the 1.9 m optical telescope of the South African Astronomical Observatory. The instrument makes use of rapidly counter-rotating (10 Hz), super-achromatic half- and quarter-waveplates, a fixed Glan-Thompson beam splitter and two photomultiplier tubes that record the modulated O and E beams. Each modulated beam permits an independent measurement of the polarisation and therefore simultaneous two-filter observations. All Stokes parameters are recorded every 0.1 sec and photometry every 1 millisecond. Post-binning of data is possible in order to improve the signal. This is ideal for applications such as measuring the rapid variability of the optical polarisation from magnetic Cataclysmic Variable stars. First light was obtained in February 2008.

The prime new optical application is direct imaging of exoplanets (which will be linearly polarized several tens of percent) against the (speckled) residual seeing image of the parent star (which, if within a few tens of parsec, will generally be polarized less than 0.1%). For a properly designed exoplanet polarimeter, this gives an advantage over photometry of a factor of order 1000. Various projects are underway and preliminary designs and concepts were presented at the 2008 SPIE conference in France and the Astronomical Polarimetry 2008 conference (see Section 5) in Canada. Examples are Planetpol (Hough *et al.* 2006), the Extreme Polarimeter (ExPo) (Jeffers *et al.* 2007), and the Gemini Planet Imager (Graham *et al.* 2007).

As the new large telescopes often have Nasmyth foci, the classical prescription for obtaining accurate optical and IR polarimetry needs to be modified. Tinbergen (2007) recently published a tutorial paper in PASP on the principles of Nasmyth focus polarimetry needed for such instruments.

5. Conferences

In May 2006, a meeting on *The Future of Photometric, Spectrophotometric and Polarimetric Standardization* was held in Blankenberge, Belgium. The meeting demonstrated convincingly the ongoing need for careful calibration for all ground-based as well as space-based projects that involve photometric measurement. Organized by former Commission 25 and Division IX president Chris Sterken, the meeting contained descriptions of many contemporary projects and the plans to calibrate instruments. Among those described were such surveys as the LSST, SDSS, Pan-STARRS, SkyMapper Southern Sky Survey, Dark Energy Survey, SNAP, and projects such as *Gaia*, CHORIZOS, Omega-CAM and the ASTRA spectrophotometer. Although most of these projects involved the visible region, the UV and IR were not neglected, with the former represented at least down to the Lyα line in *HST* measurements, and an entire session devoted to the infrared. In addition to the calibration of instruments and a full discussion of absolute calibration, a number of photometric systems were described. These included the classic *UBVRI*, the Walraven *VBLUW*, *u'g'r'i'z'*, and the *IRWG* and Mauna Kea IR systems. Another entire session was devoted to polarimetry. Chris Sterken provided some telling examples of the need for calibration and standardization, both at the beginning and at the end of the meeting. The proceedings (Sterken 2007) were published in the ASP Conference Series and were dedicated to Arlo Landolt, former President of Division IX and of Commission 25, in recognition of his life's work of setting standards in photometry.

Pierre Bastien (University of Montreal) reports that the conference *Astronomical Polarimetry 2008 – Science from Small to Large Telescopes*, took place from 6 to 11 July 2008 at the Fairmont Le Manoir Richelieu, La Malbaie, in the wonderful Charlevoix region by the St-Lawrence River in Québec, Canada. The meeting attracted 97 keen participants and covered many topics. Refer to <mars.astro.umontreal.ca/astropol2008/> for a full list. This website also includes a text describing the scientific impact of polarimetry on theoretical astrophysics.

A whole day was spent on instrumentation for the optical and near-IR, far-IR and sub-millimetre spectral regions. Other sessions dealt with interstellar matter and molecular clouds, the galactic plane and star formation, stellar magnetic fields, circumstellar matter, the solar system and exoplanets, supernovae and extragalactic polarimetry, and CMB polarimetry. Some instrumentation highlights include an instrument that can measure polarization to 1 part in 10^6 and is now in use for exoplanets, imaging polarimetry with adaptive optics, high speed photo-polarimetry and a polarimeter for use with SCUBA-2 on the JCMT.

The need for faint polarized standard stars for large telescopes was raised. It was suggested to leave this to individual astronomers but that an effort should be made to collate all results in one place. Commission 25 was mentioned as an appropriate place to do so. The production of a database of optical stellar polarization data is currently in progress. All those interested in participating with observations are encouraged to contact Antonio Mario Magalhães, Univ. of Sao Paulo, Brazil. Polarimetry for large telescopes was a main topic of discussion. For some future telescopes, it is already too late to modify planned designs but in other cases, it is still possible to have an effect. For polarimetry, it is much better to design telescopes and instrumentation with the goal of doing polarimetry initially rather than producing add-ons to do polarimetry afterwards. A mission concept study for a future millimeter-wave satellite dedicated to CMB polarimetry was funded by NASA. Its use by astronomers working in other fields is strongly encouraged. This satellite would also be a perfect instrument for a polarization survey of the whole foreground sky. The proceedings of the Astronomical Polarimetry 2008 conference will be published in the ASP Conference Series.

As for future conferences, members of the Commission 25 are preparing a session on *Photometry: Past and Present* for the January 2009 meeting of the American Astronomical Society's History of Astronomy Division at Long Beach, California. The next astronomical polarimetry conference is planned to take place in 2012 in Nice, France.

6. Ongoing and future work of Commission 25

The various large-scale surveys are generating a flood of new standard stars and standard star observations. The Commission believes that some form of coordination among the various initiatives would be helpful to the astronomical community at large. The

Commission is thus planning to develop an IAU standard star database. The final product will be a one-stop location for all standard star data access. It would be a data depository and a gateway to the various recommended data servers via internet access. Such a database would be quality-controlled by a group of experts, so that observers could easily choose suitable and reliable standard stars that they need to use for their observations. The envisaged database should also be compatible with the Virtual Observatory (VO) data standards in order to take full advantage of the suite of data manipulating tools developed by the International VO Alliance.

Apart from providing reliable standard star data, the envisaged data base could also provide information on how to use those standards properly. It has been suggested that the Commission could also compile educational material in the form of a 'cookbook' on methods of doing and using photometry. In this regard the proceedings of the conference on *The Future of Photometric, Spectrophotometric and Polarimetric Standardization*, held in Blankenberge, Belgium in May 2006, would be particularly useful. A working group will take these ideas further at the next General Assembly. In keeping with this, it has been suggested that the Commission could consider organising a symposium on standardization topics across the electromagnetic spectrum.

With regard to polarimetry, the need for faint polarized standard stars for large telescopes was raised at the Astronomical Polarimetry 2008 meeting. It was suggested that Commission 25 would be an appropriate forum to collate all the results in one place. The OC of Commission 25 takes note of this suggestion and will liaise with the organisers of the conference to follow up on the recommendation.

Acknowledgements

P. Martinez would like to thank all the members of the Organising Committee of Commission 25 for their support during the triennium 2006 - 2009. He is also very grateful to Drs P. Bastien, S.B. Potter and J. Tinbergen for their inputs on recent developments in polarimetry, and to all Commission 25 OC colleagues who contributed to the compilation of this report. This research has made use of NASA's Astrophysics Data System.

Peter Martinez
president of the Commission

References

Adelman, S. J., Gulliver, A. F., Smalley, B., *et al.* 2007, in: C. Sterken (ed.), *The Future of Photometric, Spectrophotometric and Polarimetric Standardization ASP-CS*, 364, 255
Adelman-McCarthy, J. K., Agüeros, M. A., Allam, S. S., *et al.* 2007, *ApJS*, 172, 634
Andersson, B.-G. & Potter, S.B. 2007, *ApJ*, 665, 369
Barge, P., Baglin, A., Auvergne, M., Rauer, H., *et al.* 2008, *A&A*, 482, 17
Berdyugina, S. V., Berdyugin, A. V., Fluri, D. M., & Piirola, V. 2008, *ApJ*, 673, 83
Capak, P., Aussel, H., Ajiki, M., McCracken, H. J., *et al.* 2007, *ApJS*, 172, 99
Clem, J. L., Vanden Berg, D. A., & Stetson, P. B. 2007, *AJ*, 134, 1890
Christiansen, J. L., Charbonneau, D., A'Hearn, M. F., *et al.* 2008, in: F. Pont, D. Queloz & D. D. Sasselov (eds.), *Transiting Planets* (Cambridge: CUP), in press [arXiv:0807.2852v1]
Crotts, A. P., Consortium ALPACA, 2006, *BAAS*, 38, 1041
Fossati, L., Bagnulo, S., Mason, E., & Landi Degl'Innocenti, E. 2007, *ASP-CS*, 364, 503
Graham, J. R., Macintosh, B., Doyon, R., Gavel, D., *et al.* 2007, in press [2007arXiv0704.1454] (White paper submitted to the NSF-NASA-DOE Astronomy and Astrophysics Advisory Committee ExoPlanet Task Force)

Hamilton, N. 2006, in: S. Ryan (ed.), Proc. *Advanced Maui Optical and Space Surveillance Technologies Conference*, The Maui Economic Development Board, p. E80

Hoang, T. & Lazarian, A. 2008, *MNRAS*, 388, 117

Hough, J. H., Lucas, P. W., Bailey, J. A., Tamura, M, *et al.* 2006, *PASP*, 118, 1302

Huber, D., Saio, H., Gruberbauer, M., Weiss, W. W., *et al.* 2008, *A&A*, 483, 239

Jeffers, S. V., Keller, C. U., Rodenhuis, M., Miesen, N. 2007, in: P. Kalas (ed.), Proc. conf. *In the Spirit of Bernard Lyot: The Direct Detection of Planets and Circumstellar Disks in the 21st Century* (Berkeley: UC), p. 42

Kilkenny, D., Koen, C., van Wyk, F., Marang, F., & Cooper, D. 2007, *MNRAS*, 380, 1261

Landi Degl'Innocenti, E., Bagnulo, S., & Fossati, L. 2007, in: C. Sterken (ed.) *The Future of Photometric, Spectrophotometric and Polarimetric Standardization*, ASP-CS, 364, 495

Landolt, A. U. 2007, *AJ*, 133, 2502

Lazarian, A., & Hoang, T. 2008, *ApJ*, 676, 25

Leggett, S. K., Currie, M. J., Varricatt, W. P., Hawarden, T. G., *et al.* 2006, *MNRAS*, 373, 781

McGraw, J., Ackermann, M., Williams, T., *et al.* 2006, in: S. Ryan (ed.), Proc. *Advanced Maui Optical and Space Surveillance Technologies Conference,* The Maui Economic Development Board, p. E6

Miller-Ricci, E., Rowe, J. F., Sasselov, D., Matthews, J. M., *et al.* 2008, *ApJ*, 682, 593

Milone, E. F., & Young, A. T. 2005, *PASP*, 117, 485

Milone, E. F., & Young, A. T. 2007, in: C. Sterken (ed.) *The Future of Photometric, Spectrophotometric and Polarimetric Standardization*, ASP-CS, 364, 387

Milone, E. F., & Young, A. T. 2008, *JAAVSO*, 136, in press

Rodgers, C. T., Canterna, R., Allen, D., Hausel, E., & Smith, J. A. 2006, *BAAS*, 38, 1042

Seitzer, P., Abercromby, K., Barker, E., & Rodriguez, H. 2007, in: S. Ryan (ed.), Proc. *Advanced Maui Optical and Space Surveillance Technologies Conference*, The Maui Economic Development Board, p. E37

Smith, J. A., Allam, S. S., Tucker, D. L., & Fornal, B. 2007, AAS Meeting No. 211, Abstract No. 132.14

Sterken, C. (ed.) 2007, *The Future of Photometric, Spectrophotometric and Polarimetric Standardization*, ASP-CS, 364

Strassmeier, K. G., Briguglio, R., Granzer, T., *et al.* 2008, *A&A*, 490, 287

Tinbergen, J. 2007, *PASP*, 119, 1371

Transactions IAU, Volume XXVIIA
Reports on Astronomy 2006–2009
Karel A. van der Hucht, ed.

© 2009 International Astronomical Union
doi:10.1017/S1743921308025751

DIVISION IX / COMMISSION 25 / WORKING GROUP
INFRARED ASTRONOMY

CHAIR	Eugene F. Milone
VICE-CHAIR	Andrew T. Young
MEMBERS	Eva Bauwens, Roger A. Bell,
	Michael S. Bessell, Martin Cohen,
	Robert Garrison, Ian S. Glass,
	John A. Graham, Arne A. Henden,
	Herman Hensberge, Lynne A. Hillenbrand,
	Steve B. Howell, Mark R. Kidger,
	Robert L. Kurucz, Arlo U. Landolt,
	Ian S. McLean, Matthew Mountain,
	George H. Rieke, Stephen J. Schiller,
	Douglas A. Simons, Michael F. Skrutskie,
	C. Russell Stagg, Christiaan L. Sterken,
	Roger I. Thompson, Alan T. Tokunaga,
	Kevin Volk

TRIENNIAL REPORT 2006 - 2009

1. Introduction

As we have noted before, the WG-IR was created following a Joint Commission Meeting at the IAU General Assembly in Baltimore in 1988, a meeting that provided both diagnosis and prescription for the perceived ailments of infrared photometry at the time. The results were summarized in Milone (1989). The challenges involve how to explain the failure to systematically achieve the milli-magnitude precision expected of infrared photometry and an apparent 3% limit on system transformability. The proposed solution was to re-define the broadband Johnson system, the passbands of which had proven so unsatisfactory that over time effectively different systems proliferated although bearing the same *JHKLMNQ* designations; the new system needed to be better positioned and centered in the atmospheric windows of the Earth's atmosphere, and the variable water vapour content of the atmosphere needed to be measured in real time to better correct for atmospheric extinction.

The WG-IR was formalized by Ian McLean, then president of Commission 25, at the Buenos Aires IAU General Assembly in 1991, and Milone formally appointed to the chair. A subcommittee had been formed almost immediately in 1988 to look at ways to implement the recommendations put forward in Milone (1989). It established the procedure and criteria for judging the performance of existing infrared passbands and began experimenting with passband shapes, widths, and placements within the spectral windows of the Earth's atmosphere. The method and coding were initiated and largely carried out by Andy Young, with Milone running the simulations and Stagg assisting with profiles.

By 1993, preliminary recommendations were presented at the photometry meeting in Dublin (Young, Milone, & Stagg 1993). The full details of the criteria and results of the numerical simulations were presented by Young, Milone, & Stagg (1994). Subsequent work, described in WG-IR and/or Commission 25 reports, included the use of a new MODTRAN version (3.7) to check and extend previous work. This part of the program proved so successful in minimizing the effects of water vapour on the source flux transmitted through the passband that the second stage, real-time monitoring of IR extinction, was not pursued, although this procedure remains desirable for unoptimized passbands designed for specific astrophysical purposes.

During the 2003 - 2006 triennium, the WG concentrated on gathering and presenting evidence of the usefulness of the WG-IR infrared passband set. For the near infrared portion of the WG-IR set (namely iz, iJ, iH, iK), field trials were conducted over the years 1999 - 2003 with the 1.8 m telescope at the Rothney Astrophysical Observatory of the University of Calgary. The results of those trials and the details of further work done to that date were presented in Milone & Young (2005). This paper contained, for the first time, evidence that not only were the WG-IR passbands more useful to secure precise transformations than all previous passbands, but that they were also superior in at least one measure of the signal to noise ratio. This evidence was further refined in Milone & Young (2007). As a consequence, the original purpose of the WG-IR largely has been achieved but opposition to the new passband system is still strong, and passbands that somewhat compromise the WG-IR recommendations have been advanced in order to provide more throughput, at the cost of precision and standardization. As a consequence, nonoptimized passbands are still in use at the highest altitude infrared sites. The situation is described (and decried!) in Milone & Young (2007).

2. Developments within the past triennium

The WG-IR has the had the policy of being open to input from its members at all times following the initial consultations with all segments of the infrared community. We continue this policy (Milone & Young 2007).

Thanks to previous efforts, the *need* for improved IR passbands has been accepted by the community, by far and large. The work of Salas, Cruz-González, & Tapia (2006) in making use of Padé approximants to simulate atmospheric extinction as they define a new set of unoptimized intermediate IR passbands illustrates this well. In the near infrared, the work of Simons & Tokunaga (2002) and Tokunaga, Simons, & Vacca (2002) typify the new attitude; the work by this group was summarized recently by Tokunaga & Vacca (2007).

Perhaps the most important aspect of the WG-IR's work is the promise that the near-IR WG-IR passband set holds for highly precise photometry at intermediate and even low elevation sites.

To encourage the increasingly active amateur community to consider the use of infrared passbands, and to encourage manufacturers to produce the WG-IR filters at affordable prices, we gave a presentation on the benefit of the new passbands for variable star photomery at the joint meeting of the AAVSO and the RASC in Calgary in July, 2007. The written form of that presentation is Milone & Young (2008). It stresses the achievement of smaller extinction coefficients and better signal/noise ratio with the recommended passbands, in real observational data (as well as in simulations) than is achieved with other passbands that are in current use.

3. Closing remarks

The future activities require fabrication and field testing of the remainder of the WG-IR passband set, namely the passbands iL, iL', iM, iN, in, and iQ. It is also desirable to extend the list of near infrared standard stars presented in Milone & Young (2005) to a fuller all-season set, and to extend them to fainter stars, as, for example, Landolt (1992) (and in earlier papers cited therein) has done for the visual Johnson-Cousins passbands.

Eugene F. Milone
chair of the Working Group

References

Landolt, A. 1992, *AJ*, 104, 340

Milone 1989, in: E. F. Milone (ed.), *Infrared Extinction and Standardization*, Proc. Two Sessions of IAU Commissions 25 and 9, Baltimore, MD, USA, 4 August 1988, *Lecture Notes in Physics*, 341, 1

Milone, E. F. & Young, A. T. 2005, *PASP*, 117, 485

Milone, E. F. & Young, A. T. 2007, in: C. Sterken (ed.), *The Future of Photometric, Spectrophotometric, and Polarimetric Standardization*, ASP-CS, 364, 387

Milone, E. F. & Young, A. T. 2008, *JRASC*, 36, in press

Salas, L., Cruz-González, I., & Tapia, M. 2006, *RMAA*, 42, 273

Simons, D. A. & Tokunaga, A. T. 2002, *PASP*, 114, 169

Tokunaga, A. T. & Vacca, W.D. 2007, in: C. Sterken (ed.), *The Future of Photometric, Spectrophotometric, and Polarimetric Standardization*, ASP-CS, 364, 409

Tokunaga, A. T., Simons, D. A., & Vacca, W. D. 2002, *PASP*, 114, 180

Young, A. T., Milone, E. F., & Stagg, C. R. 1993, in: C. J. Butler & I. Elliott (eds.), *Stellar Photometry – Current Techniques and Future Developments*, Proc. IAU Coll. No. 136 (Cambridge: CUP), p. 235

Young, A. T., Milone, E. F., & Stagg, C. R. 1994, *A&AS*, 105, 259

Transactions IAU, Volume XXVIIA
Reports on Astronomy 2006–2009
Karel A. van der Hucht, ed.

© 2009 International Astronomical Union
doi:10.1017/S1743921308025763

COMMISSION 30

RADIAL VELOCITIES
VITESSES RADIALES

PRESIDENT	Stephane Udry
VICE-PRESIDENT	Guillermo Torres
PAST PRESIDENT	Birgitta Nordström
ORGANIZING COMMITTEE	Francis C. Fekel, Kenneth C. Freeman, Elena V. Glushkova, Geoffrey W. Marcy, Birgitta Nordström, Robert D. Mathieu, Dimitri Pourbaix, Catherine Turon, Tomaz Zwitter

COMMISSION 30 WORKING GROUPS

Div. IX / Commission 30 WG	Radial-Velocity Standard Stars
Div. IX / Commission 30 WG	Stellar Radial Velocity Bibliography
Div. IX / Commission 30 WG	Catalogue of Orbital Elements of Spectroscopic Binaries

TRIENNIAL REPORT 2006 - 2009

1. Introduction

This three-year period has seen considerable activity in the Commission, with a wide range of applications of radial velocities as well as a significant push toward higher precision. The latter has been driven in large part by the exciting research on extrasolar planets. This field is now on the verge of detecting Earth-mass bodies around nearby stars, as demonstrated by recent work summarized below, and radial velocities continue to play a central role.

This is not to say that classical applications of RVs have lagged behind. On the contrary, this triennium has seen the release of several very large data sets of stellar radial velocities (Galactic and extragalactic) that are sure to have a significant impact on a number of fields for years to come. The era of mass-producing radial velocities has arrived. Examples include the Geneva-Copenhagen Survey, the Sloan Digital Sky Survey, and RAVE, and are described below.

Due to circumstances beyond our control, the report of Commission 30 for the previous (2003 - 2006) triennium did not appear in the printed version of the IAU Transactions XXVIA, although it did appear in the electronic version. For progress during the previous period, the reader is therefore encouraged to consult the latter, which is also available from the Commission web site.

2. Radial velocities and exoplanets (Guillermo Torres & John A. Johnson)
2.1. *Toward Earth-mass planets*

Detections of Jupiter-mass exoplanets by the radial-velocity method relying on measurements with precisions of a few $m\,s^{-1}$ are now quite routine. This technique has provided by far the majority of the more than 300 planet discoveries to date. The persistence of astronomers and the increasing precision of their instruments has led to larger and larger numbers of multi-planet systems being found. One example is the interesting case of μ Arae (Pepe *et al.* 2007), with *four* planets, one of which is as small as $10.5\,\mathrm{M_{Earth}}$. The host star also presents the signature of p-mode oscillations seen clearly in the radial velocities.

The record-holder for the most planets is the star 55 Cnc, which is orbited by no less than *five* planets (Fischer *et al.* 2008), of which the smallest has a minimum mass of $10.8\,\mathrm{M_{Earth}}$. Exciting discoveries during this period made possible by the high precision and stability of the HARPS instrument on the ESO 3.6 m telescope at La Silla (Chile) include the system Gls 581 (at only 6.3 pc), attended by at least three planets. In addition to the previously known Neptune-mass body orbiting the star with a period $P = 5.3$ d, two other low mass planets were found by Udry *et al.* (2007) with minimum masses of only $5\,\mathrm{M_{Earth}}$ ($P = 12.9$ d) and $7.7\,\mathrm{M_{Earth}}$, the latter being near the outer edge of the habitable zone of the M3V parent star ($P = 83.6$ days). The three-planet orbital solution for this case has an rms residual of only $1.2\,\mathrm{m\,s^{-1}}$.

Another system with three low-mass planets was announced by Mayor *et al.* (2008) around the nearby (13 pc) metal-poor K2V star HD 40307. The planets weighed in at 4.2, 6.9, and $9.2\,\mathrm{M_{Earth}}$, and the three-planet Keplerian orbital fit gave impressive residuals of just $0.85\,\mathrm{m\,s^{-1}}$. HARPS has demonstrated that this sort of velocity precision is achievable for 'quiet' stars that present a low level of 'jitter' in their radial velocities due to astrophysical phenomena such as p-mode oscillations, granulation, or chromospheric activity. Indications are that Neptune-mass or smaller planets are more common around solar-type (F-K) stars than previously thought (see, e.g., Mayor & Udry 2008).

2.2. *Retired A-type stars and their planets*

Most Doppler searches for planets have concentrated on main sequence stars of spectral types F or later, because the velocity precision for earlier type stars is seriously compromised by line broadening induced by rapid rotation, as well as the overall fewer number of spectral lines available. This difficulty in studying higher-mass stars introduces a bias in our understanding of planets, but it can be overcome by looking at such stars after they have left the main sequence. This is precisely the approach of an ongoing project to investigate the relationship between stellar mass and planet formation by using the HIRES instrument on the Keck 10 m telescope to search for planets in a sample of 240 intermediate-mass subgiants ($1.3 < M_*/\mathrm{M_\odot} < 2.2$). Subgiants have lower surface temperatures and rotational velocities than their main-sequence progenitors, making them ideal proxies for A- and F-type stars in Doppler studies. From a smaller sample of subgiants observed previously by Johnson *et al.* (2007a) at Lick Observatory for 4 years with a typical velocity precision of $4\,\mathrm{m\,s^{-1}}$, a strong correlation was detected between stellar mass and planet occurrence, with a detection rate of 9% within 2.5 AU among the high-mass sample, compared to 4.5% for Sun-like stars and less than 2% for M dwarfs. A paucity of planets within 1 AU of stars with masses greater than $1.5\,\mathrm{M_\odot}$ was found, indicating that stellar mass also plays a key role in planet migration (Johnson *et al.* 2007b, Johnson *et al.* 2008). The goal of the expanded Keck survey (with an increased velocity precision of about $2\,\mathrm{m\,s^{-1}}$) is to map out the relationships between stellar mass and exoplanet properties in greater detail by examining the distribution of planetary minimum masses, eccentricities, semimajor axes, and the rate of multiplicity around evolved A stars. If the 9% occurrence rate is confirmed, some 20 to 30 new planets should be found in the sample orbiting some of the most massive stars so far examined by the Doppler technique.

2.3. *Current status and prospects*

In May of 2008 NASA convened the Exoplanet Forum 2008, a meeting of experts from the US and other countries in eight different observational techniques related to exoplanet research. The purpose was to discuss paths forward for exploring and characterizing planets around other stars, and to provide specific suggestions for space missions, technology development, and observing programmes that could fulfill the recommendations of a previously held meeting of the Exoplanet Task Force (`<www.nsf.gov/mps/ast/exoptf.jsp>`). The reports resulting from these meetings are intended to provide input for consideration by various advisory committees in the USA, and in particular by the *Astronomy and Astrophysics Decadal Survey* that is currently underway.

Radial velocities was one of the eight techniques considered by the Exoplanet Forum 2008. The corresponding chapter of the report, available at `<exep.jpl.nasa.gov/exep_exfCommunityReport.cfm>`, summarized the progress in the field over the last few years, which is illustrated by a velocity precision of 1 m s^{-1} or slightly better achieved so far, led by the Swiss team using the HARPS instrument on the ESO 3.6 m telescope, and the California-Carnegie team using the HIRES instrument on the Keck 10 m telescope. The factors currently limiting the precision were discussed briefly and have been described in detail by Pepe & Lovis (2007). They include various sources of astrophysical noise (stellar oscillations, granulation, magnetic cycles, collectively known as 'stellar jitter'), guiding, the illumination of the spectrograph, and the wavelength reference. Good progress has been made in each of these areas. For example, it appears that jitter can be substantially reduced through longer exposures or binning, to the level of perhaps 10 cm s^{-1} or less. A new thorium-argon line list was developed by Lovis & Pepe (2007) that significantly improves the velocity precision when using this source as the wavelength reference. Further improvements in the velocity precision perhaps reaching a few cm s^{-1} appear possible using a dense spectrum of lines generated by a femtosecond-pulsed laser ('laser comb'), described in more detail below. The next few years will tell whether this promise can be realized in practice.

The report of the Exoplanet Forum also described recent progress in techniques to measure precise velocities in the near infrared (see, e.g., Ramsey *et al.* 2008), which are now approaching the 10 m s^{-1} level in initial tests. Longer wavelengths potentially provide a significant advantage for the Doppler detection of very small (even Earth-mass) planets, since these objects produce a larger signal when orbiting less massive stars, which emit most of their flux in the near infrared.

In addition to velocity precision, the report pointed out what is currently considered by the community to be the greatest challenge for making progress in the detection of exoplanets by the Doppler technique: the limited access to telescope time. This has a direct impact not only on the size of the samples of solar-type stars that can be studied, but also severely restricts the number of late-type (faint) stars that can be targeted to search for Earth-mass planets. The need for exposure times longer than dictated by Poisson statistics to reduce stellar jitter, as mentioned above, is a further strain on the limited resources currently available on telescopes equipped with high-precision spectrographs.

3. Toward higher radial velocity precision (Guillermo Torres)

During this period agreement has been reached for the construction of an improved copy of the very successful HARPS spectrograph, currently in operation at the ESO 3.6 m telescope at La Silla, for the northern hemisphere (HARPS-NEF). This is a high-resolution ($R \simeq 120\,000$) fibre-fed optical spectrograph with broad wavelength coverage

(3780 - 6910 Å) designed for high radial velocity precision. HARPS-NEF is a collaboration between the New Earths Facility (NEF) scientists of the Harvard Origins of Life Initiative and the HARPS team at the Geneva Observatory. It is expected to be the workhorse for follow-up of transiting planet candidates for NASA's *Kepler* mission, and should be operational perhaps in late 2010.

HARPS-NEF is a cross-dispersed echelle spectrograph that will benefit not only from updates and improvements over the original HARPS instrument, but in addition it will be installed on a larger telescope aperture in the northern hemisphere (the 4.2 m William Herschel Telescope on La Palma, Canary Islands). It is designed for ultra-high stability (10-20 cm s^{-1}), and like HARPS it will be placed in a vacuum chamber with careful temperature control.

One of the key factors that determine the precision of the RVs is the wavelength reference. Existing technologies in the optical (such as the Th-Ar technique and iodine gas absorption cell) have already reached sub-m s^{-1} precision in some cases, but further improvements are needed if the Doppler method is to reach cm s^{-1}, as is needed to detect terrestrial-mass planets. A new technology that has emerged in the last few years and that holds great promise for providing a very stable reference is that of laser 'frequency combs'. As the name suggests, a frequency comb generated from mode-locked femtosecond-pulsed lasers provides a spectrum of very narrow emission lines with a constant frequency separation given by the pulse repetition frequency, typically 1 GHz for this application. This frequency can be synchronized with an extremely precise reference such as an atomic clock. For example, using the generally available Global Positioning System (GPS), the frequencies of comb lines have long-term fractional stability and accuracy of better than 10^{-12}.

This is more than enough to measure velocity variations at a photon-limited precision level of 1 cm s^{-1} in astronomical objects (see, e.g., Murphy *et al.* 2007). This direct link with GPS as the reference allows the comparison of measurements not only between different instruments, but potentially also over long periods of time. To provide lines with separations that are well matched to the resolving powers of commonly used echelle spectrographs, a recent improvement incorporates a Fabry-Pérot filtering cavity that increases the comb line spacing to ~ 40 GHz over a range greater than 1000 Å (Li *et al.* 2008). Prototypes using a titanium-doped sapphire solid-state laser have been built that provide a reference centered around 8500 Å. In practice, of course, Doppler measurements are also affected by other instrumental problems, so that the value of this new technology for highly precise RV measurements is still to be demonstrated. Tests have been initiated during this triennium.

Plans call also for the installation of a laser comb on the HARPS-NEF spectrograph described above. Applications of this technique are not limited to stars. For example, a direct measurement of the expansion of the Universe could be made by observing *in real time* the evolution of the cosmological redshift of distant objects such as quasars. Such a measurement would require a precision in determining Doppler drifts of ~ 1 cm s^{-1}/yr (see, e.g., Steinmetz *et al.* 2008), which a laser comb can in principle deliver.

4. Radial velocities and asteroseismology (Guillermo Torres)

The significant increase in the precision of velocity measurements over the past few years, driven by exoplanet searches, has enabled important studies of the internal constitution of stars through the technique of asteroseismology. A number of spectrographs now reach the precision needed for this type of investigation. During this period Bedding *et al.* (2006) observed the metal-poor subgiant star ν Ind with the UCLES instrument on the 3.9 m Anglo-Australian Telescope, and with the CORALIE spectrograph on the 1.2 m Swiss telescope at ESO. The precision of those measurements ranged from 5.9 to 9.5 m s^{-1}, and allowed the authors to place constraints on the stellar parameters confirming that the star has a low mass and an old age. This was the first application of asteroseismology to a metal-poor star. α Cen A was observed by Bazot *et al.* (2007) with the HARPS spectrometer on the ESO 3.6 m telescope, and 34 *p*-modes were identified in the acoustic oscillation spectrum of the star. Individual observations had errors well under 1 m s^{-1}. A similar study by Mosser *et al.* (2008a) was conducted on Procyon (α CMi) using the SOPHIE spectrograph on the 1.9 m telescope at the Haute-Provence Observatory, yielding a precision of about 2 m s^{-1}. The HARPS instrument was used again by Mosser *et al.* (2008b) to study the old Galactic disk, low-metallicity star HD 203608. A total of 15 oscillation modes were identified, and the age of the star was determined to be 7.25 ± 0.07 Gyr.

5. Radial velocities in Galactic and extragalactic clusters (Elena V. Glushkova, Hugo Levato and Guillermo Torres)

Searches for spectroscopic binaries in southern open clusters have continued during this period (e.g., González & Levato 2006). These authors have reported results for the open cluster Blanco 1. Forty four stars previously mentioned in the literature as cluster candidates, plus an additional 25 stars in a wider region around the cluster were observed repeatedly during five years. Six new spectroscopic binaries have been detected and their orbits determined. All of them are single-lined spectroscopic systems with periods ranging from 1.9 to 1572 d. When considering also all suspected binaries, the spectroscopic binary frequency in this cluster amounts to 34%. Additional velocities were measured in this cluster by Mermilliod *et al.* (2008a), who obtained a rather similar binary frequency.

Results from long term radial velocity studies based on the CORAVEL spectrometers have been presented during this period for the open clusters NGC 6192, NGC 6208, and NGC 6268 (Clariá *et al.* 2006), as well as for NGC 2112, NGC 2204, NGC 2243, NGC 2420, NGC 2506, and NGC 2682 (Mermilliod & Mayor 2007). These studies were complemented with photometric observations in a variety of systems, and included membership determination and binary studies. A number of new spectroscopic binaries were discovered, and their orbital elements were determined.

Other individual cluster studies in the Milky Way, which we merely reference here without giving the details due to space limitations, include: IC 2361 (Platais *et al.* 2007), NGC 2489 (Piatti *et al.* 2007), α Per (Mermilliod *et al.* 2008b), the five distant open clusters Ru 4, Ru 7, Be 25, B 73 and Be 75 (Carraro *et al.* 2007), Tombaugh 2 (Frinchaboy *et al.* 2008), the Orion Nebula cluster (Fürész *et al.* 2008), the most massive Milky way open cluster Westerlund 1 (Mengel & Tacconi-Garman 2008), the Galactic center star cluster (Trippe *et al.* 2008), and the globular clusters M 4 (Sommariva *et al.* 2008) and ω Cen (Da Costa & Matthew 2008).

This triennium saw the publication of the final results of the 20 year efforts of J.-C. Mermilliod and colleagues to measure the radial velocities of giant stars in open clusters for a variety of studies related to their kinematics, membership, and photometric and spectroscopic properties. A catalogue of spectroscopic orbits for 156 binaries based on more than 4000 individual velocities was published by Mermilliod *et al.* (2007), based on measurements from CORAVEL and the CfA Digital Speedometers. Orbital periods range from 41 d to more than 40 yr, and eccentricities are as high as $e = 0.81$. Another 133 spectroscopic binaries were discovered but do not have sufficient observations and/or time coverage to determine orbital elements. This material provides a dramatic increase in the body of homogeneous orbital data available for red-giant spectroscopic binaries in open clusters, and should form the basis for a comprehensive discussion of membership, kinematics, and stellar and tidal evolution in the parent clusters. A companion catalogue (Mermilliod *et al.* 2008c) reports mean radial velocities for 1309 red giants in clusters based on 10 517 individual measurements, and mean radial velocities for 166 open clusters among which 57 are new. This information, combined with recent absolute proper motions, will permit a number of investigations of the galactic distribution and space motions of a large sample of open clusters.

Frinchaboy & Majewski (2008) reported on a survey of the chemical and dynamical properties of the Milky Way disk as traced by open star clusters. They used medium-resolution spectroscopy ($R \approx 15\,000$) with the Hydra multi-object spectrographs on the Cerro Tololo Inter-American Observatory 4 m and WIYN 3.5 m telescopes to derive moderately high-precision RVs ($\sigma < 3\,\mathrm{km\,s^{-1}}$) for 3436 stars in the fields of 71 open clusters within 3 kpc of the Sun. Along with the work described in the preceding paragraph, these represent the largest samples of clusters assembled thus far having uniformly determined, high-precision radial velocities.

A good deal of activity focused on kinematic analyses of globular cluster (GC) systems in other galaxies. Lee *et al.* (2008a) measured radial velocities for 748 GC candidates in M31, and Lee *et al.* (2008b) obtained radial velocities of 111 objects in the field of M60. Konstantopoulos *et al.* (2008) obtained new spectroscopic observations of the stellar cluster population of region B in the prototype starburst galaxy M 82. Schuberth *et al.* (2006) presented the first dynamical study of the GC system of NGC 4636 based on radial velocities for 174 clusters. Bridges *et al.* (2006) measured radial velocities of 38 GCs in the Virgo elliptical galaxy M 60, and Bridges *et al.* (2007) obtained new velocities for 62 GCs in M 104.

An interesting problem was discussed by Abt (2008), pointing to a possible bias in the RVs of many B-type stars. The author looked at 10 open clusters younger than about 30 million years with sufficient numbers of measured radial velocities, many of them being measured with CORAVEL, and found that in each case, the main-sequence B0–B3 stars have larger velocities than earlier- or later-type stars.

6. Radial velocities for field giants (Guillermo Torres)

A programme to measure precise radial velocities for 179 giant stars has been ongoing at the Lick Observatory, with individual errors of 5-8 m s^{-1} per measurement (Hekker *et al.* 2006). This study presented a list of 34 stable K-type giants (with RV standard deviations under 10 m s^{-1}) suitable to serve as reference stars for NASA's Space Interferometry Mission. A follow-up paper (Hekker *et al.* 2008) reported that 80% of the stars monitored show velocity variations at a level greater than 20 m s^{-1}, of which 43 exhibit significant periodicities. One of the goals was to investigate possible mechanisms that cause these variations. A complex correlation was found between the amplitude of the changes and the surface gravity of the star, in which part of the variation is periodic and uncorrelated with $\log g$, and another component is random and does correlate with surface gravity.

Massarotti *et al.* (2008) reported radial velocities made with the CfA Digital Speedometers for a sample of 761 giant stars, selected from the *Hipparcos* Catalogue to lie within 100 pc. Rotational velocities and other spectroscopic parameters were determined as well. Orbital elements were presented for 35 single-lined spectroscopic binaries and 12 double-lined binaries. These systems were used to investigate stellar rotation in field giants to look for evidence of excess rotation that could be attributed to planets that were engulfed as the parent stars expanded.

7. Galactic structure – large surveys (Birgitta Nordström and Guillermo Torres)

7.1. *The Geneva-Copenhagen Survey*

During the previous 3-year period one of the mayor surveys completed and published is the *Geneva-Copenhagen Survey of the Solar Neighbourhood* (Nordström *et al.* 2004). Unfortunately, the full description of this project and the important new science results that came out of it did not make it into the printed version of the IAU Transactions XXVIA for the triennium 2003 - 2006, so we summarize and update that information here for its significant impact for the study of Galactic structure. This survey provided accurate, multi-epoch radial velocities for a magnitude-complete, all-sky sample of 14 000 F- and G-type dwarfs down to a brightness limit of $V = 8.5$, and is volume complete to about 40 pc. The catalogue includes new mean radial velocities for 13 464 stars with typical mean errors of $0.25 \, \mathrm{km \, s^{-1}}$, based on 63 000 individual observations made mostly with the CORAVEL photoelectric cross-correlation spectrometers covering both hemispheres. Studies of this rich data set have found evidence for dynamical substructures that are probably due to dynamical perturbations induced by spiral arms and perhaps the Galactic bar. These 'dynamical streams' (Famaey *et al.* 2005) contain stars of different ages and metallicities which do not seem to have a common origin. These features, which dominate the observed UVW diagrams, make the conventional two-Gaussian decomposition of nearby stars into thin and thick disk members a highly dubious procedure. An analysis by Helmi *et al.* (2006) suggests that tidal debris from merged satellite galaxies may be found even in the solar neighbourhood.

A new release of this large catalogue with updated calibrations as well as new age and metallicity determinations was published during the present triennium by Holmberg *et al.* (2007), and is available from the CDS at <cdsweb.u-strasbg.fr/cgi-bin/qcat?J/A+A/475/51>. A follow-up paper and catalogue are expected to be available shortly, containing new kinematic data (UVW velocities) resulting from a re-analysis using the revised *Hipparcos* parallaxes (van Leeuwen 2007), and on-line updates.

7.2. *Sloan Digital Sky Survey*

This triennium saw the sixth data release of the Sloan Digital Sky Survey (Adelman-McCarthy *et al.* 2008), which now covers an area of 9583 square degrees on the sky. This release includes nearly 1.1 million spectra of galaxies, quasars, and stars with sufficient signal to be usable, along with redshift determinations, as well as effective temperature, surface gravity, and metallicity determinations for many stars. The spectra cover the wavelength region 3800–9200 Å at a resolving power R ranging from 1850 to 2200. Velocity precisions range from about $9 \, \mathrm{km \, s^{-1}}$ for A- and F-type stars to about $5 \, \mathrm{km \, s^{-1}}$ for K-type stars. The zero point of the velocities is in the process of being calibrated using spectra from the ELODIE spectrograph. These data are a valuable resource for a variety of investigations related to Galactic structure and the evolution and history of the Milky Way.

7.3. *RAVE*

The second data release of the Radial Velocity Experiment (RAVE) was published during this triennium (Zwitter *et al.* 2008). This is an ambitious spectroscopic survey to measure radial velocities as well as stellar atmosphere parameters (effective temperature, metallicity, surface gravity, rotational velocity) of up to one million stars using the 6dF multi-object spectrograph on the 1.2 m UK Schmidt telescope of the Anglo-Australian Observatory. The RAVE programme started in 2003, obtaining medium resolution spectra (median $R = 7500$) in the Ca II triplet region (8410 - 8795 Å) for southern hemisphere stars drawn from the Tycho-2 and SuperCOSMOS catalogues, in the magnitude range $9 < I < 12$. Following the first data release, the current release doubles the sample of published radial velocities, now reaching 51 829 measurements for 49 327 individual stars observed between 2003 and 2005. Comparison with external data sets indicates that the new data collected since April 2004 show a standard deviation of $1.3\,\mathrm{km\,s^{-1}}$, about twice as good as for the first data release. For the first time, this data release contains values of stellar parameters from 22 407 spectra of 21 121 individual stars. The data release includes proper motions from the STARNET 2.0, Tycho-2, and UCAC2 catalogues, and photometric measurements from Tycho-2, USNO-B, DENIS, and 2MASS. The data can be accessed via the RAVE web site at <www.rave-survey.org>. Scientific uses of these data include the identification and study of the current structure of the Galaxy and of remnants of its formation, recent accretion events, as well as the discovery of individual peculiar objects and spectroscopic binary stars. For example, kinematic information derived from the RAVE data set has been used by Smith *et al.* (2007) to constrain the Galactic escape velocity at the solar radius to $V_{\mathrm{esc}} = 536^{+58}_{-44}\,\mathrm{km\,s^{-1}}$ (90% confidence).

8. Working groups

Below are the progress reports of the three active working groups of Commission 30. Their efforts are focused on providing a service to the astronomical community at large through the compilation of a variety of information related to radial velocities.

8.1. *WG on radial velocity standard stars* (Stephane Udry)

Large radial-velocity surveys are being conducted to search for extrasolar planets around different types of stars, including A- to M-type dwarfs, and G- to K-type giants (e.g., Udry & Santos 2007). Although not aiming at establishing a set of radial-velocity standard stars, the non-variable stars in these programmes, followed over a long period of time, provide ideal candidates for our list of standards. They will moreover broaden the domain of stellar properties covered (brightness and spectral type). At this point, the results of most of those programmes are still not publicly available and we must still wait a bit in order to fine-tune and enlarge the list presently available at <obswww.unige.ch/~udry/std/std. html>. In addition to the by-product aspect of planet search programmes, a targeted observational effort, dedicated to the definition of a large sample of RV standards for *Gaia*, is being pursued with several instruments (CORALIE, SOPHIE, etc). It will provide in a few years a list of several thousand suitable standards spread over the entire sky (Crifo *et al.* 2007).

For all of the efforts above, work remains to be done to combine the data from the different instruments into a common RV system, for example through the observation of minor planets in the solar system (Zwitter *et al.* 2007). This has still to be done for most of the planet search programmes, but is already included in the *Gaia* effort.

8.2. *WG on stellar radial velocity bibliography* (Hugo Levato)

During the 2006 - 2009 triennium, the WG searched for the papers with measurements of radial velocities of stars in 33 journals. As of December 2007 113 658 entries have been catalogued. We expect to finish 2008 with more than 150,000. It is worth mentioning that at the end of 1996 there were 23 358 entries recorded, so that in 10 years the number of entries in the catalogue has expanded by a factor of five. During the triennium we have improved the search engine to search by different parameters. In the main body of the catalogue we have included information about the technical characteristics of the instrumentation used for radial velocity measurements, and comments about the nature of the objects. The catalogue can be accessed at `<www.casleo.gov.ar/catalogue/catalogue.html>`.

8.3. *WG on the catalogue of orbital elements of spectroscopic binaries (SB9)*
(Dimitri Pourbaix)

In Manchester, a WG was set up to work on the implementation of the 9th catalogue of orbits of spectroscopic binaries (SB9), superseding the 8th release of Batten *et al.* (1989) (SB8). SB9 exists in electronic format only. The web site `<sb9.astro.ulb.ac.be>` was officially released during the summer of 2001. This site is directly accessible from the Commission 26 web site, from BDB (in Besançon), and from the CDS, among others.

Since the last report, substantial progress has been accomplished, in particular in the way complex systems can be uploaded together with their radial velocities. That is the case, for instance, for triple stars with the light time effect accounted for and systems with a pulsating primary.

At the time of this writing SB9 contains 2802 systems (SB8 had 1469) and 3340 orbits (1469 in SB8). A total of 563 papers were added since August 2000, although most of them come from *outside* the WG. Many papers with orbits still await uploading into the catalogue. According to ADS, the release paper (Pourbaix *et al.* 2004) has been cited a total of 58 times since 2005. This is twice as many as the old Batten *et al.* catalogue over the same period.

Even though this work has been very well received by the community and a number of tools have been designed and implemented to make the job of entering new orbits easier (input file checker, plot generator, etc.), the WG still suffers from a serious lack of manpower. Few colleagues outside the WG spontaneously send their orbits (but they are usually pleased to send their data when we ask for them). Any help (from authors, journal editors, and others) is therefore very welcome. Uploading an orbit into SB9 also involves checking for typos. In this way we have found several mistakes in published solutions, which we have corrected. Sending orbits to SB9 prior to publication (e.g., at the proof stage) would therefore be a way to prevent some mistakes from making it into the literature.

Guillermo Torres
vice-president of the Commission

References

Abt, H. 2008, *PASP*, 120, 715

Adelman-McCarthy, J. K., Agüeros, M. A., Allam, S. S., *et al.* 2008, *ApJS*, 175, 297

Batten, A. H., Fletcher, J. M., & MacCarthy, D. G. 1989, *Eighth Catalogue of the Orbital Elements of Spectroscopic Binary Systems, Publ. DAO*, 17, 1

Bazot, M., Bouchy, F., Kjeldsen, J., *et al.* 2007, *A&A*, 470, 295

Bedding, T. R., Butler, R. P., Carrier, F., *et al.* 2006, *ApJ*, 647, 558

Bridges, T., Gebhardt, K., Sharples, R., *et al.* 2006, *MNRAS*, 373, 157

Bridges, T., Rhode, K. L., Zepf, S. E., & Freeman, K. C. 2007, *ApJ*, 658, 980

Carraro, G., Geisler, D., Villanova, S., *et al.* 2007, *A&A*, 476, 217

Claria, J. J., Mermilliod, J.-C., Piatti, A. E., & Parisi, M. C. 2006, *A&A*, 453, 91

Crifo, F., Jasniewica, G., Soubiran, C., *et al.* 2007, in: J. Bouvier, A. Chalabaev & C. Charbonnel (eds.), Proc. Ann. Meeting French Soc. Astron. Astroph., 2007, p. 459

Da Costa, G. S. & Matthew, C. G. 2008, *AJ*, 136, 506

Famaey, B., Jorissen, A., Luri, X., *et al.* 2005, *A&A*, 430, 165

Fischer, D. A., Marcy, G. W., Butler, R. P., *et al.* 2008, *ApJ*, 675, 790

Frinchaboy, P. M., Marino, A. F., & Villanova, S., *et al.* 2008, *MNRAS*, 391, 39

Frinchaboy, P. M. & Majewski, S. R. 2008, *AJ*, 136, 118

Fürész, G., Hartmann, L. W., Megeath, S. T., *et al.* 2008, *ApJ*, 676, 1109

González, J. F. & Levato, H. 2006, *RMxAA*, 26, 171

Hekker, S., Snellen, I. A. G., Aerts, C., *et al.* 2008, *A&A*, 480, 215

Hekker, S., Reffert, S., Quirrenbach, A., *et al.* 2006, *A&A*, 454, 943

Helmi, A., Navarro, J. F., Nordström, B., *et al.* 2006, *MNRAS*, 365, 1309

Holmberg, J., Nordström, B., & Andersen, J. 2007, *A&A*, 475, 519

Johnson, J. A., Marcy, G. W., Fischer, D. A., *et al.* 2008, *ApJ*, 675, 784

Johnson, J. A., Butler, R. P., Marcy, G. W., *et al.* 2007a, *ApJ*, 670, 833

Johnson, J. A., Butler, R. P., Marcy, G. W., *et al.* 2007b, *ApJ*, 665, 785

Karchenko, N. V., Scholz, R.-D., Piskunov, A. E., *et al.* 2007, *AN*, 328, 889

Konstantopoulos, I. S., Bastian, N., Smith, L. J., *et al.* 2008, *ApJ*, 674, 846

Lee, M. G., Hwang, Ho S., Kim, S. Ch., *et al.* 2008a. *ApJ*, 674, 886

Lee, M. G., Hwang, Ho S., Park, H. S., *et al.* 2008b, *ApJ*, 674, 857

Li, Ch.-H., Benedick, A. J., Fendel, P., *et al.* 2008, *Nature*, 452, 610

Lovis, C. & Pepe, F. 2007, *A&A*, 468, 1115

Massarotti, A., Latham, D. W., Stefanik, R. P., & Fogel, J. 2008, *AJ*, 135, 209

Mayor, M. & Udry, S. 2008, *PhST*, 130, 014010

Mayor, M., Udry, S., Lovis, C., *et al.* 2008, *A&A*, in press [arXiv:0806.4587]

Mengel, S. & Tacconi-Garman, L. E. 2008, in: *Young Massive Star Clusters - Initial Conditions and Environments*, Proc. meeting 2007, in press [arXiv:0803.4471]

Mermilliod, J.-C., Platais, I., James, D. J., Grenon, M., & Cargile, P. A. 2008a, *A&A*, 485, 95

Mermilliod, J.-C., Queloz, D., & Mayor, M. 2008b, *A&A*, 488, 409

Mermilliod, J.-C., Mayor, M., & Udry, S. 2008c, *A&A*, 485, 303

Mermilliod, J.-C. & Mayor, M. 2007, *A&A*, 470, 919

Mermilliod, J.-C., Andersen, J., Latham, D. W., & Mayor, M. 2007, *A&A*, 473, 829

Mosser, B., Bouchy, F., Martić, M., *et al.* 2008a, *A&A*, 478, 197

Mosser, B., Deheuvels, S., Michel, E., *et al.* 2008b, *A&A*, 488, 635

Murphy, M. T., Udem, Th., Holzwarth, R., *et al.* 2007, *MNRAS*, 380, 839

Nordström, B., Mayor, M., Andersen, J., *et al.* 2004, *A&A*, 418, 989

Pepe, F. A. & Lovis, C. 2007, in: *Physics of Planetary Systems, Nobel Symp. 135*, in press

Pepe, F., Correia, A. C. M., Mayor, M., *et al.* 2007, *A&A*, 462, 769

Piatti, A., Clariá, J. J., Mermilliod, J.-C., *et al.* 2007, *MNRAS*, 377, 1737

Platais, I., Melo, C., Mermilliod, J.-C., *et al.* 2007, *A&A*, 461, 509

Pourbaix, D., Tokovinin, A. A., Batten, A. H., *et al.* 2004, *A&A*, 424, 727

Ramsey, L. W., Barnes, J., Redman, S. L., *et al.* 2008, *PASP*, 120, 887

Schuberth, Y., Richtler, T., Dirsch, B., *et al.* 2006, *A&A*, 459, 391

Smith, M. C., Ruchti, G. R., Helmi, A., *et al.* 2007, *MNRAS*, 379, 755

Sommariva, V., Piotto, G., Rejkuba, M., *et al.* 2008, *A&A*, in press [arXiv:0810.1897]

Steinmetz, T., Wilken, T., Araujo-Hauck, C., *et al.* 2008, *Science*, 321, 1335

Trippe, S., Gillessen, S., Gerhard, O. E., *et al.* 2008, *A&A*, 492, 419

Udry, S. & Santos, N. C. 2007, *ARA&A*, 45, 397

Udry, S., Bonfils, X., Delfosse, X., *et al.* 2007, *A&A* (Letters), 469, L43

van Leeuwen, F. 2007, *A&A*, 474, 653

Zwitter, T., Mignard, F., & Crifo, F. 2007, *A&A*, 462, 795

Zwitter, T., Siebert, A., Munari, U., *et al.* 2008, *AJ*, 136, 421

Transactions IAU, Volume XXVIIA
Reports on Astronomy 2006–2009
Karel A. van der Hucht, ed.

© 2009 International Astronomical Union
doi:10.1017/S1743921308025775

DIVISION IX / COMMISSION 30 / WORKING GROUP
STELLAR RADIAL VELOCITY BIBLIOGRAPHY

CHAIR   Hugo Levato
VICE-CHAIR   Stella M. Malaroda

TRIENNIAL REPORT 2006 - 2009

1. Introduction

The Working Group *Stellar Radial Velocity Bibliography* is a very small one that was created with the purpose of continuing the cataloging of bibliography of radial velocity of stars made by Mme Barbier in successive catalogues until her retirement in 1990.

We have started the new compilation late in 1990. The first version of the catalogue after the retirement of Mme Barbier was published for the triennium 1991 - 1994 at <www.casleo.gov.ar/catalogo/>. The catalogue is updated at this page each six months.

2. Developments within the triennium

During the triennium 2006 - 2009, the WG searched for the papers with measurements of radial velocities of stars among 33 journals. By December 2007 we had cataloged 113 658 entries. We expect to finish 2008 with around 150 000. It is worth to mention that at the end of 1996 we had recorded 23 358 entries, so in 10 years the number of entries in the catalogue multiplied by five.

During the triennium we have improved the search engine to search by different parameters. In the main body of the catalogue we have included information about the technical characteristics of the instrumentation used for radial velocity measurements and comments about the nature of the objects.

3. Closing remarks and future

The future is becoming very attractive. New tools permit to increase the efficiency of the search. During the last year search engines like SCOPUS from Elsevier Publishers was under test.

Hugo Levato
chair of the Working Group

Transactions IAU, Volume XXVIIA
Reports on Astronomy 2006–2009
Karel A. van der Hucht, ed.

© 2009 International Astronomical Union
doi:10.1017/S1743921308025787

DIVISION IX / COMMISSION 30 / WORKING GROUP
CATALOG OF ORBITAL ELEMENTS OF
SPECTROSCOPIC BINARIES

CHAIR	Dimitri Pourbaix
VICE-CHAIR	Andrew T. Young
MEMBERS	Alan H. Batten, Francis C. Fekel,
	William I. Hartkopf, Hugo Levato,
	Nidia I. Morrell, Andrei A. Tokovinin,
	Guillermo Torres, Stepane Udry

TRIENNIAL REPORT 2006 - 2009

The SB9 Working Group of Commission 30 aims at compiling the *9th Catalogue of Orbits of Spectroscopic Binaries*. By definition, this is a never ending task as orbits of newly discovered systems keep appearing in the literature. Despite this, the working group tries to catch up with the delay as nothing was done in between 1989 when the 8th catalogue by Batten *et al.* and 2000 when the WG was settled. In 2006, at its business meeting, the WG decided to focus on the completeness of systems rather than on completeness of orbits. If the latter is a valuable objective, only the former is useful to any statistical investigation of spectroscopic binaries.

It was also decided that spectroscopic orbits of extrasolar planets should make it to the catalogue. Owing to the lack of manpower, to the instability of some planetary status, that task has been postponed so far. Another pending issue is the automatic grading scheme. In 2007, a student from the Université Libre de Bruxelles sought any computable quality indicator which could reproduce the original subjective grade listed in SB8 (and predecessors). No positive result has been achieved so far.

The main progress of SB9 over the past two years has focused on uploading papers with new systems, trying to be as complete as possible with some authors' series of papers (e.g. Carquillat, Griffin, Fekel, ...). 155 new systems have thus been uploaded (for 202 orbits in 44 papers). SB9 now contains 2839 systems, i.e., 1370 more than in the 8th catalogue. By the IAU XXVII General Assembly in 2009, we will try to upload the systems missing in order to double the sample of SB8.

Despite its spread in the community (assessed through ADS citations: 58 quotations to the release paper, Pourbaix *et al.* (2004, *A&A*, 424, 727) and its efficiency to identify typographic bugs in some already published orbits, SB9 does not attracts orbits yet! Besides a very limited number of authors who spontaneously send their new orbits with all the relevant data, nobody sends his/her new solution nor let us know about the coming paper. That is a very unfortunate and frustrating situation because a little help from every author would make the whole community benefit from a better and more complete catalogue.

Dimitri Pourbaix
chair of the Working Group

Transactions IAU, Volume XXVIIA
Reports on Astronomy 2006–2009
Karel A. van der Hucht, ed.

© 2009 International Astronomical Union
doi:10.1017/S1743921308025799

INTER-DIVISION IX-X WORKING GROUP ENCOURAGING THE INTERNATIONAL DEVELOPMENT OF ANTARCTIC ASTRONOMY

CHAIR   Michael Burton

MEMBERS   Carlos A. Abia, John E. Carlstrom,
                   Vincent Coudé du Foresto, Xiangqun Cui,
                   Sebastián Gurovich, Takashi Ichikawa,
                   James P. Lloyd, Mark J. McCaughrean,
                   Gino Tosti, Hans Zinnecker

TRIENNIAL REPORT 2006 – 2009

1. Overview

Two major astronomical experiments are underway at the US Amundsen-Scott South Pole Station. The first is the South Pole Telescope, a 10 m sub-millimetre telescope designed to measure primary and secondary anisotropies in the CMBR, with the aim of placing constraints on the equation of state for dark energy. The second is the IceCube neutrino observatory, which will be a cubic kilometre array designed to image sources of high energy neutrinos.

The French-Italian Concordia station at Dome C is now complete and operating through the winter. Site testing is underway through the Concordia-astro program. At Dome A, the summit of the Antarctic plateau, China are establishing a new station, and deployed the Australian-built PLATO observatory there for the austral winter of 2008, obtaining the first site testing data from that location. This was a part of an International Polar Year program.

Japan started site evaluation at Dome F in 2007.

The long duration ballooning program run from the US McMurdo Station continues to grow, with multiple programs now able to fly balloons simultaneously.

SCAR, the *Scientific Committee for Antarctic Research*, has established astronomy as an official Scientific Research Program, one of five such designated scientific programs. The IAU's application to join SCAR was recognised at the Moscow meeting of SCAR delegates in July 2008, making the IAU now the 9th scientific union member of SCAR. SCAR coordinated the International Polar Year, which ran through 2007 – 08.

2. Meetings held

• *Visit to the IRAIT Telescope*, 12–13 September 2006, University of Perugia, Italy. Progress, plans and launch for the IRAIT telescope for Dome C.

• *Large Astronomical Optical/IR Infrastructures at Concordia. Prospects and constraints*, 16-19 October 2006, Roscoff, France. First ARENA Conference.

- *Telescope and instrument robotization at Dome C*, 26-28 March 2006, Puerto Santiago, Tenerife, Spain. Conditions for successful implementation and future operation of robotic telescopes and instrumentation at Dome C.
- *Site testing at Dome C*, 11-13 June 2007, Rome, Italy. Status of Antarctic site testing.
- *Submillimetre astronomy at Dome C*, 25-27 June 2007, Saclay, France. Plans for sub-millimetre astronomy at Dome C.
- *The Astrophysical Science Cases at Dome C*, 17-21 September 2007, Potsdam, Germany. Second ARENA Conference.
- *Wide Field Telescopes*, 26-27 March 2007, University of Exeter, UK. Plans for wide-field astronomy in Antarctica.
- *Spectroscopy on Dome C: from UV to Sub-millimetric Wavelengths*, 16-18 April 2008, University of Granada, Spain. Prospects for astronomical spectroscopy in Antarctica.
- *Site Testing in Antarctica and New Antarctic Facilities*, 23-28 June 2008, Marseille, France. SPIE Meeting, Astronomical Instrumentation: Conference 7012, Session 3.
- *Astronomy in Antarctica*, St Petersburg, Russia, 8-11 July 2008. SCAR Meeting, Session 4.2.
- *Time-series Observations from Dome C*, 17-19 September 2008, Catania, Italy.
- *Interferometry at Concordia 2*, 13-14 October 2008, Garching-bei-München, Germany.
- *A Large Astrophysical Observatory at Dome C for the Next Decade*, 11-14 May 2009, Rome, Italy. Third ARENA Conference.

3. Major publications

- Burton, M. G., 2008, in: K. A. van der Hucht (ed.), *Transactions of the IAU*, Vol. XXVIB, Proc. Twenty Sixth General Assembly, Prague 2006 (Cambridge: CUP), p. 188, *Report on the Inter-Division IX-X Working Group on Encouraging the International Development of Antarctic Astronomy*
- Epchtein, N., & Candidi, M. (eds.), 2007, *Large Astronomical Infrastructures at CONCORDIA, Prospect and Constraints for Antarctic Optical/IR Astronomy*, Proc. 1st ARENA Conference, Roscoff, France, 16-19 October 2006, *EAS-PS*, 25
- Zinnecker, H., Rauer, H., & Epchtein, N. (eds.), 2008, *The Astrophysical Science Cases at Dome C*, Proc. 2nd ARENA Conference, Potsdam, 17-21 September 2007, EAS-PS, in press.

Michael G. Burton
chair of the Working Group

Transactions IAU, Volume XXVIIA
Reports on Astronomy 2006–2009
Karel A. van der Hucht, ed.

© 2009 International Astronomical Union
doi:10.1017/S1743921308025805

DIVISION X   RADIO ASTRONOMY
RADIOASTRONOMIE

Division X provides a common theme for astronomers using radio techniques to study a vast range of phenomena in the Universe, from exploring the Earth's ionosphere or making radar measurements in the Solar System, via mapping the distribution of gas and molecules in our own Galaxy and in other galaxies, to study the vast explosive processes in radio galaxies and QSOs and the faint afterglow of the Big Bang itself.

PRESIDENT	Ren-Dong Nan
VICE-PRESIDENT	Russell A. Taylor
PAST PRESIDENT	Luis F. Rodríguez
BOARD	Christopher L. Carilli, Jessica Chapman, Gloria M. Dubner, Michael Garrett, W. Miller Goss, Richard E. Hills, Hisashi Hirabayashi, Prajval Shastri, José María Torrelles

DIVISION X COMMISSION

Commission 40	Radio Astronomy

DIVISION X WORKING GROUPS

Division X WG	Interference Mitigation
Division X WG	Astrophysically Important Spectral Lines
Division X WG	Global VLBI

INTER-DIVISION WORKING GROUPS

Inter-Division X-XII WG	Historic Radio Astronomy
Inter-Division X-XI WG	Encouraging the International Development of Antarctic Astronomy
Inter-Division IX-X-IX WG	Astronomy from the Moon

TRIENNIAL REPORT 2006 - 2009

1. Introduction

We were pleased that the 2006 Nobel Prize in Physics was awarded to John C. Mather and George F. Smoot for their discovery of the black body form and anisotropy of the cosmic microwave background radiation. In addition, the URSI John Howard Dellinger Gold Medal has been awarded to Alan Rogers of MIT Haystack Observatory, and the Balthasar van der Pol Gold Medal to Jack Welch (UC/Berkeley). Moreover, Kavli Prizes in Astrophysics of 2008 were granted to Maarten Schmidt and Donald Lynden-Bell for their contributions to our understanding of the nature of quasars. It is mentioned that

Grote Reber medals were awarded for lifetime achievements in radio astronomy to Bill Erickso (2005), Bernie Mills (2006), Govind Swarup (2007), and Sandy Weinreb (2008).

During the past triennium, radio astronomy witnessed many changes and improvements in the available facilities. We have included links to all major radio astronomical facilities on the web page `<www.bao.ac.cn/IAU_COM40/2005/facilities.html>`, where the details of the recent instrumental developments can be consulted.

After the consolidation and initiation of the Atacama Large Millimeter Array(ALMA) project highlighted in the previous triennium Divisional report, perhaps international efforts associated with the Square Kilometer Array (SKA) project become the most visible event in the radio astronomical community. The SKA will have a collecting area of up to one million square meters spread over at least 3000 km, providing a sensitivity 50 times higher than existing radio telescopes, and an instantaneous FoV up to at least several tens of square degrees. Over the last few years, several SKA Pathfinder Telescopes (LOFAR, EVLA, MWA, ATA, ASKAP, MeerKAT, FAST, etc.), SKA Design work Studies (SKADS), PrepSKA and TDP have been funded, each of which is developing one or more aspects of the Reference Design technology in depth; see `<www.skatelescope.org/>`.

The scientific results have continued to flow with great continuity. In the first section of this document we list some of these results that sample the great variety of research being undertaken. By far, this report is not complete and it merely tries to convey the vitality and versatility in this field. Besides the Organizing Committee (OC) members of the Division, we would like to extend our sincere appreciation and gratitude to our colleagues, Michael Kramer, Yu H. Yan, Zhi-Qiang Shen, XueLei Chen, Wayne Orchiston, Kenneth I. Kellermann, and Masatoshi Ohishi, for their valuable contributions to the Division.

Three Divisional Working Groups have renovated their memberships and terms of reference, see `<www.bao.ac.cn/IAU_COM40/workgroup.html>`.

Three Inter-Division Working Groups are also linked to the Division X webpage: `<www.bao.ac.cn/IAU_COM40/workgroupi.htm>`.

Sadly, four distinguished astronomers died in the period under consideration.

W. N. 'Chris' Christiansen passed away in April 2007. He developed the innovative cross-type radio telescope known as the Chris Cross. He served as vice-president of the IAU 1964 - 1970, president of the International Union of Radio Science (URSI) 1978 - 1981, and was honored as its Honorary Life President in 1984.

Ronald N. Bracewell passed away in August 2007. He was Stanford University engineering professor, and internationally known for developing magnetic imaging as a tool that evolutionized medicine, and for developing radio astronomy.

As a great pioneer in Solar Radio Astronomy, Paul Wild passed away on 10 May 2008 in Canberra, Australia. He was the second winner of the Hale Prize.

The death of Barry E. Turner on 10 May 2008 was reported in National Radio Astronomy Observatory (NRAO). He pioneered the charting of the astronomical microwave spectrum, thereby establishing astrochemistry as a new field of science.

2. Scientific highlights June 2005 - June 2008

2.1. *On molecular observations*

Radio astronomy, in particular at the millimeter and sub-millimeter wavelengths, continues to be a major tool for the study of molecules in an astrophysical context. These studies extend from nearby comets to remote galaxies (e.g., Biver *et al.* 2006; Carilli *et al.* 2007), passing by rotating structures around forming stars in our Galaxy (e.g., Zapata *et al.* 2007). More than 140 molecular species have been detected in space (`<www.ph1.uni-koeln.de/vorhersagen/>`), with the most recent addition being cyanoformaldehyde (CNCHO) (Remijan *et al.* 2008). Molecular observations are a major part of the programs of existing radio telescopes as well as of the science drivers of projects like ALMA and the Large Millimeter Telescope (LMT). In what follows we briefly summarize major results on this topic obtained over the last two years.

A review on the atomic and molecular content of diffuse interstellar clouds (previously believed to be relatively devoid of molecules) was given in 2006 (Snow & McCall 2006). A complementary review on cold dark clouds, stressing the efforts toward understanding how molecular gas condenses to form stars, was presented in 2007 (Bergin & Tafalla 2007).

Magnetic fields in the Cepheus A region have been determined by Vlemmings *et al.* (2006) using linear and circular polarization observations of the 22.2 GHz H_2O masers (Vlemmings *et al.* 2006). The study of the morphology and kinematics of molecular outflows continues to be an active field of research, as evidenced by the studies in HH 211 and in the OMC 1 South Region (Zapata *et al.* 2006). Patel *et al.* (2005) presented images and kinematical data of a disk of dust and molecular gas around a high-mass protostar. This result has been confirmed by Jiménez-Serra *et al.* (2007). Molecules are also used to study possible protoplanetary disks, envelopes of evolved stars, and planetary nebulae (e.g., Qi *et al.* 2006; Fong *et al.* 2006; Tafoya *et al.* 2007). In our galaxy, an extensive survey of star-formation regions (COMPLETE Survey) has released a phase I database that includes and imaging of these regions. The origin of the large abundance of complex organic molecules in the Galactic Center has been studied, and it was concluded that all of the complex molecules that are observed were probably ejected from grain mantles by shocks (Ridge *et al.* 2006; Requena-Torres *et al.* 2006).

A new ^{12}CO (J=1-0) line survey of the Andromeda Galaxy with the highest resolution to date (23″, or 85 pc along the major axis) is reported by Nieten *et al.* (2006), who find radial gradients that suggest that either the atomic and total gas-to-dust ratios increase by a factor of 20 or the dust becomes colder towards larger radii. With increased sensitivity, it has become possible to detect and study molecules in external galaxies. Riechers *et al.* (2006) made $^{12}CO(J=1-0)$ observations of the high-redshift quasi-stellar objects (QSOs) BR 1202−0725 ($z = 4.69$), PSS J2322+1944 ($z = 4.12$), and APM 08279+5255 ($z = 3.91$) using the NRAO Green Bank Telescope (GBT) and the MPIfR Effelsberg 100 m telescope. They conclude that emission from all CO transitions is described well by a single gas component in which all molecular gas is concentrated in a compact nuclear region. Their observations and models show no indication of a luminous extended, low surface brightness molecular gas component in any of the high-redshift QSOs studied. Weiss *et al.* (2007) argue that the molecular lines and the dust continuum emission observed in the Broad Absorption Line quasar APM 08279+5255 arise from a very compact (dimensions of about 100-300 pc), highly gravitationally magnified region surrounding the central AGN.

2.2. *Pulsars and transients*

The recent years have continued to produce exciting new results in the field of pulsar astrophysics. The double pulsar discovered at Parkes in 2003 has already fulfilled its promise to provide the best tests of General Relativity in strong gravitational fields (Kramer *et al.* 2006). The second most relativistic pulsar, J1906+0746 (Lorimer *et al.* 2006), was discovered as the first of probably many remarkable discoveries of the

Pulsar ALFA Survey(P-ALFA) on-going at the Arecibo Observatory (Cordes *et al.* 2006). Surveys of globular clusters have produced an outstanding number of new discoveries (e.g., Ransom *et al.* 2005), including the discovery of the fastest rotating millisecond pulsar with a spin frequency of 716 Hz (Hessels *et al.* 2006). Further unexpected discoveries include those of several radio transient and intermittent sources, demonstrating that the dynamic radio sky is still largely unexplored. These discoveries include that of a transient source in the Galactic Centre (Hyman *et al.* 2005), the discovery of transient radio bursts from rotating neutron stars named RRATS (McLaughlin *et al.* 2006), and the discovery of an intermittent pulsar PSR 1931+24 which gives unique insight into the workings of pulsar magnetospheres (Kramer *et al.* 2006). Further unexpected radio emission was discovered from magnetars (e.g., Camilo *et al.* 2007) with the remarkable example of the anomalous X-ray pulsar XTE J1810−197, for which single pulses were detected up to 120 GHz (Camilo *et al.* 2006). The link between pulsars and supernovae and the related physics of core collapse was also further studied with the discovery of new pulsar-supernova remnant associations (e.g., Camilo *et al.* 2006) and the velocity of young pulsars (e.g., Chatterjee *et al.* 2005).

2.3. *Radio Sun*

Radio observations cover a wide range from the quiet Sun emissions to flares, CMEs and interplanetary particles. There exist some recent attempts to explore the properties at mm wavelength (e.g., Brajsa *et al.* 2007) and prepare for future ALMA observations (e.g., Wedemeyer-Böhm *et al.* 2007). The Nancay Radio Heliograph (NRH) at dual 17 and 34 GHz has covered one solar cycle, and the scientific outputs are still increasing, such as obtaining the coronal magnetic field distribution (Huang, G., Ji, H., & Wu, G. 2008). The Siberian Solar Radio Telescope (SSRT) interferometer at 5.7 GHz offers the ability to study fast fine structures (Altyntsev *et al.* 2007). High dynamic range snapshot images of the solar corona at 327 MHz were obtained by combining the Giant Metrewave Radio Telescope (GMRT) and the NRH (Mercier *et al.* 2006). Using NRH data, together with other space and ground observations, the CME features, source region of 3He-rich SEP events, electron acceleration and transport, and radio bursts of CME events were studied (e.g., Gopalswamy *et al.* 2006; Pick *et al.* 2006; Trottet *et al.* 2006; Yan *et al.* 2006).

Because of observations with high resolutions by Chinese spectrometers in the 1-7.6 GHz range, data about the detailed spectra of zebra patterns in the microwave range above 3 GHz have appeared (Chernov 2006). Among others, the radio spectrometers from cm, dm to metric wavelengths include ETH, Ondrejev, IZMIRAN, AIP, Athens, etc. An interplanetary Type II like event in 0.1-14 MHz observed by WIND/WAVES is found not to be due to plasma radiation, but instead to incoherent synchrotron radiation from near-relativistic electrons (Bastian 2007). The newly-launched S/WAVES covering the ˜1-16 MHz range can perform 3D localization and tracking of the radio emissions associated with streams of energetic electrons and shock waves associated with CMEs (Bougeret *et al.* 2008). Future major new solar-dedicated observing facilities (specifically FASR and the Chinese Radio Heliograph) will greatly extend the capabilities of achieving high spatial, temporal and spectral resolutions and high sensitivity (Hudson & Vilmer 2007).

2.4. *The Galactic Center*

The past few years have seen a lot of progress in radio observations of the Galactic Center (GC).

High-resolution Very Long Baseline Interferometry (VLBI) observations have established that the intrinsic size of Sgr A* is only about 1 AU at 3.5 mm (Shen *et al.* 2005), even though the exact structure of the interstellar scattering needs to be further examined with better longer wavelength radio observations (Bower *et al.* 2006). The size constraint, along with the lower limit to the mass of Sgr A* from the VLBA measurements of its proper motion, strongly supports the supermassive black hole nature of Sgr A*. Recently some testing observations at 1.3 m or shorter have been successfully done.

Many new observations at different radio bands with different instruments have been carried out to investigate the variability of Sgr A*. These include sub-mm observations of Sgr A* with the Caltech Submillimeter Observatory (CSO); the Submillimeter Array (SMA) monitoring observations; the Nobeyama Millimeter Array (NMA) monitoring at 100 and 140 GHz; the Australia Telescope Compact Array (ATCA) observations at 86 GHz; and the Very Large Array (VLA) monitoring at cm wavelengths, etc. As a result, intra-day variability has been detected by the Owens Valley Radio Observatory (OVRO) millimeter interferometer (Mauerhan *et al.* 2005), the NMA, the SMA and the ATCA. The cause of the variability is still in debate. Flaring activity at 43 and 22 GHz with the VLA shows the peak flare emission at 43 GHz leading the 22 GHz one by 20-40 minutes, which can be explained with the plasmon model.

Significant progress has been made in the detection of polarized emission from Sgr A* at mm and sub-mm bands. Linear polarization from Sgr A* was detected with the BIMA at 1.3 mm. Variable polarization in both magnitude and position angle on time scales down to a few hours has been measured with the SMA at 340 GHz (Marrone *et al.* 2007), although its relation to the variation of the total intensity is not clear. The inferred Faraday rotation is very low, and thus a very low accretion rate is needed. The polarized emission is also detected by the VLA at 43 and 22 GHz, which seems to suggest that the rotation measure decreases with decreasing frequency (Yusef-Zadeh *et al.* 2007).

Simultaneous spectra of Sgr A* from 90 cm to 3.8 μm showed a spectral break at 3.6 cm (An *et al.* 2005). At longer cm wavelengths it is consistent with free-free absorption due to the ionized gas. Observations at 74 and 330 MHz reveal the presence of a large-scale diffuse source of non-thermal synchrotron emission with a magnetic field several orders of magnitude below the commonly-assumed 1 mG. Recent radio and sub-mm observations have revealed that magnetic filaments are more widespread than previously thought. Observations at even higher resolution and sensitivity will be required to understand fully the origin, properties and role of the magnetic field at the GC (Lazio & LaRosa 2005).

To understand the emission from Sgr A* and establish the relationship of flaring activity across its spectrum, several campaigns of wide band (including X-ray, infrared and radio) observations towards the GC have been coordinated, with some time delays detected between various wavelengths (e.g., Eckart *et al.* 2006).

There are also many new observations of the GC such as the GBT multi-wavelength survey, the wide-area VLA continuum survey at 6 and 20 cm, the large scale CO 3-2 observations in the GC by the ASTE, Odin observations at 118 GHz, SMA observations of line and continuum emission at 1.3 mm, spectral imaging of the Sgr B2 region in multiple 3 mm molecular lines with the Mopra telescope, VLA observations of OH masers in Sgr A East and the circumnuclear disk. GC appears to be one of the best laboratories for studying the formation of complex organic molecules.

A transient source 2″.7 south of Sgr A* was detected at 330 MHz (Hyman *et al.* 2005), which was resolved by the VLA into two components (Bower *et al.* 2005). New data with the GMRT at 330 MHz indicate a very steep spectral index. A sensitive 20 cm VLA continuum survey of the GC region has resulted in a catalog of 345 discrete sources. Multi-frequency, multi-configuration VLA observations have been used to investigate the stellar winds and embedded massive star formation in the Quintuplet and Arches stellar cluster. A number of potential ultracompact and hypercompact H II regions in Sgr B2 were studied with the H 52α recombination line, consistent with an ionized outflow from high mass stars. Recently, there was a debate on the possible non-thermal nature of the radio emission arising from the Sgr B complex.

2.5. *Galactic surveys*

Advances in computing and wide-field interferometric imaging have made possible the execution of large-scale spectroscopic surveys of the Milky Way. Three such surveys are underway. The VLA Galactic Plane Survey (Stil *et al.* 2006) was completed in 2006, and the Canadian Galactic Plane Survey (Taylor *et al.* 2003), and the Southern Galactic Plane Survey (McClure- Griffiths *et al.* 2005) are nearing completion. These surveys provide arcminute-scale resolution imaging of atomic hydrogen emission covering a wide range of longitudes including the three quadrants of the Galaxy (second, first and fourth). These data make available an image of the atomic hydrogen gas in the interstellar medium at comparable angular resolution to data on the molecular gas and dust provided by sub-millimetre and infrared observations. Planned surveys at submillimetre with the James Clerk Maxwell Telescope (JCMT) and *Herschel*, promise in combination with these H I surveys to provide a complete picture of the Interstellar Medium (ISM) in the plane of the Galaxy at parsec spatial resolution. At this high angular resolution, the cold atomic hydrogen component of the multi-phase ISM is revealed by H I self absorption and continuum absorption to be ubiquitous (Gibson *et al.* 2005, Strasser *et al.* 2007). Complexes of cold H I clouds are detected, that may represent an intermediate temperature and density phase of the ISM linking the diffuse warm ISM to molecular clouds and triggered by passage of spiral arm shocks (Gibson *et al.* 2007).

Radio polarimetry of the Galaxy is enjoying a resurgence as a probe of the magneto-ionic ISM. The dense grid of rotation measure observations towards background radio sources provided by the Canadian Galactic Plane Survey (CGPS) and the Southern Galactic Plane Survey (SGPS) (e.g., Haverkorn *et al.* 2006), along with pulsar rotation measures, are being used to probe the structure of the global magnetic field of the Galaxy and the smaller-scale turbulent field. Recent developments include the discovery of oscillatory RM structures associated with spiral arm structure and the characterization of the reversals of the global field (Brown *et al.* 2007), and the relationship of fields in molecular clouds to the global-scale fields (Han & Zhang 2007).

Broad-band, multi-channel spectropolarimetry offers the possibility of carrying out rotation measure synthesis to deconvolve the three-dimensional structure of the Faraday rotating ISM, the so-called Faraday screen. New Galactic Plane surveys that are soon to be launched, will use new spectrometer systems on Arecibo (GALFACTS) and Parkes and the DRAO 26 m telescope (GMIMS) and will exploit this technique. First observations for both of these surveys will begin in 2008.

Planned surveys at submillimetre with the JCMT and *Herschel*, and new data from *Planck*, promise, in combination with these new low-frequency radio surveys of H I emission and polarimetry, to allow exploration of a more complete picture of the relationship between the phases of the ISM in the plane of the Galaxy at parsec spatial resolution.

2.6. *Active galaxies*

The study of the physics of launching AGN jets is set to make significant progress with the availability of large bodies of pc-scale total intensity, as well as linear and circular polarisation measurements at multiple frequencies and multiple epochs. Sub-milliarcsecond morphology is available for a sample of 250 relativistically-beamed AGN (Kovalev *et al.* 2005), of which proper-motion data were also earlier published for over 100 objects. Multi-frequency VLBI polarimetry with spatial resolution of a few resolution elements transverse to the direction of the jet has also been done for a significant number of AGN, which has enabled the search for and detection of transverse rotation-measure gradients across the pcscale jets (Zavala & Taylor 2005; Mahmud & Gabuzda 2007). Such a gradient has long been suggested as a signature of helical magnetic fields in the jets (Papageorgiou 2006). Homan & Lister (2006) detect significant pc-scale circular polarisation in 15133 AGN at 15 GHz. Gabuzda *et al.* (2008) detect polarisation in another seven AGN which they interpret in the helical magnetic field scenario. With VLBI at 7 mm, rotation measures have been obtained in the high frequency regime, i.e., between 2 cm and 7 mm (Mutel, Denn & Dreier 2005; Gabuzda *et al.* 2006). Measurements of Faraday Rotation can be used to get a handle on the geometry of the ordered component of the magnetic field that is parallel to the line of sight. Using global VLBI polarimetry at 8 GHz, Kharb, Shastri, & Gabuzda (2005) have made the first detections of ordered pc-scale magnetic fields in the nuclei of radio-galaxies of the Fanaroff-Riley I type. Gabuzda (2006) has reviewed the various results.

With the VLA D-array at 5 GHz Gallimore *et al.* (2006) have found that kpc-scale radio-emitting outflows are common in the low-luminosity radio-quiet Seyfert galaxies, and their deviation from the radio-infrared correlation argues for these outflows being driven by the AGN rather than a starburst.

Greenhill (2007) reviews the results on OH and water masers in active galaxies. He argues that while the results argue for OH masers being mostly driven by intense star formation, the water masers are driven by the AGN.

With the Institute for Millimetre Radio Astronomy (IRAM) 30 m telescope, Ocana-Flaquer *et al.* (2008) have searched for the $CO(1\rightarrow0)$ and $CO(2\rightarrow1)$ emission lines from a sample of 52 radio-powerful galaxies and detect these lines in 28 objects. They find that the inferred mass of the molecular gas is systematically somewhat lower than that found for FIR-selected radio galaxies by Evans *et al.* (2005) from observations of the $CO(1\rightarrow0)$ transition with the NRAO 12 m telescope. The modeled value of the dust temperature is warmer than in normal spiral galaxies, pointing to heating of the dust by the AGN. With observations from the IRAM 30 m telescope, Evans *et al.* (2006) confirm that the CO luminosity in the host galaxies of quasars is enhanced relative to the infrared luminosity compared to the infrared-luminous galaxies. This implies either that the star formation is more efficient or that the AGN results in additional dust heating; the measurements of the HCN $(1\rightarrow0)$ and HCN $(2\rightarrow1)$ transitions (in the 3 mm and mm bands respectively) support the latter possibility. Evans *et al.* (2006) have the results on molecular gas measurements in low-redshift quasars.

2.7. *Cosmology*

The Cosmic microwave background (CMB) is one of the most important sources of cosmological information. The *WMAP* team released their three year observation data in 2006 (Spergel *et al.* 2007) and five year observation data in 2008 (Komatsu *et al.* 2008). Combined with the Arcminute Cosmology Bolometer Array Receiver (ACBAR) (Reichardt *et al.* 2008), Boomerang (Jones *et al.* 2006) and the Cosmic Background

Imager (CBI) (Readhead *et al.* 2004) data, the first five peaks in the temperature angular power spectrum have been detected. The treatment of the polarization data has been improved, the strong TE correlation at small l observed in the *WMAP* first year data is believed to be due to foreground contamination, and the EE correlation has also been observed. With this update, the epoch of reionization is around $z = 11$, more in line with CDM model prediction. The CMB observations also put strong constraint on the ϕ^4 inflation model. At small l, the power is still too low compared with theoretical prediction, but with better likelihood function it is slightly less severe than the first year data indicates. There is some sign of non-Gaussianity in the data. If confirmed, it will challenge the single field slow roll models of inflation.

Experiments for probing the epoch of re-ionization by red-shifted 21 Centimeter Array (21CMA) line is being pursued by a number of teams, most notably the 21CMA (<21cma.bao.ac.cn>), LOFAR (<www.lofar.org>), and MWA (<www.haystack.mit.edu/ast/arrays/mwa>), which are all dedicated experiments. In addition, the GMRT has also been equipped with new receivers for such searches (Pen *et al.* 2008).

3. Triennial reports of the Working Groups

3.1. *Working Group on Interference Mitigation*

This Working Group, chaired by Tasso Tzioumis (Australia), continues for a third IAU triennium. At the Division X business meeting during the IAU XXVI General Assembly in Prague, 2006, there was very strong support for the continuation of the Working Group. Interference Mitigation is destined to be even more important in the era of new radio astronomical facilities like the SKA and its various path-finders (LOFAR, ATA, ASKAP, MeerKAT, etc.).

The charter of the WG was revised in 2008 and the membership was updated. Details are given in <www.bao.ac.cn/IAU_COM40/WG/WgRF.html>. The WG promotes interference mitigation activities and reports at each IAU General Assembly. The group also commissions a technical review of the current state and progress on interference mitigation and this is distributed at each IAU general assembly. Reports from the 2003 and 2006 assemblies are available on the website. A new review has been commissioned and will be presented in the IAU XXVII General Assembly in 2009.

The membership of the WG comprises: Anastasios Tzioumis (chair, Australia), Willem A. Baan (Netherlands), Darrel T. Emerson (USA), Masatoshi Ohishi (Japan), Tomas E. Gergely (USA), J. Richard Fisher (USA), John E. B. Ponsonby (UK), Albert-Jan Boonstra (Netherlands) Wim van Driel (France, IUCAF chair), Haiyan Zhang (China), and Ronald D. Ekers (Australia).

3.2. *Working Group Astrophysical Important Spectral Lines*

This Working Group, chaired by Masatoshi Ohishi (Japan), keeps working actively and continues under its previous status. The WG met during the IAU XXVI General Assembly in Prague, and presented members with a possible list of lines between 1000 and 3000 GHz. The chairman called for further input; however, nothing was submitted. The list was used to support a provisional agenda item at the World Radio Communication Conference in 2011 (WRC-11) of the International Telecommunication Union (ITU) regarding the use of frequencies between 275 and 3000 GHz. The WRC-07 adopted the provisional agenda as an agenda toward WRC-11 with the scope to update a footnote that has identified several frequency bands up to 1000 GHz of interest to the radio astronomical community. The Working Party 7D (radio astronomy) of the ITU agreed to call for further information regarding astrophysically-important spectral lines up to 3000 GHz to support the agenda above. Therefore, an update of the list between 1000 and 3000 GHz and comments on the list are needed; the revised list will be submitted to the IAU XXVII General Assembly in 2009.

4. Conferences and meetings on radio astronomy

Since the 2006 General Assembly in Prague, about sixteen international meetings intimately related to radio astronomy have been held world wide. Regarding the IAU GA and non-GA activities in 2009, we encouraged the members and OC of Division X to submit proposals for Symposia, Joint Discussions, and Special Sessions. One proposed symposium and two Special Sessions are coordinated by Division X, and eight others are supported by Division X. Some representative meetings that show the recent developments that have occurred in radio astronomy are listed below:

- *SKA Science and Engineering Committee Meeting*, (SSEC 1), Perth, Australia, 7-8 April 2008
- *SKA Pathfinder Meeting*, Perth, Australia, 31 March - 1 April 2008
- *VSOP2 Meeting: Astrophysics and Technology*, ISAS, Sagamihara, Kanagawa, Japan, 3-7 December 2007
- *World Radcommunication Conference 2007* (WRC-07), Geneva, Switzerland, 22 October - 16 November 2007
- *Frank N. Bash Symposium 2007: New Horizons in Astronomy*, Austin, TX, USA , 14-16 October 2007
- From Planets to Dark Energy: the Modern Radio Universe, University of Manchester, UK, 1-5 October 2007
- *18th Meeting International SKA Steering Committee*, Manchester, UK, 6-7 October 2007
- *SKA2007 Inter-WG meeting*, Manchester, UK, 27-29 October 2007
- *European Radio Interferometry School*, Bonn, Germany, 10-15 September 2007
- *40 Years of Pulsars, Millisecond Pulsars, Magnetars*, Montreal, Canada, 12-18 August 2007
- *Frontiers of Astrophysics: A Celebration of NRAO's 50th Anniversary*, Charlottesville, VA, USA, 17-21 June 2007
- *LMT Inauguration and Related Radio Astronomy Symposium*, Mexico, 15-17 November 2006

Ren-Dong Nan
president of the Division

References

Altyntsev, A. T., Grechnev, V. V., Meshalkina, N. S., & Yan, Y. 2007, *Sol. Phys.*, 242, 111
An, T., Goss, W. M., Zhao, J.-H., *et al.* 2005, *ApJ* (Letters), 634, L49
Bastian, T. 2007, *ApJ*, 665, 805
Bergin, E. A. & Tafalla, M. 2007, *ARAA*, 45, 339
Biver, N., Bockelée-Morvan, D., Crovisier, J., *et al.* 2006, *A&A*, 449, 1255
Bougeret, J. L., Goetz, K., Kaiser, M. L., *et al.* 2008, *SSRv*, 136, 487
Bower, G. C., Roberts, D. A., Yusef-Zadeh, F., *et al.* 2005, *ApJ*, 633, 218
Bower, G. C., Goss, W. M., Falcke, H., *et al.* 2006, *ApJ* (Letters), 648, L127
Brajša, R., Benz, A. O., Temmer, M., *et al.* 2007, *Sol. Phys.*, 245, 167
Brown, J. C., Haverkorn, M., Gaensler, B. M., *et al.* 2007, *ApJ*, 663, 258

Camilo F., Ransom, S. M.; Gaensler, B. M.; *et al.* 2006a, *ApJ*, 637, 456

Camilo, F., Ransom, S. M., Halpern, J. P., *et al.* 2006b, *Nature*, 442, 892

Camilo, F., Ransom, S. M., Halpern, J. P., & Reynolds, J. 2007, *ApJ* (Letters), 666, L93

Carilli, C. L., Neri, R., Wang, R., *et al.* 2007, *ApJ* (Letters), 666, L9

Chatterjee, S., Vlemmings, W. H. T., Brisken, W. F., *et al.* 2005, *ApJ* (Letters), 630, L61

Chernov, G. P. 2006, *SSRv*, 127, 195

Cordes, J., Freire, P. C. C., Lorimer, D. R., *et al.* 2006, *ApJ*, 637, 446

Eckart, A., Baganoff, F. K., Schdel, R., *et al.* 2006, *A&A*, 450, 535

Evans, A. S., Mazzarella, J. M., Surace, J. A., *et al.* 2005, *ApJS*, 159, 197

Evans, A. S. 2006, *NewAR*, 50, 657

Evans, A. S., Solomon, P. M., Tacconi, L. J., Vavilkin, T., & Downes, D. 2006, *AJ*, 132, 2398

Fong, D., Meixner, M., Sutton, E. C., Zalucha, A., & Welch, W. J. 2006, *ApJ*, 652, 1626

Gabuzda, D. C., Rastorgueva, E. A., Smith, P. S., & O'Sullivan, S. P. 2006, *MNRAS*, 369, 1596

Gabuzda, D. C. 2006, in: W. Baan *et al.* (eds.), Proc. *8th European VLBI Network Symposium*, 2006, Torun, Poland, p. 11

Gabuzda, D. C., Vitrishchak, V. M., Mahmud, M., & O'Sullivan, S. P. 2008, *MNRAS*, 384, 1003

Gallimore, J. F., Axon, D. J.; O'Dea, C. P., Baum, S. A., & Pedlar, A. 2006, *AJ*, 132, 546

Gibson, S. J., Taylor, A. R., Higgs, L. A., Brunt, C. M., & Dewdney, P. E. 2005, *ApJ*, 626, 195

Gibson, S. J., Taylor, A. R., Stil, J. M.. *et al.* 2007, in: B. G. Elmegreen & J. Palouš (eds.), in: *Triggered Star Formation in a Turbulent ISM*, Proc. IAU Symposium No. 237 (Cambridge: CUP), p. 363

Gopalswamy, N., Mikić, Z., Maia, D., *et al.* 2006, *SSRv*, 123, 303

Greenhill, L. 2007, in: J. M. Chapman & W. A. Baan (eds.), *Astrophysical Masers and their Environments*, Proc. IAU Symposium No. 242 (Cambridge: CUP), p. 381

Han, J. L. & Zhang, J. S. 2007, *A&A*, 464, 609

Haverkorn, M., Gaensler, B. M., McClure-Griffiths, N. M., *et al.* 2006, *ApJS*, 167, 230

Hessels, J. W. T., Ransom, S. M.. Stairs, I. H., *et al.* 2006, *Science*, 311, 1901

Homan, D. C. & Lister, M. L. 2006, *AJ*, 131, 1262

Huang, G., Ji, H., & Wu, G. 2008, *ApJ* (Letters), 672, L131

Hudson, H. & Vilmer, N., 2007, *LNP*, 725, 81

Hyman, S. D., Lazio, T. J.. W., Kassim, N. E.. *et al.* 2005, *Nature*, 434, 50

Jiménez-Serra, I., Martín-Pintado, J., *et al.* 2007, *ApJ* (Letters), 661, L187

Jones, W. C., Ade, P. A. R., Bock, J. J., *et al.* 2006, *ApJ*, 647, 823

Kharb, P., Shastri, P., & Gabuzda, D. C. 2005, *ApJ* (Letters), 632, L69

Komatsu, E., Dunkley, J., Nolta, M. R., *et al.*, *ApJS*, in press [arXiv:0803.0547]

Kovalev, Y. Y., Kellermann, K. I., Lister, M. L., *et al.* 2005, *AJ*, 130, 2473

Kramer, M., Lyne, A. G., O'Brien, J. T., *et al.* 2006a, *Science*, 312, 549

Kramer, M., Stairs, I. H., Manchester, R. N., *et al.* 2006b, *Science*, 314, 97

Lazio, T. J. W. & LaRosa, T. N. 2005, *Science*, 307, 686

Lorimer, D., Stairs, I. H., Freire, P. C., *et al.* 2006, *ApJ*, 640, 428

Mahmud, M. & Gabuzda, D. C. 2008, in: T. A. Rector & D. S. De Young (eds.), *Extragalactic Jets: Theory and Observation from Radio to Gamma Ray*, ASP-CS, 386, 494

Marrone, D. P., Moran, J. M., Zhao, J.-H., Rao, R. 2007, *ApJ* (Letters), 654, L57

Mauerhan, J. C., Morris, M., Walter, F., & Baganoff, F. K. 2005, *ApJ* (Letters), 623, L25

McClure-Griffiths, N. M., Dickey, John M., Gaensler, B. M., *et al.* 2005, *ApJS*, 158, 178

McLaughlin, M., Lyne, A. G., Lorimer, D. R., *et al.* 2006, *Nature*, 439, 817

Mercier, C., Subramanian, P., Kerdraon, A., *et al.* 2006, *A&A*, 447, 1189

Mutel, R. L., Denn, G. R., & Dreier, C. 2005, in: J. Romney & M. Reid (eds.), *Future Directions in High Resolution Astronomy: The 10th Anniversary of the VLBA*, ASP-CS, 340, 155

Nieten, C., Neininger, N., Guélin, M., *et al.* 2006, *A&A*, 453, 459

Ocana-Flaquer, B., Leon, S., Lim, J., Combes, F., & Dinh-V-Trung 2008, in: *The Evolution of Galaxies through the Neutral Hydrogen Window*, in press [arXiv:0803.4443]

Papageorgiou, A. 2006, in: T. Hovatta, E. Nieppola & I. Torniainen (eds.), Proc. *8th ENIGMA meeting*, p. E6

Patel, N., Curiel, S., Sridharan, T. K., *et al.* 2005, *Nature*, 437, 109

Pen U.-L., Chang, T.-C., Peterson, J. B., *et al.* 2008, in: *The Evolution of Galaxies through the Neutral Hydrogen Window AIP-CP*, 1035, 75

Pick, M., Mason, G. M., Wang, Y.-M., *et al.* 2006, *ApJ*, 648, 1247

Qi, C., Wilner, D. J., Calvet, N., *et al.* 2006, *ApJ* (Letters), 636, L157

Ransom, S. M., Hessels, J. W. T., Stairs, I. H., *et al.* 2005, *Science*, 307, 892

Readhead, A. C. S., Mason, B. S., Contaldi, C. R., *et al.* 2004, *ApJ*, 609, 498

Reichardt, C. L., Ade, P. A. R., Bock, J. J., *et al.* 2008, *ApJ*, submitted [arXiv:0801.1491]

Remijan, A. J., Hollis, J. M., Lovas, F. J., *et al.* 2008, *ApJ* (Letters), 675, L85

Requena-Torres, M. A., Martín-Pintado, J., Rodríguez-Franco, A., *et al.* 2006, *A&A*, 455, 971

Ridge, N. A., Di Francesco, J., Kirk, H., *et al.* 2006, *AJ*, 131, 2921

Riechers, D. A., Walter, F., Carilli, C. L.. *et al.* 2006, *ApJ*, 650, 604

Shen, Z.-Q., Lo, K. Y., Liang, M.-C., Ho, Paul T. P., Zhao, J.-H. 2005, *Nature*, 438, 62

Snow, T. P. & McCall, B. J. 2006, *ARAA*, 44, 367

Spergel, D. N., Bean, R., Doré, O., *et al.* 2007, *ApJS*, 170, 377

Stil, J. M., Taylor, A. R., Dickey, J. M., *et al.* 2006, *AJ*, 132, 1158

Strasser, S. T., Dickey, J. M., Taylor, A. R., *et al.* 2007, *AJ*, 134, 2252

Tafoya, D., Gómez, Y., Anglada, G., *et al.* 2007, *AJ*, 133, 364

Taylor, A. R., Gibson, S. J., Peracaula, M., *et al.* 2003, *AJ*, 125, 3145

Trottet, G., Correia, E., Karlický, M., *et al.* 2006, *Sol. Phys.*, 236, 75

Vlemmings, W. H. T., Diamond, P. J., van Langevelde, H. J., *et al.* 2006, *A&A*, 448, 597

Wedemeyer-Böhm, S., Ludwig, H. G., Steffen, M., *et al.* 2007, *A&A*, 471, 977

Weiß, A., Downes, D., Neri, R., *et al.* 2007, *A&A*, 467, 955

Yan, Y., Yan, Y., Pick, M., Wang, M., Krucker, S., & Vourlidas, A. 2006, *Sol. Phys.*, 239, 277

Yusef-Zadeh, F., Wardle, M., Cotton, W. D., *et al.* 2007, *ApJ* (Letters), 668, L47

Zapata, L. A., Ho, P. T. P., Rodríguez, L. F., *et al.* 2006, *ApJ*, 653, 398

Zapata, L. A., Ho, P. T. P., Rodríguez, L. F., *et al.* 2007, *A&A* (Letters), 471, L59

Zavala, R. T. & Taylor, G. B., 2005, *ApJ* (Letters), 626, L73

Transactions IAU, Volume XXVIIA
Reports on Astronomy 2006–2009
Karel A. van der Hucht, ed.

© 2009 International Astronomical Union
doi:10.1017/S1743921308025817

DIVISION X / COMMISSION 40 / WORKING GROUP
GLOBAL VERY LONG BASELINE INTERFEROMETRY

CHAIR   Jonathan D. Romney
PAST CHAIR   Richard T. Schilizzi
MEMBERS   Simon T. Garrington, Leonid I. Gurvits,
                    Xiaoyu Hong, David L. Jauncey,
                    Hideyuki Kobayashi, Richard Porcas,
                    Robert A. Preston, Christopher John Salter,
                    Arpad Szomoru, Masato Tsuboi,
                    James S. Ulvestad, Alan R. Whitney

TRIENNIAL REPORT 2006 - 2009

1. Introduction

This triennium began with an action to re-create the Terms of Reference for the Working Group Global VLBI (WG-GV). These had been lost over the years since the Group was established in 1990. Fortunately, the personal archive of one long-term member yielded a copy of the original memorandum by R. D. Ekers, which was found to coincide quite well with current practice and areas of interest. New Terms of Reference, based on modern conditions, were drafted and accepted by both IAU and URSI.

2. Terms of Reference, d.d. 12 December 2006

The WG-GV was established in 1990 as a Working Group of Commission J at the URSI General Assembly in Prague, and recognized in 1991 at the IAU General Assembly in Buenos Aires as a Division X Working Group. The mandate of the WG-GV, its membership and chair, are reviewed at Commission J business sessions during URSI General Assemblies. The current mandate of the WG-GV comprises the following tasks: 1. To develop a concept for an International VLBI Network, comprising existing or future national and regional networks. 2. To promote compatibility of technology in VLBI instrumentation. 3. To serve as a liaison between ground-based observatories and national or international space agencies, for coordination of participation by ground radio telescopes in Space VLBI missions. The WG-GV carries out its tasks in conjunction with the organizations concerned, and presents summaries of its activities to URSI Commission J and IAU Division X at their respective General Assemblies. The current chair of the WG-GV is Jonathan Romney (NRAO, USA); membership is being updated.

3. Membership

Simultaneously, WG-GV membership was renewed through a poll of all previous members, who were asked either to confirm their continued membership or to propose replacements. The new membership was selected with an eye toward balanced representation in both a geographic sense, and in terms of expertise. Geographic balance was based on three longitude regions: Asia & Australia, Europe & Africa, and the Americas. Expertise available among current WG-GV members spans areas including scientific studies in astrophysics, astrometry, and geodesy; and techniques including e-VLBI and Space VLBI.

4. Working Group meetings

The WG-GV did not meet as a group during the 2006 IAU XXVI General Assembly, because the chair was not able to travel to that meeting. The WG met during the recent URSI General Assembly, held 7-16 August 2008 in Chicago, (IL, USA). Unfortunately, however, that meeting had to be held during a 'splinter meeting' session on the final day of the GA, when a number of members were not able to attend. It is not clear whether the past practice of holding WG-GV meetings during the General Assemblies of these international unions remains tenable, but other options are also problematic. The diversity of interests among the membership, and the graduation of VLBI beyond the specialized-technique category for which periodic international 'VLBI meetings' were appropriate, make it difficult to identify other venues at which a reasonable fraction of members can attend.

5. Basic tasks

Among the three basic tasks specified in the WG-GV's Terms of Reference, work was concentrated in the Space VLBI area during this triennium, in preliminary work related to the VSOP-2 mission. This effort was carried out by the chair individually at this early stage. Guidance, developed in consultation with the chair of URSI Commission J (who is a past chair of WG-GV), was provided to the mission in two areas: on compiling a suitable list of ground radio telescopes (GRTs) whose participation would be valuable to the mission; and on issuing invitations to the institutions operating those telescopes. As of this date, a substantial list of suitable GRTs exists, but the invitations have not yet been issued by the mission.

The chair was appointed, on an ex-officio basis, as a member of the VISC-2 mission advisory committee, and participated in that group's first face-to-face meeting on 13 May 2008 in Bonn, Germany, in a limited meeting during the recent URSI GA, and in a number of VISC-2 teleconferences.

The WG-GV's role in the VSOP-2 mission has been called into question recently by two major potential GRT networks. The European VLBI Network, principally, has expressed a strong preference to negotiate and schedule their participation bilaterally with the VSOP-2 mission. And, with NASA having decided not to fund the 'Samurai' proposal for US VSOP-2 mission support, the participation of the VLBA and other instruments of the US National Radio Astronomy Observatory will also have to be negotiated directly with the mission, on the basis of whatever funding arrangement may be developed. The mission itself strongly favors GRT coordination through the WG-GV. An overall resolution of this issue will have to be a major focus for the WG in the coming years.

6. The chair

The chair tendered his resignation to the president of IAU Division X and the chair of URSI Commission J, effective as of the end of the URSI GA. Selection of a successor chair is still under way. Therefore, this report covers only the approximately two-year subset of the triennium to date. Future directions will be determined by the new chair.

Jonathan D. Romney
chair of the Working Group

Transactions IAU, Volume XXVIIA
Reports on Astronomy 2006–2009
Karel A. van der Hucht, ed.

© 2009 International Astronomical Union
doi:10.1017/S1743921308025829

INTER-DIVISION IV-V-IX / WORKING GROUP
HISTORIC RADIO ASTRONOMY

CHAIR   **Wayne Orchiston**
VICE-CHAIR   **Kenneth I. Kellermann**
MEMBERS   **Rodney D. Davies, Suzanne V. Débarbat,**
                                   Masaki Morimoto, Slava Slysh,
                                   Govind Swarup, Hugo van Woerden,
                                   Jasper V. Wall, Richard Wielebinski

TRIENNIAL REPORT 2006 - 2009

1. Introduction

The Working Group was formed at the IAU XXV General Assembly in Sydney, 2003, as a joint initiative of Commissions 40 *Radio Astronomy* and Commission 41 *History of Astronomy*, in order to assemble a master list of surviving historically-significant radio telescopes and associated instrumentation found worldwide, and document the technical specifications and scientific achievements of these instruments. In addition, it would maintain an on-going bibliography of publications on the history of radio astronomy, and monitor other developments relating to the history of radio astronomy (including the deaths of pioneering radio astronomers).

Wayne Orchiston (Australia) and Ken Kellermann (USA) were appointed as the WG chair and vice-chair, respectively, at the WG Business Meeting at IAU XXVI General Assembly in Prague, 2006, where the new WG Committee and terms of reference for 2006 - 2009 were finalized.

The membership list of the WG contains the names of about one hundred astronomers who are active in the history of radio astronomy field or interested in it. More than 30 WG members actively researched aspects of radio astronomical history during 2005 - 2007.

2. National master lists of surviving historically-significant radio telescopes

WG members actively worked on national master lists for Australia, France, Germany, the Netherlands, the United Kingdom and the USA, and a number of research papers were prepared documenting individual instruments or instruments and research associated with specific radio astronomy field stations.

3. Preservation of historically-significant radio telescopes and associated relics

Stanford University recently demolished the five 60 ft antennas at their field station off Highway 280 (California), but the concrete piers there that many well-known radio astronomers carved their names on were preserved for posterity.

4. Publications on the history of radio astronomy

A list of papers and books on the history of radio astronomy published during 2005 - 2008 is scheduled to appear in the July 2008 issue of the *Journal of Astronomical History and Heritage*. Meanwhile, this journal continues to include history of radio astronomy papers in every issue.

5. The IAU XXVI General Assembly, Prague 2006

At the Prague General Assembly of the IAU the WG Committee held a Business Meeting, a one-day science meeting devoted to *The History of European Radio Astronomy* and a half-day Science Meeting on *Radio Astronomy Fifty Years Ago: From Field Stations to 'Big Science'*. Papers presented at the European Radio Astronomy meeting subsequently were published in a special issue of *Astronomische Nachrichten*, edited by Richard Wielebinski, Ken Kellermann & Wayne Orchiston.

6. The end of an era

The WG continued to record the deaths of pioneering radio astronomers and report these in its Progress Reports published in the *Journal of Astronomical History and Heritage* (Orchiston et al. 2004, 2005, 2006).

7. Graduate studies in history of radio astronomy

A notable recent development has been the introduction of an off-campus part-time PhD program in history of astronomy at James Cook University, Townsville, Australia. Currently, six students are researching aspects of early Australian and U.S. radio astronomy, supervised by Wayne Orchiston, Bruce Slee, Richard Strom and Richard Wielebinski. In addition, one student (supervised by Alastair Gunn and Wayne Orchiston) completed a U.K. history of radio astronomy Masters research project.

Wayne Orchiston
chair of the Working Group

References

Orchiston, W., Davies, R., Denisse, J.-F., *et al.* 2004, *JAHH*, 7, 53
Orchiston, W., Bracewell, R., Davies, R., *et al.* 2005, *JAHH*, 8, 65
Orchiston, W., Davies, R., Kellermann, K., *et al.* 2006, *JAHH*, 9, 203

Transactions IAU, Volume XXVIIA
Reports on Astronomy 2006–2009
Karel A. van der Hucht, ed.

© 2009 International Astronomical Union
doi:10.1017/S1743921308025830

DIVISION XI   SPACE & HIGH-ENERGY ASTROPHYSICS

ASTROPHYSIQUE SPATIALE & DES HAUTES ÉNERGIES

Division XI connects astronomers using space techniques or particle detectors for an extremely large range of investigations, from *in-situ* studies of bodies in the solar system to orbiting observatories studying the Universe in wavelengths ranging from radio waves to γ-rays, to underground detectors for cosmic neutrino radiation.

PRESIDENT   **Günther Hasinger**
VICE-PRESIDENT   **Christine Jones**
PAST PRESIDENT   **Haruyuki Okuda**
BOARD   **João Braga, Noah Brosch,**
                       Thijs de Graauw, Leonid I. Gurvits,
                       George Helou, Ian D. Howarth,
                       Hideyo Kunieda, Thierry Montmerle,
                       Marco Salvati, Kulinder Pal Singh

DIVISION XI COMMISSION

Commission 44   **Space & High-Energy Astrophysics**

INTER-DIVISION WORKING GROUP

Division IX-X-XI WG   **Astronomy from the Moon**

TRIENNIAL REPORT 2006 - 2009

1. Introduction

Division XI was born by merging Commission 44 *Space and High-Energy Astrophysics* and Commission 48 *High Energy Astrophysics* by the decision at the IAU General Assembly in Den Haag (1994). While historically, space astrophysics started with the high-energy range only accessible form space, in modern astronomy almost all subfields are utilizing or planning to build space observatories. On the other hand, during the last few years, ground-based very-high-energy (TeV) γ-ray astronomy has emerged as a truly observational astrophysical discipline, and other ground-based facilities like gravitational wave antennas, neutrino detectors high-energy cosmic ray arrays stand by to join the era of multi-messenger astrophysics.

The IAU Division XI therefore has a quite broad range of scientific interests and observing methods, which is however naturally biased towards high-energy astrophysics. This situation is also reflected in the Division membership, which is quite large, but the majority of the members are high-energy astronomers. Other astronomy communities often choose their Division by astrophysical interest rather than wavelength. This difficulty, which is also existing in some of the other more wavelength oriented Divisions, has been addressed in the previous triennial report of Division XI (2003 - 2006) and is a matter of an active debate in the IAU Executive Committee, which hopefully will lead to some adjustments to the Division structure.

Scientifically, the Division XI is enjoying very exciting and prosperous years, many missions in a variety of fields are in orbit and being operated actively. Several new missions have been launched recently, others are in preparation and waiting for launches in the near future. The missions cover almost the whole range of wavelengths, from radio astronomy to gamma ray astronomy, as well as even facilities related to particle astrophysics and gravitational wave detections. Their productivity is enormous and it is beyond the scope of this report to cover them completely, so only a brief summary of major activities will be given with help of the board members.

2. X-ray and γ-ray astronomy

2.1. *Facilities in operation*

2.1.1. *Chandra*

Since its launch in July 1999, the *Chandra* X-ray Observatory has been NASA's flagship mission for X-ray astronomy. With its arcsecond angular resolution, *Chandra*'s capabilities have contributed to a wide variety of astronomical studies including mapping the distribution of heavy elements and obtaining evidence for the acceleration of cosmic rays in supernova remnants, detecting X-ray flares from Sgr A* and a large population of nearby stellar mass black holes, measuring the size and intensity variability of knots in the jets from AGN, determining the power and age of AGN outbursts by measuring the cavities created in the hot gas in galaxies and clusters of galaxies, using clusters of galaxies to constrain cosmological parameters including the equation of state for dark energy, providing strong evidence for the existence of dark matter through X-ray and lensing observations of the Bullet cluster which show the positional offset of the luminous material from the dark matter, and through very deep observations, resolving much of the extragalactic X-ray background and of the Galactic ridge emission into discrete sources. Recently, *Chandra* could identify the youngest supernova remnant in our Milky Way, which must have exploded only 140 years ago. GoogleSky now includes *Chandra* images.

2.1.2. *XMM-Newton*

With its three co-aligned X-ray telescopes providing large collecting area and good high energy response up to 12 keV, ESA's *XMM-Newton* excels in obtaining high throughput X-ray imaging, timing and spectroscopy for a wide range of astrophysical sources from comets and planets to quasars and clusters of galaxies. Recent notable accomplishments include constraints on the spin as well as the mass of black holes in Active Galactic Nuclei, the discovery of quasi-periodic oscillations (QPOs) in several ULXs, the demonstration that cool gas is being uplifted by bubbles from the Virgo cluster core, and the release of the second XMM Serendipitous Source Catalog, or 2XMM, which contains 192,000 unique sources drawn from 3491 *XMM-Newton*-EPIC observations and covers a sky area of 330 square degrees. It is the largest X-ray source catalog ever produced. One of the recent highlights from the mission is the analysis of the slew observations, when the satellite moves from one target to another. This way a nova explosion could be identified recently, which although in principle visible to the naked eye, was missed by all optical observations. In a similar fashion, *XMM-Newton* could identify new candidates for for tidal disruption events from stars swallowed by black holes in the center of normal galaxies.

2.1.3. SUZAKU

The 5th Japanese X-ray astronomy satellite *SUZAKU* (*Astro-E2*) was launched in July 2005. With its excellent CCD energy-resolution in the soft X-ray range, its comparatively low intrinsic background, and its unprecedented sensitivity in the hard X-ray band, *SUZAKU* now works as a powerful wideband X-ray observatory yielding exciting results. Science highlights concern imaging spectroscopy of hot plasmas in SNRs, our Galactic center, and clusters of galaxies, revealing the distribution of density, temperature, and abundance, as well as broad Fe-K lines from black holes and neutron stars which reveal strong gravitational effects and black hole spins. *SUZAKU* broadband observations of AGN detected in the *Swift*-BAT hard X-ray survey indicate a new class of as yet unrecognized, heavily absorbed Seyfert galaxies. One of the recent highlights, e.g., concerns the discovery of solar-wind charge-exchange X-ray emission from the Earth's magnetosheath, which can have fundamental consequences for the interpretation of the soft X-ray background.

2.1.4. RXTE

The *Rossi X-Ray Timing Explorer* (*RXTE*) celebrated its 12th launch anniversary in 2007 and continues to make important scientific discoveries. *RXTE* observations have been crucial to the study of millisecond timing properties of accreting neutron stars as well as finding fast spinning neutron stars in low mass X-ray binaries, some of which may be intermittent pulsars. *RXTE*'s throughput and fast timing have enabled new results on the micro-quasar GRS 1915+105 that argue for a maximally spinning black hole. The QPO in XTE J1650−500 led to determine the currently smallest black hole mass of only 3.8 ± 0.5 $M_\odot$. *RXTE* observations of 4U 1636−53 directly showed the separation between unstable and marginally stable nuclear burning, as the QPO frequency drifted down to about 8 mHz and ceased, when a burst occurred. Another major accomplishment has been the demonstration that Anomalous X-ray Pulsars and Soft Gamma-ray Repeaters are magnetars, neutron stars with very strong magnetic fields. *RXTE*'s long lifetime has allowed recurrent transients such as the black hole system GX 339−4 to be studied during successive outbursts. Another recent highlight was the analysis of the Galactic ridge X-ray emission constructed from the 3-20 keV *RXTE* scan and slew observations. This emission follows closely the Galactic stellar mass distribution. These observations ended a decade long debate about the nature of the ridge emission, which is now interpreted as the sum of weak X-ray sources, mostly cataclysmic variables and coronal active binaries.

2.1.5. Swift and HETE-2

After the successful launch and operation of *HETE-2*, which provides γ-ray burst triggers on a regular basis, the field has been transformed by the launch of the *Swift* γ-ray burst mission. Since its launch in November 2004, *Swift* has detected more than 300 Gamma-ray bursts (GRBs), including nearly 30 short GRBs. A recent highlight was the discovery and identification of an exceptionally luminous burst at a redshift $z = 6.3$, comparable to the most distant galaxies and QSOs. The burst detected on 19 March 2008 made headlines as the first 'naked eye' GRB when the optical emission from this redshift $z = 0.937$ source reached 5.5 mag. In the burst GRB 060729, *Swift* observed the power-law decay of the X-ray afterglow for more than 100 d following the burst, suggesting that the jet may have a large opening angle. *Swift* also observed the actual supernova explosion of a giant star in January 2008, which was followed-up with extensive multi-wavelength observations. An ongoing *Swift*-BAT survey has detected 250 active galactic nuclei, with about 10-20% of them being heavily obscured by dust. Also in non-GRB observations, in April 2008, *Swift* detected a giant flare from the young star EV Lacertae that was thousands of times more powerful than any observed from our middle-aged Sun.

2.1.6. *INTEGRAL*

The *INTErnational Gamma-Ray Astrophysics Laboratory* (*INTEGRAL*), an ESA mission in cooperation with Russia and the United States, was launched in October 2002 to study gamma-ray sources in the energy range 15-10 MeV. A recent highlight is the discovery of a significant asymmetry of the spatial distribution of the 511 keV annihilation emission line in the Galactic disk, which resembles an asymmetry in the distribution of hard low mass X-ray binaries and possibly eliminates the need for more exotic explanations, such as those involving dark matter. *INTEGRAL* observations of Al^{26} at 1.8 MeV detect the signature of Doppler shifted emission caused by the rotation of the Milky Way. *INTEGRAL* observations have also discovered a population of obscured high-mass X-ray binaries with OB supergiants as the companion stars. Recently a long *INTEGRAL* observation of the binary pulsating X-ray source Her X-1 showed a significant increase in its spin that was several times greater than the average spin-up rate. The third *INTEGRAL* catalog lists 421 sources mostly identified with black hole or neutron star binaries or with active galaxies, but with 25% of the sources still unidentified.

2.1.7. *AGILE and GLAST*

Two new high-energy (GeV) γ-ray missions have been launched recently. *AGILE*, a small Italian satellite, was launched in April 2007 into an equatorial orbit with a very low particle background and is providing first scientific results. The *Gamma-ray Large Area Space Telescope* (*GLAST*) is an international and multi-agency mission designed to observe objects from 10 keV to 300 GeV. *GLAST* carries a Large Area Telescope (LAT), as well as a Gamma Burst Monitor (GBM). *GLAST* was successfully launched on 11 June 2008. Although just completing its two month check-out period, it has already observed a dozen γ-ray bursts and detected the blazar PKS 1502+106. With its unprecedented sensitivity and angular resolution, the *GLAST*-LAT will transform our knowledge of the γ-ray sky.

2.1.8. *H.E.S.S. and MAGIC*

During the last years, very-high-energy (TeV) γ-ray astronomy has emerged as an observational discipline, largely driven by the European-led High-Energy Stereoscopic System (H.E.S.S.) and the Major Atmospheric Gamma-ray Imaging Cherenkov (MAGIC) telescope. More than 70 TeV γ-ray sources have been detected, representing different galactic and extragalactic source populations such as young shell type supernova remnants, pulsar wind nebulae, giant molecular clouds, Wolf-Rayet stars, binary pulsars, micro-quasars, the Galactic Center, Active Galactic Nuclei, and large number of yet unidentified Galactic objects. Of particular cosmological importance is the upper limit on the extragalactic IR background light provided by the TeV observations of nearby AGN.

2.2. *Missions in preparation*

2.2.1. *Astrosat*

India's first Astronomy Satellite, *Astrosat*, is getting ready for launch by ISRO in late 2009. A general purpose multi-wavelength satellite, it will cover the X-ray spectral band from 0.3 to 150 keV with three co-aligned X-ray instruments, and provide simultaneous UV and optical coverage with two 40 cm telescopes. A scanning Survey Monitor will look for transient X-ray sources in the sky. The three X-ray instruments are (*i*) Large Area high pressure Xenon Proportional Counters (LAXPC) with a total area of 6000 cm^2; (*ii*) a soft X-ray focusing Telescope (SXT) of 2 m focal length using conical foil mirrors with gold surface and a CCD detector; and (*iii*) a CdznTe Imager (CZTI) having a coded aperture mask.

2.2.2. *NUStar*

The *Nuclear Spectroscopic Telescope Array (NuSTAR)* is a NASA SMEX mission with a launch foreseen in 2011 that will open the high-energy X-ray sky for sensitive imaging studies for the first time. By focusing X-rays at energies up to ~ 80 keV, *NuSTAR* will answer fundamental questions, e.g., about black holes and the formation of the elements.

2.2.3. *Spektr-Röntgen-Gamma (SRG)*

The Russian *Spektrum-Röntgen-Gamma (SRG)* platform is planned for a launch in 2011. It will perform a new sensitive X-ray all-sky survey in the energy band up to 12 keV. One of the main goals of the EROSITA telescope aboard *SRG* is to study Dark Energy through X-ray observations of $\sim 100\,000$ clusters of galaxies.

2.2.4. *Astro-H (NeXT)*

The X-ray observatory mission NeXT was approved by JAXA for a launch in 2013 and was named *Astro-H*. The two major new capabilities of this observatory will be high-resolution spectroscopy of $E/dE > 1000$ at 6 keV and hard X-ray imaging up to 60 keV.

2.2.5. *SVOM*

The *Space multi-band Variable Object Monitor (SVOM)* is a mission prepared in a Chinese/French collaboration aimed at continuing the investigation of γ-ray bursts. The launch is scheduled for 2013.

2.2.6. *The International X-ray Observatory (IXO)*

In May 2008 ESA and NASA established a coordination group involving ESA, NASA and JAXA, with the intent of exploring a joint mission merging the ongoing *XEUS* and *Constellation-X* into developing an *International X-ray Observatory (IXO)*. The starting configuration for the *IXO* study will be a mission featuring a single large X-ray mirror and an extendable optical bench with a 20-25 m focal length, with several instruments on an interchangeable focal plane. This plan establishes an *IXO* study, which will be the input to the US decadal process and to the ESA selection for the Cosmic Vision Plan. The *IXO* study supersedes the ongoing *XEUS* and *Constellation-X* activities.

3. UV, Optical, infrared/submillimeter and radio astronomy

3.1. *Missions in operation*

3.1.1. *GALEX*

The UV mission *GALEX* (*Galaxy Evolution Explorer*), a successor of *FUSE*, is making extensive surveys of galaxies with its wide field of view UV cameras. With its capabilities complementary to *HST*, the global structure of star formation activity in the galaxies is well contrasting the general stellar population.

3.1.2. *HST*

In August 2008, in its 18th year of operation, NASA's *Hubble Space Telescope* has celebrated its 100 000th orbit with a spectacular image of a star forming region near the star cluster NGC 2074. By a continuous restoration and addition of new instruments, *HST* keeps a leading role at the forefront of astronomical discoveries. The Advanced Camera for Surveys (ACS), installed in 2002, provided a major breakthrough in imaging capability and was at the heart of many new scientific results. It, e.g., helped to pin-point a likely intermediate mass black hole in the core of the spectacular globular cluster ω Centauri. The camera was used to image the GOODS North and South fields, as well as the COSMOS field to unprecedented depth and solid angle. It was also used to discover and study more than 20 new distant Type Ia supernovae and thus to trace the history of cosmic expansion over the last 10 billion yr. Unfortunately, ACS failed in 2007, but NASA decided to pursue the fourth *Hubble* Servicing Mission (SM4), which is planned in October 2008 and will bring new instruments to the telescope. In addition to the Wide Field Camera-3 (WFC3), the Cosmic Origins Spectrograph (COS) will be inserted. It will also aim to repair the Space Telescope Imaging Spectrograph (STIS) as well as ACS. New batteries and gyros will maintain the telescope's power and pointing systems and should hopefully extend *Hubble*'s lifetime to at least 2013, by which time NASA will be getting close to launching *JWST*.

3.1.3. *Spitzer Space Telescope (SST)*

Following the great success of *ISO*, the *Spitzer Space Telescope* was launched in 2003 into an Earth-trailing solar orbit, and is expected to continue operating at cryogenic temperatures until spring 2009. The combination of low background and new detector technology have enabled superb sensitivity and mapping speed. The spectroscopic coverage of *Spitzer* (5-38 μm) has encompassed phenomena ranging from planets orbiting other stars to galaxies at $z \simeq 3$, a mere 2 Gyr from the Big Bang, while *Spitzer* deep imaging (3.6, 4.5, 5.8, 8.0, 24, 70 and 160 μm) has reached the most distant objects known to date, galaxies at $z \simeq 7$. This has made *Spitzer* an essential element in the fabric of international astronomical research. More than 2 700 individuals from over 24 countries have become active *Spitzer* PIs or co-Is, and more than 1 100 refereed papers have been published based on 4.25 years of *Spitzer* operation.

Among its many discoveries and results, *Spitzer* demonstrated that mature galaxies exist at redshifts $z > 6$, i.e., less than 1 Gyr after the Big Bang. Some of these galaxies may be sufficiently massive to challenge the accepted theoretical framework of galaxy formation. It resolved the infrared background into discrete sources, galaxies and obscured AGN, and enabled a census of these objects during the peak of star and galaxy formation activity at redshifts $1 < z < 2$. *Spitzer* generated arcsecond-resolution maps of more than 200 deg^2 of the Milky Way, equivalent to a full accounting of all star formation in our Galaxy. It revealed the signature of dust formation in the Cas A supernova remnant. *Spitzer* data have led to an understanding of the building of planetesimals in the circumstellar disks of young stellar objects. It has shown that comets have a mineralogy similar to that of dust debris disks surrounding nearby stars. *Spitzer* went beyond the first direct detection of exoplanets, and has collected data to characterize their physical properties such as density and gas/rock structure and their surface temperature distribution. It has offered the first detection of water vapor absorption in an exoplanet atmosphere. After *Spitzer* runs out of its Helium cryogen, it is expected to continue operating in the two shortest bands at 3.6 and 4.5 μm with essentially the same sensitivity; this phase of warm operations with the focal plane between 25 and 30 K might last several years. This mode of operation will be open to the international astronomical community with the same peer-review philosophy followed until now.

3.1.4. *AKARI (ASTRO-F) and SPICA*

The first Japanese infrared satellite *AKARI* launched in February 2006, has been operated very successfully. It completed its all sky survey by August 2007, covering 94% of the whole sky with better sensitivity and angular resolution than *IRAS*, in the whole infrared region. The northern and southern ecliptic polar regions have been extensively mapped, providing extremely deep surveys, including the Lockman Hole and LMC areas. Besides the general survey, about 5 000 pointing observations have been made for individual sources, covering planetary sources, young and old stars, galaxies and ULIRGs. The capability of near infrared grism spectroscopy is particularly valuable to observe spectra of brown dwarfs and AGNs. *AKARI* is still continuing its post helium phase in the near infrared range. The bright source catalogue will be released late in 2008 for the team members and to the public one year later.

Following the success of the *AKARI*, a more advanced infrared mission *SPICA*, a cryogenically cooled 3.5 m telescope for mid- and far-infrared observations is under planning in ISAS/JAXA. The implementation of the mission is under development in collaboration with ESA, aiming at a launch in 2017. Collaboration with the US and Korean communities is also in discussion.

3.1.5. *Wilkinson Microwave Anisotropy Probe (WMAP)*

The first major cosmology mission *COBE* took a great step towards establishing the Big Bang cosmology, with the perfect confirmation of the black body nature of the cosmic microwave background radiation and the first evidence of anisotropy in its brightness distribution. These pioneering milestones were awarded the Nobel Prize in 2007. The advanced mission *WMAP*, launched in 2001, has confirmed *COBE*s results perfectly with higher sensitivity and angular resolution. A simple cosmological model with only six parameters fits not only the *WMAP* temperature and polarization data, but also small-scale CMB data, light element abundances, large-scale structure observations, and the supernova luminosity/distance relationship. Therefore, *WMAP* and the other precision cosmology measurements have provided us with a highly accurate reference frame for further astrophysical studies of the early Universe.

3.2. *Facilities in preparation*

3.2.1. *TAUVEX*

The Israeli TAUVEX three-telescope array for wide-field imaging in the ultraviolet has been adapted for a mission on-board the geo-synchronous technological demonstrator satellite GSAT-4 of the Indian Space Research Organization. The Flight Model was undertaking the final calibrations in summer 2008, prior to shipping to Bangalore to be integrated with the satellite. The launch is planned for 2008/2009 for a >3 yr mission.

3.2.2. *Herschel*

Herschel is an ESA Corner-Stone Mission with instrument contributions by NASA, expected for launch in early 2009 into an L2 orbit, on the same rocket carrying *Planck*. It will be the first versatile far-infrared and submillimeter observatory, and the largest IR-capable telescope ever flown into space. Herschel offers imaging and spectroscopy from 55 to 672 μm on a passively cooled 3.5 m telescope equipped with three instruments (HIFI, PACS and SPIRE). *Herschel* will build on *Spitzer* results, pursuing unsolved questions with higher spatial resolution, and with spectroscopy at longer wavelengths, including very high spectral resolution with the heterodyne instrument HIFI. *Herschel* is well equipped to advance our understanding of interstellar chemistry and kinematics, critical processes in star and planet formation, and the structure and nature of powerful extragalactic sources known as Ultra-Luminous InfraRed Galaxies (ULIRG).

3.2.3. *Planck*

The *Planck Surveyor* (*Planck*), an ESA mission with NASA contributions, is an all-sky survey covering with nine bands the range from $350\,\mu$m to 1 cm. Expected to launch in early 2009 and to operate for at least 15 months, it is aimed at improving dramatically our measurement of the intensity and polarization of the Cosmic Microwave Background (CMB) and of astrophysical foregrounds at a resolution down to a few arcminutes. *Planck*s longest wavelengths overlap *WMAP*, and the shortest wavelengths extend far into the submillimeter in order to improve the separation between galactic foregrounds and the CMB. *Planck* will yield exquisitely sensitive maps of the Milky Way at wavelengths (1.4, 2.1, 3.0, 4.3, 6.8, & 10.0 mm and 350, 550 and $850\,\mu$m) and polarization maps at a few arcminute resolution in some of these bands and open up new vistas of research in Milky Way emission mechanisms, ISM structure, dust properties, and star formation processes. These data will be even more valuable when combined with the complementary maps produced by *IRAS*, 2MASS and *AKARI*, and those expected from *WISE*.

3.2.4. *SOFIA*

The atmospheric window for infrared observations opens up substantially at stratospheric altitudes, even though background emission remains a limiting factor. The *Stratospheric Observatory for Far-Infrared Astronomy* (*SOFIA*) is designed to exploit this opening by flying a 2.5 m telescope on board a modified Boeing 747SP airplane; it is a joint venture of NASA and DLR, the German Aerospace Agency. Besides the large aperture, the most exciting prospects of *SOFIA* derive from the concept of an ever-evolving instrument complement, with continual technology updates, allowing for breakthroughs on *SOFIA* and a path to space adaptation. *SOFIA* has flown many hours with the telescope cavity door closed, and is expected to observe first light while airborne in 2009.

3.2.5. *WISE*

The *Wide-field Infrared Survey Explorer* (*WISE*) is a NASA MIDEX mission expected to launch in late 2009, and map the sky at 3.3, 4.7, 12 and $23\,\mu$m to a sensitivity ranging from 0.1 to a few mJy. With its cooled 40 cm telescope, it is expected to detect a hundred million objects, including the most luminous galaxies and quasars in the Universe. *WISE* will be able to complete the nearby census of stars and brown dwarfs down to extremely low temperatures within several tens of parsecs of the Sun.

3.2.6. *Astro-G (VSOP-2)*

The space radio VLBI mission *VSOP-2* has been approved by JAXA and is now named *Astro-G*. It is planned for launch in 2012. One of the major scientific aims is to resolve supermassive black hole shadows by very long baseline radio interferometry with ground based radio telescopes. It will achieve an angular resolution of $40\,\mu$m-arcsec.

3.2.7. *JWST*

The introduction of adaptive optics and interferometric technique has made ground based telescopes competitive to *HST* in resolution, particularly in infrared observations.

However, the superiority of the observational conditions in space, in particular regarding the infrared sky background and atmospheric transmission is irreplaceable. This is the motivation of implementing the 6.5 m *James Webb Space Telescope* (*JWST*), which is now under development and to be launched in 2013.

4. Other missions

4.1. *Gravitational waves. LISA*

Complementary to the operating ground-based gravitational wave detectors, as TAMA-300, GEO-600, Virgo and LIGO, a space mission, the *Laser Interferometer Space Antenna* (*LISA*) is under planning both in ESA and NASA. It is a long baseline laser interferometer for detection of gravitational waves, to be launched into an L2 orbit and a candidate for the ESA Cosmic Vision L1 slot.

4.2. *Ultra high-energy cosmic rays. Auger*

Cosmic rays of ultra-high energy remain one of the least understood phenomena in the Universe. A new international facility, the Pierre Auger observatory, a huge 3000 km^2 particle array combined with four wide-angle optical telescopes of atmospheric fluorescence light, located in Argentina, is now delivering its first, highly tantalizing results. These results demonstrate the existence of a statistically significant spectral feature (steepening or cutoff) at around 5×10^{19} eV. Also, a possible correlation between the arrival directions of the highest-energy cosmic rays, $\geqslant 6 \times 10^{19}$ eV, and the positions of nearby AGN has been reported. A Space Station borne mission called *JEM-EUSO* has been proposed as an international collaborative endeavor under Japanese leadership.

5. Summary

Space missions have been expanded to almost every field of astronomy and relevant to every kind of astronomical objects or phenomena. Dozens of missions are now in orbit and another dozen of missions are under development, awaiting launch in the coming years. The golden age of space observations will continue and the advanced missions will provide great strides of progress, in a variety of fields of astronomy. Many of the missions are carried out by large consortia, with multi-discipline participants. Sometimes, collaborations are organized among multiple missions and with groundbased observatories. Many missions are international in hardware developments and data analysis. As the scale of missions becomes larger, international collaboration is indispensable to afford the high cost and work force needed.

These developments should increase the importance of the activity of Division XI, but, at the same time, raise the complexity of its activities.

Günther Hasinger
president of the Division

Transactions IAU, Volume XXVIIA
Reports on Astronomy 2006–2009
Karel A. van der Hucht, ed.

© 2009 International Astronomical Union
doi:10.1017/S1743921308025842

INTER-DIVISION IX-X-XI WORKING GROUP ASTRONOMY FROM THE MOON

PRESIDENT	Sallie L. Baliunas
VICE-PRESIDENT	Yoji Kondo
PAST PRESIDENT	Norio Kaifu
MEMBERS	Oddbjørn Engvold, Norio Kaifu, Haruyuki Okuda, Yervant Terzian

TRIENNIAL REPORT 2006 - 2009

1. Introduction

During the period the Working Group had proposed and was granted renewed status by Division XI *Space and High-Energy Astrophysics*. Additionally the Working Group requested to be extended to Division IX *Optical and Infrared Techniques*, Division X *Radio Astronomy*, as well as Division XI .

We recall the memory of a member of the Organizing Committee Willem Wamsteker († 24 November 2005). Norio Kaifu, past chair of the Working Group, is appreciated for his past outstanding service.

Haruyuki Okuda had previously remarked, in the Business Meeting of the WG at the IAU XXVI General Assembly in Prague, 2006, that new opportunities are emerging across the world for lunar research. Haruyuki Okuda's remarks remain in the forefront of scientific goals. We are mindful of Haruyuki Okuda's parallel comment, namely, that such projects necessarily will be large in scope and infrastructure, requiring international cooperation.

2. Working Group Business Meeting at the IAU XXVII GA in 2009

Interested participants were asked to discuss and generate proposals for the 2009 IAU General Assembly concerning astronomy from the moon.

A Business Meeting that will include scientific reports has been requested to be scheduled at the IAU XXVII General Assembly in Rio de Janeiro, Brazil, August 2009.

The website <www.cfa.harvard.edu/moon> is being re-structured for the distribution of WG material.

Sallie L. Baliunas
chair of the Working Group

Transactions IAU, Volume XXVIIA
Reports on Astronomy 2006–2009
Karel A. van der Hucht, ed.

© 2009 International Astronomical Union
doi:10.1017/S1743921308025854

DIVISION XII   UNION-WIDE ACTIVITIES

ACTIVITÉS D'INTÉRÊT GÉNÉRAL DE L'UAI

Division XII consists of Commissions that formerly were organized under the Executive Committee, that concern astronomers across a wide range of scientific sub-disciplines and provide interactions with scientists in a wider community, including governmental organizations, outside the IAU.

PRESIDENT	Malcolm G. Smith
VICE-PRESIDENT	Françoise Genova
PAST PRESIDENT	Johannes Andersen
BOARD	Steven R. Federman, Alan C. Gilmore,
	Il-Seong Nha, Raymond P. Norris,
	Ian E. Robson, Magda G. Stavinschi,
	Virginia L. Trimble, Richard J. Wainscoat,
	Lars Lindberg Christensen

DIVISION XII COMMISSIONS

Commission 5	Documentation and Astronomical Data
Commission 6	Astronomical Telegrams
Commission 14	Atomic and Molecular Data
Commission 41	History of Astronomy
Commission 46	Astronomy Education and Development
Commission 50	Protection of Existing and Potential Observatory Sites
Commission 55	Communicating Astronomy with the Public

INTER-DIVISION X-XII WORKING GROUP

Inter-Division X-XII WG Historic Radio Astronomy

TRIENNIAL REPORT 2006 - 2009

1. Introduction

After the addition of Commission 55 at the IAU XXVI General Assembly in Prague, 2006, Division XII now consists of seven IAU Commissions. The first six had remained outside the IAU Division structure when it was created in 1994, because their activities did not fit comfortably into any of the eleven initial Divisions which are organised by scientific or technical discipline.

The activities of these seven Commissions cut across the full range of scientific sub-disciplines – comets to cosmology – but also deal with important policy interfaces to the outside world which are important for astronomy and for the IAU as a whole – e.g., data access rights, science education and outreach, and protection of the environment. Recognising this unifying theme, the IAU XXV General Assembly created Division XII during its meetings in Sydney, 2003, on the proposal of the Executive Committee. Light pollution control is an example of one of the Division's Commissions at work (Commission 50) and is currently featured in the 'Themes' highlighted on the IAU home page `<www.iau.org>`. The 'Themes' section comprises 'the most popular topics in astronomy and most frequently asked questions that the IAU Press Office has received over the years'. The work of Commission 55 is a particularly significant, Union-Wide activity during the *International Year of Astronomy 2009*, along with the underpinning role of Commission 46 and its Program Groups. The *IYA* also offers a special opportunity to showcase the work of Commission 41. For more on the *IYA2009*, visit `<www.astronomy2009.org>`.

For further details of Division activities, visit the Division XII web site `<http://www.iaudivisionxii.org/>`.

2. Reports of Commissions and Working Groups

The detailed reports and projected activities, for the period September 2006 - August 2009, of most of the above-listed individual Commissions, along with with those of their Working Groups and Programme Groups, follow below. These include a variety of conferences and meeting activities. An effort has been made to cross-fertilize the efforts of Commissions 41, 45, 50 and 55 in their support activities for *IYA2009*, following the lead of the IAU Executive Committee Working Group.

Malcolm G. Smith
president of the Division

Transactions IAU, Volume XXVIIA
Reports on Astronomy 2006–2009
Karel A. van der Hucht, ed.

© 2009 International Astronomical Union
doi:10.1017/S1743921308025866

COMMISSION 5

DOCUMENTATION AND ASTRONOMICAL DATA

DOCUMENTATION ET
DONNÉES ASTRONOMIQUES

PRESIDENT	**Raymond P. Norris**
VICE-PRESIDENT	**Masatoshi Ohishi**
PAST PRESIDENT	**Françoise Genova**
ORGANIZING COMMITTEE	**Uta Grothkopf, Oleg Yu. Malkov,**
	William D. Pence, Marion Schmitz,
	Robert J. Hanisch, Xu Zhou

COMMISSION 5 WORKING GROUPS

Div. XII / Commission 5 WG	**Astronomical Data**
Div. XII / Commission 5 WG	**Designations**
Div. XII / Commission 5 WG	**Libraries**
Div. XII / Commission 5 WG	**FITS**
Div. XII / Commission 5 WG	**Virtual Observatories, Data Centers and Networks**
Div. XII / Commission 5 TF	**Preservation and Digitization of Photographic Plates**

TRIENNIAL REPORT 2006 - 2009

1. Introduction

IAU Commission 5 deals with data management issues, and its working groups and task groups deal specifically with information handling, with data centres and networks, with technical aspects of collection, archiving, storage and dissemination of data, with designations and classification of astronomical objects, with library services, editorial policies, computer communications, ad hoc methodologies, and with various standards, reference frames, etc., *FITS*, astronomys *Flexible Image Transport System*, the major data exchange format, is controlled, maintained and updated by the Working Group *FITS*.

2. Highlights

Highlights of the 2006 - 2009 triennium have so far included:

• WG-AD: Three astronomers, including the WG-AD chair, participated in ICSUs *Strategic Committee for Information and Data* (SCID), resulting in a proposal to restructure the *World Data Centre System* and the *Federation for Astronomical and Geophysical data analysis Systems* (FAGS).

• WG-FITS: A New FITS Standard Document has been approved in July 2008 following a lengthy period of consultation and debate. The new standard is available on the FITS Support Office web site at `<fits.gsfc.nasa.gov>`.

3. Future Commission 5 activities

Planned activities include:

• A one-day session on *Astronomical data and the Virtual Observatory*, at the CO-DATA GA in Kiev, October 2008.

• A meeting of FAGS and WDC representatives at the 2008 CODATA GA in Kiev to discuss the establishment of the new WDS.

• An electronic discussion led by the WG-AD on *Data Challenges of Next-generation Astronomical Instrumentation*, in the months leading up to the IAU XXVII General Assembly in Rio de Janeiro, Brazil, August 2009.

• A three-day Special Session on Accelerating the Rate of Progress in Astronomy at the IAU XXVII General Assembly in Rio de Janeiro, Brazil, August 2009.

4. Task Force PDPP

As well as its Working Groups, whose reports are published separately in this volume, Commission 5 includes a Task Force on *Preservation and Digitization of Photographic Plates*, whose short report is included here. A full report is at `<www.atnf.csiro.au/people/rnorris/IAUC5/TF_PDPP2008.htm>`. Many Task Force members are involved in studies of long-term variability, which require monitoring over a time exceeding that available from digital observations alone. Historic data reproduced accurately from archived plates are thus central to such research. Several projects involve *Carte du Ciel* (CdC) plates; in 2006 the WG-CdC was merged with the TF-PDPP.

4.1. *Plate preservation*

A North American Astronomical Plate repository has been established at Pisgah Astronomical Research Institute, North Carolina, USA. The Cambridge (UK) plate store was dismantled and the contents repatriated or disseminated.

4.2. *Direct plates*

Measurements of several collections of astrograph plates with the USNO StarScan plate measuring machine have been completed with a repeatability of $0.2\,\mu$m. Positions and magnitudes from century-old Sydney Observatory Galactic Survey plates have now been cataloged, and early analyses are already contributing new science. A comparison between *Gaia* and CdC measurements along an equatorial belt is being proposed to search for changes, and evidence for the interstellar medium has also been detected from CdC plates. A catalogue of Vatican Observatory Schmidt plates (now digitized), plus 'thumbnail' scans, will be posted on the web. However, the preservation and digitization of the substantial historic collections of direct and objective-prism plates from the National, Qing Dao and Purple Mountain Observatories is hampered by lack of resources.

4.3. *Spectroscopy*

The digitizing of selected spectrograms borrowed from various US observatories for telluric ozone research has re-started at the DAO, including a collaboration with the Carnegie Institution to digitize a subset of Mount Wilson spectra. The 10-10 microdensitometer from KPNO has been brought in as a first step in setting up an international scanning laboratory. The major objective-prism survey of the Byurakan Observatory has now been digitized, and is in the public domain; early scientific results are already impressive.

Raymond P. Norris
president of the Commission

Transactions IAU, Volume XXVIIA
Reports on Astronomy 2006–2009
Karel A. van der Hucht, ed.

© 2009 International Astronomical Union
doi:10.1017/S1743921308025878

DIVISION XII / COMMISSION 5 / WORKING GROUP
ASTRONOMICAL DATA

CHAIR	Raymond P. Norris
MEMBERS	57

TRIENNIAL REPORT 2006 - 2009

1. The Working Group

Astronomers are well aware of the changing nature, volume, and complexity of astronomical data. Most of us are aware that next generation instruments, with Terabyte databases, are going to present enormous challenges to the way that we process data, and our current ways of managing astronomical databases will probably no longer work. So there are a number of initiatives within the astronomical community, most notably the Virtual Observatory, which aim to address these. However, many astronomers are not aware that similar challenges are being met in other disciplines (e.g., geosciences, life sciences, etc.) and that similar solutions are being sought. As a result, astronomy has much to gain from cross-fertilisation with other disciplines.

The Working Group on *Astronomical Data* (WG-AD) has two key roles: to act as a communication channel between the IAU and CODATA (ICSU *Committee on Scientific Data*), and to provide a forum for discussion of data issues. Through CODATA, the WG-AD also contributes to strategic ICSU activities, such as ICSUs *Strategic Committee on Information and Data* (SCID), of which the WG-AD chair is a member. SCID has recently released for review a report which proposes, amongst other things, a re-structuring of the *World Data Centre* system and the *Federation for Astronomical and Geophysical data analysis Services* (FAGS).

At the 2006 IAU XXVI General Assembly in Prague, 2006, WG-AD held a Special Session on *Astronomical Data Management*, an important component of which was an electronic discussion in the months leading up to the GA. The WG-AD plans to conduct a similar e-discussion in the months leading up to the 2009 IAU XXVII General Assembly in Rio de Janeiro, Brazil, 2009, focussing particularly on the challenges posed by new instrumentation demands, new ways of managing data, including the Virtual Observatory, and the advent of Petabyte astronomical databases. To ensure that this discussion is relevant and timely, the WG-AD initiated a membership drive to enlist the expertise of those individuals actively engaged in managing next-generation instruments and their data management challenges. As a result of this the membership of the WG-AD has increased in 2008 from 34 to 58. See: See <`http://www.atnf.csiro.au/people/rnorris/WGAD/members.htm`>.

2. CODATA

CODATA, the ICSU *Committee on Data for Science and Technology*, is the cross-disciplinary forum in which such issues are addressed. Given the rapidly-changing nature of data and its management, it is unsurprising that CODATA is in a process of reinvigoration, and is exploring ways in which it can deliver greater value to science.

The CODATA General Assembly in Beijing, October 2006, included a session on *Astronomical Data Management*, with talks given by several WG-AD members. In its recently-released Draft Strategic Plan, CODATA identifies three strategic CODATA initiatives, including the *Global Information Commons for Science Initiative* (GICSI) , and identifies a number of ways of achieving greater engagement with the broader scientific community. It also recognises the success of existing CODATA activities such as the *Data Science Journal*, the Task Groups, and the biennial CODATA conferences, and proposes ways in which these might be made even more effective. CODATA has also adopted a new Mission Statement to reflect these changes.

In conjunction with these new initiatives, CODATA membership is growing. Australia, Ireland, and the UK are in the process of becoming members of CODATA, as has the International Union of Geodesy and Geophysics (IUGG).

Another important development has been the release of the Draft Report of ICSUs Strategic Committee on Information and Data (SCID), whose members included three astronomers. Recommendations of this report include:

• Support for the CODATA Strategic Plan, and a recommendation that CODATA focus on its strategic initiatives.

• A proposed restructuring of the *Federation of Astronomical and Geophysical data analysis Services* (FAGS) and *World Data Center* system (WDC) into a new body, the *World Data Services* (WDS). A measure of success of this new body will be the value that it delivers to bodies such as the existing astronomical data centres and the Virtual Observatory.

• A proposed restructuring of ICSU activities, to enable ICSU to assert a strategic leadership role on behalf of the global scientific community in relation to the policies, management and stewardship of scientific data and information, including the establishment of a new body to work closely with CODATA and WDS.

The next CODATA Conference will take place in Kiev, Ukraine, in October 2008. The conference will include a one-day session on *Astronomical Data and the Virtual Observatory* and also a meeting of FAGS and WDC representatives to discuss the establishment of the new WDS. Other challenges include the bottleneck in transferring data from journals to data centres. This difficult challenge is being faced by several disciplines, and we have much to learn from each other. Another potential example is that ICSU may be useful in helping these centres achieve their ambitious goals. But it is early days, and we are only just starting to explore potential ways forward.

Raymond P. Norris
chair of the Working Group

Transactions IAU, Volume XXVIIA
Reports on Astronomy 2006–2009
Karel A. van der Hucht, ed.

© 2009 International Astronomical Union
doi:10.1017/S174392130802588X

DIVISON XII / COMMISSION 5 / WORKING GROUP
LIBRARIES

CHAIR   Uta Grothkopf
CO-CHAIR   Fionn Murtagh
MEMBERS   Christina Birdie, Marsha Bishop,
                     Laurence Bobis, Donna J. Coletti,
                     Brenda G. Corbin, Monique Gomez,
                     Halima Naimova

TRIENNIAL REPORT 2006 - 2009

1. Introduction

The IAU Working Group on *Libraries* was officially recognized for the first time in the Transactions of the IAU XXIA *Reports on Astronomy* (McNally 1991), but librarians were involved in Commission 5 activities before that (largely due to the efforts of George A. Wilkins, president of the Commission from 1982 to 1988, see Transactions of the IAU XXA *Reports on Astronomy*, Swings 1988). Wayne Warren (NASA GSFC, MD, USA) and Helen Knudsen (Calech, CA, USA) were the group's first chairpersons, Brenda Corbin (U.S. Naval Observatory) became Helen Knudsen's successor in 1990. Since 1996, Fionn Murtagh (School of Computer Science, Queen's University of Belfast, Northern Ireland) and Uta Grothkopf (European Southern Observatory, Germany) have been co-chairs.

The WG on *Libraries* aims to facilitate better cooperation between astronomers and librarians. The publication paradigm continues to shift from printed material to electronic formats for scientific literature, leading to an evolved concept of library services and information access. Librarians are taking on new and diversified roles. While demands for traditional services continue, the importance of accessibility and archiving of older materials is widely recognized. Long-term solutions, developed cooperatively by publishers, scientists and librarians, are needed in order to guarantee future access to today's knowledge.

Some projects of interest to both librarians and astronomers (though not developed under the auspices of this group) are the International Astronomy Meetings List, a database of Reports of Observatories, listings of IAU colloquia and astronomical newsletters, the Directory of Astronomy Librarians and Libraries, a book reviews database, digitization projects of historical documents, maintenance of core lists of astronomy books and astronomy and physics journals and the mailing lists Astrolib and EGAL. The IAU Thesaurus, compiled by R.M. & R.R. Shobbrook under the auspices of the IAU, may prove to be of use as a tool for automated information retrieval, provided that necessary updates can be implemented. The IAU funded the first *Library and Information Services in Astronomy* (LISA) conference, held in 1988 in Washington, DC, USA (IAU Colloquium No. 110) and supported the second (IAU Technical Workshop, Garching, Germany, 1995). Since then, LISA has developed into a series of conferences which are held at regular intervals; LISA VI will be held in Pune, India, in February 2010.

2. Membership developments

In 2007 and 2008, we were happy to welcome two new members to our Working Group: Marsha Bishop, (National Radio Astronomy Observatory, USA), and Christina Birdie (Indian Institute of Astrophysics, India). The role of librarians among the IAU is becoming ever more visible: by now, five members (Brenda Corbin, Donna Coletti, Peter Hingley, Monique Gomez and Uta Grothkopf) have been accepted by the IAU as full members rather than consultants, which has previously been the typical status of librarians among the IAU.

3. Activities within the past triennium

During the IAU XXVI General Assembly in Prague, Czech Republic, August 2006, the Working Group on *Libraries*, organized a Business Meeting jointly with the Working Group on *Publishing*.

- Françoise Genova (CDS) described the *Use and Validation of the IAU Astronomy Thesaurus in Ontologies*, a collaboration of French computing departments in the frame of the *Massive Data in Astronomy* project.
- Uta Grothkopf (ESO) and Brenda Corbin (USNO) presented a summary of the *Library and Information Services in Astronomy. V* (LISA V) conference,held in Cambridge, MA, USA, June 2006.
- Brenda Corbin also gave an *Overview of Observatory Archives* with a special focus on the setup and maintenance of the archives of Lowell Observatory, the National Radio Astronomy Observatory and Yerkes Observatory in the USA.
- Guenther Eichhorn (ADS) presented an update on the ADS *Abstract Service* and reported on the (close) *Cooperation between the ADS and Libraries*, both in the early days of the ADS as well as nowadays.
- Karel A. van der Hucht (IAU Assistant General Secretary) explained *The New Structure of the IAU Proceedings Series*: after the reorganization of IAU symposia and colloquia (with the colloquia having been merged into the symposia series), the WG Publishing is now pleased to report much shorter production times for proceedings volumes.
- Uta Grothkopf reported on the *Use of Bibliometrics by Observatories*, based on a questionnaire distributed among major observatories.

Slides of the individual presentations can be found at <www.eso.org/sci/libraries/ IAU-WGLib/iau06/>.

Subsequent to the colloquium *Future professional communication in Astronomy*, held in Brussels, Belgium, in June 2007 (Heck & Houziaux 2007), Terry Mahoney (Instituto de Astrofisica de Canarias, Spain), and a group of librarians, among them Karen Moran (Royal Observatory Edinburgh, UK), Shireen Davis (South African Astronomical Observatory, Cape Town), and Monique Gomez (Insituto de Astrofisica de Canarias, Spain) wrote a *Declaration Concerning the Evolving Role of Libraries in Research Centres*, also called the *Manifesto*. This document "reflects the concerns of its drafters at the increasing invisibility of research libraries *vis-á-vis* recent accelerated changes in publishing and reader-access technology". The document is meant to raise discussion about these topics and, ultimately, the authors hope that it might be considered for adoption as official IAU policy. The declaration has been published in the December 2007 issue of The Observatory (Mahoney *et al.* 2007), and in the July 2008 issue of the *IAU Information Bulletin* (Mahoney *et al.* 2008).

In January 2008, the ESO librarians Uta Grothkopf and Christopher Erdmann compiled a document on *Open Access – state of the art*. The paper gives a brief introduction to open access (OA), helps readers to distinguish between open access and open access publishing, summarizes variations in publishers' approaches towards OA, and reports on the situation in astronomy. Recent developments in Europe and the U.S. are briefly explained. The document has been published in the July 2008 issue of the *IAU Information Bulletin* (Grothkopf & Erdmann 2008).

4. Closing remarks

The Working Group on *Libraries* intends to get involved at the IAU XXVII General Assembly, to be held in Rio de Janeiro, Brazil, August 2009. Sessions may focus on data access and preservation, science metrics, and the role of librarians in fostering research.

The web site of the Working Group can be found at <`www.eso.org/libraries/IAU-WGLib/`>.

Uta Grothkopf & Fionn Murtagh
chair & co-chair of the Working Group

References

Grothkopf, U. & Erdmann, C. 2008, *IAU Information Bulletin*, No. 102, p. 64

Heck, A. & Houziaux, L. (eds.) 2007, *Future Professional Communication in Astronomy*, Proc. Coll. Brussels, 10-13 June 2007 (Bruxelles: Acad. Royale de Belgique)

Mahoney, T. J., *et al.* 2007, *The Observatory*, 127, 401

Mahoney, T. J., *et al.* 2008, *IAU Information Bulletin*, No. 102, p. 70

McNally, D. (ed.) 1991, *Reports on Astronomy*, Transactions of the IAU XXIA (Dordrecht: Kluwer Acad. Publ.)

Swings, J.-P. (ed.) 1988, *Reports on Astronomy*, Transactions of the IAU XXA (Dordrecht: Kluwer Acad. Publ.)

Transactions IAU, Volume XXVIIA
Reports on Astronomy 2006–2009
Karel A. van der Hucht, ed.

© 2009 International Astronomical Union
doi:10.1017/S1743921308025891

DIVISION XII / COMMISSION 5 / WORKING GROUP
FITS DATA FORMAT

CHAIR   William D. Pence
VICE-CHAIR   François Ochsenbein
PAST CHAIR   Donald C. Wells
MEMBERS   Steven W. Allen, Mark R. Calabretta,
                    Lucio Chiappetti, Daniel Durand,
                    Thierry Forveille, Carlos Gabriel,
                    Eric Greisen, Preben J. Grosbol,
                    Robert J. Hanisch, Walter J. Jaffe,
                    Osamu Kanamitsu, Oleg Yu. Malkov,
                    Clive G. Page, Arnold H. Rots,
                    Richard A. Shaw, Elizabeth Stobie,
                    William T. Thompson, Douglas C. Tody,
                    Andreas Wicenec

TRIENNIAL REPORT 2006 - 2009

1. Introduction

The Working Group *FITS* (WG-FITS) is the international control authority for the *Flexible Image Transport System* (FITS) data format. The WG-FITS was formed in 1988 by a formal resolution of the IAU XX General Assembly in Baltimore (MD, USA), 1988, to maintain the existing FITS standards and to approve future extensions to FITS.

2. Developments within the past triennium

The two most significant activities of the WG during the 2006 - 2009 triennium were to produce a new version of the formal FITS standard document, which precisely defines the requirements of the FITS format, and to create a new registry for documenting existing FITS conventions. These activities are described in the following two subsections.

2.1. *New FITS standard document*

The requirements of the FITS format are formally defined in the FITS standard document which is controlled by the WG-FITS. The last major revision to this document was made in 1999, so in early 2007, a small subcommittee (L. Chiappetti, C. Page, W. Pence, R. Shaw, and E. Stobie) was appointed to consider ways to update the document. After several months of work, this subcommittee produced a new draft of the FITS standard which, while not proposing any major changes to the FITS requirements, did contain dozens of small technical changes, as well as hundreds of other editorial changes to better organize and clarify many sections in the document. One significant change was the addition of a new section that summarizes the conventions for representing world coordinate information, such as in defining the position of an image on the celestial sphere. This information was previously only available in three separately published papers.

This proposed new FITS standard document went through a rigorous review process starting with a public comment period on the official FITS newsgroup, `<sci.astro.fits>`, from July through October 2007. A slightly revised version of the document was then formally endorsed by the four regional FITS committees (representing North America, Europe, Japan, and Australia & New Zealand) in February 2008. This was then followed by a final review by the WG-FITS which formally approved this new version of the FITS standard in July 2008. It is now publicly available on the FITS Support Office web site at `<fits.gsfc.nasa.gov>`.

2.2. *Registry of FITS conventions*

The other main activity of the WG-FITS during this period was to create a registry on the FITS Support Office Web site for documenting the many existing FITS conventions that are in use within the FITS community. These conventions have been independently developed by many different groups. The registry provides a convenient central location for documenting the various FITS keywords or other FITS data structures that are used by each convention. The registration process is intentionally quite simple, so unlike changes to the FITS standard which can take years to complete, a convention can be entered into the registry in a matter of weeks. There is a required public comment period before the convention is registered, but this is mainly just to ensure that the documentation about the convention is sufficiently clear and complete.

As of July 2008, the following conventions have been registered:

• CONTINUE Long String Keyword convention for writing string keyword values that are longer than the 68-character limit of a single FITS keyword.

• FOREIGN file encapsulation for wrapping other types of files in FITS.

• CHECKSUM keyword convention for verifying the integrity of FITS HDUs.

• INHERIT keyword which indicates that a HDU should inherit the primary header keywords.

• Column Limits keywords (TLMINn/TLMAXn and TDMINn/TDMAXn) which define the legal and actual minimum and maximum values in a table column.

• Tiled Image Compression convention for dividing an image into a grid of tiles and then storing each compressed tile in a variable length array column of a binary table.

• ESO HIERARCH Keyword convention uses a hierarchical structure to define the keyword name. This convention can be generalized to support keyword names longer than 8 characters or containing characters that would not be allowed in a standard FITS keyword name.

• Euro3D interchange data format for integral field spectroscopy in which 1-dimensional spectra are obtained at multiple positions over a 2-dimensional spatial field of view.

• OIFITS data format for optical interferometry.

• Multi-Beam FITS (MBFITS) data format for single-dish mm/submm telescopes.

• Hierarchical Grouping convention for defining hierarchical associations of HDUs.

Several more conventions are pending, and it is expected that the registry will continue to grow each year.

3. Future plans

There are a number of other important FITS issues that need to be addressed by the WG-FITS in the next triennium. Some of these are

- Develop standard conventions for specifying time and date coordinates.
- Complete the draft specification for representing distortions in world coordinate systems.
- Rewrite the FITS Users Guide which is currently obsolete.
- Consider elevating some of the conventions that are documented in the FITS registry to the status of a full FITS standard.

These activities are certain to keep the WG-FITS busy for the foreseeable future.

William D. Pence
chair of the Working Group

Transactions IAU, Volume XXVIIA
Reports on Astronomy 2006–2009
Karel A. van der Hucht, ed.

© 2009 International Astronomical Union
doi:10.1017/S1743921308025908

DIVISION XII / COMMISSION 5 / WORKING GROUP
VIRTUAL OBSERVATORIES, DATA CENTERS & NETWORKS

CHAIR   Robert J. Hanisch
MEMBERS   Beatriz Barbuy, Robert D. Bentley,
           Piero Benvenuti, Daniel Egret,
           Toshio Fukushima, Françoise Genova,
           Preben J. Grosbol, George Helou,
           Raymond P. Norris, Peter Quinn

TRIENNIAL REPORT 2006 - 2009

1. Introduction

The Working Group *Virtual Observatories, Data Centers, and Networks* was established under Commission 5 at the Prague General Assembly in 2006. The purpose of the WG is to provide IAU oversight of the activities of the *International Virtual Observatory Alliance* (IVOA, <www.ivoa.net/>), to encourage data centers and other data providers to archive and publish data according to IVOA standards, and to help assure that astronomical research facilities are electronically linked with current network technologies. The WG coordinates activities closely with the WG-*FITS*, as the IVOA uses FITS as its primary format for binary data exchange, and the WG on *Astronomical Data*.

2. Activities

To date the WG has been inactive, the bulk of standards developments being managed within the IVOA context. As various virtual observatory projects transition from their development phases to operational facilities, it will be the purview of the WG to review the IVOA standards (primarily from a process perspective, to assure that astronomy community interests are being served) and to act as a liaison with astronomical data providers and encourage their adoption of IVOA standards. The WG will convene as part of Commission 5 meetings during the IAU XXVII General Assembly in Rio de Janeiro, Brazil, in 2009.

Robert J. Hanisch
chair of Working Group

Transactions IAU, Volume XXVIIA
Reports on Astronomy 2006–2009
Karel A. van der Hucht, ed.

© 2009 International Astronomical Union
doi:10.1017/S174392130802591X

COMMISSION 6

ASTRONOMICAL TELEGRAMS
TÉLÉGRAMMES ASTRONOMIQUES

PRESIDENT	Alan C. Gilmore
VICE-PRESIDENT	Nicolay N. Samus
PAST PRESIDENT	Kaare Aksnes
ORGANIZING COMMITTEE	Daniel W. E. Green, Brian G. Marsden, Syuichi Nakano, Elizabeth Roemer, Jana Tichá, Hitoshi Yamaoka

TRIENNIAL REPORT 2006 - 2009

1. Introduction

From Director Dan Green's report, following this report, it is obvious that the *Central Bureau for Astronomical Telegrams* (CBAT) continues its excellent work. The *Electronic Telegrams* (CBETs), established in the previous triennium, have become the regular means for fast communication, with the *Circulars* providing the official and archival record of discoveries and designations. It is regretted that subscriptions to the printed *Circulars* continue to decline, but inevitable in this age of electronic communication.

2. Funding

The need for the Director to continually seek external funding for the *Bureau* is unfortunate so the Commission is grateful to the U.S. National Science Foundation for its support. The Commission is the also very grateful to Dan Green for his efforts to obtain funding, on top of the extraordinary work-load of running the *Bureau*.

3. CBAT and the Minor Planet Center

The co-operation between the *Bureau* and the *Minor Planet Center*, where CBAT Assistant Director Gareth Williams works, provides very efficient processing and notification of comet and other solar-system discoveries.

Alan C. Gilmore
president of the Commission

Transactions IAU, Volume XXVIIA
Reports on Astronomy 2006–2009
Karel A. van der Hucht, ed.

© 2009 International Astronomical Union
doi:10.1017/S1743921308025921

DIVISION XII / COMMISSION 6 / SERVICE
CENTRAL BUREAU FOR ASTRONOMICAL TELEGRAMS

DIRECTOR   **Daniel W. E. Green**
ASSISTANT DIRECTOR **Gareth V. Williams**

TRIENNIAL REPORT 2006 - 2009

1. Introduction

There were 401 *IAU Circulars* and 1248 *Central Bureau Electronic Telegrams* (*CBETs*) issued during the triennium 2005-2008:

Dates	Circulars	CBETs
2005 July-Dec.	Nos. 8556-8652	Nos. 175-345
2006 Jan.-June	Nos. 8653-8727	Nos. 346-565
2006 July-Dec.	Nos. 8728-8789	Nos. 566-791
2007 Jan.-June	Nos. 8790-8853	Nos. 792-989
2007 July-Dec.	Nos. 8854-8906	Nos. 990-1189
2008 Jan.-June	Nos. 8907-8956	Nos. 1190-1422

2. CBETs

This was the second Triennium in which the Central Bureau issued electronic-only *CBETs* to aid in the rapid dissemination of reports. Only 174 *CBETs* had been issued in the 2002 - 2005 Triennium, but they evolved into a full supplemental and complemental publication to the *Circulars* during this past Triennium, reflecting the move toward electronic publication and away from printed publication.

3. CBETs and Circulars

It is the intention that all items requiring designation by the Central Bureau (super-novae, novae, comets, solar-system satellites) continue to be be noted on the printed *Circulars*; supernovae are now published almost entirely on *CBETs*, due to their great numbers and to the unfortunate lack of financial support from many in the supernova community (though a few supernova researchers continue to be long-time supporters of the CBAT through their paid subscriptions), but summaries of newly announced super-novae have been published later on the *Circulars* with citation to the specific *CBETs*.

4. Distribution

Subscribers may receive the *Circulars* in printed and/or electronic form, the latter being available by e-mail or by logging in to the Computer Service, either directly on the Bureau's computers or via the World Wide Web. Since 1997, the *Circulars* have been made freely available at the CBAT website, but following complaints by paying subscribers, the general delay in posting for non-subscribers is now about one year (expanded in late 2004 from the previous 4-6 weeks). The *CBETs* are also posted at the CBAT website, with the earlier *CBETs* also available freely (and, like the *Circulars*, are indexed via the web-based bibliographic *Astrophysics Data System* (ADS), the ostensible replacement to the now-defunct *Astronomy and Astrophysics Abstracts* series).

5. Funding

Funding is currently being sought to allow the CBAT to post all electronic *Circulars* and *CBETs* freely. A step in this direction was thankfully accomplished with the help of the U.S. National Science Foundation's accepting a proposal to fund the CBAT at a 50% level beginning in February 2008. Great additional effort is being expended by the CBAT Director to seek alternate – and more extensive and long-lasting – sources of income, including from international sources.

The number of subscriptions to the printed *Circulars* was around 125 at the end of June 2008, continuing the slow decline (which peaked around 800 in the 1990s). The number of subscribers to the joint Computer Service of the CBAT and the *Minor Planet Center* (MPC) has remained relatively stable, though it too slowly declined by about 10% in the last triennium (to around 435 in June 2008). There, nonetheless, continues to be considerable interest to maintain having a printed copy of the *Circulars*, though subscription rates may need to increase to cover the increasing costs of printing and postage.

6. Supernovae and comets

Supernovae and comets have continued to dominate the activities of the *Bureau*, as related in the annual reports of the Bureau as published in the *IAU Information Bulletins* and made available at the Commission 6 website `<cfa-www.harvard.edu/iau/Commission6.html>`. The pattern continues regarding increasing numbers of comet discoveries from NEO surveys being first reported as objects of asteroidal appearance, where they are often posted on the MPC's 'NEO Confirmation' web page because of their unusual motion – follow-up observations then showing some of the objects to be of cometary appearance. The working link between the CBAT and MPC continues to be highly useful in the joint announcement of such objects.

7. The CBAT web site

The CBAT has expanded its website over the past triennium to include much information regarding supernovae, novae, comets, satellites of solar-system planets, etc., – a snapshot of which can be gleaned from the 'Headlines' web page `<cfa-www.harvard.edu/iau/Headlines.html>`. Plans are underway to complete the availability of all *IAUCs* back to No. 1 at the CBAT website, and to provide better indexing of the on-line *Circulars* and to provide flexibility for subscribers in terms of what sorts of *CBETs* and/or *IAUCs* they would like to receive by subject or type of object. The CBAT Director interacts with members of numerous other Commissions at the triennial IAU General Assemblies and during the course of the triennium by e-mail in efforts to increase the value of the CBAT to all astronomers, and all scientists are encouraged to dialogue with the CBAT Director regarding how the work of the CBAT can be more useful to their own work.

8. Amateur astronomers

Amateur astronomers in numerous countries are finding more supernovae during dedicated CCD surveys, and amateurs still dominate in finding Galactic novae. Their help with follow-up observations (astrometry, photometry) of new comets, novae and supernovae is also appreciated by the CBAT. The *IAU Circulars* have continued to serve as the official announcement medium for the annual Edgar Wilson Award for amateur discoveries of comets.

9. Refereeing

Since the *IAU Circulars* are in fact refereed (to a greater extent than many contributors realize), the CBAT benefits from consultation with members of Commission 6 in their various areas of astronomical expertise, as well as referees from the general astronomical community.

Daniel W. E. Green
director of the Bureau

Transactions IAU, Volume XXVIIA
Reports on Astronomy 2006–2009
Karel A. van der Hucht, ed.

© 2009 International Astronomical Union
doi:10.1017/S1743921308025933

COMMISSION 14

ATOMIC AND MOLECULAR DATA
DONNÉES ATOMIQUES ET MOLÉCULAIRES

PRESIDENT	**Steven R. Federman**
VICE-PRESIDENT	**Glenn M. Wahlgren**
PAST PRESIDENT	**Sveneric Johansson**
ORGANIZING COMMITTEE	**Milan Dimitrijevic, Alain Jorissen, Lyudmilla I. Mashonkina, Farid Salama, Jonathan Tennyson, Ewine F. van Dishoeck**

COMMISSION 14 WORKING GROUPS

Div. XII / Commission 14 WG	**Atomic Data**
Div. XII / Commission 14 WG	**Collision Processes**
Div. XII / Commission 14 WG	**Molecular Data**
Div. XII / Commission 14 WG	**Solids and Their Surfaces**

TRIENNIAL REPORT 2006 - 2009

1. Introduction

The main purpose of Commission 14 is to foster interactions between the astronomical community and those conducting research on atoms, molecules and solid state particles. One way the Commission accomplishes this goal is through triennial compilations on recent relevant research in atomic, molecular and solid state physics, as well as related chemical fields. The most recent compilations appear in the following set of WG Triennial Reports, which were produced by members of the Working Groups and the Organizing Committee of Commission 14. Before presenting the Reports, we highlight the meetings supported by the Commission.

2. Supported meetings within the past triennium

Past meetings:
- IAU Symposium No. 234, *Planetary Nebula in our Galaxy and Beyond*, April 2006, Waikoloa Beach, HI, USA
- IAU Symposium No. 251, *Organic Matter in Space*, February 2008, Hong Kong, China

Future meetings, in the IAU XXVII General Assembly, Rio de Janeiro, August 2009:
- JD 13, *Eta Carinae in the Context of the Most Massive Stars*
- JD 4, *Progress in Understanding the Physics of Ap and Related Stars*
- SpS 1, *IR and Sub-mm Spectroscopy – a New Tool for Studying Stellar Evolution*

Steven R. Federman
president of the Commission

Transactions IAU, Volume XXVIIA
Reports on Astronomy 2006–2009
Karel A. van der Hucht, ed.

© 2009 International Astronomical Union
doi:10.1017/S1743921308025945

DIVISION XII / COMMISSION 14 / WORKING GROUP ATOMIC DATA

CHAIR   **Gillian Nave**
VICE-CHAIRS   **Glenn M. Wahlgren, Jeffrey R. Fuhr**

TRIENNIAL REPORT 2006 - 2009

1. Energy levels, wavelengths, line classifications, and line structure

The references cited in this section are mostly papers on original laboratory research; compilations and data bases are covered in another section. The references, ordered by atomic number and spectrum, are given in parentheses following the spectral notations. References including experimental data on line structure, hyperfine structure (HFS) or isotope structure (IS) are also included.

Li I [181], C I [136, 247], N I HFS,IS [109], Ne II IS [182, 125, 128], Ne III [182, 129], Ne IV [92], Mg I IS [212], Al VI HFS [45], S VII-XIV [146], K II [193], Sc I HFS [135], Cr I IS [87], Mn I HFS [39], Mn II HFS [40], Fe XV-XIX [161], Kr I [214], Sr I HFS,IS [53], Nb I HFS [133, 134 Cs I HFS [55], La I HFS [22, 89], La II HFS [220], Pr I HFS [88], Pr II HFS [90, 91], Nd II HFS,IS [122, 157, 198], Nd III [199], Nd IV [256, 257], Sm II HFS [156], Eu III [258], Gd I IS [107], Tm I HFS [21], Tm IV [165], Yb I HFS [56], Yb III [183], Hf II [155], Ta I HFS [106], Pb I IS [252].

The references for elements heavier than Ni ($Z > 28$) are limited to the first three or four spectra only, these data being of most interest for solar and stellar spectroscopy. The references of the lighter elements are also incomplete, the selection being limited to those of highest astrophysical interest. The data in a number of the references above include and/or supersede all or most of the previously data for the indicated spectrum.

Current analyses of high-resolution laboratory spectra (energy levels, wavelengths) is ongoing at Lund, Sweden (transition and rare-earth elements), London, UK (iron-group elements), NIST, USA (transition and rare-earth elements, HFS), Troitsk, Russia (heavy elements), and Meudon & Orsay, France (transition and rare-earth elements, theory).

2. Wavelength standards

Ritz wavelengths of forbidden lines of [Fe II], [Ti II], and [Cr II] have been measured using energy levels derived using Fourier transform spectroscopy (FTS) [10]. Accurate wavelengths of spectral lines in iron-group elements have been measured using FTS with uncertainties of around 10^{-5} nm [11]. The most accurate frequency standards are now being measured using laser spectroscopy with calibration from a laser frequency comb. Frequencies with uncertainties of less than 1 MHz have been measured for H I [99], Mg I [212], K I [73], Rb I [160], Sr I [53], Cs I [77, 94], and In II [250] using this technique.

Wavelengths of spectral lines from a Th/Ar hollow cathode lamp suitable for calibration of astronomical spectrographs have been published by various authors [117, 154, 172]. A correction to the wavelength scale of Ar I published by Whaling *et al.* in 2002 [253] is given in [213].

3. Transition probabilities

The references listed in section 7 are for the period 2005 - 2008. The transition-probability data in these references were obtained by both theoretical and experimental methods. The references for elements heavier than Ni ($Z > 28$) are limited to the first three or four spectra only. All papers contain a significant amount of numerical data, normally for more than ten spectral lines. Extensive results of atomic structure calculations are also given in reference [81].

4. Larger compilations, reviews, conference proceedings

The following compilations on wavelengths and energy levels have been published by the Atomic Spectroscopy group at NIST: Be II [124], B I [130], Ne II [128], Ne III [129], Ne VII [126], Ne VIII [127], K II-XIX [217], Ga I-XXXI [227], Kr I-XXXVI [211], Rb I-XXXVII [215], W I [131], W II [131], Hg I [210], Fr I [216]

In addition to these comprehensive compilations, the *Handbook of basic atomic spectroscopic data* [218] contains a selection of the strongest spectral lines of neutral and singly-ionized atoms of all elements from hydrogen to einsteinium. Compilations by other groups include HE I [69, 169], B II [201], and a compilation of coronal lines [76].

The following additional major compilations of transition probability data have been published during the latest three year period: Na-like to Ar-like sequences [81], ^{4}He I [70], Fe I and Fe II [85], C I, C II, N I and N II [254], (Erratum: [255]). Additional data can be found in *NIST Atomic Transition Probabilities,*86.

A number of papers on atomic spectroscopic data are included in proceedings of the *Ninth International Colloquium on Atomic Spectra and Oscillator Strengths for Astrophysical and Laboratory Plasmas*, held in Lund, Sweden, August 2007. Invited papers are published in the *Physica Scripta* T Series (Wahlgren, Wiese & Beiersdorfer 2008) and poster papers are published in the on-line open access *Journal of Physics Conference Series* (Wahlgren, Wiese & Beiersdorfer 2008).

Papers on astrophysical data needs are included in the proceedings from the *International Conferences on Atomic and Molecular Data* (ICAMDATA, Meudon, France, October 2006; Beijing, China, October 2008) and the *Fifteenth International Conference on Atomic Processes in Plasmas* (NIST, March 2007). These proceedings contain review papers as well as descriptions of atomic and molecular databases.

5. Atomic spectroscopic data on the internet

The following databases of atomic spectra are available at NIST. Most of these have received major updates since the last triennial report.
- *Atomic Spectra Database*
<physics.nist.gov:/PhysRefData/ASD/index.html>
- *Handbook of Basic Atomic Spectroscopic Data*
<physics.nist.gov/PhysRefData/Handbook/index.html>
- *Energy Levels of Hydrogen and Deuterium*
<physics.nist.gov/PhysRefData/HDEL/index.html>
- *Ground Levels and Ionization Energies for the Neutral Atoms*
<physics.nist.gov/PhysRefData/IonEnergy/ionEnergy.html>
- *Spectral Data for the Chandra X-ray Observatory*
<physics.nist.gov/PhysRefData/Chandra/index.html>

- *NIST Atomic Spectra Bibliographic Databases*
`<physics.nist.gov/PhysRefData/ASBib1/index.html>`, consists of three databases of publications on atomic transition probabilities, atomic energy levels and spectra, and atomic spectral line broadening. All three are updated on a frequent basis.

Additional on-line databases including significant quantities of atomic data include:
- The *MCHF/MCDHF Collection on the Web* (C. Froese Fischer *et al.*), `<www.vuse.vanderbilt.edu/~cff/mchf_collection/>` contains results of multi-configuration Hartree-Fock (MCHF) or multi-configuration Dirac-Hartree-Fock (MCDHF) calculations for hydrogen and Li-like through Ar-like ions, mainly for $Z \leqslant 30$. Data for fine-structure transitions are included.
- The *TOPbase and Opacity Projects* include transition probability and oscillator strength data for astrophysically abundant ions for $Z \leqslant 26$). A database is available at `<vizier.u-strasbg.fr/topbase/topbase.html>`. Revised opacities for stellar astrophysics have been made available during the current reporting period.
- The *Database on Rare Earths at Mons university* (DREAM) database `<w3.umh.ac.be/~astro/dream.shtml>` continues to be a relevant source of data for wavelengths, energy levels, oscillator strengths and radiative lifetimes for neutral, singly and doubly-ionized rare earth elements. New data have not been added to this database during the past three years.
- *CHIANTI*, an atomic database for spectroscopic diagnostics of astrophysical plasmas `<wwwsolar.nrl.navy.mil/chianti.html>`, contains atomic data and programs for computing spectra from astrophysical plasmas, with the emphasis on highly-ionized atoms. During the current reporting period additions to the database (v.5) include new physical processes and atomic data.
- The *Vienna Atomic Line Database* (VALD) web site `<ams.astro.univie.ac.at/~vald/>` allows users to extract atomic line data from a compilation of numerous sources according to element or presence in stellar spectra.
- The *bibl database* is a comprehensive bibliographic database on atomic spectroscopy with a search engine on various atomic parameters is available at the Institute of Spectroscopy, Troitsk, `<das101.isan.troitsk.ru/bibl.htm>`.

6. List of References

The references are identified by a running number. This refers to the general reference list at the end of this report, where the literature is ordered alphabetically according to the first author. Each reference contains one or more code letters indicating the method applied by the authors, defined as follows:

Theoretical methods

Q	quant. mech. calc.	QF	quant. mech. calc. forbidden lines

Experimental methods

El	energy levels	Wl	wavelength
HFS	hyperfine structure	IS	isotope structure
L	lifetime	M	miscellaneous
TE	emission transition probabilities		

Other methods

CP	data compilation	CM	comments
R	relative value only	F	forbidden line

7. References on lifetimes and transition probabilities

Ac I: 194
Ac II: 194

Ag II: 44

Al XIII: 8

Ar I: 57, 123, 271
Ar II: 206
Ar VII: 239
Ar X: 41
Ar XI-Ar XVIII: 170
Ar XI: 139
Ar XIV: 143, 142
Ar XVII: 1

Au I: 84
Au II: 30, 84

B II: 49, 201, 219, 274
B IV: 110

Ba I: 72, 222
Ba II: 207

Be I: 95, 274
Be III: 110

Bi III: 195

Br I: 191, 276
Br II: 191

C I: 79, 254
C II: 249, 254
C III: 192, 219, 274

Ca II: 132, 166, 207
Ca III: 14, 237
Ca X: 97
Ca XIII: 137
Ca XIV: 138
Ca XIX: 119

Cd I: 101
Cd II: 167

Cl I: 184, 228, 231, 276

Co XI: 7, 242, 244
Co XIII: 242, 246
Co XVI: 9
Co XVII: 266

Cr I: 230
Cr II: 180
Cr XII: 68
Cr XIII: 238

Cs I: 65

Cu II: 185

F I: 276
F VI: 49, 219, 248, 274
F VIII: 110

Fe I-Fe XVI: 96
Fe I: 35, 85
Fe II: ?, 52, 85, 102
Fe III: 25, 64, 240
Fe IV: 174
Fe VII: 265
Fe VIII: 273
Fe IX: 6, 245, 273
Fe X: 229
Fe XI: 229
Fe XII: 270, 246
Fe XIII: 14, 115
Fe XIV: 42, 68, 229
Fe XV: 9, 171
Fe XVI: 4, 2
Fe XVII-Fe XXIII: 141
Fe XVII: 152
Fe XVIII: 267, 175
Fe XXII: 112
Fe XXIII: 269, 219
Fe XXIV: 268
Fe XXV: 111, 119

Ga I: 202
Ga II: 16, 114, 164

Gd II: 100

H I: 36, 103

He I: ?, 54, 69, 104, 110

Hf I: 155
Hf II: 144, 155

In I: 205
In II: 113

Ir I: 262
Ir II: 262

K V - KVII: 32

Kr II: 59, 159

Li II: 110

Lu III: 29

Mg I: 12, 95, 116
Mg II - Mg XI: 116
Mg V: 28, 63
Mg VI: 140
Mg VIII: 177
Mg IX: ?

Mn I: 37
Mn II: 190
Mn XIII: 98

Mo I: 190

N I: 17, 18, 19, 47, 46, 224, 234, 254
N II: 101, 173, 226, 243, 254
N IV: 49, 219, 241, 248, 274

Na I-Na X: 116

Nd II: 31
Nd IV: 257

Ne I: 60
Ne II: 43
Ne III: 129
Ne VII: 101, 219, 248, 274

Ni II: 108
Ni VI: 118
Ni XI: 5
Ni XIV: 246

Ni XVI: 68
Ni XVII: 9
Ni XVIII: 266
Ni XIX: 3
Ni XXVI: 277

O I: 20, 48, 223
O II: 178, 225, 235
O III: 204
O IV: 233
O V: 15, 219, 241, 248, 274
O V-O VIII: 93

Os I: 196
Os II: 196

P II: 14, 50, 75

Pb II: 195, 209, 208

Pd I: 261

Pm II: 83

Pr II: 148
Pr IV: 51

Ra I: 34, 72, 194, 222
Ra II: 194

Rb I - XXXVII: 215

Re I: 188
Re II: 190, 189

S I: 62, 272
S II: 80
S III: 14
S X: 232
S XII: 177
S XIII: 71, 263
S XV: 153, 119

Sc XIX: 251

Si I: 35
Si II: 236
Si III: 74
Si IX: 149
Si X: 149, 177
Si XI: 27, 149
Si XIII: 186

Sm II: 145, 197

Sn II: 205

Sr I: 150, 264
Sr II: 147, 179, 207

Ta I: 82
Tb III: 260

Tc I: 187
Tc II: 190

Th IV: 203

Ti I: 38
Ti II: 23
Ti IV: 120, 200
Ti VI: 168
Ti XVIII: 275
Ti XIX: 221
Ti XX: 176

Tl I: 33

V I: 259
V II: 259
V III: 105
V V: 66, 121

W I: 131
W II: 131

Xe I: 61, 67
Xe II: 58

Yb II: 179
Yb III: 183

Zn I: 78, 95
Zn II: 163

Zr II: 151, 158
Zr III: 162

References

[1] Aggarwal, K. M. & Keenan, F. P. 2005, *A&A*, 441, 831 [Q,QF]
[2] _____ 2006, *A&A*, 450, 1249 [Q,QF]
[3] _____ 2006, *A&A*, 460, 959 [Q,QF]
[4] _____ 2007, *A&A*, 463, 399 [Q,QF]
[5] _____ 2007, *A&A*, 475, 393, [Q,QF]
[6] Aggarwal, K. M., Keenan, F. P., Kato, T., & Murakami, I. 2006, *A&A*, 460, 331 [Q,QF]
[7] Aggarwal, K. M., Keenan, F. P., & Msezane, A. Z. 2007, *A&A*, 473, 995 [Q,QF]
[8] Aggarwal, K. M., Keenan, F. P., & Rose, S. J. 2005, *A&A*, 432, 1151 [Q]
[9] Aggarwal, K. M., Tayal, V. G., et al. 2007, *At. Data Nucl. Data Tables*, 93, 615 [Q,QF]
[10] Aldenius, M. & Johansson, S. G. 2007, *A&A*, 467, 753 [El,Wl]
[11] Aldenius, M., Johansson, S. G., & Murphy, M. T. 2006, *MNRAS*, 370, 444 [El,Wl]
[12] Aldenius, M., Tanner, J. D., Johansson, S., Lundberg, H., et al. 2007, *A&A*, 461, 767 [TE,L]
[13] Alexander, S. A. & Coldwell, R. L., 2006, *J. Chem. Phys.*, 124, 054104, [Q]
[14] Andersson, M. & Brage, T. 2007, *J. Phys. B*, 40, 709 [Q]

[15] Andersson, M., Brage, T., Hutton, R., Kink, I., *et al.* 2006, *J. Phys. B*, 39, 2815 [Q]

[16] Andersson, M., Jönsson, P., & Sabel, H. 2006, *J. Phys. B*, 39, 4239 [Q]

[17] Bacławski, A. & Musielok, J. 2007, *Eur. Phys. J. Spec. Top.*, 144, 221 [CM,TE,R]

[18] Bacławski, A., Wujec, T., & Musielok, J. 2006, *Eur. Phys. J. D*, 40, 195 [TE]

[19] ‒‒‒‒‒‒ 2007, *Eur. Phys. J. D*, 44, 427 [TE,R]

[20] Barklem, P. S. 2007, *A&A*, 462, 781 [Q]

[21] Başar, G., Başar, G., Öztürk, İ. K., Acar, F. G., & Kröger S. 2005, *Phys. Scr.*, 71, 159 [HFS]

[22] Başar, G., Başar, G., Er, A., & Kröger, S. 2007, *Phys. Scr.*, 75, 572 [HFS]

[23] Bautista, M. A., Hartman, H., Gull, T. R., *et al.* 2006, *MNRAS*, 370, 1991 [Q,QF]

[24] Beck, D. R. 2007 *J. Phys. B*, 40, 651 [Q]

[25] ‒‒‒‒‒‒ 2007, *J. Phys. B*, 40, 3505 [Q]

[26] Bhatia, A. K. & Landi, E. 2007, *At. Data Nucl. Data Tables*, 93, 742 [Q,QF]

[27] ‒‒‒‒‒‒ 2007, *At. Data Nucl. Data Tables*, 93, 275 [Q,QF]

[28] Bhatia, A. K., Landi, E., & Eissner, W. 2006, *At. Data Nucl. Data Tables*, 92, 105 [Q,QF]

[29] Biémont, É. 2005, *Phys. Scr.*, T119, 55 [Q]

[30] Biémont, É., Blagoev, K., Fivet, V., *et al.* 2007, *MNRAS*, 380, 1581 [TE,R,Q]

[31] Biémont, É., Ellmann, A., Lundin, P. *et al.* 2007, *Eur. Phys. J. D*, 41, 211 [L,F,QF]

[32] Biémont, É., Garnir, H.-P., Lefèbvre, P.-H., *et al.* 2006, *Eur. Phys. J. D*, 40, 91 [Q]

[33] Biémont, É., Palmeri, P., Quinet, P., *et al.* 2005, *J. Phys. B*, 38, 3547 [L,Q]

[34] Bieroń, J.,Indelicato, P., & Jönsson,P. 2007, *Eur. Phys. J. Spec. Top.*, 144, 75 [Q,QF]

[35] Bigot, L. & Thévenin, F. 2006, *MNRAS*, 372, 609 [CM,M]

[36] Binh, D. Hoang, 2005, *Comput. Phys. Commun.*, 166, 191 [Q]

[37] Blackwell-Whitehead, R. J. & Bergemann, M. 2007, *A&A* (Letters), 472, L43 [TE]

[38] Blackwell-Whitehead, R. J., Lundberg, H., Nave, G., *et al.* 2006, *MNRAS* 373, 1603 [TE,L]

[39] Blackwell-Whitehead, R. J., Pickering, J. C., Pearse, O., *et al.* 2005, *ApJS*, 157, 402 [HFS]

[40] Blackwell-Whitehead, R. J., Toner, A., Hibbert, A., *et al.* 2005, *MNRAS*, 364, 705 [HFS]

[41] Bogdanovich, P., *et al.* 2005, *Nucl. Instrum. Methods Phys. Res. B*, 235, 174–179, [Q]

[42] Brenner, G., Crespo López-Urrutia, *et al.* 2007, *Phys. Rev. A*, 75, 032504, [L,F]

[43] Bridges,J. M. & Wiese, W. L. 2007, *Phys. Rev. A*, 76, 022513 [TE]

[44] Campos, J., Ortiz, M., Mayo, R., *et al.* 2005, *MNRAS*, 363, 905 [TE,Q]

[45] Casassus, S., Storey, P. J., Barlow, M. J., & Roche, P. F. 2005, *MNRAS* 359, 1386 [HFS]

[46] Çelik, G., Akin, E., & Kiliç, H. Ş. 2006, *Eur. Phys. J. D*, 40, 325 [Q]

[47] ‒‒‒‒‒‒ 2007, *Int. J. Quantum Chem.*, 107, 495 [Q]

[48] Çelik, G. & Ateş, Ş. 2007, *Eur. Phys. J. D*, 44, 433 [Q]

[49] Chen, M.-K. 2005, *Nucl. Instrum. Methods Phys. Res. B*, 235, 165 [Q]

[50] Choudhury, K. B., Deb, N. C., Roy, K., *et al.* 2005, *Indian J. Phys.*, 79, 1353 [Q]

[51] Churilov, S. S., Ryabtsev, A. N., Wyart, J.-F., *et al.* 2005, *Phys. Scr.*, 71, 589, [Q]

[52] Corrégé, G. & Hibbert, A. 2006, *ApJ*, 636, 1166 [Q]

[53] Courtillot, I., Quessada-Vial, A., *et al.* 2005, *Eur. Phys. J. D*, 33, 161 [Wl,IS,HFS]

[54] Dall, R. G., Baldwin, K. G. H., Byron, L. J., *et al.* 2008, *Phys. Rev. Lett.*, 100, 023001, [L]

[55] Das, D., Banerjee, A., Barthwal, S., & Natarajan, V. 2006, *Eur. Phys. J. D*, 38, 545 [HFS]

[56] Das, D., Barthwal, S., Banerjee, A., & Natarajan, V. 2005, *Phys. Rev. A*, 72, 032506 [HFS,IS]

[57] Das, M. B. & Karmakar, S. 2005, *Phys. Scr.*, 71, 266 [L]

[58] ‒‒‒‒‒‒ 2006, *Eur. Phys. J. D*, 40, 339 [L]

[59] ‒‒‒‒‒‒ 2006, *JQSRT*, 102, 387 [L]

[60] Das, M. B., Mitra, D., & Karmakar, S. 2005, *Phys. Scr.*, 71, 599 [L]

[61] Dasgupta, A., Apruzese, J. P., Zatsarinny, O., *et al.* 2006, *Phys. Rev. A*, 74, 012509 [Q]

[62] Deb, N. C. & Hibbert, A. 2006, *J. Phys. B*, 39, 4301 [Q]

[63] ‒‒‒‒‒‒ 2007, *At. Data Nucl. Data Tables*, 93, 585 [Q]

[64] ‒‒‒‒‒‒ 2007, *J. Phys. B*, 40, F251 [Q]

[65] Derevianko, A. & Porsev, S. G. 2007, *Eur. Phys. J. A*, 32, 517 [Q]

[66] Dixit, G., Sahoo, B. K., Deshmukh, P. C., *et al.* 2007, *ApJS*, 172, 645 [Q]

[67] Dong, C. Z., Fritzsche, S., & Fricke, B. 2006, *Eur. Phys. J. D*, 40, 317 [Q,QF]

[68] Dong, C. Z., Kato, T., Fritzsche, S., & Koike, F. 2006, *MNRAS*, 369, 1735 [Q,QF]

[69] Drake, G. W. F. & Morton, D. C. 2007, *ApJS*, 170, 251 [CP]

[70] ——— 2007, *ApJS*, 170, 251 [Q]

[71] Du, S.-B., Yang, Z.-H., Chang, H.-W., & Su, H. 2005, *Chin. Phys. Lett.*, 22, 1638 [L]

[72] Dzuba, V. A. & Ginges, J. S. M. 2006, *Phys. Rev. A*, 73, 032503 [Q,QF]

[73] Falke, S., Tiemann, E., Lisdat, C., *et al.* 2006, *Phys. Rev. A*, 74, 032503 [Wl,IS]

[74] Fan, J., Zhang, T. Y., Zheng, N. W., *et al.* 2007, *Chin. J. Chem. Phys.*, 20, 265 [Q]

[75] Federman, S. R., Brown, M., Torok, S., *et al.* 2007, *ApJ*, 660, 919 [TE,L]

[76] Feldman, U. & Doschek, G. A. 2007, *At. Data Nucl. Data Tables*, 93, 779 [El,Wl]

[77] Fendel, P., Bergeson, S. D., Udem, Th., & Hänsch, T. W. 2007, *Opt. Lett.*, 32, 701 [Wl,HFS]

[78] Fischer, C. F. & Zatsarinny, O. 2007, *Theor. Chem. Accounts*, 118, 623 [Q]

[82] Fivet, V., Palmeri, P., Quinet, P., *et al.* 2006, *Eur. Phys. J. D*, 37, 29 [L,Q]

[83] Fivet, V., Quinet, P., Biémont, É., *et al.* 2007, *MNRAS*, 380, 771 [Q]

[84] Fivet, V., Quinet, P., Biémont, E., & Xu, H. L. 2006, *J. Phys. B*, 39, 3587 [L,Q]

[79] Froese Fischer, C. 2006, *J. Phys. B*, 39, 2159 [Q,QF]

[80] Froese Fischer, C., Gaigalas, G., *et al.* 2006, *Comput. Phys. Commun.*, 175, 738 [CM]

[81] Froese Fischer, C., Tachiev, G., *et al.* 2006, *At. Data Nucl. Data Tables*, 92, 607 [Q,QF]

[85] Fuhr, J. R. & Wiese, W. L. 2006, *J. Phys. Chem. Ref. Data*, 35, 1669 [CP,CPF]

[86] Fuhr, J. R. & Wiese, W. L. 2006, in: D. R. Lide (ed.), *CRC Handbook of Chemistry & Physics*, (Boca Raton: CRC Press, 88th Edition) [CP,CPF]

[87] Furmann, B., Jarosz, A., Stefańska, D., *et al.* 2005, *Spectrochim. Acta, Part B*, 60, 33 [HFS]

[88] Furmann, B., Krzykowski, A., Stefańska, D., *et al.* 2006, *Phys. Scr.*, 74, 658 [El,HFS,Wl]

[89] Furmann, B., Stefanska, D., & Dembczynski, J. 2007, *Phys. Scr.*, 76, 264 [HFS]

[90] Furmann, B., Stefańska, D., Dembczyński, J., *et al.* 2005, *Phys. Scr.*, 72, 300 [El,HFS,Wl]

[91] ——— 2007, *At. Data Nucl. Data Tables*, 93, 127 [El,Wl]

[92] Gallardo, M., Raineri, M., Reyna Almandos, *et al.* 2007, *Spectrosc. Lett.*, 40, 879 [El,Wl]

[93] García, J., Mendoza, C., Bautista, M. A., *et al.* 2005, *ApJS*, 158, 68 [Q]

[94] Gerginov, V., Calkins, K., Tanner, C. E., *et al.* 2006, *Phys. Rev. A*, 73, 032504 [Wl,HFS]

[95] Głowacki, L. & Migdałek, J. 2006, *J. Phys. B*, 39, 1721 [Q]

[96] Gu, M. F., Holczer, T., Behar, E., & Kahn, S. M. 2006, *ApJ*, 641, 1227 [Q]

[97] Gupta, G. P. & Msezane, A. Z. 2006, *J. Phys. B*, 39, 4977 [Q]

[98] ——— 2007, *Phys. Scr.*, 76, 225 [Q]

[99] Hänsch, T. W., Alnis, J., *et al.* 2005, *Phil. Trans. R. Soc. London, Ser. A*, 363, 2155 [Wl]

[100] den Hartog, E. A., Lawler, J. E., Sneden, C., & Cowan, J. J. 2006, *ApJS*, 167, 292 [TE,L]

[101] Hibbert, A. 2005, *Phys. Scr.*, T120, 71[Q]

[102] Hibbert, A. & Corrégé, G. 2005, *Phys. Scr.*, T119, 61 [Q]

[103] Horbatsch, M. W., Horbatsch, M., & Hessels, E. A. 2005, *J. Phys. B*, 38, 1765, [Q]

[104] Hussain, S., Saleem, M., & Baig, M. A. 2007, *Phys. Rev. A*, 76, 012701 [M]

[105] Irimia, A. 2007, *J. Astrophys. Astron.*, 28, 157 [QF]

[106] Jaritz, N., Windholz, L., Zaheer, U., *et al.* 2006, *Phys. Scr.*, 74, 211 [Wl,EL,HFS]

[107] Jelvani, S., Khodadoost, B., *et al.* 2005, *Opt. Spektrosk.*, 98, 875 (in Russian) [IS]

[108] Jenkins, E. B. & Tripp, T. M. 2006, *ApJ*, 637, 548 [M]

[109] Jennerich, R. M., Keiser, A. N., & Tate, D. A. 2006, *Eur. Phys. J. D*, 40, 81 [IS]

[110] Jitrik, O. & Bunge, C. F. 2005, *Nucl. Instrum. Methods Phys. Res. B*, 235, 105 [Q,QF]

[111] Johnson, W. R. & Safronova, U. I. 2007, *At. Data Nucl. Data Tables*, 93, 139 [Q]

[112] Jonauskas, V., Bogdanovich, P., Keenan, F. P., *et al.* 2006, *A&A*, 455, 1157 [Q,QF]

[113] Jönsson, P. & Andersson, M. 2007, *J. Phys. B*, 40, 2417 [Q]

[114] Jönsson, P., Andersson, M., Sabel, H., & Brage, T. 2006, *J. Phys. B*, 39, 1813 [Q]

[115] Keenan, F. P., Jess, D. B., Aggarwal, K. M., *et al.* 2007, *MNRAS*, 376, 205 [Q,QF]

[116] Kelleher, D. E. & Podobedova, L. I. 2008, *J. Phys. Chem. Ref. Data*, 37, 267 [CP,CPF]

[117] Kerber, F., Nave, G., & Sansonetti, C. J. 2008, *ApJ*, 178, 374 [Wl]

[118] Kildiyarova, R. R. Churilov, S. S. Joshi, Y. N., *et al.* 2006, *Phys. Scr.*, 73, 249 [Q]

[119] Kimura, E., Nakazaki, S., Berrington, K. A., *et al.* 2000, *J. Phys. B*, 33, 3449 [Q]

[120] Kingston, A. E. & Hibbert, A. 2006, *J. Phys. B*, 39, 2217 [Q]

[121] _____ 2007, *J. Phys. B*, 40, 3389 [Q]

[122] Koczorowski, W., Stachowska, E., *et al.* 2005, *Spectrochim. Acta, Part B*, 60, 447 [IS]

[123] Koryukina, E. V. 2005, *J. Phys. D*, 38, 3296 [Q]

[124] Kramida, A. E. 2005, *Phys. Scr.*, 72, 309 [CP]

[125] Kramida, A. E., Brown, C. M., Feldman, U., & Reader, J. 2006, *Phys. Scr.*, 74, 156 [El,Wl]

[126] Kramida, A. E. & Buchet-Poulizac, M.-C. 2006, *Eur. Phys. J. D*, 38, 265 [El,Wl,CP]

[127] _____ 2006, *Eur. Phys. J. D*, 39, 173 [El,Wl,CP]

[128] Kramida, A. E. & Nave, G. 2006, *Eur. Phys. J. D*, 39, 331 [El,Wl,CP]

[129] _____ 2006, *Eur. Phys. J. D*, 37, 1 [El,Wl,CP]

[130] Kramida, A. E. & Ryabtsev, A. N. 2007, *Phys. Scr.*, 76, 544 [CP]

[131] Kramida, A. E. & Shirai, T. 2006, *J. Phys. Chem. Ref. Data*, 35, 423 [CP]

[132] Kreuter, A., Becher, C., Lancaster, G. P. T., *et al.* 2005, *Phys. Rev. A*, 71, 032504 [L,F,QF]

[133] Kröger, S. 2007, *Eur. Phys. J. D*, 41, 55 [HFS]

[134] Kröger, S., Öztürk, I. K., Acar, F. G., *et al.* 2007, *Eur. Phys. J. D*, 41, 61 [El,HFS]

[135] Krzykowski, A. & Stefańska, D. 2008, *J. Phys. B*, 41, 055001 [HFS]

[136] Labazan, I., Reinhold, E., Ubachs, W., *et al.* 2005, *Phys. Rev. A*, 71, 040501 [Wl]

[137] Landi, E. & Bhatia, A. K. 2005, *A&A*, 444, 305 [Q,QF]

[138] _____ 2005, *At. Data Nucl. Data Tables*, 90, 177 [Q,QF]

[139] _____ 2006, *At. Data Nucl. Data Tables*, 92, 305 [Q,QF]

[140] _____ 2007, *At. Data Nucl. Data Tables*, 93, 1 [Q,QF]

[141] Landi, E. & Gu, M. F. 2006, *ApJ*, 640, 1171, [Q,QF]

[142] Lapierre, A., Jentschura, U. D., *et al.* 2005, *Phys. Rev. Lett.*, 95, 183001 [L,F]

[143] Lapierre, A., Crespo López-Urrutia, J. R., *et al.* 2006, *Phys. Rev. A*, 73, 052507 [L,F]

[144] Lawler, J. E., den Hartog, E. A., Labby, Z. E., *et al.* 2007, *ApJS*, 169, 120 [TE,L]

[145] Lawler, J. E., den Hartog, E. A, Sneden, C., *et al.* 2006, *ApJS*, 162, 227 [TE,L]

[146] Lepson, J. K., Beiersdorfer, P., Behar, E., & Kahn, S. M. 2005, *ApJ*, 625, 1045 [Wl]

[147] Letchumanan, V., Wilson, M. A., Gill, P., *et al.* 2005, *Phys. Rev. A*, 72, 012509 [L,F]

[148] Li, R., Chatelain, R., Holt, R. A., *et al.* 2007, *Phys. Scr.*, 76, 577 [TE]

[149] Liang, G., Zhao, G., & Zeng, J. 2007, *At. Data Nucl. Data Tables*, 93, 375 [Q]

[150] Liu, Y., Andersson, M., Brage, T., *et al.* 2007, *Phys. Rev. A*, 75, 014502 [Q,QF]

[151] Ljung, G., Nilsson, H., Asplund, M. & Johansson, S. 2006, *A&A*, 456, 1181 [TE,Q]

[152] Loch, S. D., Pindzola, M. S., Ballance, C. P., & Griffin, D. C. 2006, *J. Phys. B*, 39, 85 [Q]

[153] López-Urrutia, C. J. R., Beiersdorfer, P., *et al.* 2006, *Phys. Rev. A*, 74, 01250 [L,F]

[154] Lovis, C. & Pepe, F. 2007, *A&A*, 468, 1115 [Wl]

[155] Lundqvist, M., Nilsson, H., Wahlgren, G. M., *et al.* 2006, *A&A*, 450, 407 [Wl,E,L]

[156] Lundqvist, M., Wahlgren, G. M., & V. Hill 2007, *A&A*, 463, 693 [IS]

[157] Ma, H. L. 2005, *Chin. Phys.*, 14, 511 [IS]

[158] Malcheva, G., Blagoev, K., Mayo, R., *et al.* 2006, *MNRAS*, 367, 754 [L,Q]

[159] Mar, S., del Val, J. A., Rodríguez, F., *et al.* 2006, *J. Phys. B*, 39, 3709 [TE]

[160] Maric, M., McFerran, J. J., & Luiten, A. N. 2008, *Phys. Rev. A*, 77, 032502 [Wl,IS]

[161] May, M. J., Beiersdorfer, P., Dunn, J., *et al.* 2005, *ApJS*, 158, 230 [Wl]

[162] Mayo, R., Campos, J., Ortiz, M., *et al.* 2006, *Eur. Phys. J. D*, 40, 169 [L,Q]

[163] Mayo, R., Ortiz, M., & Campos, J. 2006, *Eur. Phys. J. D*, 37, 181 [TE,Q]

[164] McElroy, T. & Hibbert, A. 2005, *Phys. Scr.*, 71, 479 [Q]

[165] Meftah, A., Wyart, J.-F., Champion, N., *et al.* 2007, *Eur. Phys. J. D*, 44, 35 [El,Wl]

[166] Meléndez, M., Bautista, M. A., & Badnell, N. R. 2007, *A&A*, 469, 1203 [Q,QF]

[167] Moehring, D. L., Blinov, B. B., Gidley, D. W., *et al.* 2006, *Phys. Rev. A*, 73, 023413, [L]

[168] Mohan, M., Singh, A. K., *et al.* 2007, *At. Data Nucl. Data Tables*, 93, 105 [Q]

[169] Morton, D. C., Wu, Q., & Drake, G. W. F. 2006, *Can. J. Phys.*, 84, 83–105, [El,Q]

[170] Mulye, Y. G. & Natarajan, L. 2005, *JQSRT*, 94, 477 [Q]

[171] Murakami, I., Kato, T., Kato, D., *et al.* 2006, *J. Phys. B*, 39, 2917 [Q]

[172] Murphy, M. T., Tzanavaris, P.,Webb, J. K., *et al.* 2007, *MNRAS*, 378, 221 [Wl]

[173] Musielok, J. 2005, *Acta Phys. Pol. A*, 108, 449 [CM]

[174] Nahar, S. N. 2006, *A&A*, 448, 779 [Q,QF]

[175] ―――― 2006, *A&A*, 457, 721 [Q,QF]

[176] Nandi, T., Ahmad, N., Wani, A. A., & Marketos, P.2006, *Phys. Rev. A*, 73, 032509 [L,F]

[177] Nataraj, H. S., Sahoo, B. K., Das, B. P., *et al.* 2007, *J. Phys. B*, 40, 3153 [Q,QF]

[178] Natarajan, L. 2006, *JQSRT*, 97, 267 [Q]

[179] Nayak, M. K. & Chaudhuri, R. K. 2006, *Eur. Phys. J. D*, 37, 171 [Q]

[180] Nilsson, H., Ljung, G., Lundberg, H., & Nielsen, K. E. 2006, *A&A*, 445, 1165 [TE,L]

[181] Noble, G. A., Schultz, B. E., Ming, H., *et al.* 2006, *Phys. Rev. A*, 74, 012502 [Wl,IS]

[182] Öberg, K. J. 2007, *Eur. Phys. J. D*, 41, 25 [Wl,IS]

[183] Öberg, K. J. & Lundberg, H. 2007, *Eur. Phys. J. D*, 42, 15 [El,Wl,L]

[184] Oliver, P. & Hibbert, A. 2007, *J. Phys. B*, 40, 2847 [Q]

[185] Ortiz, M., Mayo, R., Biémont, E., *et al.* 2007, J. Phys. B, 40, 167 [TE,Q]

[186] Özdemir, L. & Ürer, G. 2007, *JQSRT*, 103, 281 [Q]

[187] Palmeri, P., Froese Fischer, C., Wyart, J.-F., *et al.* 2005, *MNRAS*, 363, 452 [Q]

[189] Palmeri, P., Quinet, P., Biémont, E., *et al.* 2005, *MNRAS*, 362, 1348 [L,Q]

[188] ―――― 2006, *Phys. Scr.*, 74, 297 [L,Q]

[190] Palmeri, P., Quinet, P., Biémont, E., *et al.* 2007, *MNRAS*, 374, 63 [Q]

[191] Pan, L. & Beck, D. R. 2006, *J. Phys. B*, 39, 4581 [Q]

[192] Peng, Y.-L., Han, X.-Y., Wang, M.-S., & Li, J.-M. 2005, *J. Phys. B*, 38, 3825 [Q]

[193] Pettersen, K., Ekberg, J. O., Martinson, I., & Reader, J. 2007, *Phys. Scr.*, 75, 702 [El,Wl]

[194] Quinet, P., Argante, C., Fivet, V., *et al.* 2007, *A&A*, 474, 307 [Q]

[195] Quinet, P., Biémont, E., Palmeri, P., & Xu, H. L. 2007, *J. Phys. B*, 40, 1705 [Q]

[196] Quinet, P., Palmeri, P., Biémont, E., *et al.* 2006, *A&A*, 448, 1207 [L,Q]

[197] Rehse, S. J., Li, R., Scholl, T. J., 2006, *Can. J. Phys.*, 84, 723 [TE]

[198] Rosner, S. D., Masterman, D., Scholl, T. J., *et al.* 2005, *Can. J. Phys.*, 83, 841, [HFS,IS]

[199] Ryabchikova, T., Ryabtsev, A., Kochukhov, O., & Bagnulo, S. 2006, *A&A*, 456, 329 [El,Wl]

[200] Ryabtsev, A. N., Churilov, S. S., *et al.* 2005, *Opt. Spektrosk.*, 98, 568 (in Russian) [Q]

[201] Ryabtsev, A. N., Kink, I., Awaya, Y., *et al.* 2005, *Phys. Scr.*, 71, 489 [El,Wl,CP]

[202] Safronova, U. I., Cowan, T. E., & Safronova, M. S. 2006, *J. Phys. B*, 39, 749 [Q]

[203] Safronova, U. I., Johnson, W. R., *et al.* 2006, *Phys. Rev. A*, 74, 042511 [Q,QF]

[204] Safronova, U. I., Ralchenko, Yu., Murakami, I., *et al.* 2006, *Phys. Scr.*, 73, 143 [Q]

[205] Safronova, U. I., Safronova, M. S., & Kozlov, M. G. 2007, *Phys. Rev. A*, 76, 022501 [Q]

[206] Saha, B. & Fritzsche, S. 2005, *J. Phys. B*, 38, 1161 [Q,QF]

[207] Sahoo, B. K., Islam, Md. R., Das, B. P., *et al.* 2006, *Phys. Rev. A*, 74, 062504

[208] Sahoo, B. K., Majumder, S., Chaudhuri, R. K., *et al.* 2004, *J. Phys. B*, 37, 3409 [QF]

[209] Sahoo, B. K., Majumder, S., Merlitz, H. R., *et al.* 2006, *J. Phys. B*, 39, 355 [Q]

[210] Saloman, E. B. 2006, *J. Phys. Chem. Ref. Data*, 35, 1519 [CP]

[211] ―――― 2007, *J. Phys. Chem. Ref. Data*, 36, 215 [CP]

[212] Salumbides, E. J., Hannemann, S., Eikema, K. S. E., *et al.* 2006, *MNRAS* 373, L41 [Wl,IS]

[213] Sansonetti, C. J. 2007, *J. Res. Natl. Inst. Sc. & Technol.*, 112, 297 [Wl]

[214] Sansonetti, C. J. & Greene, M. B. 2007, *Phys. Scr.*, 75, 577 [El,Wl]

[215] Sansonetti, J. E. 2006, *J. Phys. Chem. Ref. Data*, 35, 301 [CP]

[216] ―――― 2007, *J. Phys. Chem. Ref. Data*, 36, 497 [CP]

[217] ―――― 2008, *J. Phys. Chem. Ref. Data*, 37, 7 [CP]

[218] Sansonetti, J. E. & Martin, W. C. 2005, *J. Phys. Chem. Ref. Data*, 34, 1559 [CP]

[219] Savukov, I. M. 2004, *Phys. Rev. A*, 70, 042502 [Q]

[220] Schef, P., Björkhage, M., Lundin, P., & Mannervik, S. 2006, *Phys. Scr.*, 73, 217 [HFS]

[221] Schippers, S., Schmidt, E. W., *et al.* 2007, *Phys. Rev. Lett.*, 98, 033001 [L,F]

[222] Scielzo, N. D., Guest, J. R., Schulte, E. C., *et al.* 2006, *Phys. Rev. A*, 73, 010501 [L]

[223] Sharpee, B. D. & Slanger, T. G. 2006, *J. Phys. Chem. A*, 110, 6707 [M,F,R]

[224] Sharpee, B. D., Slanger, T. G., *et al.* 2005, *Geophys. Res. Lett.*, 32, L12106 [M,F,R]

[225] Sharpee, B. D., Slanger, T. G., Huestis, D. L., & Cosby, P. C. 2004, *ApJ*, 606, 605 [M,F,R]

[226] Shen, X.-Z., Yuan, P., Zhang, H.-M., & Wang, J. 2007, *Chin. Phys.*, 16, 2934 [M]

[227] Shirai, T., Reader, J., Kramida, A. E., *et al.* 2007, *J. Phys. Chem. Ref. Data*, 36, 509 [CP]

[228] Singh, N., Jha, A. K. S., & Mohan, M. 2006, *Eur. Phys. J. D*, 38, 285, [Q]

[229] Smith, S. J., Chutjian, A., & Lozano, J. A. 2005, *Phys. Rev. A*, 72, 062504 [L,F]

[230] Sobeck, J. S., Lawler, J. E., & Sneden, C. 2007, *ApJ*, 667, 1267 [TE]

[231] Sonnentrucker, P., Friedman, S. D., & York, D. G. 2006, *ApJ (Letters)*, 650, L115 [M]

[232] Tayal, S. S. 2005, *ApJS*, 159, 167 [Q]

[233] _____ 2006, *ApJS*, 166, 634 [Q]

[234] _____ 2006, *ApJS*, 163, 207 [Q]

[235] _____ 2007, *ApJS*, 171, 331 [Q]

[236] _____ 2007, *J. Phys. B*, 40, 2551 [Q]

[237] Tayal,V. & Gupta, G. P. 2007, *Phys. Scr.*, 75, 331 [Q]

[238] _____ 2007, *Eur. Phys. J. D*, 44, 449 [Q]

[239] Tayal, V., Gupta, G. P., & Tripathi, A. N. 2005, *Indian J. Phys.*, 79, 1243 [Q]

[240] Toner, A. & Hibbert, A. 2005, *MNRAS*, 364, 683 [Q,QF]

[241] Träbert, E., Knystautas, E. J., Saathoff, G., & Wolf, A. 2005, *J. Phys. B*, 38, 2395 [L]

[242] Träbert, E., Reinhardt, S. , Hoffmann, J., & Wolf, A. 2006, *J. Phys. B*, 39, 945 [L,F]

[243] Träbert, E., Saathoff, G., & Wolf, A. 2005, *Phys. Scr.*, 72, 35 [L,F]

[244] Verma, N., Jha, A. K. S., & Mohan, M., 2005, *J. Phys. B*, 38, 3185 [Q]

[245] _____ 2006, *ApJS*, 164, 297 [Q]

[246] Vilkas, M. J. & Ishikawa, Y. 2004, *J. Phys. B*, 37, 4763 [Q]

[247] Wallace, L. & Hinkle, K. 2007, *ApJS*, 169, 159 [El,Wl]

[248] Wang, F. & Gou, B. C. 2006, *At. Data Nucl. Data Tables*, 92, 176 [Q]

[249] Wang, G.-L. & Zhou, X.-X. 2007, *Chin. Phys.*, 16, 2361, [Q]

[250] Wang, Y. H., Dumke, R., Zhang, J., et al. 2007, *Eur. Phys. J. D*, 44, 307 [Wl,IS]

[251] Wang, Z.-W., Yang, D., Hu, M.-H., et al. 2005, *Chin. Phys.*, 14, 1559 [Q]

[252] Wąsowicz, T. J. & Kwela, J. 2008, *Phys. Scr.*, 77, 025301 [IS]

[253] Whaling, W., et al. 2002, *J. Res. Natl. Inst. Sc. & Technol.*, 107, 149 [Wl]

[254] Wiese, W. L. & Fuhr, J. R. 2007, *J. Phys. Chem. Ref. Data*, 36, 1287 [CP,CPF]

[255] _____ 2007, *J. Phys. Chem. Ref. Data*, 36, 1737 [CP,CPF]

[256] Wyart, J.-F., Meftah, A., Bachelier, A., et al. 2006, *J. Phys. B*, 39, L77 [El,WL]

[257] Wyart, J.-F., Meftah, A., Tchang-Brillet, W.-Ü. L., et al. 2007, *J. Phys. B*, 40, 3957 [El,Wl]

[258] Wyart, J.-F., Tchang-Brillet, W.-Ü. L., Churilov, S. S., et al. 2008, *A&A*, 483, 339 [El,Wl]

[259] Xu, H., Jiang, Z., & Lundberg, H. 2006, *J. Opt. Soc. Am. B*, 23, 2597 [L]

[260] Xu, H.-L. & Jiang, Z.-K. 2005, *Chin. Phys. Lett.*, 22, 2798 [L]

[261] Xu, H. L., Sun, Z. W., Dai, Z. W., et al. 2006, *A&A*, 452, 357 [L,Q]

[262] Xu, H. L., Svanberg, S., Quinet, P., et al. 2007, *JQSRT*, 104, 52 [L,Q]

[263] Yang, Z.-H., Du, S.-B., Chang, H.-W., et al. 2005, *Chin. Phys. Lett.*, 22, 1099 [L]

[264] Yasuda, M. & Katori, H. 2004, *Phys. Rev. Lett.*, 92, 153004 [L,F]

[265] Young, P. R., Berrington, K. A., & Lobel, A. 2005, *A&A*, 432, 665 [MR]

[266] Younis, W. O., Allam, S. H., et al. 2006, *At. Data Nucl. Data Tables*, 92, 187 [Q]

[267] Del Zanna, G. 2006, *A&A*, 459, 307 [Q,QF]

[268] _____ 2006, *A&A*, 447, 761 [Q]

[269] Del Zanna, G., Chidichimo, M. C., & Mason, H. E. 2005, *A&A*, 432, 1137 [Q,QF]

[270] Del Zanna, G. & Mason, H. E. 2005, *A&A*, 433, 731 [Q,QF]

[271] Zatsarinny, O. & Bartschat, K. 2006, *J. Phys. B*, 39, 2145 [Q]

[272] _____ 2006, *J. Phys. B*, 39, 2861 [Q]

[273] Zeng, J.-L., Wang, Y.-G., Zhao, G., & Yuan, J.-M. 2006, *Chin. Phys.*, 15, 1502 [Q]

[274] Zhang, M. & Gou, B.-C. 2005, *Chin. Phys.*, 14, 1554 [Q]

[275] Zhong, J. Y., Zeng, J. L., Zhao, G., Bari, M. A., & Zhang, J. 2005, *PASJ*, 57, 835 [Q,QF]

[276] Zon, B. A., Kretinin, Y. I., & Chernov, V. E. 2006, *Opt. Spectrosc.*, 101, 501 [Q]

[277] Zou, Y., Hutton, R., et al. 2005, *Nucl. Instrum. Methods Phys. Res. B*, 235, 192 [L]

Transactions IAU, Volume XXVIIA
Reports on Astronomy 2006–2009
Karel A. van der Hucht, ed.

© 2009 International Astronomical Union
doi:10.1017/S1743921308025957

DIVISION XII / COMMISSION 14 / WORKING GROUP COLLISION PROCESSES

CO-CHAIRS **Gillian Peach,**
Milan S. Dimitrijevic,
Phillip C. Stancil

TRIENNIAL REPORT 2006 - 2009

1. Introduction

Research in atomic and molecular collision processes and spectral line broadening has been very active since our last report (Schultz & Stancil 2007, Allard & Peach 2007). Given the large volume of the published literature and the limited space available, we have attempted to identify work most relevant to astrophysics. Since our report is not comprehensive, additional publications can be found in the databases at the web addresses listed in the final section. Elastic and inelastic collisions among electrons, atoms, ions, and molecules are included and reactive processes are also considered, but except for charge exchange, they receive only sparse coverage.

Numerous meetings on collision processes and line broadening have been held throughout the report period. Important international meetings that provide additional sources of data through their proceedings are: the XXIV *International Conference on Photonic, Electronic, and Atomic Collisions* (ICPEAC) (Fainstein *et al.* 2006), XXV ICPEAC (Becker *et al.* 2007), the NASA *Laboratory Astrophysics Workshop* (Weck *et al.* 2006), the 18th *International Conference on Spectral Line Shapes* (ICSLS) (Oks & Pindzola 2006) and the VIth *Serbian Conference on Spectral Line Shapes in Astrophysics* (SC-SLSA)(Popović & Dimitrijević 2007). The 19th ICSLS has just taken place in June 2008.

2. Electron collisions with atoms, ions, molecules, and molecular ions

Collisions of electrons with atoms, ions, molecules, and molecular ions are the major excitation mechanism for a wide range of astrophysical environments. In addition, electron collisions play an important role in ionization and recombination, contribute to cooling and heating of the gas, and may contribute to molecular fragmentation and formation. In the following sections we summarize recent work on electron collisions with astrophysically relevant species, including elastic scattering, excitation, dissociation, ionization, recombination, and electron detachment from negative ions.

2.1. *Electron-atom scattering*

New work on elastic scattering from neutral atoms is limited to Xe (Linert *et al.* 2007), Cs (Zatsarinny & Bartschat 2008), In (Rabasović *et al.* 2008) and Au (Maslov *et al.* 2008). The excitation of atomic oxygen has been investigated (Wang & Zhou 2006, Barklem 2007), while new work on ionization has been carried out for Mg (Bolognesi *et al.* 2008).

2.2. *Electron-ion scattering*

For atomic ions, new work has primarily focused on excitation and includes: C^+ (Wilson *et al.* 2005), N^+ (Hudson & Bell 2005), O^+ (Tayal 2007), Al^{12+} (Aggarwal *et al.* 2005,

Aggarwal *et al.* 2008), S^{4+} (Hudson & Bell 2006), Ar^{16+} (Aggarwal & Keenan 2005b), Ca^+ (Meléndez *et al.* 2007), Kr^{6+} (Ishikawa & Vilkas 2008), Fe^+ (Ramsbottom *et al.* 2007), Fe^{3+} (McLaughlin *et al.* 2006), Fe^{4+} (Ballance *et al.* 2007), Fe^{6+} (Witthoeft & Badnell 2008), Fe^{9+} (Aggarwal & Keenan 2005a), Fe^{11+} (Storey *et al.* 2005), Fe^{15+} (Aggarwal & Keenan 2006), Fe^{17+} (Witthoeft *et al.* 2006), Fe^{19+} (Witthoeft *et al.* 2007), Fe^{22+} (Chidichimo *et al.* 2005), Fe^{25+} (Aggarwal *et al.* 2008), Ni^{3+} (Meléndez & Bautista 2005), and for Si, Cl, and Ar isonuclear sequences (Colgan *et al.* 2008).

New elastic data exists for He^+ and Li^{2+} (Bhatia 2008). Ionization and detachment have been studied for H^- (Jung 2008); C^{2+}, N^{3+}, and O^{4+} (Fogle *et al.* 2008); Ne^{4+} and Au^{47+} (Pindzola *et al.* 2008); and Si, Cl, and Ar isonuclear sequences (Colgan *et al.* 2008).

Another important process is recombination for which a number of new works including radiative and dielectronic recombination have appeared: C^{2+}, N^{3+}, and O^{4+} (Fogle *et al.* 2005); Si^{3+} (Schmidt *et al.* 2007); Mg^{2+} (Fu *et al.* 2008); and Fe^{16+} (Chen 2008).

2.3. *Electron-molecule scattering*

For molecules, new elastic scattering references have appeared as follows: C_3 (Munjal & Baluga 2006); CH_4 (Cho *et al.* 2008); H_2CO (Kaur & Baluja 2005); C_6H_6, C_6F_6, C_6H_{12}, C_6H_{14}, C_6F_{14}, C_8H_{16}, C_8H_{18}, and C_8F_{18} (Shi *et al.* 2008); SF_4 (Szmytkowski *et al.* 2005); SO_2Cl_2 (Szmytkowski *et al.* 2006); SiN_2, $SiCO$, and $CSiO$ (Fujimoto *et al.* 2007); pyrazine (Winstead 2007); propane (Bettega, da Costa, & Lima 2008); methanol and ethanol (Khakoo *et al.* 2008); propene and cyclopropane (Makochekanwa *et al.* 2008); and NeF (Kaur *et al.* 2008).

For excitation, new references include: H_2 (da Costa *et al.* 2005, Kato *et al.* 2008); C_3 (Munjal & Baluga 2006); CH_2 (Allan 2007); CH_4 (Čurik *et al.* 2008); CO and C_2H_4 (da Costa *et al.* 2007); CO_2 (Rescigno *et al.* 2007); CF_4 (Irrera & Gianturco 2005); N_2 (Tashiro & Morokuma 2007, Khakoo *et al.* 2007), ethene (Allan, Winstead, & McKoy 2008); and NeF (Kaur *et al.* 2008).

New work for dissociative processes are for dissociative electron attachment to H_2O (Haxton *et al.* 2007), C_2H_2 (Chourou & Orel 2008, May *et al.* 2008), and C_4H_2 (May *et al.* 2008). Research on molecular ionization has been limited to H_2 and D_2 (Martín 2007) and H_2O (Kaiser *et al.* 2007).

2.4. *Electron-molecular ion scattering*

References on dissociative processes have appeared for: H_2^+ (Motapon *et al.* 2008), He_2^+ (Buhr *et al.* 2008), NeH^+ and NeD^+ (Ngassam *et al.* 2008), H_3^+ (Kokoouline & Greene 2005), D_2H^+ (Zhaunerchyk *et al.* 2008b), HCO^+ and DCO^+ (Douguet *et al.* 2008), O_3^+ (Zhaunerchyk *et al.* 2008a), and $HCNH^+$ (Ngassam *et al.* 2005).

For excitation, a study has been conducted for H_3^+ (Faure *et al.* 2006). New results for ionization include H_2^+ (Pindzola *et al.* 2005), while detachment has been investigated for Si_2^- (Lindahl *et al.* 2008). Finally, vibrational excitation due to electron impact has been studied for NeH^+ and NeD^+ (Ngassam *et al.* 2008).

3. Heavy particle collisions

3.1. *Ion-atom and atom-atom collisions*

Charge exchange has seen a substantial amount of activity over the report period as it plays an important role in a variety of environments. Studies for collisions on H include: H^+ (Bradley *et al.* 2005, Dubois *et al.* 2005, Zeng *et al.* 2008), He^{2+} (Havener *et al.* 2005), Be^{4+} (Minami *et al.* 2006), C^{6+} (Liu *et al.* 2005), N^{2+} and O^{2+} (Barragán *et al.*

2006a, Barragán *et al.* 2006c), O^{8+} (Perez & Olson 2005), F^{2+} (Dutta *et al.* 2005), Ne^{10+} (Errea *et al.* 2005a, Barragán *et al.* 2006b), Si^{3+} (Wang *et al.* 2006, Bruhns *et al.* 2008), S^{16+} (Janowicz *et al.* 2005), Cl^{7+} (Zhao *et al.* 2007), and Ar^{18+} (Errea *et al.* 2005a). A database for cross sections for all carbon ions colliding with hydrogen has been constructed by Suno & Kato (2006).

Neutral helium is also an important target for which studies have been carried out for the incident ions: He^+ (Bradley *et al.* 2005), C^{4+} (Hoshino *et al.* 2007), F^{7+} (Zouros *et al.* 2008), and $Ne^{(2-6)+}$ (Hasan 2005).

Charge exchange due to proton impact on Ca (Dutta *et al.* 2006, Pandey & Dubey 2007) and Mg (Pandey & Dubey 2007), alpha particles on Na (Lee 2006), and N^{7+} on atomic oxygen (Perez & Olson 2005) have been studied, while radiative charge transfer in Ne^{2+} collisions with He (Zhao *et al.* 2006) has been investigated.

Elastic scattering due to proton impact on He, Ne, and Ar (Ovchinnikov *et al.* 2006) have been studied. Inelastic processes involving hyperfine changing collisions have been investigated for: H + H (Zygelman 2005); $(n - n')$-mixing in $H^*(n)$ + H(1s) collisions (Mihajlov *et al.* 2005); fine structure transitions for O + H and C + H (Abrahamsson & Krems 2007), C^+ + H and Si^+ + H (Barinovs *et al.* 2005) and depolarization collisions of excited atoms and ions with hydrogen (Derouich *et al.* 2005b, Derouich *et al.* 2005a, Derouich 2007, Derouich & Barklem 2007, Sahal-Bréchot *et al.* 2007). Electronic transitions have been studied for O + H (Krems *et al.* 2006) and H impact on S^{3+}, Ar^{13+}, and Fe^{13+} (Burgess & Tully 2005).

Excitation, charge transfer, and ionization due to He^{2+} collisions with H including Debye screening in a dense plasma has been studied by Liu *et al.* (2008a), Liu *et al.* (2008b). Electron detachment in He collisions with H^- (Huang *et al.* 2005, Ogurtsov *et al.* 2006) and C^- (Huang *et al.* 2005) has also been investigated. Rate coefficients have been computed for the formation of HD (Dickinson 2005); CH^+ (Barinovs & van Hemert 2005); and SO, SO^+, and S_2 (Andreazza & Marinho 2005) by radiative association.

3.2. *Ion-, atom-, and molecule-molecule collisions*

In photo-ionized environments, multiply charged ions may coexist with neutral molecules. Examples include x-ray ionized regions and solar wind interactions with cometary gas. In these environments charge transfer plays an important role. Recent studies of ion-molecule charge transfer include He^{2+} (Dubois *et al.* 2005, Kusakabe *et al.* 2006), C^{4+} (Zarour *et al.* 2005), O^+ (Kimura *et al.* 2006), O^{5+} and Ar^{5+} (Dubois *et al.* 2005), and F^{7+} (Zouros *et al.* 2008) with H_2; H^+ (Lindsay *et al.* 2005a, Wells *et al.* 2005, Kumar *et al.* 2006, Lin *et al.* 2007) and He^{2+} (Kusakabe *et al.* 2006) with CO; He^{2+} (Abu-Haija *et al.* 2005b, Kusakabe *et al.* 2006), $Ar^{5,4+}$ (Abu-Haija *et al.* 2005a), and Kr^{8+} (Kaneyasu *et al.* 2005) with N_2; H^+ and O^+ (Luna *et al.* 2005), He^{2+} (Abu-Haija *et al.* 2005b, Kusakabe *et al.* 2006), and $Ar^{5,4+}$ (Abu-Haija *et al.* 2005a) with O_2; H^+ (Kimura *et al.* 2006), He^{2+} (Bodewits, *et al.* 2005, Seredyuk *et al.* 2005c), O^{6+} (Seredyuk *et al.* 2005b, Bodewits & Hoekstra 2007), and a range of ions (Otranto & Olson 2008) with water; H^+ (Lindsay *et al.* 2005a), He^{2+} (Abu-Haija *et al.* 2005b, Kusakabe *et al.* 2006), and O^{6+} (Seredyuk *et al.* 2005b) with CO_2; H^+ with NH_2 (Suno *et al.* 2006); He^{2+} (Abu-Haija *et al.* 2005b, Kusakabe *et al.* 2006) with NH_3; H^+ (Lindsay *et al.* 2005b), He^{2+} (Seredyuk *et al.* 2005a), and C^{4+} and O^{6+} (Seredyuk *et al.* 2005b) with CH_4; H^+ (Suzuki *et al.* 2005) and He^{2+} (Seredyuk *et al.* 2005a) with C_2H_4; and He^{2+} (Seredyuk *et al.* 2005a) with C_2H_6. Other charge exchange studies include H_2^+ + H (Errea *et al.* 2005b); N_2^+ + N_2 and O_2^+ + O_2 (Tong & Nanbu 2007); and O^+ + O_2 (Martinez *et al.*

2006), while Cornelius (2006) has developed a scaling relation for total cross sections with H_2.

For applications to x-ray emission from comets and planetary atmospheres, charge exchange has been considered for molecular targets including $C^{(3-6)+}$, $N^{(4-7)+}$, and $O^{(5-7)+}$ with methane (Djurić et al. 2008); highly charged L-shell Fe ions with various neutrals (Wargelin et al. 2005, Beiersdorfer et al. 2008); and for a range of ions and molecules (Otranto et al. 2006).

The internal level populations of molecular rovibrational states are primarily controlled through collisional excitation by atom and molecule impact. Investigations have been carried out for excitation of H_2 by H (Wrathmall & Flower 2006, Wrathmall et al. 2007), He (Mack et al. 2006), H_2 (Lee et al. 2006), and H^- (Giri & Sathyamurthy 2006); HF by He (Reese et al. 2005); CO by H (Yang et al. 2006a, Shepler et al. 2007), He (Yang et al. 2006a), and H_2 (Wernli et al. 2006, Yang et al. 2006b); CO^+ by H (Andersson et al. 2008); CN by C_2H_2 (Olkhov & Smith 2007); CS by He (Lique et al. 2006b, Lique & Spielfiedel 2007); PN by He (Tobola et al. 2007); SiO by H (Palov et al. 2006) and He (Dayou & Balança 2006); SiS by He (Vincent et al. 2007); SO by He (Lique et al. 2006a, Lique et al. 2006c); H_2O by He (Yang & Stancil 2007) and H_2 (Dubernet et al. 2006), NH_3 (Machin & Roueff 2005, Yang & Stancil 2008), NH_2D (Machin & Roueff 2006), and ND_2H by He (Machin & Roueff 2007); and HC_2N by He and H_2 (Wernli et al. 2007).

Other investigations include: collisional dissociation of highly excited H_2 by He Ohlinger et al. (2007), fragmentation of CO by slow C^{6+} and Ar^{11+} (Wells et al. 2008), and dissociation and fragmentation of N_2 by slow Xe^{q+} ions, $(q = 15\text{-}21)$, (Zhu et al. 2005). Mutual neutralization in H_2^+ collisions with H^- has been studied by Liu et al.(2006) and the formation of HeH_2^+ via radiative association by Mrugala & Kraemer (2005).

4. Reactive scattering and chemistry

Due to space limitations, we cannot review the many advances in reactive scattering and chemical processes relevant to astrophysics. One noteworthy and relevant study involves a quasi-classical trajectory investigation of $H + CH_4 \rightarrow H_2 + CH_3$ (Xie et al. 2006), and updates to the UMIST Astrochemistry database which gives rate coefficient fits for 4572 reactions, has been completed recently (Woodall et al. 2007).

5. Stark broadening

Knowledge of line widths and shifts for atomic transitions is very important for the interpretation of stellar spectra and also for circumstellar conditions and galactic H II regions.

The *Critical Review of Selected Data on Experimental Stark Widths and Shifts for Spectral Lines of Neutral and Ionized Atoms* for the period 2001 - 2007, see Lesage (2008), contains tables where measured values are listed and compared with semi-classical calculations. A book entitled *Stark Broadening of Hydrogen and Hydrogenlike Spectral Lines in Plasmas* has been published by Oks (2006) and contains many useful references.

5.1. *Developments in line broadening theory*

Poquérusse & Alexiou (2006) have extended standard semi-classical impact theory for hydrogen-like ions to include penetrating collisions. For transitions in highly excited hydrogen-like atoms and ions Stambulchik & Maron (2008) have produced a simple analytical method for calculating line shapes and Gigosos et al. (2007) have developed an exact expression for the impact broadening operator of hydrogen. The asymmetry of hydrogen lines has been studied by Demura *et al.* (2008a) and Demura *et al.* (2008b).

Stambulchik & Maron (2008) have studied broadening of lines subject to external electric and magnetic fields and Godbert-Mouret *et al.* (2006) have developed new code to calculate their lineshapes. Dubau *et al.* (2007) have carried out quantum mechanical calculations of electron impact broadening for XUV lines in plasmas.

5.2. *Isolated lines*

For isolated lines, Stark broadening is dominated by collisions with plasma electrons. Broadening parameters have been determined theoretically for:

Ar II 476.5 nm, 480.6 nm and Kr II 469.4 nm lines (Dimitrijević & Csillag 2006), Cd I 33 singlets and 37 triplets (Simić *et al.* 2005a), 26 Ne V multiplets (Hamdi *et al.* 2007) and 15 Si VI multiplets (Hamdi *et al.* 2008).

Also for: 4 N IV, 3 O V, 1 F VI and 4 Ne IV (Elabidi *et al.* 2008a), 2 N II, O III, F IV and Ne V (Ivković *et al.* 2005), O II (Mahmoudi *et al.* (2005)), F II (Srećković *et al.* 2005b), 1 F III (Simić *et al.* 2005b), 6 Ar I (Dimitrijević *et al.* 2007a), 9 Cr I (Dimitrijević *et al.* 2005), 3 Mn II (Popović *et al.* 2008), 3 Te I (Dimitrijević *et al.* 2007b, Dimitrijević *et al.* 2008), F III (Simić *et al.* 2005b), Ne VII, Ne VIII and Si XI (Elabidi *et al.* 2008b), S II, S III and S IV (Milovanović N. & Dimitrijević 2007), Cu III, Zn III and Se III (Simić *et al.* 2006), Ga II (N'Dollo & Donga-Passi 2006), Sn II (Colón & Alonso-Medina 2006) and Pb III (Alonso-Medina *et al.* 2008) lines.

Broadening parameters have been obtained experimentally for:

C I 247.8561 nm (Djeniže *et al.* 2006b), Mg II 448.1 nm (Djeniže *et al.* 2005b), Fe I 381.58 nm (Bengoechea *et al.* 2006) and Fe I 538.34 nm (Bengoechea *et al.* 2005) lines.

Also for: 2 N I (Bartecka *et al.* 2005), 3 O II (Bukvić *et al.* 2005), 23 O III (Srećković *et al.* 2005a), 10 N II and 8 O III (Ivković *et al.* 2005), F II (Srećković *et al.* 2005b), 4 Ne I (Dzierżega *et al.* 2006), 13 Ne I (Jovićević *et al.* 2005), 7 Ar I (Milosavljević *et al.* 2006), 6 Ar II (Iglesias *et al.* 2006), 6 Mn I (Srećković *et al.* 2007), 11 Mn II and 3 Mn III (Djeniže *et al.* 2006e), 17 Ni II (Mayo *et al.* 2008), 35 Kr II (del Val *et al.* 2008), 2 Ag I and 2 Au I (Djeniže *et al.* 2006c), 2 Ag I, 11 Ag II and 3 Ag III (Djeniže *et al.* 2005a), 26 Au II (Ortiz & Mayo 2005), 43 Sn I and 27 Sn II (Alonso-Medina & Colón 2008), 12 Sn I and 16 Sn II (Djeniže *et al.* 2006d), 16 In III (Djeniže *et al.* 2006a), 10 Pb III lines (Alonso-Medina & Colón 2007), 31 Pb II (Colón & Alonso-Medina 2006), 38 Xe III (Peláez *et al.* 2006a) and shifts of 110 Xe and 42 Xe III (Ćirišan *et al.* 2006) lines.

Djurović *et al.* (2006) have presented a review of experimental work on Stark broadening of 80 singly ionized xenon lines and Purić *et al.* (2008) have used published Stark widths for spectral lines originating from 3s-3p transition arrays of multiply charged ions, to establish trends from which Stark widths are predicted for Mg VII, Mg IX, Mg X, Na VII, Na VIII, Al VIII, Al IX, Si XI, Ti XI, Cr XIII, Cr XIV, Fe XV, Fe XVI, Fe XXIII and Ni XVIII.

Stehlé *et al.* (2005) have examined current Stark broadening theory as a basis for diagnostics of low-temperature plasmas and Mahmoudi *et al.* (2008) have provided new expressions for diagonal multiplet factors of complex configurations, required for studies of isolated lines. Zmerli *et al.* (2008) have proposed an improved interpolation method for widths as a function of temperature.

5.3. *Transitions in hydrogenic and helium-like systems*

New quantum mechanical calculations of the broadening of Lyβ, Lyγ and Lyδ have been carried out by de Kertanguy *et al.* (2005). Transitions in the Balmer series have been studied by Gigosos & González (2006), and by Stambulchik *et al.* (2007) for $n \leqslant 15$. Broadening of high-n transitions are considered by Lisgo *et al.* (2006) and benchmarked against electron density measurements. New experiments for Hβ for the wide range of plasma parameters have been carried out by Djurović *et al.* (2005) and Griem *et al.* (2005) compare Hα profiles measured at high electron densities with theoretical results.

Broadening of the radio recombination lines of hydrogen has been studied theoretically by Watson (2006) and Gavrilenko & Oks (2007) and lines of hydrogen-like and helium-like ions of C, Si and Ar have been examined by Stambulchik & Maron (2006) who find that ion dynamics is very important.

New theoretical calculations of broadening have been reported for He I 667.8 nm and 587.6 nm lines (Ben Chaouacha *et al.* (2007)) and He I 728.1, 706.5, 504.8, 492.2 and 471.3 nm lines in a dense plasma (Omar *et al.* (2006)). Peláez *et al.* (2006b) have carried out experiments for He I 318.8 nm and 402.6186 nm lines.

6. Broadening by neutral atoms and molecules

The analysis of experimental molecular spectra in order to extract line shape parameters is often very difficult. Line shapes can be affected by collisional narrowing and the dependence of collisional broadening and shifting on molecular speed. When these effects are sufficiently important, fitting Voigt profiles to experimental spectra produces systematic errors in the parameters retrieved.

A collection of papers concerning the status of the molecular spectroscopic database, HITRAN 2000, has been published by Rothman et al. (2003) and this has recently been updated for the current version HITRAN 2004 by Rothman *et al.* (2005).

6.1. *Broadening of atomic lines*

Some theoretical work has been published in the period 2005-2008 and the transitions with the perturbing atoms or molecules are listed below.

Li; 2s-3s transition broadened by Ar, Kr and Xe (Rosenberry *et al.* 2007).
Li; wings of the resonance line broadened by He and H$_2$ (Allard *et al.* 2005).
Na and K; wings of the resonance lines broadened by He (Zhu *et al.* 2006).
Li, Na and K; impact widths for the resonance lines broadened by He (Mullamphy *et al.* 2007).
Rb and Cs; resonance line profiles, including far line wings, broadened by He and H$_2$ (Allard & Spiegelman 2006).
Fe II; 24188 lines broadened by collisions with H (Barklem & Aspelund-Johansson 2005).
Sr: $5s^2\,^1S_0 \rightarrow 5s5p^3P_1$ and $5s5p^3P_{0,1,2} \rightarrow 5s6s^3S_1$ transitions broadened by the rare gases (Holtgrave & Wolf 2005).

6.2. *Broadening and shift of molecular lines*

Much new data have been published since the last report was prepared. The molecules are listed below with their perturbing atomic or molecular species and are labelled by 'E' and 'T' to indicate experimental work and theoretical analysis respectively.

H$_2$-H$_2$ collision-induced absorption (T) (Orton *et al.* 2007); in binary mixtures H$_2$-N$_2$ and H$_2$-CO (E) (Abu-Kharma *et al.* 2006).
H$_2$ lines broadened and shifted by He (T) (Ma *et al.* 2007).
HDO lines broadened and shifted by N$_2$ (E) (Bach *et al.* 2005).
HCN lines broadened and shifted by HCN and air (E) (Devi *et al.* 2005); N$_2$ (E) (Smith *et al.* 2008); H$_2$, N$_2$, O$_2$, CH$_3$CN and rare gases (E) (Rohart *et al.* 2007).
HC$_3$N lines broadened by H$_2$, He and N$_2$ (E) (Colmont *et al.* 2007b).

H_2CO lines broadened by H_2CO, N_2 and O_2 (E) (Staak *et al.* 2005).

HCO^+ lines broadened by He and Ar (T) (Buffa 2007).

HNO_3 lines broadened by air (E) (Cazzoli *et al.* 2005).

HO_2 lines broadened by air (E) (Ibrahim *et al.* 2007); H_2O and N_2 (E) (Kanno *et al.* 2005).

HI lines broadened by HI (E) (Bulanin *et al.* 2005, Hartmann *et al.* 2005); He (E) (Flaud *et al.* 2006).

HI, HBr lines broadened by rare gases (E) (Domanskaya *et al.* 2007).

H_2O lines broadened by H_2O (T) (Tolchenov & Tennyson 2005, Ptashnik *et al.* 2005, Antony & Gamache 2007, Antony *et al.* 2007); air (E) (Liu *et al.* 2007, Seta *et al.* 2008); H_2O and air (E+T) (Toth 2005, Jenouvrier *et al.* 2007, Ibrahim *et al.* 2008); N_2 (E+T) (Aldener *et al.* 2005, Bandyopadhyay *et al.* 2007, Tran *et al.* 2007, Bykov *et al.* 2008); N_2 and air (E+T) (Aldener *et al.* 2005, Bandyopadhyay *et al.* 2007, Tran *et al.* 2007, Bykov *et al.* 2008, Hodges *et al.* 2008); N_2 and O_2 (E) (Golubiatnikov *et al.* 2008, Hoshina *et al.* 2008); H_2O, N_2 and O_2 (E) (Cazzoli *et al.* 2007, Koshelev *et al.* 2007, Cazzoli *et al.* 2008); H_2O and Ar (E) (Li *et al.* 2008); H_2O, H_2, N_2, O_2, CO_2 and rare gases (E) (Golubiatnikov 2005); H_2O, H_2, N_2, O_2, CO_2, He and air (E) (Brown *et al.* 2005).

CH_4 lines broadened by air (E) (Predoi-Cross *et al.* 2006); CH_4 and air (E) (Predoi-Cross *et al.* 2007a); CH_4, air, He and H_2 (E) (Lucchesini & Gozzini 2007); CH4 (E) (Lepère 2006, Wishnow *et al.* 2007); H_2 and He (T) (Tran *et al.* 2006); N_2 (E) (Mondelain *et al.* 2007, Martin & Lepère 2008); CH_4 and N_2 (E) (Menard-Bourcin *et al.* 2007); N_2 and O_2 (E) (Mondelain *et al.* 2005, Lepère *et al.* 2005); N_2, O_2 and air (T) (Antony *et al.* 2008).

C_2H_2 self-broadened lines (E+T) (Lepère *et al.* 2007, Nguyen *et al.* 2008, Lyulin *et al.* 2008); broadened by He (T) (Thibault 2005, Nguyen *et al.* 2006); H_2, D_2, N_2, air and rare gases (E) (Arteaga *et al.* 2007); CO_2 (E+T) (Martin *et al.* 2006).

C_2H_4 lines broadened by N_2 (E) (Blanquet *et al.* 2005).

C_3H_2 lines self broadened (E) (Achkasova *et al.* 2006).

CH_3Br lines broadened by N_2 (E+T) (Jacquemart *et al.* 2007, Tran *et al.* 2008); CH_3Br and N_2 (E) (Jacquemart & Tran 2008).

CH_3F lines broadened by H_2 (E) (Lerot *et al.* 2006b); N_2, O_2 (E) (Lerot *et al.* 2006a); CH_3F (Lerot *et al.* 2005).

CH_3CN broadened by CH_3CN and N_2 (E) (Rinsland *et al.* 2008).

CO lines broadened by He (E) (Thibault *et al.* 2007); CO_2 (E) (Sung & Varanasi 2005); Ar (E+T) (Wehr *et al.* 2006a, Wehr *et al.* 2006b); N_2, O_2, CO_2 and rare gases (Colmont *et al.* 2007a).

CO_2 lines broadened by air (E) (Predoi-Cross *et al.* 2007c, Toth *et al.* 2007); self-broadened lines (E) (Hikida *et al.* 2005, Le Barbu *et al.* 2006, Predoi-Cross *et al.* 2007b), (T) (Toth *et al.* 2006); air and CO_2 (E) (Devi *et al.* 2007a, Toth *et al.* 2008, Joly *et al.* 2008, Devi *et al.* 2007a); air and Ar (E) (Li *et al.* 2008); N_2 and O_2 (E) (Hikida & Yamada 2006).

Cs_2 lines broadened by N_2 (E+T) (Misago *et al.* 2006); O_2 (E+T) (Misago *et al.* 2007); Ar and air (E+T) (Misago *et al.* 2008).

N_2 self-broadened lines (E) (El-Kader & Moustafa 2005, Hashimoto & Kanamori 2006); N_2-H_2 collision-induced absorption (E) (Boissoles *et al.* 2005).

NH_3 lines broadened by NH_3 (E) (Leary *et al.* 2008); N_2, O_2 and air (E) (Dhib *et al.* 2007).

N_2O lines broadened by air (E) (Grossel *et al.* 2008).

O_2 self-broadened lines (E) (Tretyakov *et al.* 2007, Predoi-Cross *et al.* 2008a); lines broadened by O_2 and N_2 (Tretyakov *et al.* 2005); N_2 (E+T) (Predoi-Cross *et al.* 2008b).

O_3 lines broadened by O_3 (E) (Yamada & Amano 2005); N_2 and O_2 (E+T) (Rohart *et al.* 2008).

OCS lines broadened by OCS (E) (Matton *et al.* 2006); N_2 and O_2 (E) (Koshelev *et al.* 2006).

PH_3 lines broadened by N_2 (E+T) (Bouanich *et al.* 2005, Bouanich & Blanquet 2007).

SO_2 self-broadened lines (E+T) (Zéninari *et al..* 2007, Henningsen *et al.* 2008).

7. Databases

A database for atomic and molecular processes is maintained at the Oak Ridge National Laboratory Controlled Fusion Atomic Data Center (CFADC) at the address `<cfadc.phy.ornl.gov>` and a useful on-line database of rovibrational collisional excitation data, BASECOL, can be found at `<basecol.obspm.fr>`.

Some collisional data are also available on the Leiden Atomic and Molecular Database `<www.strw.leidenuniv.nl/~moldata>` and the UMIST Astrochemistry database is at `<www.udfa.net>`.

A 'virtual observatory' for astronomers can be found at `<cdsarc.u-strasbg.fr>` and the latest version of the database High resolution Transmission, HITRAN 2004, is at `<www.hitran.com>`.

The current version of the database Gestion et Etude des Informations Spectroscopiques Atmosphériques (GEISA-03) is at `<ara.lmd.polytechnique.fr>` and the Spherical Top data System (STDS) has address `<icb.u-bourgogne.fr/OMR/SMA/SHTDS/STDS>`.

The National Institute for Standards and Technology (NIST) maintains a database at `<www.physics.nist.gov/PhysRefData>`. which contains the Bibliography on Atomic Line Shapes and Shifts up to 2008 and the database at the Observatoire de Paris, `<amrel.obspm.fr/balss>` contains a Bibliography up to 2007.

The Vienna Atomic Line Database (VALD) can be found at `<ams.astro.univie.ac.at/~vald>` and the Belgrade database at `<www.aob.bg.ac.yu/BELDATA>`.

Acknowledgements

We thank Ms. Fay Ownby and Dr. David Schultz (Oak Ridge National Laboratory) for assistance with bibliographic data. Philip C. Stancil acknowledges support from NASA grant NAG NNG05GD98G.

References

Abrahamsson, E. & Krems, R. V. 2007, *ApJ*, 654, 1171
Abu-Haija, O., Faify S. A., Olmez, G., *et al.* 2005a, *Nucl. Inst. Meth. B*, 241, 109
Abu-Haija, O., Kamber, E. Y., Ferguson, S. M., *et al.* 2005b, *Phys. Rev. A*, 72, 042701
Abu-Karma, M., Stamp, C., Varghese, G., & Reddy, S. P. 2006, *JQSRT*, 97, 332
Achkasova, E., Araki, M., Denisov, A., & Maier, J. P. 2006, *JMoSp*, 237, 70
Aggarwal, K. M. & Keenan, F. P. 2005a, *A&A*, 439, 1215
Aggarwal, K. M. & Keenan, F. P. 2005b, *A&A*, 441, 831
Aggarwal, K. M. & Keenan, F. P. 2006, *A&A*, 450, 1249
Aggarwal, K. M., Hamada, K., Igarashi, A., *et al.* 2008, *A&A*, 484, 879
Aggarwal, K. M., Igarashi, A., Keenan, F. P., & Nakazaki, S. 2008, *A&A*, 479, 585
Aggarwal, K. M., Keenan, F. P., & Rose, S. J. 2005, *A&A*, 432, 1151
Aldener, M., Brown, S. S., Stark, H., *et al.* 2005, *JMoSp*, 232, 223
Allan, M. 2007, *J. Phys. B: At. Mol. Opt. Phys.*, 40, 3531
Allan, M., Winstead, C., & McKoy, V. 2008, *Phys. Rev. A*, 77, 042715

Allard, N. F., Allard, F., & Kielkopf, J. F. 2005, *A&A*, 440, 1195

Allard, N. F. & Peach, G. 2007, in: O. Engvold (ed.), *Reports on Astronomy 2002-2005, IAU Transactions XXVIA* (Cambridge: CUP), 149

Allard, N. F. & Spiegelman, F. 2006, *A&A*, 452, 351

Alonso-Medina, A. & Colón, C. 2007, *A&A*, 466, 399

Alonso-Medina, A. & Colón, C. 2008, *ApJ*, 672, 1286

Alonso-Medina, A., Colón, C., & Zanón, A. 2008, *MNRAS*, 385, 261

Andersson, S., Barinovs, G., & Nyman, G. 2008, *ApJ*, 678, 1042

Andreazza, C. M. & Marinho, E. P. 2005, *ApJ*, 624, 1121

Antony, B. K. & Gamache, R. R. 2007, *JMoSp*, 243, 113

Antony, B. K., Neshyba, S., & Gamache, R. R. 2007, *JQSRT*, 105, 148

Antony, B. K., Niles, D. L., Wroblewski, S. B., *et al.* 2008, *JMoSp*, 251, 268

Arteaga, S. W., Bejger, C. M., Gerecke, J. L., *et al.* 2007, *JMoSp*, 243, 253

Bach, M., Fally, S., Coheur, P.-F, *et al.* 2005, *JMoSp*, 232, 341

Ballance, C. P., Griffin, D. C., *et al.* 2007, *JPhB: At. Mol. Opt. Phys.*, 40, F327

Bandyopadhyay, A., Ray, B., Ghosh, P. N., *et al.* 2007, *JMoSp*, 242, 10

Barinovs, G. & van Hemert, M. C. 2005, *ApJ*, 636, 923

Barinovs, G., Hemert, M. C., Krems, R., & Dalgarno, A. 2005, *ApJ*, 620, 537

Barklem, P. S. 2007, *A&A*, 462, 781

Barklem, P. S. & Aspelund-Johansson, J. 2005, *A&A*, 435, 373

Barragán, P., Errea, L. F., Méndez, L., Rabadán, I., & Riera, A. 2006a, *ApJ*, 636, 544

Barragán, P., Errea, L. F., Méndez, L., Rabadán, I., & Riera, A. 2006b, *PhRvA*, 74, 024701

Barrag'an, P., Le, A.-T.., & Lin, C. D. 2006c, *PhRvA*, 74, 012720

Bartecka, A., Baclawski, A., Wujec, T., & Musielok, J. 2005, *EPJD*, 37, 163

Becker, A., Mosshammer, R., Mokler, P., & Ullrich, J., eds. 2007, Proc. XXV *International Conference on Photonic, Electronic, & Atomic Collisions*, JP-CS, 88

Beiersdorfer, P., Schweikhard, L., Liebisch, P., & Brown, G. V. 2008, *ApJ*, 672, 726

Ben Chaouacha, H., Sahal-Bréchot, S., & Ben Nessib, N. 2007, *A&A*, 465, 651

Bengoechea, J., Aragón, C., & Aguilera, J. A. 2005, *Spectrochim. Acta B*, 60, 897

Bengoechea, J., Aguilera, J. A., & Aragón, C. 2006, *Spectrochim. Acta B*, 61, 69

Bettega, M. H. F., da Costa, R. F.,, & Lima, M. A. P. 2008, *PhRvA*, 77, 052706

Bhatia, A. K. 2008, *PhRvA*, 77, 052707

Blanquet, G., Bouanich, J.-P., Walrand, J., & Lepèere, M. 2005, *JMoSp*, 229, 198

Bodewits, D., Tielens, A. G. G. M., Morgenstern, R., & Hoekstra, R. 2005, *NIMPB*, 235, 358

Bodewits, D., Hoekstra, R., Seredyuk, B., *et al.* 2006, *ApJ*, 642, 593

Bodewits, D. & Hoekstra, R. 2007, *PhRvA*, 76, 032703

Boissoles, J., Domanskaya, A., Boulet, C., Tipping, R. H., & Ma, Q. 2005, *JQSRT*, 95, 489

Bolognesi, P., Pravica, L., Veronesi, S., *et al.* 2008, *PhRvA*, 77, 054704

Bouanich J.-P. & Blanquet, G. 2007, *JMoSp*, 241, 186

Bouanich, J.-P., Walrand, J., & Blanquet, G. 2005, *JMoSp*, 232, 40

Bradley, J., O'Rourke, S. F. C., & Crothers, D. S. F. 2005, *PhRvA*, 71, 032706

Brown, L. R., Benner, D. C., Devi, V. M., *et al.* 2005, *J. Mol. Struct.*, 742, 111

Bruhns, H., Kreckel, H., Savin, D. W., Seely, D. G., & Havener, C. C. 2008, *PhRvA*, 77, 064702

Buffa, G. 2007, *PhRvA*, 76, 042509

Bulanin, M. O., Domanskaya, A. V., Kerl, K., & Maul, C. 2005, *JMoSp*, 230, 87

Buhr, H., Pedersen, H. B., Altevogt, S., *et al.* 2008, *PhRvA*, 77, 032719

Bukvić, S., Srećković, A., & Djeniže, S. 2005, *Publ. Astron. Soc. 'Rudjer Bošković'*, 5, 135

Burgess, A. & Tully, J. A. 2005, *JPhB: At. Mol. Opt. Phys.*, 38, 2629

Bykov, A. D., Lavrentieva, N. N., Mishina, T. P., *et al.* 2008, *JQSRT*, 109, 1834

Cazzoli, G., Dore, L., Puzzarini, C., *et al.* 2005, *JMoSp*, 229, 158

Cazzoli, G., Puzzarini, C., Buffa, G., & Tarrini, O. 2007, *JQSRT*, 105, 438

Cazzoli, G., Puzzarini, C., Buffa, G., & Tarrini, O. 2008, *JQSRT*, 109, 1563

Chen, G.-X. 2008, *PhRvA*, 77, 022703

Chidichimo, M. C., Del Zanna, G., Mason, H. E., *et al.* 2005, *A&A*, 430, 331

Cho, H., Park, Y. S., y Castro, E. A., *et al.* 2008, *JPhB: At. Mol. Opt. Phys.*, 41, 045203
Chourou, S. T. & Orel, A. E. 2008, *PhRvA*, 77, 042709
Ćirišan, M., Peláez, R. J., Djurović, S., *et al.* 2006, *JPhB: At. Mol. Opt. Phys.*, 40, 3477
Colgan, J., Zhang, H. L., & Fontes, C. J. 2008, *PhRvA*, 77, 062704
Colmont, J.-M., Nguyen, L., Rohart, F., & Włodarczak, G. 2007a, *JMoSp*, 246, 86
Colmont, J.-M., Rohart, F., & Włodarczak, G. 2007b, *JMoSp*, 241, 119
Colmont, J.-M., Rohart, F., Włodarczak, G., & Bouanich, J.-P. 2006, *JMoSp*, 238, 98
Colón, C. & Alonso-Medina, A. 2006, *Spectrochim. Acta B*, 61, 856
Colón, C., Alonso-Medina, A., Rivero, C., & Fernández, F. 2006, *PhyS*, 73, 410
Cornelius, K. R. 2006, *PhRvA*, 73, 032710
da Costa, R. F., Bettega, M. H. F., Ferreira, L. G., *et al.* 2007, *JPh-CS*, 88, 012028
da Costa, R. F., da Paixão, F. J ., & Lima, M. A. P. 2005, *JPhB: At. Mol. Opt. Phys.*, 38, 4363
Čurik, R., Čársky P., & Allan, M. 2008, *JPhB: At. Mol. Opt. Phys.*, 41, 115203
Dayou, F. & Balança, C. 2006, *A&A*, 459, 297
Demura, A. V., Demchenko, G. V., & Nikolić, D. 2008a, *EPJD*, 46, 203
Demura, A. V., Demchenko, G. V., & Nikolić, D. 2008b, *EPJD*, 46, 111
Derouich, M. 2007, *A&A*, 466, 683
Derouich, M. & Barklem, P. S. 2007, *A&A*, 462, 1171
Derouich, M., Barklem, P. S., & Sahal-Bréchot, S. 2005a, *A&A*, 441, 395
Derouich, M., Sahal-Bréchot, S., & Barklem, P. S. 2005b, *A&A*, 434, 779
Devi, V. M., Benner, D. C., Smith, M. A. H., *et al.* 2005, *JMoSp*, 231, 66
Devi, V. M., Benner, D. C., Brown, L. R., Miller, C. E., & Toth, R. A. 2007a, *JMoSp*, 242, 90
Devi, V. M., Benner, D. C., Brown, L. R., Miller, C. E., & Toth, R. A. 2007b, *JMoSp*, 245, 52
Dhib, M., Ibrahim, N., Chelin, P., *et al.* 2007, *JMoSp*, 242, 83
Dickinson, A. S. 2005, *JPhB: At. Mol. Opt. Phys.*, 38, 4329
Dimitrijević, M. S., Christova, M., & Sahal-Bréchot, S. 2007a, *PhyS*, 75, 809
Dimitrijević, M. S., Ryabchikova, T., Popović, L.Č., *et al.* 2005, *A&A*, 435, 1191
Dimitrijević, M. S., Simić, S., Kovačević, A., *et al.* 2007b, in: *Flows, Boundaries, and Interaction Workshop, AIP-CP*, 934, 202
Dimitrijević, M. S., Simić, S., *et al.* 2008, *Contr. Obs. Astron. Skalnate Pleso*, 38, 403
Dimitrijević, M. S. & Csillag, L. 2006, *JApSp*, 73, 458
Djeniže, S., Srećković, A., & Bukvić, S. 2005a, *Spectrochim. Acta B*, 60, 1552
Djeniže, S., Srećković, A., & Bukvić, S. 2005b, *Jap. J. Appl. Phys.*, 44, 1450
Djeniže, S., Srećković, A., & Bukvić, S. 2005c, *Z. Naturforsch. A*, 60, 282
Djeniže, S., Srećković, A., & Bukvić, S. 2006a, *Spectrochim. Acta B*, 61, 588
Djeniže, S., Srećković, A., & Bukvić, S. 2006b, *Z. Naturforsch. A*, 61, 91
Djeniže, S., Srećković, A., Bukvić, S., & Vitas, N. 2006c, *Z. Naturforsch. A*, 61, 491
Djeniže, S., Srećković, A., & Nikolić, Z. 2006d, *JPhB: At. Mol. Opt. Phys.*, 39, 3037
Djeniže, S., Bukvić, S., Srećković, A., & Nikolić, Z. 2006e, *New Astronomy*, 11, 256
Djurić, N., Smith, S.J., Simcić, J., & Chutjian, A. 2008, *ApJ*, 679, 1661
Djurović, S., Nikolić, D., Savić, I., Sörge, S., & Demura, A. V. 2005, *PhRvE*, 71, 036407
Djurović, S., Peláez, R. J., Ćirišan, M., *et al.* 2006, *JPhB: At. Mol. Opt. Phys.*, 39, 2901
Domanskaya, A. V., Bulanin, M. O., Kerl, K., & Maul, C. 2007, *JMoSp*, 243, 155
Douguet, N., Kokoouline, V., & Greene, C. H. 2008, *PhRvA*, 77, 064703
Dubau, J., Blancard, C., & Cornille, M. 2007, *High Energy Density Physics*, 3, 76
Dubernet, M.-L., Daniel, F., Grosjean, A., *et al.* 2006, *A&A*, 460, 323
Dubois, A., Caillet, J., Hansen, J. O., *et al.* 2005, *NIMPB*, 241, 48
Dutta, C.M., Gu, J.P., Hirsch, G., *et al.* 2005, *PhRvA*, 72, 052715
Dutta, C.M., Oubre, C., Nordlander, P., Kimura, M., & Dalgarno, A. 2006, *PhRvA*, 73, 032714
Dzierżega, K., Musiol, K., Pokrzywka, B., & Zawadzki, W. 2006, *Spectrochim. Acta B*, 61, 850
Elabidi, H., Ben Nessib, N., Cornille, M., *et al.* 2008a, *JPhB: At. Mol. Opt. Phys.*, 41, 025702
Elabidi, H., Ben Nessib, N., & Dimitrijević, M. S. 2008b, *New Astronomy*, 12, 64
El-Kader, M. S. A. & Moustafa, S. I. 2005, *CP*, 318, 199
Errea, L.F., Illescas, C., Méndez, L., Pons, B., Riera, A. , & Suárez, J. 2005a, *NIMPB*, 235, 315

Errea, L. F., Macias, A., Méndez, L., Rabadán, I., & Riera, A. 2005b, *NIMPB*, 235, 362

Fainstein, P. D., Lima, M. A., P., Miraglia, J. E., *et al.* (eds.) 2006, Proc. XXIV Intern. Conf. *Photonic, Electronic, & Atomic Collisions* (Singapore: World Scientific Publ. Co.)

Faure, A., Kokoouline, V., Greene, C. H., *et al.* 2006, *JPhB: At. Mol. Opt. Phys.*, 39, 4261

Flaud, P.-M., Orphal, J., Boulet, C., & Hartmann, J.-M. 2006, *JMoSp*, 235, 149

Fogle, M., Badnell, N. R., Glans, P., *et al.* 2005, *A&A*, 442, 757

Fogle, M., Bahati, E. M., Bannister, M. E., *et al.* 2008, *ApJS*, 175, 543

Fu, J., Gorczyca, T. W., Nikolić, D., *et al.* 2008, *PhRvA*, 77, 032713

Fujimoto, M. M., Michelin, S. E., Mazon, K. T., *et al.* 2007, *PhRvA*, 76, 012709

Gavrilenko, V. P. & Oks, E. 2007, *PhyS*, 76, 43

Gigosos, M. A. & González, M. A. 2006, in: *Physics of Ionized Gases*, AIP-CP, 876, 294

Gigosos, M. A., González, M.Á., Talin, B., & Calisti, A. 2007, *A&A*, 466, 1189

Giri, K. & Sathyamurthy, N. 2006, *JPhB: At. Mol. Opt. Phys.*, 39, 4123

Godbert-Mouret, L., Capes, H., Koubiti, M., *et al.* 2006, in: *PLASMA 2005*, AIP-CP, 812, 423

Golubiatnikov, G. Yu. 2005, *JMoSp*, 230, 196

Golubiatnikov, G. Yu., Koshelev, M. A., & Krupnov, A. F. 2008, *JQSRT*, 109, 1828

Griem, H. R., *et al.* 2005, *JPhB: At Mol. Opt. Phys.*, 38, 975; erratum, 2006, 39, 3705

Griem, H. R. & Ralchenko, Y. 2006, in: *Spectral Line Shapes*, AIP-CP, 874, 14

Grossel, A., Zéninari, V., Parvitte, B., *et al.* 2008, *JQSRT*, 109, 1845

Hamdi, R., Ben Nessib, N., Dimitrijević, M. S., & Sahal-Bréchot, S. 2007, *ApJS*, 170, 243

Hamdi R., Ben Nessib, N., Milovanović, N., *et al.* 2008, *MNRAS*, 387, 871

Hartmann, J.-M., Bouanich, J.-P., Boulet, C., *et al.* 2005, *JQSRT*, 95, 151

Hashimoto, T. & Kanamori, H. 2006, *JMoSp*, 235, 104

Hasan, A. T. 2005, *EJPhD*, 35, 461

Havener, C. C., Rejoub, R., Kristić, P.S., & Smith, A. C. H. 2005, *PhRvA*, 71, 042707

Haxton, D. J., Rescigno, T. N., & McCurdy, C. W. 2007, *PhRvA*, 75, 012711

Henningsen, J., Barbe, A., & De Backer-Barilly, M.-R. 2008, *JQSRT*, 109, 2491

Hikida, T. & Yamada, K. M. T. 2006, *JMoSp*, 239, 154

Hikida, T., Yamada, K. M. T., Fukabori, M., Aoki, T., & Watanabe, T. 2005, *JMoSp*, 232, 202

Hodges, J. T., Lisak, D., Lavrentieva, N., *et al.* 2008, *JMoSp*, 249, 86

Holtgrave, J. C. & Wolf, P. J. 2005, *PhRvA*, 72, 012711

Hoshina, H., Seta, T., Iwamoto, T., *et al.* 2008, *JQSRT*, 109, 2303

Hoshino, M., Pichl, L., Kanai, Y., *et al.* 2007, *PhRvA*, 75, 012716

Huang, Y., Wu, S., Yang, E., Gao, M., Zhang, X., Li, G., & Lu, F. 2005, *NIMPB*, 229, 46

Hudson, C. E. & Bell, K. L. 2005, *A&A*, 430, 725

Hudson, C. E. & Bell, K. L. 2006, *A&A*, 452, 1113

Ibrahim, N., Chelin, P., Orphai, J., & Baranov, Y. I. 2008, *JQSRT*, 109, 2523

Ibrahim, N., Thiebaud, J., Orphal, J., & Fittschen, C. 2007, *JMoSp*, 242, 64

Irrera, S. & Gianturco, F. A. 2005, *New J. Phys.*, 7, 1

Ishikawa, Y. & Vilkas, M. J. 2008, *PhRvA*, 77, 052701

Iglesias, E. J., Ghosh, J., Elton, R. C., & Griem, H. R. 2006, *JQSRT*, 98, 101

Ivković, M., Ben Nessib, N., & Konjević, N. 2005, *JPhB: At. Mol. Opt. Phys.*, 38, 715

Jacquemart, D., Tchana, F. K., Lacome, N., & Kleiner, I. 2007, *JQSRT*, 105, 264

Jacquemart, D. & Tran, H. 2008, *JQSRT*, 109, 569

Janowicz, M., Słabkowska, K., Matuszak, P., & Polasik, M. 2005, *NIMPB*, 235, 337

Jenouvrier, A., Daumont, L., Régalia-Jarlot, L., *et al.* 2007, *JQSRT*, 105, 326

Joly, L., Gibert, F., Grouiez, B., *et al.* 2008, *JQSRT*, 109, 426

Jovićević, S., Ivković, M., Žikić, R., & Konjević, N. 2005, *JPhB: At. Mol. Opt. Phys.*, 38, 1249

Jung, Y.-D. 2008, *ApJ*, 674, 1207

Kaiser, C., Spieker, D., Gao, J., *et al.* 2007, *JPhB: At. Mol. Opt. Phys.*, 40, 2563

Kaneyasu, T., Azuma, T., & Okuno, K. 2005, *JPhB: At. Mol. Opt. Phys.*, 38, 1341

Kanno, N., Tonokura, K., Tezaki, A., & Koshi, M. 2005, *JMoSp*, 229, 193

Kato, H., Kawahara, H., Hoshino, M., *et al.* 2008, *PhRvA*, 77, 062708

Kaur, S. & Baluja, K. L. 2005, *JPhB: At. Mol. Opt. Phys.*, 38, 3917

Kaur, S., Baluja, K. L., & Tennyson, J. 2008, *PhRvA*, 77, 032718

de Kertanguy, A., Feautrier N., & Motapon O. 2005, *A&A*, 432, 1131

Khakoo, M.A., Wang, S., Laher, R., *et al.* 2007, *JPhB: At. Mol. Opt. Phys.*, 40, F167

Khakoo, M. A., Blumer, J., Keane, K., *et al.* 2008, *PhRvA*, 77, 042705

Kimura, M., Pichl, L., Li, Y., *et al.* 2006, *EPJD*, 38, 85

Kokoouline, V. & Greene, C. H. 2005, *JPh-CS*, 4, 74

Koshelev, M. A., Tretyakov, M. Yu., Golubiatnikov, G. Yu., *et al.* 2007, *JMoSp*, 241, 101

Koshelev, M. A., Tretyakov, M. Yu., Lees, R. M., & Xu, L.-H. 2006, *JMoSt*, 780, 7

Krems, R., Jamieson, M. J., & Dalgarno, A. 2006, *ApJ*, 647, 1531

Kumar, T. J. D., Saieswari, A., & Kumar, S. 2006, *JCP*, 124, 034314

Kusakabe, T., Miyamoto, Y., Kimura, M., & Tawara, H. 2006, *PhRvA*, 73, 022706

O'Leary, D. M., Orphal, J., Ruth, A. A., *et al.* 2008, *JQSRT*, 109, 1004

Le Barbu, T., Zéninari, V., Parvitte, B., *et al.* 2006, *JQSRT*, 98, 264

Lee, T.-G. 2006, *JPhB: At. Mol. Opt. Phys.*, 39, 3665

Lee, T.-G., Balakrishnan, N., & Forrey, R. C. 2006, *JCP*, 125, 114302

Lepère, M. 2006, *JMoSp*, 238, 193

Lepère, M., Blanquet, G., Walrand, J., *et al.* 2007, *JMoSp*, 242, 25

Lepère, M., Valentin, A., Henry, A., *et al.* 2005, *JMoSp*, 233, 86

Lerot, C., Blanquet, G., Bouanich, J.-P., Walrand, J., & Lepère, M. 2005, *JMoSp*, 230, 153

Lerot, C., Blanquet, G., Bouanich, J.-P., Walrand, J., & Lepère, M. 2006a, *JMoSp*, 235, 196

Lerot, C., Blanquet, G., Bouanich, J.-P., Walrand, J., & Lepère, M. 2006b, *JMoSp*, 238, 224

Lesage, A. 2008, *New Astron. Revs*, in press

Li, H., Farooq, A., Jeffries, J. B., & Hanson, R. K. 2008, *JQSRT*, 109, 132

Li, J. S., Liu, K., Zhang, W. J., Chen, W. D., & Gao, X. M. 2008, *JQSRT*, 109, 1575

Lin, C. Y., Stancil, P. C., Li, Y., *et al.* 2007, *PhRvA*, 76, 012702

Lindahl, A. O., Andersson, P., Collins, G. F., *et al.* 2008, *PhRvA*, 77, 022710

Lindsay, B. G., Yu, W. S., & Stebbings, R. F. 2005a, *PhRvA*, 71, 032705

Lindsay, B. G., Wu, W. S., & Stebbings, R. F. 2005b, *JPhB: At. Mol. Opt. Phys.*, 38, 1977

Linert, I., Mielewska, B., King, G. C., & Zubek, M. 2007, *PhRvA*, 76, 032715

Lique, F., Dubernet, M.-L., Spielfiedel, A., & Feautrier, N. 2006a, *A&A*, 450, 399

Lique, F., Spielfiedel, A., & Cernicharo, J. 2006b, *A&A*, 451, 1125

Lique, F., Spielfiedel, A., Dhont, G., & Feautrier, N. 2006c, *A&A*, 458, 331

Lique, F. & Spielfiedel, A. 2007, *A&A*, 462, 1179

Lisgo, S., Brooks, N., Oks, E., *et al.*D. 2006, in: *Spectral Line Shapes, AIP-CP*, 874, 253

Liu, C.-N., Cheng, S.-C., Le, A.-T., & Lin, C. D. 2005, *PhRvA*, 72, 012717

Liu, C. L., Wang, J. G., & Janev, R. K. 2006, *JPhB: At. Mol. Opt. Phys.*, 39, 1223

Liu, C. L., Wang, J. G., & Janev, R. K. 2008a, *PhRvA*, 77, 032709

Liu, C. L., Wang, J. G., & Janev, R. K. 2008b, *PhRvA*, 77, 042712

Liu, X., Zhou, X., Jeffries, J. B., & Hanson, R. K. 2007, *JQSRT*, 103, 565

Lucchesini, A. & Gozzini, S. 2007, *JQSRT*, 103, 209

Luna, H., McGrath, C., Shah, M. B., *et al.* 2005, *ApJ*, 628, 1086

Lyulin, O. M., Jacquemart, D., Lacome, N., *et al.* 2008, *JQSRT*, 109, 1856

Ma, Q., Tipping, R. H., Boulet, C., *et al.* 2007, *JMoSp*, 243, 105

Machin, L. & Roueff, E. 2005, *JPhB: At. Mol. Opt. Phys.*, 38, 1519

Machin, L. & Roueff, E. 2006, *A&A*, 460, 953

Machin, L. & Roueff, E. 2007, *A&A*, 465, 647

Mack, A., Clark, T. K., Forrey, R. C., *et al.* 2006, *PhRvA*, 74, 052718

McLaughlin, B. M., Hibbert, A., Scott, M. P., *et al.* 2006, *A&A*, 446, 1185

Mahmoudi, W. F., Ben Nessib, N., & Dimitrijević, M. S. 2005, *A&A*, 434, 773

Mahmoudi, W. F., Ben Nessib, N., & Sahal-Bréchot, S. 2008, *EPJD*, 47, 7

Makochekanwa, C., Hoshino, M., Kato, H., *et al.* 2008, *PhRvA*, 77, 042717

Martin, B. & Lepère, M. 2008, *JMoSp*, 250, 70

Martin, B., Walrand, J., Blanquet, G., Bouanich, J.-P., & Lepère, M. 2006, *JMoSp*, 236, 52

Martín, F. 2007, *JPh-CS*, 88, 012001

Martinez, H., Hernandez, C. L., & Yousif, F. B. 2006, *JPhB: At. Mol. Opt. Phys.*, 39, 2535

Maslov, M., Brunger, M. J., Teubner, P. J. O., *et al.* 2008, *PhRvA*, 77, 062711

Matton, S., Rohart, F., Bocquet, R., *et al.* 2006, *JMoSp*, 239, 182

May, O., Fedor, J., Ibänescu, B. C., & Allan, M. 2008, *PhRvA*, 77, 040701

Mayo, R., Ortiz, M., & Plaza, M. 2008, *JPhB: At. Mol. Opt. Phys.*, 41, 095702

Meléndez M. & Bautista, M. A. 2005, *A&A*, 436, 1123

Meléndez, M., Bautista, M. A., & Badnell, N. R. 2007, *A&A*, 469, 1203

Menard-Bourcin, F., Menard, J., & Boursier, C. 2007, *JMoSp*, 242, 55

Mihajlov, A. A., Ignjatović, M. Lj., & Dimitrijević, M. S. 2005, *A&A*, 437, 1023

Milosavljević, V., Ellingboe, A. R., & Djeniže, S. 2006, *Spectrochim. Acta B*, 61, 81

Milovanović, N. & Dimitrijević, M. S. 2007, in: *Spectral Line Shapes in Astrophysics*, Proc. VI Serbian Conf., *AIP-CP*, 938, 258

Minami, T., Pindzola, M. S., Lee, T.-G., *et al.* 2006, *JPhB: At. Mol. Opt. Phys.*, 39, 2877

Misago, F., Lepère, M., Bouanich, J.-P., & Blanquet, G. 2008, *JMoSp*, 248, 14

Misago, F., Lepère, M., Walrand, J., Bouanich, J.-P., & Blanquet, G. 2006, *JMoSp*, 237, 46

Misago, F., Lepère, M., Walrand, J., Bouanich, J.-P., & Blanquet, G. 2007, *JMoSp*, 241, 61

Mondelain, D., Chelin, P., Valentin, A., *et al.* 2005, *JMoSp*, 233, 23

Mondelain, D., Payan, S., Deng, W., *et al.* 2007, *JMoSp*, 244, 130

Motapon, O., Tamo, F. O. W., Urbain, X., & Schneider, I. F. 2008, *PhRvA*, 77, 052711

Mrugala, F. & Kraemer, W. P. 2005, *JCP*, 122, 224321

Mullamphy, D. F. T., Peach, G., Venturi, V., *et al.* 2007, *JPhB: At. Mol. Opt. Phys.*, 40, 1141

Munjal, H. & Baluja, K. L. 2006, *JPhB: At. Mol. Opt. Phys.*, 39, 3185

N'Dollo, M. & Donga-Passi, J. 2006, *PhyS*, 74, 208

Ngassam, V., Orel, A. E., & Suzor-Weiner, A. 2005, *JPh-CS*, 4, 224

Ngassam, V., Florescu-Mitchell, A. I., & Orel, A. E. 2008, *PhRvA*, 77, 042706

Nguyen, L., Blanquet, G., Anwera, J. V., & Lepère, M. 2008, *JMoSp*, 249, 1

Nguyen, L., Ivanov, S. V., Buzykin, O. G., & Buldyreva, J. 2006, *JMoSp*, 239, 101

Ogurtsov, G. N., Mikoushkin, V. M., Ovchinnikov, S. Yu., *et al.* 2006, *PhRvA*, 74, 042720

Ohlinger, L., Forrey, R. C., Lee, T.-G., & Stancil, P. C. 2007, *PhRvA*, 76, 042712

Olkhov, R. V. & Smith, I. W. M. 2007, *JCP*, 126, 134314

Oks, E., *et al.* (eds.) 2006, Proc. 18th Intern. Conf. *Spectral Line Shapes*, *AIP-CP*, 874

Oks, E. 2006, *Stark Broadening of Hydrogen, & Hydrogenlike Spectral Lines in Plasmas* (Oxford: Alpha Science International)

Omar, B., Günter, S., Wierling, A., & Röpke, G. 2006, *PhRvE*, 73, 056405

Ortiz, M. & Mayo, R. 2005, *JPhB: At. Mol. Opt. Phys.*, 38, 3953

Orton, G. S., Gustafsson, M., Burgdorf, M., & Meadows, V. 2007, *Icarus*, 189, 544

Otranto, S. & Olson, R. E. 2008, *PhRvA*, 77, 022709

Otranto, S., Olson, R. E., & Beiersdorfer, P. 2006, *PhRvA*, 73, 022723

Ovchinnikov, S. Yu, Krstić, P. S., & Macek, J. H. 2006, *PhRvA*, 74, 042706

Palov, A. P., Gray, M. D., Field, D., & Balint-Kurti, G. G. 2006, *ApJ*, 639, 204

Pandey, M. K. & Dubey, R. K. 2007, *EPJD*, 41, 275

Peláez, R. J., Ćirišan, M., Djurović, S., *et al.* 2006a, *JPhB: At. Mol. Opt. Phys.*, 39, 5013

Peláez, R. J., González, V. R., Rodríguez, F., Aparicio, J. A., & Mar, S. 2006b, *A&A*, 453, 751

Perez, J. A. & Olson, R. E. 2005, *NIMPB*, 241, 134

Pindzola, M. S., Loch, S. D., Colgan, J., & Fontes, C. J. 2008, *PhRvA*, 77, 062707

Pindzola, M. S., Robicheaux, F., & Colgan, J. 2005, *JPhB: At. Mol. Opt. Phys.*, 38, L285

Popović, L.Č., & Dimitrijević, M.S. (eds, 2007, *Spectral Line Shapes in Astrophysics*, Proc. VIth Serbian Conf., *AIP-CP*, 938

Popović, L.Č., Dimitrijević, M. S., Simić, S., *et al.* 2008, *New Astron.*, 13, 85

Poquérusse, A. & Alexiou, S. 2006, *JQSRT*, 99, 493

Predoi-Cross, A., Brawley-Tremblay, M., Brown, L. R., *et al.* 2006, *JMoSp*, 236, 201

Predoi-Cross, A., Hambrook, K., Keller, R., *et al.* 2008a, *JMoSp*, 248, 85

Predoi-Cross, A., Holladay, C., Heung, H., *et al.* 2008b, *JMoSp*, 251, 159

Predoi-Cross, A., Unni, A. V., Heung, H., *et al.* 2007a, *JMoSp*, 246, 65

Predoi-Cross, A., Unni, A. V., Liu, W., *et al.* 2007b, *JMoSp*, 245, 34

Predoi-Cross, A., Liu, W., Holladay, C., *et al.* 2007c, *JMoSp*, 246, 98

Ptashnik, I. V., Smith, K. M., & Shine, K. P. 2005, *JMoSp*, 232, 186

Purić, J., Dojčinović, I. P., Nikolić, M., *et al.* 2008, *ApJ*, 680, 803

Rabasović, M. S., Kelemen, V. I., Tošić, S. D., *et al.* 2008, *PhRvA*, 77, 062713

Rabli, D. & McCarroll, R. 2005, *JPhB: At. Mol. Opt. Phys.*, 38, 3311

Ramsbottom, C. A., Hudson, C. E., Norrington, P. H., & Scott, M. P. 2007, *A&A*, 475, 765

Reese, C., Stoecklin, T., Voronin, A., & Rayez, J.C. 2005, *A&A*, 430, 1139

Rescigno, T. N., McCurdy, C. W., Haxton, D. J., *et al.* 2007, *JPh-CS*, 88, 012027

Rinsland, C. P., Devi, V. M., Benner, D. C., *et al.* 2008, *JQSRT*, 109, 974

Rohart, F., Nguyen, L., Buldyreva, J., *et al.* 2007, *JMoSp*, 246, 213

Rohart, F., Włodarczak, G., Colmont, J.-M., *et al.* 2008, *JMoSp*, 251, 282

Rosenberry, M. A., Burgess, K. M., & Stewart, B. 2007, *JPhB: At. Mol. Opt. Phys.*, 40, 177

Rothman, L. S., Jacquemart, D., Barbe, A., *et al.* 2005, *JQSRT*, 96, 139

Sahal-Bréchot, S., Derouich, M., Bommier, V., & Barklem, P.S. 2007, *A&A*, 465, 667

Schmidt, E. W., Bernhardt, D., Müller, A., *et al.* 2007, *PhRvA*, 76, 032717

Schultz, D. R. & Stancil, P. C. 2007, in: O. Engvold (ed.), *Reports on Astronomy 2002 - 2005, IAU Transactions XXVIA* (Cambridge: CUP), 134

Seredyuk, B., McCullough, R. W., & Gilbody, H. B. 2005a, *PhRvA*, 71, 022713

Seredyuk, B., McCullough, R. W., & Gilbody, H. B. 2005b, *PhRvA*, 72, 022710

Seredyuk, B., McCullough, R. W., Tawara, H., *et al.* 2005c, *PhRvA*, 71, 022705

Seta, T., Hoshina, H., Kasai, Y., *et al.* 2008, *JQSRT*, 109, 144

Shepler, B. C., Yang, B. H., Kumar, T. J. D., *et al.* 2007, *A&A* (Letters), 475, L15

Shi, D., Sun, J., Liu, Y., & Zhu, Z. 2008, *JPhB: At. Mol. Opt. Phys.*, 41, 025205

Simić, Z., Dimitrijević, M. S., Milovanović, N., & Sahal-Bréchot, S. 2005a, *A&A*, 441, 391

Simić, Z., Dimitrijević, M. S., Popović, L.Č., & Dačić, M. D. 2005b, *JApSp*, 72, 443

Simić, Z., Dimitrijević, M. S., Popović, L.Č., & Dačić, M. D. 2006, *New Astron.*, 12, 187

Smith, M. A. H., Rinsland, C. P., Blake, T. A., *et al.* 2008, *JQSRT*, 109, 922

Sobocinski, P., Pešić R., Hellhammer, R., *et al.* 2005, *NIMPB*, 233, 207

Srećković, A., Bukvić, S., & Djeniže, S. 2005a, *PhSc*, 71, 218

Srećković, A., Bukvić, S., Djeniže, S., *et al.* 2005b, *Publ. Astron. Soc. "Rudjer Bošković"*, 5, 275

Srećković, A., Nikolić, Z., Bukvić, S., & Djeniže, S. 2007, *JQSRT*, 105, 536

Staak, M., Gash, E. W., Venables, D. S., & Ruth, A. A. 2005, *JMoSp*, 229, 115

Stambulchik, E., Alexiou, S., Griem, H. R., & Kepple, P. C. 2007, *PhRvE*, 75, 016401

Stambulchik, E. & Maron, Y. 2006, *JQSRT*, 99, 730

Stambulchik, E. & Maron, Y. 2008, *JPhB: At. Mol. Opt. Phys.*, 41, 095703

Stehlé, C., Busquet, M., Gilles, D., & Demura, A.V. 2005, *Laser & Particle Beams*, 23, 357

Storey, P. J., Del Zanna, G., Mason, H. E., & Zeippen, C. J. 2005, *A&A*, 433, 717

Sung, K. & Varanasi, P. 2005, *JQSRT*, 91, 319

Suno, H. & Kato, T. 2006, *At. Data Nucl. Data Tables*, 92, 407

Suno, H., Rai, S. N., Liebemann, H.-P., *et al.* 2006, *PhRvA*, 74, 012701

Suzuki, R., Rai, S. N., Liebemann, H.-P., *et al.* 2005, *PhRvA*, 71, 032710

Szmytkowski, C., Domaracka, A., Możejko, P., *et al.* 2005, *JPhB: At. Mol. Opt. Phys.*, 38, 745

Szmytkowski, C., Możejko, P., Kwitnewski, S., *et al.* 2006, *JPhB: At. Mol. Opt. Phys.*, 39, 2571

Tashiro, M. & Morokuma, K. 2007, *PhRvA*, 75, 012720

Tayal, S. S. 2007, *ApJS*, 171, 331

Thibault, F. 2005, *JMoSp*, 234, 286

Thibault, F., Mantz, A. W., Claveau, C., *et al.* 2007, *JMoSp*, 246, 118

Tobola, R., Klos, J., Lique, F., Chalasinski, G., & Alexander, M. H. 2007, *A&A*, 468, 1123

Tolchenov, R. N. & Tennyson, J. 2005, *JMoSp*, 231, 23

Tong, L. & Nanbu, K. 2007, *Vacuum*, 81, 1119

Toth, R. A. 2005, *JQSRT*, 94, 1

Toth, R. A., Brown, L. R., Miller, C. E., Devi, V. M., & Benner, D. C. 2006, *JMoSp*, 239, 243

Toth, R. A., Brown, L. R., Miller, C. E., Devi, V. M., & Benner, D. C. 2008, *JQSRT*, 109, 906

Toth, R. A., Miller, C. E., Devi, V. M., Benner, D.C., & Brown, L.R. 2007, *JMoSp*, 246, 133

Tran, H., Bermejo, D., Domenech,J.-L., *et al.* 2007, *JQSRT*, 108, 126

Tran, H., Flaud, P.-M., Flouchet, T., Gabard, T., & Hartmann, J.-M. 2006, *JQSRT*, 101, 306

Tran, H., Jacquemart, D., Mandin, J.-Y., & Lacome, N. 2008, *JQSRT*, 109, 119

Tretyakov, M. Yu., Koshelev, M. A., Dorovskikh, V. V., *et al.* 2005, *JMoSp*, 231, 1

Tretyakov, M. Yu., Koshelev, M. A., Koval, I. A., *et al.* 2007, *JMoSp*, 241, 109

del Val, J. A., Peláez, R. J., Mar, S., *et al.* 2008, *PhRvA*, 77, 012501

Valognes, J. C., Bardet, J. P., Vitel, Y., & Flih, S.A. 2005, *JQSRT*, 95, 113

Vincent, L. F. M., Spielfiedel, A., & Lique, F. 2007, *A&A*, 472, 1037

Wang, J. G., He, B., Ning, Y., Liu, C. L., Yan, J., *et al.* 2006, *PhRvA*, 74, 052709

Wang, Y. & Zhou, Y. 2006, *JPhB: At. Mol. Opt. Phys.*, 39, 3009

Wargelin, B. J., Beiersdorfer, P., Neill, P. A., Olson, R. E., & Scofield, J. H. 2005, *ApJ*, 634, 687

Watson ,J. K. G. 2006, *JPhB: At. Mol. Opt. Phys.*, 39, 1889

Weck, P. F., Kwong, V. H. S., & Salama, F. 2006, Proc. *NASA Laboratory Astrophysics Workshop*, *NASA-CP* 2006-214549

Wehr, R., Ciuryło, R., Vitcu, A., *et al.* 2006a, *JMoSp*, 235, 54

Wehr, R., Vitcu, A., Thibault, F., *et al.* 2006b, *JMoSp*, 235, 69; erratum, 237, 126

Wells, E., Krishnamurthi, V., Carnes, K. D., *et al.* 2005, *PhRvA*, 72, 022726

Wells, E., Nishide, T., Tawara, H., *et al.* 2008, *PhRvA*, 77, 064701

Wernli, M., Valiron, P., Faure, A., *et al.* 2006, *A&A*, 446, 367

Wernli, M., Wiesenfeld, L., Faure, A., & Valiron, P. 2007, *A&A*, 464, 1147

Wilson, N. J., Bell, K. L., & Hudson, C. E. 2005, *A&A*, 432, 731; erratum 461, 765

Winstead, C. & McKoy, V. 2007, *PhRvA*, 76, 012712

Wishnow, E. H., Orton, G. S., Ozier, I., & Gush, H. P. 2007, *JQSRT*, 103, 102

Witthoeft, M. C. & Badnell, N. R. 2008, *A&A*, 481, 543

Witthoeft, M. C., Badnell, N. R., Del Zanna, G., *et al.* 2006, *A&A*, 446, 361

Witthoeft, M. C., Del Zanna, G., & Badnell, N. R. 2007, *A&A*, 466, 763

Woodall, J., Agúndez, M., Markwick-Kemper, A. J., & Millar, T. J. 2007, *A&A*, 466, 1197

Wrathmall, S. A. & Flower, D. R. 2006, *JPhB: At. Mol. Opt. Phys.*, 39, L249

Wrathmall, S. A., Gusdorf, A., & Flower, D. R. 2007, *MNRAS*, 382, 133

Xie, Z., Bowman, J. M., & Zhang, X. 2006, *JChPh*, 125, 133120

Yamada, M. M. & Amano, T. 2005, *JQSRT*, 95, 221

Yang, B. & Stancil, P. C. 2007, *JCP*, 126, 154306

Yang, B. & Stancil, P. C. 2008, *EuPhJD*, 47, 351

Yang, B., Perera, H., Balakrishnan, N., *et al.* 2006a, *JPhB: At. Mol. Opt. Phys.*, 39, S1229

Yang, B., Stancil, P. C., Balakrishnan, N., & Forrey, R. C. 2006b, *JCP*, 124, 104304

Zarour, B., Champion, C., Hanssen, J., & Lasri, B. 2005, *NIMPB*, 235, 374

Zatsarinny, O. & Bartschat, K. 2008, *PhRvA*, 77, 062701

Zeng, S. L., Liu, L., Wang, J. G., & Janev, R. K. 2008, *JPhB: At. Mol. Opt. Phys.*, 41, 135202

Zéninari, V., Joly, L., Grouiez, B., Parvitte, B., & Barbe, A. 2007, *JQSRT*, 105, 312

Zhao, L. B., Wang, J. G., Stancil, P. C., *et al.* 2006, *JPhB: At. Mol. Opt. Phys.*, 39, 5151

Zhao, L.B., Watanabe, A., Stancil, P. C., & Kimura, M. 2007, *PhRvA*, 76, 022701

Zhaunerchyk, V., Geppert, W. D., Österdahl, F., *et al.* 2008a, *PhRvA*, 77, 022704

Zhaunerchyk, V., Thomas, R. D., Geppert, W. D., *et al.* 2008b, *PhRvA*, 77, 034701

Zhu, C., Babb, J. F., & Dalgarno, A. 2006, *PhRvA*, 73, 012506

Zhu, X. L., Ma, X., Wei, B., Liu, H. P., Wang, Z. L., *et al.* 2005, *NIMPB*, 235, 387

Zmerli, B., Ben Nessib, N., & Dimitrijević, M. S. 2008, *EPJD*, 48, 389

Zouros, T. J. M., Sulik, B., Gulyás, L., & Tökési, K. 2008, *PhRvA*, 77, 050701

Zygelman, B. 2005, *ApJ*, 622, 1356

Transactions IAU, Volume XXVIIA
Reports on Astronomy 2006–2009
Karel A. van der Hucht, ed.

© 2009 International Astronomical Union
doi:10.1017/S1743921308025969

DIVISION XII / COMMISSION 14 / WORKING GROUP
SOLIDS AND THEIR SURFACES

CHAIR   Gianfranco Vidali

SEXENNIAL REPORT 2003 - 2009

1. Introduction

In the last decade there has been a tremendous increase of interest in studying processes occurring on IS dust. In part this is due to the availability of ground-based and space-borne high quality instruments which have been used to detect molecules in diverse astrophysical environments, from protoplanetary disks to hot cores and dense clouds. It has also been recognized that IS dust has an important role in the formation of molecules, from molecular hydrogen to methanol. Therefore, it is necessary not to study only properties of dust, but also understand how atoms and molecules interact with and on dust.

This has prompted a number of laboratories with a tradition of working in surface science to study the processes associated with dust. Besides the standard probes that have been used in the past, now there are available techniques that can give precise information at the atomic/molecular level about the formation of molecules on dust. For instance, Thermal Programmed Desorption (TPD), Reflection Absorption Infrared Spectrometry (RAIRS), Resonant Enhanced Multiphoton Ionization (REMPI), and Atom Force Microscopy (AFM) give information about the kinetics and energetics of diffusion of atoms/molecules on and desorption from surfaces, the products of reaction, the ro-vibrational state of ejected products, and the morphology of the solid surfaces, respectively. One of the consequences of the interest by surface science laboratories in studying physical/chemical properties of dust analogues and reactions occurring on them is that works of interest to astrochemistry are now regularly published in chemical physics/ surface science journals such as J. Chem. Phys., J. Phys. Chem., Phys. Chem. Chem. Phys., Surface Science, and others.

While in the past there has been a large number of laboratory studies of the interaction of charged particles and radiation with ice-covered dust grain analogues, most recent work points at new directions of research that will likely continue to be studied in the near future, i.e. the formation of molecules in/on ices by hydrogenation reactions, the properties of mixed ices, and the formation and properties of dust particles, including nanoparticles. Observations with ALMA, *SOFIA* and *Herschel* will yield more detailed information on dust and molecules, and theoretical studies will need to sort out the role of dust particles in molecule formation.

2. Meetings

Sessions about atomic/molecular interaction with surfaces are often featured at regularly scheduled COSPAR, AAS and Lunar and Planetary Institute meetings. For more information about these meetings, visit the Web sites of the respective organizations. For information about the meetings below, visit the web site of the Canadian Astronomy Data Centre (Web link: `<www1.cadc-ccda.hia-iha.nrc-cnrc.gc.ca/meetings/>`). Unfortunately, a number of meetings' official web sites have been taken down.

Most important meetings (listed in inverse chronological order):
- *Cosmic Dust – Near And Far*, Heidelberg, Germany, 8-12 September 2008
- *Bridging the Laboratory and Astrophysics*, 212th AAS, St. Louis, MO, USA, 1-5 June 2008
- *The Molecular Universe: International Meeting on the Physics and Chemistry of the IS Medium*, Arcachon, France, 5-8 May 2008
- *AbSciCon 2008: Fifth Astrobiology Science Conference*, Santa Clara, CA, USA, 15-17 April 2008
- *Titan Observations, Experiments, Computations, and Modeling*, Miami, FL, USA, 24-26 March 2008
- *Organic Matter in Space*, IAU Symposium No. 251, Hong Kong, 18-22 February 2008
- *The Evolving Insterstellar Medium in the Milky Way and Nearby Galaxies*, Pasadena, CA, USA, 2-5 December 2007
- *Bioastronomy 2007: Molecules, Microbes, and Extraterrestrial Life*, San Juan, PR, USA, 16-20 July 2007
- *Origins of Solar Systems*, 2007 Gordon Conference, South Hadley, MA, USA, 8-13 July 2007
- *Molecules in Space and Laboratory*, Paris, France, 14-18 May 2007
- *Astronomy in the Submillimeter and Far Infrared Domains with the Herschel Space Observatory*, Les Houches Winter School, France, 23 April - 4 May 2007
- *Titan Observations, Experiments, Computations, and Modeling*, Honolulu, HI, USA, 5-7 February 2007
- *Science with ALMA: a New Era for Astrophysics*, Madrid, Spain, 13-16 November 2006
- *IS Medium*, Heidelberg Summer School, Germany, 25-29 September 2006
- *From Dust to Planetesimals*, Ringberg Castle, Bavaria, Germany, 11-15 September 2006
- *Cosmic Chemistry and Molecular Astrophysics*, Nobel Symposium, Sudertuna, Sweden, 10-15 June 2006
- *Complex Molecules in Space – present Status and Prospects with ALMA*, Fuglsocentret, Denmark, 8-11 June 2006
- *Carbon in Space*, International workshop, Lago di Como, Italy, 22-25 June 2006
- NASA *Laboratory Astrophysics Workshop*, Las Vegas, NV, USA, 14-16 February 2006
- *Astrochemistry - A Molecular Approach*, Honolulu, HI, USA, 17-18 December 2005
- *Hunt for Molecules*, Paris, France, 19-20 September 2005
- *Protostars and Planets. V*, Big Island, HI, USA, 24-28 October 2005
- *5th. European Workshop on Astrobiology*, Budapest, Hungary, 10-12 October 2005
- *Astrochemistry throughout the Universe: Recent Successes and Current Challenges*, IAU Symposium No. 231, Monterey, CA, USA, 29 August - 2 September 2005
- *Astrobiology and the Origins of Life*, Hamilton, Canada, 24- May -10 June 2005
- *The Spitzer Space Telescope: New Views of the Cosmos*, Pasadena, CA, USA, 9-12 November 2004
- *The Dusty and Molecular Universe: A prelude to Herschel and ALMA*, Paris, France, 27-29 October 2004

- *Effects of Space Radiation on Solar System Ices*, AOGS 2004 Session SP2, Singapore, 5-9 July 2004
- *Astrophysics of Dust*, Estes Park, CO, USA, 2003

Published works in the area of molecular reactions on solid surfaces have been sorted in 4 sections:

(*a*) reviews

(*b*) observations of dust and ices in the ISM

(*c*) dust (formation, properties, and exposure to space environment)

(*d*) interactions of atoms and molecules with solids in simulated ISM conditions

(*e*) interaction of radiation and charged particles with ices in simulated ISM conditions

Obviously, there is a certain degree of arbitrariness in the sorting. Several papers could be entered in more than one section. The papers listed here are the ones that appeared in print since the last review by the Working Group on Molecular Reactions on Solid Surfaces in 2002; therefore, this report covers a six-year period. Works are listed in inverse chronological order.

3. Reviews

References

Charnley, S. B. & Rodgers, S. D. 2008, *SSRv*, 40. *IS reservoirs of cometary matter*

Williams, D. A., Brown, W. A., Price, S. D., *et al.* 2007, *Astron. & Geophys.*, 48, 25. *Molecules, ices and astronomy*

Tothill, N. F. H. 2007, *EAS-PS*, 25, 327. *AST/RO: lessons from a decade of sub-mm astronomy at the South Pole*

Slavin, J. D. & Frisch, P. C. 2007, *SSRv*, 130, 409. *The chemical composition of IS matter at the solar location*

Herbst, E. & Cuppen, H. M. 2006, *PNAS*, 103, 12257. *IS chemistry special feature: Monte Carlo studies of surface chemistry and nonthermal desorption involving IS grains*

Dartois, E. 2005, *SSRv*, 119, 293. *The ice survey opportunity of ISO.*

Abergel, A., Verstraete, L., Joblin, C., *et al.* 2005, *SSRv*, 119, 247. *The cool IS medium*

Molster, F. & Kemper, C. 2005, *SSRv*, 119, 3. *Crystalline silicates*

Grün, E., Srama, R., Krüger, H., *et al.* 2005, *Icarus*, 174, 1. *Dust astronomy. 2002 Kuiper Prize lecture*

van Dishoeck, E. F. 2004, *ARAA* 42, 119. *ISO spectroscopy of gas and dust – from molecular clouds to protoplanetary disks*

Clayton, D. D., Nittler, L. R. 2004, *ARAA*, 42, 39. *Astrophysics with presolar stardust*

Dorschner, J. 2003, *Astromineralogy* 609, 1. *From dust astrophysics towards dust mineralogy - a historical review*

Draine, B. T. 2003, *ARAA* 41, 241. *IS dust grains*

Williams, D. A. & Viti, S. 2002, *Ann. Rep. Prog. Chem. Sect. C* 98, 87. *Recent progress in astrochemistry*

4. Observations of dust and ices in the ISM

References

Öberg, K. I., Boogert, A. C. A., Pontoppidan, K. M., *et al.* 2008, *ApJ*, 678, 1032. *The c2d Spitzer spectroscopic survey of ices around low-mass young stellar objects. III. CH_4*

Pontoppidan, K. M., Boogert, A. C. A., Fraser, H. J., *et al.* 2008, *ApJ*, 678, 1005. *The c2d Spitzer spectroscopic survey of ices around low-mass young stellar objects. II. CO_2*

Boogert, A. C. A., Pontoppidan, K. M., Knez, C., *et al.*, 2008, *ApJ*, 678, 985. *The c2d Spitzer spectroscopic survey of ices around low-mass young stellar objects. I. H_2O and the 5-8 μm bands*

Sonnentrucker, P., Neufeld, D. A., Gerakines, P. A., *et al.* 2008, *ApJ*, 672, 361-370. *Fully sampled maps of ices and silicates in front of Cepheus A East with the Spitzer Space Telescope*

Li, Y., Li, A., & Wei, D. M. 2008, *ApJ*, 678, 1136. *Determining the dust extinction of gamma-ray burst host galaxies: a direct method based on optical and X-ray photometry*

Hough, J. H., Aitken, D. K., Whittet, D. C. B., *et al.* 2008, *MNRAS*, 387, 797. *Grain alignment in dense IS environments: spectropolarimetry of the 4.67-μm CO-ice feature in the field star Elias 16 (Taurus dark cloud)*

Sirocky, M. M., Levenson, N. A., Elitzur, M., *et al.* 2008, *ApJ*, 678, 729. *Silicates in ultraluminous infrared galaxies*

Berné, O., Joblin, C., Rapacioli, M., *et al.* 2008, *A&A* (Letters), 479, L41. *Extended red emission and the evolution of carbonaceous nanograins in NGC 7023*

Li, Y., Hopkins, P. F., Hernquist, L., *et al.* 2008, *ApJ*, 678, 41. *Modeling the dust properties of $z \simeq 6$ quasars with ART^2 – All-wavelength Radiative Transfer with Adaptive Refinement Tree*

Nozawa, T., Kozasa, T., Habe, A., *et al.* 2008, in: *Origin of Matter and Evolution of Galaxies*, AIP-CP, 1016, 55. *Evolution of dust in primordial supernova remnants and its influence on the elemental composition of hyper-metal-poor stars*

Dwek, E., Arendt, R. G.; Bouchet, P., *et al.* 2008, *ApJ*, 676, 1029. *Infrared and X-Ray evidence for circumstellar grain destruction by the blast wave of Supernova 1987A*

Draine, B. T., Dale, D. A., Bendo, G., *et al.* 2007, *ApJ*, 663, 866. *Dust masses, PAH sbundances, and starlight intensities in the SINGS galaxy sample*

Angeloni, R., Contini, M., Ciroi, S., & Rafanelli, P. 2007, *AJ*, 134, 205. *Silicates in D-Type symbiotic stars: an Infrared Space Observatory overview*

Mason, R. E., Wright, G. S., Adamson, A., & Pendleton, Y. 2007, *ApJ*, 656, 798. *Spectropolarimetry of the 3.4 μm absorption feature in NGC 1068*

Terada, H., Tokunaga, A. T., Kobayashi, *et al.* 2007, *ApJ*, 667, 303. *Detection of water ice in edge-on protoplanetary disks: HK Tauri B and HV Tauri C*

Quanz, S. P., Henning, T., Bouwman, J., *et al.* 2007, *ApJ*, 668, 359. *Evolution of dust and ice features around FU Orionis objects*

Whittet, D. C. B., Shenoy, S. S., Bergin, E. A., *et al.* 2007. *ApJ*, 655, 332. *The abundance of carbon dioxide ice in the quiescent intracloud medium*

Markwick-Kemper, F., Gallagher, S. C., Hines, D. C., & Bouwman, J. 2007, *ApJ* (Letters), 668, L107. *Dust in the wind: crystalline silicates, corundum, and periclase in PG 2112+059*

Bottinelli, S., Boogert, A. C. A., van Dishoeck, E. F., Oberg, K., *et al.* 2007, in: J. L. Lemaire & F. Combes (eds.), *Molecules in Space and Laboratory* (Paris: S. Diana), p. 11. *NH_3 and CH_3OH in ices surrounding low-mass YSOs*

Bai, L., Rieke, G. H., & Rieke, M. J. 2007, *ApJ* (Letters), 668, L5. *A search for infrared emission from intracluster dust in Abell 2029*

Davis, S. S. 2007, *ApJ*, 660, 1580. *Ice formation in radiated accretion disks*

Ellison, S. L., Prochaska, J. X., & Lopez, S. 2007, *MNRAS*, 380, 1245. *The Galactic deuterium abundance and dust depletion: insights from an expanded Ti/H sample*

Vollmer, C., Hoppe, P., Brenker, F. E., & Holzapfel, C. 2007, *ApJ* (Letters), 666, L49. *Stellar $MgSiO_3$ perovskite: a shock-transformed stardust silicate found in a meteorite*

Bacmann, A., Lefloch, B., Parise, B., *et al.* 2007, in: J. L. Lemaire & F. Combes (eds.), *Molecules in Space and Laboratory* (Paris: S. Diana), p. 9. *Methanol and deuterium fractionation in pre-stellar cores*

Maiolino, R. 2007, in: K. A. van der Hucht (ed.), *Highlights of Astronomy*, 14, 262. *Dust at $z > 6$. Observations and theory*

Andersson, B.-G. & Potter, S. B. 2007, *ApJ* 665, 369. *Observational constraints on IS grain alignment*

Sujatha, N. V., Murthy, J., Shalima, P., & Henry, R. C. 2007, *ApJ*, 665, 363. *Measurement of dust optical properties in the Coalsack nebula*

Kulkarni, V. P., York, D. G., Vladilo, G., & Welty, D. E. 2007, *ApJ* (Letters), 663, L81. *9.7 μm silicate absorption in a damped Lyα absorber at $z = 0.52$*

Bethell, T. J., Chepurnov, A., Lazarian, A., & Kim, J. 2007, *ApJ*, 663, 1055. *polarization of dust emission in clumpy molecular clouds and cores*

Dwek, E., Galliano, F., & Jones, A. P. 2007, *ApJ*, 662, 927. *The evolution of dust in the early Universe with applications to the galaxy SDSS J1148+5251*

Dessauges-Zavadsky, M., Combes, F., & Pfenniger, D. 2007, *A&A* 473, 863. *Molecular gas in high-velocity clouds: revisited scenario*

Bisschop, S. E., Jørgensen, J. K., van Dishoeck, E. F., & de Wachter, E. B. M. 2007, *A&A*, 465, 913. *Testing grain-surface chemistry in massive hot-core regions*

Inoue, A. K., Buat, V., Burgarella, D., et al. 2006, *MNRAS*, 370, 380. *Effects of dust scattering albedo and 2175-Å bump on ultraviolet colours of normal disc galaxies*

Colangeli, L. 2006, *MemSAIS*, 9, 161. *Measurement of dust properties in different solar system environments*

Nozawa, T., Kozasa, T., & Habe, A. 2006, *ApJ*, 648, 435. *Dust destruction in the high-velocity shocks driven by supernovae in the early Universe*

Spoon, H. W. W., Tielens, A. G. G. M., Armus, L., et al. 2006, *ApJ*, 638, 759. *The detection of crystalline silicates in ultraluminous infrared galaxies*

Sofia, U. J., Gordon, K. D., Clayton, G. C., et al. 2006, *ApJ*, 636, 753. *Probing the dust responsible for Small Magellanic Cloud extinction*

Huard, T. L., Pontoppidan, K. M., Boogert, A., et al. 2006, *BAAS*, 38, 1055. *Variations in the extinction law, ice abundance, and dust grains in molecular cloud cores*

Chiar, J. E., Pendleton, Y., Ennico, K., et al. 2006, *BAAS*, 38, 1013. *The non-linear relationship between silicate absorption depth and IR extinction in dense clouds*

Huebner, W. F. & Snyder, L. E. 2006, in: *Comets and the Origin and Evolution of Life* (Springer), p. 113. *Macromolecules: from star-forming regions to comets to the origins of life*

Prieto-Ballesteros, O., Kargel, J. S., Fernández-Sampedro, M., et al. 2005, *Icarus*, 177, 491. *Evaluation of the possible presence of clathrate hydrates in Europa's icy shell or seafloor*

Knez, C., Boogert, A. C. A., Pontoppidan, K. M., Kessler-Silacci, J., et al. 2005. *ApJ* (Letters), 635, L145. *Spitzer mid-infrared spectroscopy of ices toward extincted background stars.*

Velusamy, T., Langer, W. D., & Willacy, K. 2005, in: D. C. Lis, G. A. Blake & E. Herbst (eds.), *Astrochemistry throughout the Universe: Recent Successes and Current Challenges*, Proc. IAU Symposium No. 231 (Cambridge: CUP), p. 137. *CO_2 ice in cold dense dark cloud cores: abundance vs. visual extinction*

Lee, J. C. & Ravel, B. 2005, in: *X-ray Diagnostics of Astrophysical Plasmas: Theory, Experiment, and Observation AIP-CP*, 774, 255. *Prospects for determining the grain composition of the IS medium with Chandra and Astro E2.*

Bernstein, M. P., Sandford, S. A., & Allamandola, L. J. 2005, *ApJS*, 161, 53. *The mid-infrared absorption spectra of neutral polycyclic aromatic hydrocarbons in conditions relevant to dense IS clouds*

Ruiterkamp, R., Cox, N. L. J., Spaans, M., et al. 2005, *A&A*, 432, 515. *PAH charge state distribution and DIB carriers: implications from the line of sight toward HD 147889*

Massey, P., Plez, B., Levesque, E. M., et al. 2005, *ApJ*, 634, 1286. *The reddening of red supergiants: when smoke gets in your eyes*

Andrews, S. M. & Williams, J. P. 2005, *ApJ*, 631, 1134. *Circumstellar dust disks in Taurus-Auriga: the sub-millimeter perspective*

Stratta, G., Perna, R., Lazzati, D., et al. 2005, *NCimC* 28, 693. *Dust extinction properties of a sample of bright X-rays afterglows*

del Burgo, C. & Laureijs, R. J. 2005, *MNRAS*, 360, 901. *New Insights into the dust properties of the Taurus molecular cloud TMC-2 and its surroundings*

Joblin, C., Abergel, A., Bernard, J.-P., et al. 2005, in: D. C. Lis, G. A. Blake & E. Herbst (eds.), *Astrochemistry throughout the Universe: Recent Successes and Current Challenges*, Proc. IAU Symposium No. 231 (Cambridge: CUP), p. 194. *Very small particles and chemistry in photo-dissociation regions: from ISO to Spitzer*

Remijan, A. J., Hollis, J. M., Lovas, F. J., et al. 2005, *ApJ*, 632, 333. *IS isomers: the importance of bonding energy differences*

Gerakines, P. A., Bray, J. J., Davis, A., & Richey, C. R. 2005, *ApJ*, 620, 1140. *The strengths of near-infrared absorption features relevant to IS and planetary ices*

Snell, R. L., Hollenbach, D., Howe, J. E., et al. 2005, *ApJ*, 620, 758. *Detection of water in the shocked gas associated with IC 443: constraints on shock models*

Pontoppidan, K. M. & the c2d Team 2005, in: D. C. Lis, G. A. Blake & E. Herbst (eds.), *Astrochemistry throughout the Universe: Recent Successes and Current Challenges*, Proc. IAU Symposium No. 231 (Cambridge: CUP), p. 319. *The spatial distribution of ices in star-forming regions*

van Boekel, R., Min, M., Waters, L. B. F. M., *et al.* 2005, *A&A*, 437, 189. *A 10 μm spectroscopic survey of herbig Ae star disks: grain growth and crystallization*

Duchêne, G., McCabe, C., Ghez, A. M., & Macintosh, B. A. 2004, *ApJ*, 606, 969. *A multiwavelength scattered light analysis of the dust grain population in the GG Tauri circum-binary ring*

Jones, A. P., d'Hendecourt, L. B., Sheu, S.-Y., *et al.* 2004, *A&A*, 416, 235. *Surface C-H stretching features on meteoritic nanodiamonds*

Pendleton, Y. J. 2004, in: A. N. Witt, G. C. Clayton & B. T. Draine (eds.), *Astrophysics of Dust ASP-CS*, 309, 573. *Hydrocarbons in meteorites, the Milky Way, and other galaxies*

Bot, C., Boulanger, F., Lagache, *et al.* 2004, *A&A*, 423, 567. *Multi-wavelength analysis of the dust emission in the Small Magellanic Cloud*

Ehrenfreund, P., Fraser, H. J., Blum, J., *et al.* 2003, *P&SS*, 51, 473. *Physics and chemistry of icy particles in the Universe: answers from microgravity*

Johnstone, D., Fiege, J. D., Redman, R. O., *et al.* 2003, *ApJ (Letters)*, 588, L37. *The G11.11-0.12 infrared-dark cloud: anomalous dust and a non-magnetic isothermal model*

Wolf, S., Padgett, D. L., & Stapelfeldt, K. R. 2003, *ApJ*, 588, 373. *The circumstellar disk of the butterfly star in Taurus.*

Kimura, H., Mann, I., & Jessberger, E. K. 2003, *ApJ*, 582, 846. *Elemental abundances and mass densities of dust and gas in the Local IS Cloud*

Lu, N., Helou, G., Werner, M. W., *et al.* 2003, *ApJ*, 588, 199. *Infrared emission of normal galaxies from 2.5 to 12 μm: ISO spectra, near-IR continuum, and mid-IR emission features*

Savaglio, S., Fall, S. M., & Fiore, F. 2003, *ApJ*, 585, 638. *Heavy-element abundances and dust depletions in the host galaxies of three γ-ray bursts*

Weingartner, J. C. & Murray, N. 2002, *ApJ*, 580, 88. *X-ray vs. optical observations of active galactic nuclei: evidence for large grains?*

Dupac, X., Giard, M., Bernard, J.-P., *et al.* 2002, *A&A*, 392, 691. *Sub-millimeter dust emission of the M 17 complex measured with PRONAOS*

Ishii, M., Nagata, T., Chrysostomou, A., & Hough, J. H. 2002, *AJ*, 124, 2790. *3.4 μm feature on the shoulder of ice-band absorptions in three luminous young stellar objects: IRAS 18511+0146, IRAS 21413+5442, and IRAS 04579+4703*

Boogert, A. C. A., Blake, G. A., Tielens, A. G. G. M. 2002, *ApJ*, 577, 271. *High-resolution 4.7 μm Keck/NIRSPEC spectra of protostars. II. Detection of the ^{13}CO isotope in icy grain mantles*

Chiar, J. E., Adamson, A. J., Pendleton, Y. J., *et al.* 2002, *ApJ*, 570, 209. *Hydrocarbons, ices, and 'XCN' in the line of sight toward the Galactic Center*

Pendleton, Y. J. & Allamandola, L. J. 2002, *ApJS*, 138, 75. *The organic refractory material in the diffuse IS medium: mid-IR spectroscopic constraints*

5. Properties of dust

References

Thompson, S. P. 2008, *A&A*, 484, 251. *Structural signatures of medium-range order in annealed laboratory silicates*

Mason, N. J., Drage, E. A., Webb, S. M., *et al.* 2008, *Faraday Discuss.*, 137, 367. *The spectroscopy and chemical dynamics of micro-particles explored using an ultrasonic trap*

Whittet, D. C. B., Hough, J. H., Lazarian, A., & Hoang, T. 2008, *ApJ*, 674, 304. *The efficiency of grain alignment in dense IS clouds: a re-assessment of constraints from near-IR polarization*

Iatì, M. A., Saija, R., Borghese, F., *et al.* 2008, *MNRAS*, 384, 591. *Stratified dust grains in the IS medium. I. An accurate computational method for calculating their optical properties*

Lazzati, D. 2008, *MNRAS*, 384, 165. *Non-LTE dust nucleation in sub-saturated vapours*

Rosenberg, J. L., Wu, Y., Le Floc'h, E., *et al.* 2008, *ApJ*, 674, 814. *Dust properties and star-formation rates in star-forming dwarf galaxies*

Engelbracht, C. W., Rieke, G. H., Gordon, K. D., *et al.* 2008, *ApJ*, 678, 804. *Metallicity effects on dust properties in starbursting galaxies*

Voshchinnikov, N. V., & Henning, T. 2008, *A&A* (Letters), 483, L9. *Is the silicate emission feature only influenced by grain size?*

Jiménez-Serra, I., Caselli, P., Martín-Pintado, J., Hartquist, T. W. 2008, *A&A*, 482, 549. *Parametrization of C-shocks. Evolution of the sputtering of grains*

Pitman, K. M., Hofmeister, A. M., Corman, A. B., & Speck, A. K. 2008, *A&A*, 483, 661. *Optical properties of silicon carbide for astrophysical applications. I. New laboratory IR reflectance spectra and optical constants*

Calura, F., Pipino, A., & Matteucci, F. 2008, *A&A*, 479, 669. *The cycle of IS dust in galaxies of different morphological types*

Weingartner, J. C. & Jordan, M. E. 2008, *ApJ*, 672, 382. *Torques on spheroidal silicate grains exposed to anisotropic IS radiation fields*

Fabian, A. C., Vasudevan, R. V., & Gandhi, P. 2008, *MNRAS* (Letters), 385, L43. *The effect of radiation pressure on dusty absorbing gas around active galactic nuclei*

Lazarian, A. & Hoang, T. 2008, *ApJ* (Letters), 676, L25. *Alignment of dust with magnetic inclusions: radiative torques and superparamagnetic Barnett and nuclear relaxation*

Nozawa, T., Kozasa, T., Habe, A., *et al.* 2008, in: *First Stars III AIP-CP*, 990, 426. *Dust evolution in Pop. III supernova remnants*

Zhukovska, S., Gail, H.-P., & Trieloff, M. 2008, *A&A*, 479, 453. *Evolution of IS dust and stardust in the solar neighbourhood*

Guillet, V., Pineau Des Forêts, G., & Jones, A. P. 2007, *A&A*, 476, 263. *Shocks in dense clouds. I. Dust dynamics*

Dartois, E. 2007, in: D. C. Lis, G. A. Blake & E. Herbst (eds.), *Astrochemistry throughout the Universe: Recent Successes and Current Challenges*, Proc. IAU Symposium No. 231 (Cambridge: CUP), p. 54. *IS dust grains: the hydrogenated amorphous carbon contribution*

Wada, K., Tanaka, H., Suyama, T., *et al.* 2007, *ApJ*, 661, 320. *Numerical simulation of dust aggregate collisions. I. Compression and disruption of 2-D aggregates*

Draine, B. T., Li, A. 2007, *ApJ*, 657, 810. *Infrared emission from IS dust. IV. The silicate-graphite-PAH model in the post-Spitzer era*

Ferguson, J. W., Heffner-Wong, A., Penley, J. J., *et al.* 2007, *ApJ*, 666, 261. *Grain physics and Rosseland-mean opacities*

Vaidya, D. B., Gupta, R., & Snow, T. P. 2007, *MNRAS*, 379, 800. *Composite IS grains*

Miville-Deschênes, M.-A., Lagache, G., Boulanger, F., & Puget, J.-L. 2007, *A&A*, 469, 595. *Statistical properties of dust far-IR emission*

Maheswar, G., Muthu, C., Sujatha, N. V., *et al.* 2007, *BASI*, 35, 233. *IS dust studies with TAUVEX*

Djouadi, Z., Gattacceca, J., D'Hendecourt, L., *et al.* 2007, *A&A* (Letters), 468, L9. *Ferromagnetic inclusions in silicate thin films: insights into the magnetic properties of cosmic grains*

Huss, G. R. & Draine, B. T. 2007, in: K. A. van der Hucht (ed.), *Highlights of Astronomy*, 14, 353. *What can pre-solar grains tell us about the Solar nebula?*

Nguyen, A. N., Stadermann, F. J., Zinner, E., *et al.* 2007, *ApJ*, 656, 1223. *Characterization of pre-solar silicate and oxide grains in primitive carbonaceous chondrites*

Ott, U., Hoppe, P. 2007. Pre-solar Grains in Meteorites and Interplanetary Dust: an Overview. *Highlights of Astronomy* 14, 341-344

Min, M., Waters, L. B. F. M., de Koter, A., *et al.* 2007, *A&A*, 462, 667. *The shape and composition of IS silicate grains*

Pelkonen, V.-M., Juvela, M., & Padoan, P. 2007, *A&A*, 461, 551. *Simulations of polarized dust emission*

Ormel, C. W., Spaans, M., Tielens, A. G. G. M. 2007, *A&A*, 461, 215. *Dust coagulation in protoplanetary disks: porosity matters*

Patil, M. K., Pandey, S. K., Sahu, D. K., & Kembhavi, A. 2007, *A&A*, 461, 103. *Properties of dust in early-type galaxies*

Vaillancourt, J. E. 2007, *EAS-PS*, 23, 147. *Polarized emission from IS dust*

Lazarian, A. 2007, *JQSRT*, 106, 225. *Tracing magnetic fields with aligned grains*

Ehrenfreund, P., Ruiterkamp, R., Peeters, Z., *et al.* 2007, *P&SS*, 55, 383. *The ORGANICS Experiment on BIOPAN V: UV and space exposure of aromatic compounds*

Li, M. P., Zhao, G., & Li, A. 2007, *MNRAS* (Letters), 382, L26. *On the crystallinity of silicate dust in the IS medium*

Bianchi, S. & Schneider, R. 2007, *MNRAS*, 378, 973. *Dust formation and survival in supernova ejecta*

Lazarian, A. & Hoang, T. 2007, *MNRAS*, 378, 910. *Radiative torques: analytical model and basic properties*

Abbas, M. M., Tankosic, D., Craven, P. D., *et al.* 2006, *ApJ*, 645, 324. *Photoelectric emission measurements on the analogs of individual cosmic dust grains*

Duley, W. W. 2006, *Faraday Discuss.*, 133, 415. *Polycyclic aromatic hydrocarbons, carbon nanoparticles and the diffuse IS bands*

Ehrenfreund, P., Ruiterkamp, R., Peeters, Z., *et al.* 2006, *COSPAR*, 36, 3635. *The ORGANICS Experiments on BIOPAN V: UV and space exposure of aromatic compounds*

Freund, M. M. & Freund, F. T. 2006, *ApJ*, 639, 210. *Solid solution model for IS dust grains and their organics*

Altobelli, N., Grün, E., & Landgraf, M. 2006, *A&A*, 448, 243. *A new look into the Helios dust experiment data: presence of IS dust inside the Earth's orbit*

Stroud, R. M. 2005, in: A. N. Krot, E. R. D. Scott & B. Reipurth (eds.), *Chondrites and the Protoplanetary Disk*, ASP-CS, 341, 645. *Micro-structural investigations of the cosmochemical histories of pre-solar grains*

Kemper, F., Vriend, W. J., & Tielens, A. G. G. M. 2005, *ApJ* 633, 534. *Erratum: 'The absence of crystalline silicates in the diffuse IS medium', ApJ, 609, 826, 2004*

Ruiterkamp, R., Cox, N. L. J., Spaans, M., *et al.* 2005, *A&A*, 432, 515. *PAH charge state distribution and DIB carriers: implications from the line of sight toward HD 147889*

Tielens, A. G. G. M., Waters, L. B. F. M., & Bernatowicz, T. J. 2005, in: A. N. Krot, E. R. D. Scott & B. Reipurth (eds.), *Chondrites and the Protoplanetary Disk*, ASP-CS, 341, 605. *Origin and evolution of dust in circumstellar and IS environments*

Dwek, E. 2005, in: *Planetary Nebulae as Astronomical Tools*, AIP-CP, 804, 197. *The chemical evolution of IS dust*

Grün, E., Srama, R., Krüger, H., *et al.* 2005, *Icarus*, 174, 1. *2002 Kuiper Prize Lecture: Dust Astronomy*

Draine, B. T. 2005, *ESA-SP*, 577, 251. *Infrared emission and models of IS dust*

Voshchinnikov, N. V., Il'in, V. B., & Henning, T. 2005, *A&A*, 429, 371. *Modelling the optical properties of composite and porous IS grains*

Bréchignac, P. & Schmidt, M. 2005, in: D. C. Lis, G. A. Blake & E. Herbst (eds.), *Astrochemistry throughout the Universe: Recent Successes and Current Challenges*, Proc. IAU Symposium No. 231 (Cambridge: CUP), p. 255. *From cold gas phase coronene clusters to hydrocarbonated nanograins*

Rapacioli, M., Joblin, C., Calvo, F., *et al.* 2005, in: D. C. Lis, G. A. Blake & E. Herbst (eds.), *Astrochemistry throughout the Universe: Recent Successes and Current Challenges*, Proc. IAU Symposium No. 231 (Cambridge: CUP), p. 200. *Theoretical properties of polycyclic aromatic hydrocarbon clusters of astrophysical interest*

Mennella, V. 2005, *JPhCS*, 6, 197. *Dust evolution from the laboratory to the IS medium*

van Breugel, W., Bajt, S., Bradley, J., *et al.* 2005, in: A. Wilson (ed.), *The Dusty and Molecular Universe: a Prelude to Herschel and ALMA*, ESA-SP, 577, 91. *Star formation in high-pressure, high-energy density environments: laboratory experiments of ISM dust analogs*

Wooden, D. H., Harker, D. E., & Brearley, A. J. 2005, in: A. N. Krot, E. R. D. Scott & B. Reipurth (eds.), *Chondrites and the Protoplanetary Disk*, ASP-CS, 341, 774. *Thermal processing and radial mixing of dust: evidence from comets and primitive chondrites*

Zinov'Eva, T. V. 2005, *AstL*, 31, 458. *Modeling infrared absorption bands with non-spherical particles*

Biennier, L., Hammond, M., Elsila, J., *et al.* 2005, in: D. C. Lis, G. A. Blake & E. Herbst (eds.), *Astrochemistry throughout the Universe: Recent Successes and Current Challenges*, Proc. IAU Symposium No. 231 (Cambridge: CUP), p. 214. *From organic molecules to carbon particles: implications for the formation of IS Dust*

Keane, J. V., Pendleton, Y. J., & Allamandola, L. J. 2005, in: D. C. Lis, G. A. Blake & E. Herbst (eds.), *Astrochemistry throughout the Universe: Recent Successes and Current Challenges*,

Proc. IAU Symposium No. 231 (Cambridge: CUP), p. 242. *Refractory carbonaceous material in luminous galaxies: mid-IR spectroscopic constraints*

Salama, F. & Biennier, L. 2004, *COSPAR*, 35, 3067. *Formation and destruction processes of IS dust: from organic molecules to carbonaceous grains*

Zubko, V., Dwek, E., & Arendt, R. G. 2004, *ApJS*, 152, 211. *IS dust models consistent with extinction, emission, and abundance constraints*

Li, A. 2004. in: A. N. Witt, G. C. Clayton & B. T. Draine (eds.), *Astrophysics of Dust, ASP-CS*, 309, 417. *Interaction of nanoparticles with radiation*

Iatì, M. A., Giusto, A., Saija, R., *et al.* 2004, *ApJ*, 615, 286. *Optical properties of composite IS grains: a morphological analysis*

Jones, A. P. & D'Hendecourt, L. B. 2004, in: A. N. Witt, G. C. Clayton & B. T. Draine (eds.), *Astrophysics of Dust, ASP-CS*, 309, 589. *IS nano-diamonds*

Shenoy, S. S., Whittet, D. C. B., Chiar, J. E., *et al.* 2003, *ApJ*, 591, 962. *A test case for the organic refractory model of IS dust*

Hegmann, M. & Kegel, W. H. 2003, *MNRAS*, 342, 453. *Radiative transfer in clumpy environments: absorption and scattering by dust*

Gibson, S. J. & Nordsieck, K. H. 2003, *ApJ*, 589, 362. *The Pleiades reflection nebula. II. Simple model constraints on dust properties and scattering geometry*

Stepnik, B., Abergel, A., Bernard, J.-P., *et al.* 2003, *A&A*, 398, 551. *Evolution of dust properties in an IS filament*

Juvela, M. & Padoan, P. 2003, *A&A*, 397, 201. *Dust emission from inhomogeneous IS clouds: radiative transfer in 3-D with transiently heated particles*

Mann, I. & Jessberger, E. K. 2003, *Astromineralogy*, 609, 189. *The in-situ study of solid particles in the Solar system*

Hanner, M. S. 2003, *Astromineralogy* 609, 171. *The mineralogy of cometary dust*

Lazarian, A. 2003, *JQSRT* 79, 881. *Magnetic fields via polarimetry: progress of grain alignment theory*

Mathis, J. S., Whitney, B. A., & Wood, K. 2002, *ApJ*, 574, 812. *Can reflection from grains diagnose the albedo?*

Stepnik, B., Abergel, A., Bernard, J. P., *et al.* 2002, in: M. Giard *et al.* (eds.), *Infrared and Sub-millimeter Space Astronomy, EAS-PS*, 4, 309. *Evolution of IS dust properties from diffuse medium to a dense cloud*

Stepnik, B., Jones, A. P., Abergel, A., *et al.* 2002. in: M. Giard *et al.* (eds.), *Infrared and Sub-millimeter Space Astronomy, EAS-PS*, 4, 31. *Grain-grain co-agulation in the ISM*

Ruiterkamp, R., Ehrenfreund, P., *et al.* 2002, in: F. Salama (ed.), *NASA Laboratory Astrophysics Workshop, NASA/CP-2002-21186*, p. 149. *Laboratory calibration studies in support of ORGANICS on the International Space Station: evolution of organic matter in space*

Cox, N., Ehrenfreund, P., Cami, J., *et al.* 2002, in: H. Lacoste (ed.), Proc. *First European Workshop on Exo-Astrobiology, ESA-SP* 518, 447. *Complex carbon chemistry and the diffuse IS bands in the Magellanic Clouds*

6. Interactions of atoms and molecules with solids in simulated ISM conditions

References

Luna, R., Millán, C., Domingo, M., & Satorre, M. Á. 2008, *Ap&SS*, 314, 113. *Thermal desorption of CH_4 retained in CO_2 ice*

Cazaux, S., Caselli, P., Cobut, V., & Le Bourlot, J. 2008, *A&A*, 483, 495. *The role of carbon grains in the deuteration of H_2*

Vidali, G., Pirronello, V., Li, L., *et al.* 2007, *JPhChA*, 111, 12611. *Analysis of molecular hydrogen formation on low temperature surfaces in temperature programmed desorption experiments*

Li, L., Manico, G., Congiu, E., *et al.* 2007, in: J. L. Lemaire & F. Combes (eds.), *Molecules in Space and Laboratory* (Paris: S. Diana), p. 58. *Formation of molecular hydrogen on amorphous silicate surfaces*

Al-Halabi, A., van Dishoeck, E. F., 2007, *MNRAS*, 382, 1648. *Hydrogen adsorption and diffusion on amorphous solid water ice*

Roberts, J. F., Rawlings, J. M. C., Viti, S., & Williams, D. A. 2007, *MNRAS*, 382, 733. *Desorption from IS ices*

Perets, H. B., Lederhendler, A., Biham, O., *et al.* 2007, *ApJ* (Letters), 661, L163. *Molecular hydrogen formation on amorphous silicates under IS conditions*

Bouwman, J., Ludwig, W., Awad, Z., *et al.* 2007, *A&A*, 476, 995. *Band profiles and band strengths in mixed H_2O:CO ices*

Öberg, K. I., Fuchs, G. W., Awad, Z., *et al.* 2007, *ApJ* (Letters), 662, L23. *Photodesorption of CO ice*

Chaabouni, H., Amiaud, L., Dulieu, F., *et al.* 2007, in: J. L. Lemaire & F. Combes (eds.), *Molecules in Space and Laboratory* (Paris: S. Diana), p. 67. *Sticking of deuterium molecules onto non-porous water ice surface: temperature dependence of the impinging molecules*

Chang, Q., Cuppen, H. M., & Herbst, E. 2007, *A&A*, 469, 973. *Gas-grain chemistry in cold IS cloud cores with a microscopic Monte Carlo approach to surface chemistry*

Watanabe, N., Mouri, O., *et al.* 2007, *ApJ*, 668, 1001. *Laboratory simulation of competition between hydrogenation and photolysis in the chemical evolution of H_2O-CO ice mixtures*

Canto, J., Gomis, O., Vilaplana, R., & Domingo, M. 2007, in: J. L. Lemaire & F. Combes (eds.), *Molecules in Space and Laboratory* (Paris: S. Diana), p. 65. *A laboratory for studying simple ices and their mixtures in the far-IR region*

Bisschop, S. E., Fuchs, G. W., van Dishoeck, E. F., & Linnartz, H. 2007, *A&A*, 474, 1061. *H-atom bombardment of CO_2, HCOOH, and CH_3CHO containing ices*

Xie, H.-B., Shao, C.-B., & Ding, Y.-H., 2007, *ApJ*, 670, 449. *Radical-molecule reaction C_3H + H_2O on amorphous water ice: a promising route for IS propynal*

Amiaud, L., Dulieu, F., Fillion, J.-H., *et al.* 2007, *JChPh*, 127, 4709. *Interaction of atomic and molecular deuterium with a non-porous amorphous water ice surface between 8 and 30 K*

Goldsmith, P.F. 2007, in: J. L. Lemaire & F. Combes (eds.), *Molecules in Space and Laboratory* (Paris: S. Diana), p. 93. *Conversion of HI to H_2 and the age of molecular clouds*

Goumans, T. P. M., Wander, A., Catlow, C. R. A., & Brown, W. A. 2007, *MNRAS*, 382, 1829. *Silica grain catalysis of methanol formation*

Elsila, J. E., Dworkin, J. P., Bernstein, M. P., *et al.* 2007, *Mechanisms of amino acid formation in IS ice analogs ApJ*, 660, 911

Bar-Nun, A., Notesco, G., & Owen, T. 2007, *Icarus*, 190, 655. *Trapping of N_2, CO and Ar in amorphous ice application to comets*

Öberg, K. I., Fraser, H. J., Boogert, A. C. A., *et al.* 2007, *A&A*, 462, 1187. *Effects of CO_2 on H_2O band profiles and band strengths in mixed H_2O:CO_2 ices*

Brown, W. A. & Bolina, A. S. 2007, *MNRAS*, 374, 1006. *Fundamental data on the desorption of pure IS ices*

Hidaka, H., Kouchi, A., & Watanabe, N. 2007, *JChPh*, 126, 204707. *Temperature, composition, and hydrogen isotope effect in the hydrogenation of CO on amorphous ice surface at 10-20 K*

Gálvez, O., Ortega, I. K., Maté, B., *et al.* 2007, *A&A*, 472, 691. *A study of the interaction of CO_2 with water ice*

Acharyya, K., Fuchs, G. W., Fraser, H. J., *et al.* 2007, *A&A*, 466, 1005. *Desorption of CO and O_2 IS ice analogs*

Garrod, R. T., Wakelam, V., & Herbst, E. 2007, *A&A*, 467, 1103. *Non-thermal desorption from IS dust grains via exo-thermic surface reactions*

Xie, H.-B., Shi, G.-S., & Ding, Y.-H. 2007, *ApJ*, 662, 758. *Chemical behavior of polycyanoacetylene radicals on gaseous and ice water: a computational perspective*

Linnartz, H., Acharyya, K., Awad, Z., *et al.* 2007, in: J. L. Lemaire & F. Combes (eds.), *Molecules in Space and Laboratory* (Paris: S. Diana), p. 47. *Solid state astrophysics and -Chemistry four Questions- four answers*

Bisschop, S. E., Fuchs, G. W., Boogert, A. C. A., *et al.* 2007, *A&A*, 470, 749. *Infrared spectroscopy of HCOOH in IS ice analogues*

Bernstein, M. P., Sandford, S. A., Mattioda, A. L., & Allamandola, L. J. 2007, *ApJ*, 664, 1264. *Near- and mid-IR laboratory spectra of PAH cations in solid H_2O*

Barzel, B. & Biham, O. 2007, *ApJ* (Letters), 658, L37. *Efficient simulations of IS gas-grain chemistry using moment equations*

Cuppen, H. M. & Herbst, E. 2007, *ApJ*, 668, 294. *Simulation of the formation and morphology of ice mantles on IS grains*

Xie, H.-B., Ding, Y.-H., & Sun, C.-C. 2006, *ApJ*, 643, 573. *Reaction mechanism of oxygen atoms with cyanoacetylene in the gas phase and on water ice*

Brown, W. A., Viti, S., Wolff, A. J., & Bolina, A. S. 2006, *Faraday Discuss.*, 133, 113. *Laboratory investigations of the role of the grain surface in astrochemical models*

Ehrenfreund, P. & Sephton, M. A. 2006, *Carbon molecules in space: from astrochemistry to astrobiology Faraday Discuss.* 133, 277

Garrod, R., Park, I. H., Caselli, P., & Herbst, E. 2006, *Faraday Discuss.*, 133, 51. *Are gas-phase models of IS chemistry tenable? The case of methanol*

Madzunkov, S., Shortt, B. J., MacAskill, J. A., *et al.* 2006, *PRA*, 73, 020901. *Measurements of polyatomic molecule formation on an icy grain analog using fast atoms*

Hiraoka, K., Mochizuki, N., & Wada, A. 2006, in: *Astrochemistry - from Laboratory Studies to Astronomical Observations*, AIP-CP, 855, 86. *How are CH_3OH, HNC/HCN, and NH_3 formed in the IS Medium?*

van Dishoeck, E. F., Acharyya, K., Al-Halabi, A., *et al.* 2006, in: *Astrochemistry - from Laboratory Studies to Astronomical Observations AIP-CP*, 855, 113. *Spectroscopy and processing of IS ice analogs*

Peeters, Z., Rodgers, S. D., Charnley, S. B., *et al.* 2006, *A&A*, 445, 197. *Astrochemistry of dimethyl ether*

Bisschop, S. E., Fraser, H. J., Öberg, K. I., *et al.* 2006, *A&A*, 449, 1297. *Desorption rates and sticking coefficients for CO and N_2 IS ices*

Collings, M. P., Chen, R., & McCoustra, M. R. S. 2006, in: *Astrochemistry - from Laboratory Studies to Astronomical Observations AIP-CP*, 855, 62. *Probing the morphology of IS ice analogues*

Amiaud, L., Fillion, J. H., Baouche, S., *et al.* 2006, *JChPh*, 124, 4702. *Interaction of D_2 with H_2O amorphous ice studied by temperature-programed desorption experiments*

Vidali, G., Roser, J. E., Li, L., *et al.* 2006, *Faraday Discuss.*, 133, 125. *The formation of IS molecules via reactions on dust grain surfaces*

Watanabe, N., Hidaka, H., & Kouchi, A. 2006, in: *Astrochemistry - from Laboratory Studies to Astronomical Observations AIP-CP*, 855, 122. *Relative reaction rates of hydrogenation and deuteration of solid CO at very low temperatures*

Woon, D. E. 2006, in: *Astrochemistry - from Laboratory Studies to Astronomical Observations AIP-CP*, 855, 305. *Ab initio quantum chemical studies of reactions in astrophysical ices - reactions involving CH_3OH, CO_2, CO, and $HNCO$ in $H_2CO/NH_3/H_2O$ ices*

Fraser, H. J., Bisschop, S. E., Pontoppidan, K. M., *et al.* 2005, *MNRAS*, 356, 1283. *Probing the surfaces of IS dust grains: the adsorption of CO at bare grain surfaces*

Lipshtat, A. & Biham, O. 2005, *MNRAS*, 362, 666. *The effect of grain size distribution on H_2 formation rate in the IS medium*

Collings, M. P., Dever, J. W., McCoustra, & M. R. S. 2005, *CPL*, 415, 40. *Sub-monolayer coverages of CO on water ice*

Öberg, K. I., van Broekhuizen, F., Fraser, H. J., *et al.* 2005, *ApJ* (Letters), 621, L33. *Competition between CO and N_2 desorption from IS ices*

Bisschop, S. E., Fraser, H. J., Fuchs, G., *et al.* 2005, in: D. C. Lis, G. A. Blake & E. Herbst (eds.), *Astrochemistry throughout the Universe: Recent Successes and Current Challenges*, Proc. IAU Symposium No. 231 (Cambridge: CUP), p. 168. *The behavior of N_2 and O_2 in pure, mixed or layered CO ices*

Roberts, H. 2005, in: D. C. Lis, G. A. Blake & E. Herbst (eds.), *Astrochemistry throughout the Universe: Recent Successes and Current Challenges*, Proc. IAU Symposium No. 231 (Cambridge: CUP), p. 27. *Modelling of deuterium chemistry in star-forming regions*

Charnley, S. B. & Rodgers, S. D. 2005, in: D. C. Lis, G. A. Blake & E. Herbst (eds.), *Astrochemistry throughout the Universe: Recent Successes and Current Challenges*, Proc. IAU Symposium No. 231 (Cambridge: CUP), p. 237. *Pathways to molecular complexity*

Hornekaer, L., Baurichter, A., Petrunin, V. V., & Luntz, A. C. 2005, *ESA-SP*, 577, 369. *The influence of dust grain morphology on H_2 formation and desorption in the IS medium*

Fraser, H. J., Bisschop, S. E., Pontoppidan, K. M., *et al.* 2005, *MNRAS*, 356, 1283. *Probing the surfaces of IS dust grains: the adsorption of CO at bare grain surfaces*

Woon, D. E. & Park, J.-Y. 2005, in: D. C. Lis, G. A. Blake & E. Herbst (eds.), *Astrochemistry throughout the Universe: Recent Successes and Current Challenges*, Proc. IAU Symposium

No. 231 (Cambridge: CUP), p. 89. *A density functional theory study of the formation and spectroscopy of the formate (HCOO$^-$) and ammonium (NH$_4^+$) ions in IS ices*

Cuppen, H. M. & Herbst, E. 2005, in: D. C. Lis, G. A. Blake & E. Herbst (eds.), *Astrochemistry throughout the Universe: Recent Successes and Current Challenges*, Proc. IAU Symposium No. 231 (Cambridge: CUP), p. 34. *Molecular hydrogen formation on IS grains*

Fillion, J.-H., Amiaud, L., Dulieu, F., *et al.* 2005, in: D. C. Lis, G. A. Blake & E. Herbst (eds.), *Astrochemistry throughout the Universe: Recent Successes and Current Challenges*, Proc. IAU Symposium No. 231 (Cambridge: CUP), p. 101. *Experimental studies of H$_2$ and D$_2$ interaction with water ice films*

Wolff, A. J. & Brown, W. A. 2005, in: D. C. Lis, G. A. Blake & E. Herbst (eds.), *Astrochemistry throughout the Universe: Recent Successes and Current Challenges*, Proc. IAU Symposium No. 231 (Cambridge: CUP), p. 57. *Mixed methanol/water ice on cosmic dust grain analogues*

Vidali, G., Roser, J. E., Manicó, G., & Pironello, V. 2005, in: D. C. Lis, G. A. Blake & E. Herbst (eds.), *Astrochemistry throughout the Universe: Recent Successes and Current Challenges*, Proc. IAU Symposium No. 231 (Cambridge: CUP), p. 355. *Molecular hydrogen formation on dust grains: a summary of experimental results on molecular hydrogen formation on dust grain analogues*

Perets, H. B., Biham, O., Manicó, G., *et al.* 2005, *ApJ*, 627, 850. *Molecular hydrogen formation on ice under IS conditions*

Vidali, G., Roser, J. E., Manicó, *et al.* 2005, *JPhCS*, 6, 36. *Formation of molecular hydrogen on analogues of IS dust grains: experiments and modelling*

Fraser, H. J., Bisschop, S. E., Pontoppidan, K. M., *et al.* 2005, in: D. C. Lis, G. A. Blake & E. Herbst (eds.), *Astrochemistry throughout the Universe: Recent Successes and Current Challenges*, Proc. IAU Symposium No. 231 (Cambridge: CUP), p. 31. *CO chemisorbed on bare grain surfaces: the potential for heterogeneous chemistry*

Amiaud, L., Baouche, S., Dulieu, F., *et al.* 2004, in: F. Combes *et al.* (eds.), *SF2A-2004: Semaine de l'Astrophysique Francaise* (EdP-CS), p. 487. *D$_2$ sticking coefficient and desorption rate on various forms of water ice films under IS conditions*

Al-Halabi, A., Fraser, H., Kroes, G. J., & van Dishoeck, E. F. 2004, *A&A*, 422, 777. *Adsorption of CO on amorphous water-ice surfaces*

Sandford, S. A., Bernstein, M. P., & Allamandola, L. J. 2004, *ApJ*, 607, 346. *The mid-IR laboratory spectra of naphthalene (C$_{10}$H$_8$) in Solid H$_2$O*

Vidali, G., Roser, J. E., Manicó, G., & Pironello, V. 2004, *JGRE*, 109, E07S14. *Laboratory studies of formation of molecules on dust grain analogues under ISM conditions*

Lipshtat, A., Biham, O., & Herbst, E. 2004, *MNRAS*, 348, 1055. *Enhanced production of HD and D$_2$ molecules on small dust grains in diffuse clouds*

Morisset, S., Aguillon, F., Sizun, M., & Sidis, V. 2003, *PCCP*, 5, 506. *The dynamics of H$_2$ formation on a graphite surface at low temperature*

Lipshtat, A. & Biham, O., 2003, *A&A*, 400, 585. *Moment equations for chemical reactions on IS dust grains*

Meierhenrich, U. J., Muñoz Caro, G. M., Schutte, W. A., *et al.* 2002, *Exo-Astrobiology*, 518, 25. *The prebiotic synthesis of amino acids – IS vs. atmospheric mechanisms*

Woon, D. E. 2002, *ApJ*, 569, 541. *Modeling gas-grain chemistry with quantum chemical cluster calculations. I. Heterogeneous hydrogenation of CO and H$_2$CO on icy grain mantles*

Basiuk, V. A. & Bogillo, V. I. 2002, *AdSpR*, 30, 1439. *Theoretical study of amino acid precursor formation in the IS medium. I. Reaction of methylenimine with hydrogen cyanide*

7. Interaction of radiation and charged particles with ices in simulated ISM conditions

References

Chen, Y.-J., Nuevo, M., Yih, T.-S., *et al.* 2008, *MNRAS*, 384, 605. *Amino acids produced from the UV/EUV irradiation of naphthalene in a H$_2$O+NH$_3$ ice mixture*

Nuevo, M., Auger, G., Blanot, D., & D'Hendecourt, L. 2008, *OLEB*, 38, 37. *A detailed study of the amino acids produced from the vacuum UV irradiation of IS ice analogs*

Zheng, W., Jewitt, D., Osamura, Y., & Kaiser, R. I. 2008, *ApJ*, 674, 1242. *Formation of nitrogen and hydrogen-bearing molecules in solid ammonia and implications for Solar system and IS ices*

Thrower, J. D., Burke, D. J., Collings, M. P., et al. 2008, *ApJ*, 673, 1233. *Desorption of hot molecules from photon irradiated IS ices*

Davoisne, C., Leroux, H., Frère, M., et al. 2008, *A&A*, 482, 541. *Chemical and morphological evolution of a silicate surface under low-energy ion irradiation*

Chen, Y.-J., Nuevo, M., Hsieh, J.-M., et al. 2007, *A&A*, 464, 253. *Carbamic acid produced by the UV/EUV irradiation of IS ice analogs*

Bringa, E. M., Kucheyev, S. O., Loeffler, M. J., et al. 2007, *ApJ*, 662, 372. *Energetic processing of IS silicate grains by cosmic rays*

Zheng, W. & Kaiser, R. I. 2007, *CP*, 450, 55. *On the formation of carbonic acid (H_2CO_3) in Solar system ices*

Bennett, C. J. & Kaiser, R. I. 2007, *ApJ*, 660, 1289. *The formation of acetic acid (CH_3COOH) in IS ice analogs*

Bennett, C. J. & Kaiser, R. I. 2007, *ApJ*, 661, 899. *On the formation of glycolaldehyde ($HCOCH_2OH$) and methyl formate ($HCOOCH_3$) in IS ice analogs*

Sivaraman, B., Jamieson, C. S., Mason, N. J., & Kaiser, R. I. 2007, *ApJ*, 669, 1414. *Temperature-dependent formation of ozone in solid oxygen by 5 keV electron irradiation and implications for Solar system ices*

Wada, A., Mochizuki, N., & Hiraoka, K. 2006, *ApJ*, 644, 300. *Methanol formation from electron-irradiated mixed H_2O/CH_4 ice at 10 K*

Ricca, A., Bakes, E. L. O., Bauschlicher, C. W. 2007, *ApJ*, 659, 858. *The energetics for hydrogen addition to naphthalene cations*

Öberg, K. I., Fuchs, G. W., Awad, Z., et al. 2007, *ApJ (Letters)*, 662, L23. *Photodesorption of CO ice*

Schriver, A., Schriver-Mazzuoli, L., Ehrenfreund, P., & D'Hendecourt, L. 2007, *CP*, 334, 128. *One possible origin of ethanol in IS medium: photochemistry of mixed $CO_2C_2H_6$ films at 11 K. A FTIR study*

Mason, N. J., Dawes, A., Holtom, P. D., et al. 2006, in: *Astrochemistry - From Laboratory Studies to Astronomical Observations AIP-CP*, 855, 128. *VUV spectroscopy of extraterrestrial ices*

Loeffler, M. J., Raut, U., Vidal, R. A., et al. 2006, *Icarus*, 180, 265. *Synthesis of hydrogen peroxide in water ice by ion irradiation*

Jamieson, C. S., Mebel, A. M., & Kaiser, R. I. 2006, *ApJS*, 163, 184. *Understanding the kinetics and dynamics of radiation-induced reaction pathways in carbon monoxide ice at 10 K*

Palumbo, M. E. 2006, *A&A*, 453, 903. *Formation of compact solid water after ion irradiation at 15 K*

Jamieson, C. S., Guo, Y., Gu, X., et al. 2006, in: F. Salama (ed.), *NASA Laboratory Astrophysics Workshop, NASA/CP-2002-21186*, p. 68. *Laboratory studies on the formation of carbon-bearing molecules in extraterrestrial environments: from the gas phase to the solid state*

Wooden, D. H., Harker, D. E., & Brearley, A. J. 2005, in: A. N. Krot, E. R. D. Scott & B. Reipurth (eds.) *Chondrites and the Protoplanetary Disk ASP-CS*, 341, 774. *Thermal processing and radial mixing of dust: evidence from comets and primitive chondrites*

Bernstein, M. P., Sandford, S. A., & Allamandola, L. J. 2005, *ApJS*, 161, 53. *The mid-IR absorption spectra of neutral PAHs in conditions relevant to dense IS clouds*

Moore, M. H. & Hudson, R. L. 2005, in: D. C. Lis, G. A. Blake & E. Herbst (eds.), *Astrochemistry throughout the Universe: Recent Successes and Current Challenges*, Proc. IAU Symposium No. 231 (Cambridge: CUP), p. 247. *Production of complex molecules in astrophysical ices*

Farenzena, L. S., Iza, P., Martinez, R., et al. 2005, *EM&P*, 97, 311. *Electronic sputtering analysis of astrophysical ices*

Hudson, R. L., Khanna, R. K., & Moore, M. H. 2005, *ApJS*, 159, 277. *Laboratory evidence for solid-phase protonation of HNCO in IS ices*

Holtom, P. D., Bennett, C. J., Osamura, Y., et al. 2005, *ApJ*, 626, 940. *A combined experimental and theoretical study on the formation of the amino acid glycine (NH_2CH_2COOH) and its isomer ($CH_3NHCOOH$) in extraterrestrial ices*

Bernstein, M. P., Mattioda, A. L., Sandford, S. A., & Hudgins, D. M. 2005, *ApJ*, 626, 909. *Laboratory IR spectra of polycyclic aromatic nitrogen heterocycles: quinoline and phenanthridine in solid argon and H_2O*

Shalabiea, O. M., Awad, Z., Chigai, T., & Yamamoto, T. 2005, in: D. C. Lis, G. A. Blake & E. Herbst (eds.), *Astrochemistry throughout the Universe: Recent Successes and Current Challenges*, Proc. IAU Symposium No. 231 (Cambridge: CUP), p. 64. *New rate constants of hydrogenation on IS grains and their astrophysical implications*

Creighan, S. C. & Price, S. D. 2005, in: D. C. Lis, G. A. Blake & E. Herbst (eds.), *Astrochemistry throughout the Universe: Recent Successes and Current Challenges*, Proc. IAU Symposium No. 231 (Cambridge: CUP), p. 55. *Studies of hydrogen formation on IS grain analogues*

Strazzulla, G. & Moroz, L. 2005, *A&A*, 434, 593. *Ion irradiation of asphaltite as an analogue of solid hydrocarbons in the IS medium*

Mukerji, R. J., Dawes, A., Holtom, P. D., *et al.* 2005, in: D. C. Lis, G. A. Blake & E. Herbst (eds.), *Astrochemistry throughout the Universe: Recent Successes and Current Challenges*, Proc. IAU Symposium No. 231 (Cambridge: CUP), p. 188. *Studies of the temperature dependence of the photo-absorption spectrum of solid ammonia*

Strazzulla, G., Leto, G., LaDelfa, S., *et al.* 2005, *MemSAIS*, 6, 51. *Oxidants produced after ion bombardment of water/carbon dioxide icy mixtures*

Holtom, P. D., Dawes, A., Davis, M. P., *et al.* 2005, in: D. C. Lis, G. A. Blake & E. Herbst (eds.), *Astrochemistry throughout the Universe: Recent Successes and Current Challenges*, Proc. IAU Symposium No. 231 (Cambridge: CUP), p. 130. *VUV photo-absorption spectroscopy of amorphous and crystalline sulphur dioxide films*

Davis, M. P., Dawes, A., Holtom, P. D., *et al.* 2005, in: D. C. Lis, G. A. Blake & E. Herbst (eds.), *Astrochemistry throughout the Universe: Recent Successes and Current Challenges*, Proc. IAU Symposium No. 231 (Cambridge: CUP), p. 128. *Vacuum UV spectroscopy of laboratory-simulated astrophysical ices*

Dawes, A., Holtom, P. D., Mukerji, R. J., *et al.* 2005, in: D. C. Lis, G. A. Blake & E. Herbst (eds.), *Astrochemistry throughout the Universe: Recent Successes and Current Challenges*, Proc. IAU Symposium No. 231 (Cambridge: CUP), p. 131. *Low-energy singly and multiply charged ion irradiation of astrophysical ices*

Gomis, O. & Strazzulla, G. 2005, *Icarus*, 177, 570. *CO_2 production by ion irradiation of H_2O ice on top of carbonaceous materials and its relevance to the galilean satellites*

Hudson, R. L., Moore, M. H., & Cook, A.M. 2005, *AdSpR*, 36, 184. *IR characterization and radiation chemistry of glycolaldehyde and ethylene glycol ices*

Kroes, G. J. & Andersson, S. 2005, in: D. C. Lis, G. A. Blake & E. Herbst (eds.), *Astrochemistry throughout the Universe: Recent Successes and Current Challenges*, Proc. IAU Symposium No. 231 (Cambridge: CUP), p. 427. *Theory of molecular scattering from and photo-chemistry at ice surfaces*

Palumbo, M. E. 2005, in: D. C. Lis, G. A. Blake & E. Herbst (eds.), *Astrochemistry throughout the Universe: Recent Successes and Current Challenges*, Proc. IAU Symposium No. 231 (Cambridge: CUP), p. 59. *Formation of compact solid water after cosmic ion irradiation*

Mennella, V., Baratta, G. A., Palumbo, M. E., & Bergin, E. A. 2006, *ApJ*, 643, 923. *Synthesis of CO and CO_2 molecules by UV irradiation of water ice-covered hydrogenated carbon grains*

Loeffler, M. J., Raut, U., Vidal, R. A., *et al.* 2006, *Icarus*, 180, 265. *Synthesis of hydrogen peroxide in water ice by ion irradiation*

Hudson, R. L. & Moore, M. H. 2004, *Icarus*, 172, 466. *Reactions of nitriles in ices relevant to Titan, comets, and the IS medium: formation of cyanate ion, ketenimines, and isonitriles*

Palumbo, M. E., 2005, *JPhCS*, 6, 211. *The morphology of IS water ice*

Moore, M. H. & Hudson, R. L. 2005, in: D. C. Lis, G. A. Blake & E. Herbst (eds.), *Astrochemistry throughout the Universe: Recent Successes and Current Challenges*, Proc. IAU Symposium No. 231 (Cambridge: CUP), p. 247. *Production of complex molecules in astrophysical ices*

Tatischeff, V. & Kiener, J. 2004, *NwARv*, 48, 99. *Gamma-ray lines from cosmic-ray interactions with IS dust grains*

Mennella, V., Palumbo, M. E., & Baratta, G. A. 2004, *ApJ*, 615, 1073. *Formation of CO and CO_2 molecules by ion irradiation of water ice-covered hydrogenated carbon grains*

Gerakines, P. A., Moore, M. H., Hudson, R. L. 2004, *Icarus*, 170, 202. *UV photolysis and proton irradiation of astrophysical ice analogs containing hydrogen cyanide*

Gomis, O., Leto, G., & Strazzulla, G. 2004, *A&A*, 420, 405. *Hydrogen peroxide production by ion irradiation of thin water ice films*

Sandford, S. A., Bernstein, M. P., & Allamandola, L. J. 2004, *ApJ*, 607, 346. *The mid-IR laboratory spectra of naphthalene ($C_{10}H_8$) in solid H_2O*

Muñoz Caro, G. M., Meierhenrich, U., Schutte, W. A., et al. 2004, *A&A*, 413, 209. *UV-photo-processing of IS ice analogs: detection of hexamethylenetetramine-based species*

Baratta, G. A., Brunetto, R., Leto, G., et al. 2004, *MemSAIS*, 5, 33. *Ion irradiation of ices relevant to astrophysics*

Woon, D. E. 2004, *AdSpR*, 33, 44. *Photoionization in UV processing of astrophysical ice analogs at cryogenic temperatures*

Mennella, V., Baratta, G. A., Esposito, A., et al. 2003, *The effects of ion irradiation on the evolution of the carrier of the 3.4 μm IS absorption band ApJ*, 587, 727

Chilton, D. L., Mohr, R., & Gerakines, P. 2003, *BAAS*, 35, 1269. *Simulating the effect of UV photolysis on IS ices*

Muñoz Caro, G. M., & Schutte, W. A. 2003, *A&A*, 412, 121. *UV-photo-processing of IS ice analogs: new IR spectroscopic results*

Cooper, J. F., Christian, E. R., Richardson, J. D., & Wang, C. 2003, *EM&P*, 92, 261. *Proton irradiation of Centaur, Kuiper Belt, and Oort Cloud objects at plasma to cosmic ray energy*

Wu, C. Y. R., Judge, D. L., Cheng, B.-M., et al. 2003, *JGRE*, 108, 5032. *EUV photolysis of CO_2-H_2O mixed ices at 10 K.*

Mennella, V., Baratta, G. A., Esposito, A., et al. 2003, *ApJ*, 587, 727. *The effects of ion irradiation on the evolution of the carrier of the 3.4 μm IS absorption band*

Muñoz Caro, G. M., Meierhenrich, U.J., Schutte, W.A., et al. 2002, *Nature*, 416, 403. *Amino acids from UV irradiation of IS ice analogues*

Carrez, P., Demyk, K., Cordier, P., et al. 2002, *Science*, 37, 1599. *Low-energy helium ion irradiation-induced amorphization and chemical changes in olivine: insights for silicate dust evolution in the IS medium, meteoritics and planetary*

Mennella, V., Brucato, J. R., Colangeli, L., & Palumbo, P. 2002, *AdSpR*, 30, 1451. *Hydrogenation of carbon grains by exposure to hydrogen atoms: implications for the 3.4 μm IS absorption band*

Transactions IAU, Volume XXVIIA
Reports on Astronomy 2006–2009
Karel A. van der Hucht, ed.

© 2009 International Astronomical Union
doi:10.1017/S1743921308025970

COMMISSION 41	**HISTORY OF ASTRONOMY**
	HISTOIRE DE L'ASTRONOMIE

PRESIDENT	Nha Il-Seong
VICE-PRESIDENT	Clive L.N. Ruggles
PAST PRESIDENT	Alexander A. Gurshtein
SECRETARY	Rajesh K. Kochhar
ORGANIZING COMMITTEE	David H. DeVorkin, Teije de Jong,
	Tsuko Nakamura, Wayne Orchiston,
	Antonio A.P. Videira, Brian Warner

COMMISSION 41 WORKING GROUPS

Div. XII / Commission 41 WG	Archives
Div. XII / Commission 41 WG	Historical Instruments
Div. XII / Commission 41 WG	Transits of Venus
Div. XII / Commission 41 WG	Astronomy and World Heritage

INTER-DIVISION X-XII WORKING GROUP

Inter-Division X-XII WG	Historic Radio Astronomy

TRIENNIAL REPORT 2006 - 2009

1. Introduction

Commission 41 of the International Astronomical Union deals with all aspects of astronomical history and heritage from ancient sky knowledge to developments in modern astronomy that have occurred within living memory. It encourages and supports research in the history of astronomy and related fields such as archaeoastronomy and is also concerned with the identification, documentation and preservation of vital aspects of our astronomical heritage such as sites, artifacts, instruments and archives. Commission 41 is one of the largest Commissions in the Union, and is a member of Division XII on *Union-Wide Activities*.

All Commission 41 members are also, ipso facto, members of the Inter-Union Commission for the History of Astronomy (ICHA). This is a joint Commission of the IAU and the International Union for the History and Philosophy of Science (IUHPS). Membership is open to IAU members as well as to scholars who, being primarily historians rather than astronomers, are not members of the IAU.

Commission 41's activities during this period have been primarily geared toward the 400th anniversary of Galileo's use of telescope. Luckily, the year 2009 happens to be the year of the IAU General Assembly. The UN has also declared 2009 to be the *International Year of Astronomy* and has asked the IAU to serve as the facilitating body. Being a dedicated Commission on history of astronomy Commission 41 has been fully alive to its responsibilities.

2. Co-sponsored meetings in 2009

Commission 41 is co-sponsoring three scientific meetings for 2009, two of them being Special Sessions at the IAU XXVII General Assembly in Rio de Janeiro, 2006. The main aim is to integrate history of astronomy with issues of contemporary significance.

(*i*) SpS4 on *Astronomy education between Past and Future*
SOC co-chairs: Rajesh K. Kochhar, Jean-Pierre de Greve and Edward F. Guinan

(*ii*) Sp5 on *Accelerating the rate of astronomical discovery*
SOC co-chairs: Raymond P. Norris and Clive L.N. Ruggles

(*iii*) Commission 41 is co-sponsoring a joint symposium on *Astronomy and its instrumentation before and after Galileo*, being organized by INAF Astronomical Observatory of Padova, in Venice 28 September - 3 October 2009. Co-chairs: Luisa Pigatto and Clive L.N. Ruggles)

3. UNESCO Astronomy and World Heritage Initiative

In 2005, UNESCO approved the *Astronomy and World Heritage Initiative* (AWHI) for the recognition, promotion, protection and preservation of places of exceptional cultural value and significance relating to astronomy. Here astronomy is broadly interpreted to include world-wide perceptions of the sky through the ages. The first task is to identify the most exceptional astronomical site and properties that can be included in the world list. Formal criteria for inclusion in the final list should emerge from this exercise. On behalf of the IAU, Commission 41 is coordinating its efforts on the AWHI with UNESCO. As an important first step, a tentative preliminary list of such sites was prepared through received suggestions and submitted. The next step, now in progress, is to set up a representative Working Group for which inputs from individuals have already been received.

At the occasion of the official signing of a Memorandum of Understanding between UNESCO and the IAU on 30 October 2008 in Paris, the following press releases have been issued.

3.1. *UNESCO Flah Info No. 151-2008. UNESCO and International Astronomical Union sign a Memorandum of Understanding*

Office of the Spokesperson
La Porte-parole

On 30 October 2008 the Director-General of UNESCO, Mr Koïchiro Matsuura, and the General Secretary of the International Astronomical Union (IAU), Professor Karel A. van der Hucht, signed a Memorandum of Understanding (MoU) formalizing their intention to expand their collaboration in the development of UNESCOs Astronomy and World Heritage thematic initiative.

Also present at the ceremony were Professor Clive Ruggles, Vice-President of the IAU Commission on the History of Astronomy, Professor David Valls-Gabaud, Deputy Director of the Observatoire de Paris and Mr Cipriano Marin, Coordinator of the Starlight Initiative, Spanish National Commission for UNESCO.

Under the MoU, the two organizations will undertake joint activities focusing on establishing the link between science and culture by identifying properties related to astronomy located around the world, preserving their memory and saving them from progressive deterioration.

In his remarks, the Director-General noted that very few of the 878 World Heritage properties currently inscribed on the List had been nominated for their astronomical values. Referring to the World Heritage Committees Global Strategy for a Balanced Representative and Credible World Heritage List, Mr Matsuura underscored the importance of diversifying the types of sites on the List, as well as the number of countries whose heritage featured in it. The work plan foreseen under this MoU would make an important contribution to these efforts.

Continuing, the Director-General expressed his gratitude for the IAUS involvement in the development of the Astronomy and World Heritage initiative since 2003 and expressed his pleasure that this would be expanded through this important framework for co-operation in the fields of research, education and awareness about our astronomical heritage, noting that its formalization on the eve of 2009, International Year of Astronomy was most fitting. Speaking of his pride that UNESCO had been designated lead agency of the International Year by the 62nd UN General Assembly, Mr Matsuura noted that many important events were being co-organized with the IAU, the Years facilitator. He looked forward to welcoming Professor van der Hucht and his delegation to UNESCO on 15-16 January 2009 for the inauguration of the International Year.

UNESCO
7, Place de Fontenoy
75007 Paris
France
Website: www.unesco.org/dg

– End of Flash Info –

3.2. *IAU Press Release – IAU0808: UNESCO and the IAU sign key Agreement on Astronomy and World Heritage Initiative*

Oct 30, 2008, Paris

A Memorandum of Understanding is to be signed today, 30 October 2008, between UNESCO and the International Astronomical Union (IAU). The IAU will be integrally involved in the process of developing UNESCOs Astronomy and World Heritage Initiative, helping to promote astronomical sites of 'Outstanding Universal Value'.

The world-famous UNESCO World Heritage Convention is renowned for its work protecting and promoting sites that celebrate the heritage of humanity. Examples include the Pyramids of Giza in Egypt, the Mayan city of Chichen Itza in Mexico, and the Stonehenge in the United Kingdom.

However, astronomical heritage is currently under-represented. All too often, neglect and mistreatment cause irreversible harm. The new Memorandum will place the Astronomy and World Heritage Initiative in a better position to reverse this trend by raising awareness of the cultural importance of astronomical sites, both ancient and modern.

Adopting the successful strategy previously applied to architectural and natural sites, the new UNESCO Astronomy and World Heritage initiative will officially recognise, promote and preserve astronomical sites that are of outstanding significance to humankind. The places in question include landmarks whose design or location relate to celestial events, whether with symbolic or direct connection with astronomy. Historic sites, instruments and representations help to broaden and enhance our perception of the sky. This theme is integral to the upcoming International Year of Astronomy 2009. The initiative is therefore designated as one of the Year's key Cornerstone projects, which are being organised to increase public understanding and appreciation of astronomy throughout the coming year and around the world.

In order to fulfil its commitment to the UNESCO Initiative, the IAU has set up a new Working Group on *Astronomy and World Heritage* under the chairmanship of Professor Clive Ruggles, Emeritus Professor of Archaeoastronomy at the University of Leicester, UK. Ruggles, who is also Vice-President of the IAU's Commission 41 on the *History of Astronomy*, has already worked with UNESCO in the early stages of developing the initiative. He said: "The globalisation of human culture is proceeding at a relentless pace, and it is becoming increasingly urgent to preserve some of the more fragile aspects of our common cultural heritage. 'Fine', you might say, 'but why worry about astronomy in this regard?' The main reason, I think, is that every human culture has a sky, and strives to interpret what people perceive there. The understanding they develop inevitably comes to form a vital part of their fundamental knowledge concerning the cosmos and their place within it. Astronomy is not just a modern science but a fundamental reflection of how all people, past and present, understand themselves in relation to the Universe".

At present, States Parties to the World Heritage Convention may nominate sites for inscription on the World Heritage List for a variety of reasons; but until now, there have been few precedents and no guidelines for nominations relating principally to astronomy. Identifying and defining criteria that demonstrate 'Outstanding Universal Value' in relation to astronomy is not a straightforward task. They must encompass a wide range of sites, from prehistoric monuments to modern observatories. Helping to establish such criteria is the IAU Working Group's top priority. As Ruggles says, "without such guidelines member states of UNESCO will have little motivation to put forward astronomical sites for the World Heritage List, since they will have very little idea of their chances of success".

The agreement between UNESCO and the IAU is designed to set the wheels in motion. As a result, astronomical heritage will become much better represented in the World Heritage List.

– End of Press Release –

4. Dictionary of Historical Astronomical Instruments

Commission 41's *Historical Instruments* Working Group (chair: Luisa Pigatto) is engaged in compiling a dictionary of historical instruments. The text will provide authentic information on various instruments, their correct name, their inventors, and use. This dictionary is expected to be of assistance in the AWHI mentioned above.

5. IAU archives

Commission 41's *Archives* Working Group (Chair: Ileana Chinnici) is currently engaged in a survey of existing archival documents in IAU founder countries (see WG Report).

6. Exhibition in collaboration with the Vatican Observatory

Some Italian members of Commission 41 are engaged in the organization of an exhibition *Astrum 2009* to be opened in October 2009 as a joint initiative by INAF Italian Institute for Astrophysics and the Vatican Observatory. The exhibition will display historical instruments, books and documents kept in the Italian Observatories and will be held in Rome at the Vatican Museums.

7. Observatoire de Paris

A group on history of astronomy is part of the Department *Systèmes de Reference Temps-Espace* (SYRTE). It is located at 61 avenue de l'Observatoire, F-75014, Paris, and is under the charge of Michel-Pierre Lerner. During the triennium the history of astronomy group comprised eight permanent researchers and as many associated researchers, working part-time or after retirement. The group's researches are briefly described below.

(*i*) Astronomy and cosmology during Renaissance and at the beginning of modern times - mostly under *Project Copernic*. A critical edition of the *De Revolutionibus Orbium Coelestium* is planned for 2009.

(*ii*) Astronomical studies for the period 17th-18th centuries – under *Project D'Alembert* – in the light of the publication of his *Euvres complètes*. Two volumes have already been published. The third and the fourth are expected to appear by the end of 2008 and the beginning of 2009. A *Colloque Grandjean de Foucy* was organized in 2007 at the Observatoire de Paris.

(*iii*) Astronomical studies for the period 19th and 20th centuries. A biography of Janssen was published in 2008 and four related seminars were organized. A book had earlier been published in 2005 on Arago and the speed of light.

(*iv*) Researches have been carried out, based mostly on the collections in the archives of Paris Observatory. These include inventories of Chinese documents, historical astrometry, history of the Observatory and related subjects, such as the metric system, time, etc.

(*v*) In addition to research papers, many activities as well as publications have been directed at laypersons. For list of publications, please contact Suzanne V. Débarbat.

8. ICOA conferences

In July 2008 the ICOA-6 Conference was held at James Cook University (Australia), with papers presented in sessions on Applied Historical Astronomy, Ethnoastronomy & Archaeoastronomy, Islamic Astronomy, The Emergence of Astrophysics, The History of Radio Astronomy, and 'Other Recent Research'.

During the triennium, two proceedings from earlier ICOA conferences were published:
- Chen, K. Y., & Sun, X. (eds.) 2006, *Frontiers of Oriental Astronomy*, Proc. ICOA-2, (Chinese Science and Technology Press),
- Chen, K.-Y., Orchiston, W., Soonthornthum, B., & Strom, R. (eds.) 2006, Proc. *Fifth International Conference on Oriental Astronomy* (Chiang Mai University).

9. History of astronomy graduate studies at James Cook University

Since 2006, Commission 41 Committee members, Wayne Orchiston (Australia) and Brian Warner (South Africa) and WG *Historic Radio Astronomy* member Richard Wielebinski (Germany), along with Ian Glass (South Africa), Kim Malville (USA), Bruce Slee (Australia), Richard Stephenson (UK) and Richard Strom (Netherlands), have been involved in developing off-campus internet-based part-time doctoral and masters programs in history of astronomy at James Cook University (JCU), Townsville, Australia. Eighteen students are currently enrolled in doctorates, and eight students have completed masters degrees in history of astronomy. During 2006-2008 the Centre for Astronomy at JCU ran three invitation-only mini-conferences on history of astronomy. The Centre also continued to produce the Journal of Astronomical History and Heritage.

Rajesh K. Kochhar
secretary of the Commission

Transactions IAU, Volume XXVIIA
Reports on Astronomy 2006–2009
Karel A. van der Hucht, ed.

© 2009 International Astronomical Union
doi:10.1017/S1743921308025982

DIVISION XII / COMMISSION 41 / WORKING GROUP ARCHIVES

CHAIR	**Ileana Chinnici**
PAST CHAIR	**Brenda G. Corbin**
MEMBERS	**Suzanne V. Débarbat, Daniel W.E. Green,**
	Lee Jung-Bok, Oscar T. Matsuura,
	Wayne Orchiston, Adam Perkins
	Irakli A. Simonia

TRIENNIAL REPORT 2006 - 2009

1. Introduction

The Working Group *Archives* deals with all aspects of the identification and preservation of astronomical archives. In 2009 the IAU will celebrate its 90th anniversary, and on this occasion the WG is taking action toward preserving the archival materials related with the history of IAU. An institution must keep memory of its own past and, as the centenary of IAU is approaching, for the 2006 - 2009 triennium the WG *Archives* has started evaluating the archival collections related to the establishment of IAU, in order to check their extent and the current conditions of preservation and conservation of such documents.

2. Working plan

Following the suggestions from S. Débarbat, a Working Plan, entitled *IAU Archives* has been proposed by I. Chinnici to the WG members. The document can be downloaded from the Commission 41 web site `<www.le.ac.uk/has/c41/WGArchivesPlan2006-09.pdf>`.

The archival documents linked to the establishment of the IAU has been divided into two classes:

A. Documents held at the IAU Secretariat in Paris

B. Documents held in the countries which founded the IAU.

No urgent action is required for class A documents, thanks to the sorting already made by Blaauw (1994), even though it is room for improvements, as suggested by Chinnici (see Working Plan).

In contrast, the lack of information on the extent of class B collections calls for a stronger effort in this direction, before the uncatalogued documents are lost. Archives of the astronomical institutions located in the nineteen countries founders of the IAU must be searched for the 1919 - 1922 period. It has also been considered the possibility to check for IAU documents in countries which initially were not IAU members.

Moreover, in 1919 thirty-two Commissions were created within the IAU, and it is possible that more archival documents can be found in the fonds of the Commission presidents.

3. Ongoing action

In order to distribute the work, Chinnici has proposed that each WG *Archives* members identifies reference persons in existing archives and make contact with them, in order to set up survey works in the following geographical areas:

Débarbat: France, Germany, Netherlands, Denmark, Switzerland

Chinnici: Italy, Greece, Spain, Portugal, Vatican

Corbin and Green: USA, Canada

Jung-Bok: Far East Countries

Matsuura: Latin America

Orchiston: Australia, New Zealand

Perkins: UK, Ireland, Norway, Sweden, Finland

Simonia: Eurasian Countries and Middle East Countries

As the African countries are not covered, a WG member for this area will be decided in the near future.

It is important to say that the WG met with difficulties in contacting people in some countries and/or in obtaining collaboration for the survey work; this could be due to the voluntary nature of this kind of research, which appears as a lengthy task. Therefore, the WG has recently decided to create some sub-committees, formed by people corresponding with the WG members in the several countries, to stimulate their collaboration. The creation of these sub-committees is ongoing.

4. Current results

At present, the most advanced survey work has been carried out in Latin America, thanks to the intense and fruitful effort of Oscar Matsuura. Countries in this area which affiliated to IAU in chronological order are: Mexico (1921), Brazil (1922), Argentina (1927), Chile (1947), Venezuela (1953), Uruguay (1970), Peru (1988), Bolivia (1998), Cuba (2001).

Matsuura has presented two reports to the WG, in 2007 and 2008, illustrating the results of his actions:

(*i*) in founder countries (Mexico and Brazil), the identification of class B documents is advancing and, particularly in Brazil, the WG task has been embedded in a research program of history of astronomy - what seems to be a good model to reproduce elsewhere.

(*ii*) in non-founder countries, the work has progressed in Venezuela, Cuba and Uruguay; a contact has been recently established in Argentina, while other contacts are still needed in Chile, Peru and Bolivia. See Matsuura's report for 2008 in the web site, <www.le.ac.uk/has/c41/>.

A report about selected astronomical archives in USA and Canada has also been prepared by B. Corbin. In these countries, as in many others, astronomical archives are not centralized in one or two locations but held in many places. This is one of the main difficulties of a survey work.

Therefore, Corbin asked for summary guides from seven major institutions in the US and one in Canada in order to give a first representation of archives in those countries. These institutions are: Harvard University (Cambridge, MA); National Radio Astronomy Observatory (Charlottesville, VA); Niels Bohr Library and Archives (American Institute of Physics, College Park, MD); the Observatories of the Carnegie Institution of Washington (Pasadena, CA); Smithsonian Institution (including Smithsonian Astrophysical Observatory, Washington, DC); U. S. Naval Observatory (Washington, DC); Yerkes

Observatory (Archives now at The University of Chicago Library, Chicago, IL);and, in Canada, the University of Toronto. In USA and Canada the search for IAU archival documents is ongoing.

In Europe, the situation shows similar difficulties and complexity, because of the spreading of these documents among many institutions and private archives. To give the WG some idea of what the Greenwich Observatory collection holds, relating to the IAU, as well as other Cambridge institutions including the Churchill College Archives Centre and St. John's College, A. Perkins has made a Boolean search of on-line catalogues available for 'International+Astronomical+Union' after 1900. The result is a fifteen-page word file which does reflect how dispersed and general in nature the material is.

S. Débarbat has sent a short report about IAU Archives in France. There, the work appears to be simplified because of the classification work already done by Blaauw at the IAU Secretariat, at the Institut d'Astrophysique and at the Observatoire de Paris. Nevertheless, other documents concerning IAU are to be searched in the fonds of astronomers occupying positions at the IAU in the past.

Work is also in progress in Italy, while in other countries no contacts have yet been established.

All reports of the WG members, at authors' agreement, are coming soon on-line on the Commission 41 web site `<www.le.ac.uk/has/c41/>`.

5. Next steps

In the near future, the WG goals are the full establishment of the sub-committees, the identification of possible contacts in other countries and the reminder for the missing reports. Such an important work is inevitably slow and it is only the beginning of an effort which hopefully could be achieved – and published – before the centenary of IAU.

The WG-Archives therefore invites all astronomers and historians wishing to collaborate in the surveying of IAU archival documents in their own countries, to contact the corresponding WG members above indicated.

Ileana Chinnici
Chair of the Working Group

References

Blaauw, A. 1994, *History of the IAU. The Birth and First Half-Century of the International Astronomical Union*, pp. 281-284.

Transactions IAU, Volume XXVIIA
Reports on Astronomy 2006–2009
Karel A. van der Hucht, ed.

© 2009 International Astronomical Union
doi:10.1017/S1743921308025994

DIVISION XII / COMMISSION 41 / WORKING GROUP
HISTORICAL INSTRUMENTS

CHAIR   **Luisa Pigatto**
PAST CHAIR   **Nha Il-Seong**
MEMBERS   **Jürgen Hamel, Kevin Johnson,**
                     Rajesh K. Kochhar, Tsuko Nakamura,
                     Wayne Orchiston, Bjørn R. Pettersen
                     Sara J. Schechner, Shi Yunli

TRIENNIAL REPORT 2006 - 2009

1. IYA2009 initiative

The *Historical Instruments* Working Group (WG-HI) and Commission 41 started planning an interdisciplinary conference titled *Astronomy and its instruments before and after Galileo* since January 2007. This conference, as an IYA2009 initiative, aims "to highlight mankind's path toward an improved knowledge of the sky using mathematical and mechanical tools as well as monuments and buildings, giving rise, in doing so, to scientific astronomy". Commission 46 and Commission 55 also support this conference, to be held on the Isle of San Servolo, Venice (Italy), 27 September – 3 October 2009. As a fact of history, it was in Venice that Galileo was advised and got material (glass) to make his telescope, and in Venice he presented an working instrument to Venetian Doge in August 1609. The conference is co-sponsored by IAU as a Joint Symposium with the INAF - Astronomical Observatory of Padova, Italy.

2. The thesaurus of historical instruments

The WG-HI thesaurus is now better defined as far as its structure and content are concerned. Like a dictionary, it will be a list of terms – single-word or multi-word nouns plus adjectives and adverbs when useful – to give a correct identification of historical instruments and their use. Terms, including variants and synonyms from each country, have to be introduced to identify instruments and parts of them. Etymology, general definition, particular definition related to adjectives or adverbs, of all listed terms have to be given in a synthetic way, as well as a bibliography related to earliest written sources (primary sources), and images when available to show a prototype of each instrument and parts of it. A first list is now circulating among the WG-HI members as a starting point in compiling the thesaurus. The WG-HI thesaurus aims to provide interested people with correct information about historical instruments, their names and use in astronomy and related disciplines as well as their inventors, in each country. It aims to help with the identification of astronomical heritage in line with 'The Astronomy and World Heritage Initiative'.

Luisa Pigatto
chair of the Working Group

Transactions IAU, Volume XXVIIA
Reports on Astronomy 2006–2009
Karel A. van der Hucht, ed.

© 2009 International Astronomical Union
doi:10.1017/S1743921308026008

COMMISSION 46

ASTRONOMY EDUCATION AND DEVELOPMENT

ÉDUCATION ASTRONOMIQUE ET DÉVELOPPEMENT

PRESIDENT	Magdalena G. Stavinschi
VICE-PRESIDENT	Rosa M. Ros
PAST PRESIDENT	Jay M. Pasachoff
ORGANIZING COMMITTEE	Johannes Andersen, Susana E. Deustua Jean-Pierre de Greve, Edward F. Guinan, Hans J. Haubold, John B. Hearnshaw, Barrie W. Jones, Rajesh K. Kochhar, Kam-Ching Leung, Laurence A. Marschall, John R. Percy Silvia Torres-Peimbert

COMMISSION 46 WORKING GROUPS

Div. XII / Commission 46 PG	World Wide Development of Astronomy
Div. XII / Commission 46 PG	Teaching for Astronomy Development
Div. XII / Commission 46 PG	International Schools for Young Astronomers
Div. XII / Commission 46 PG	Collaborative Programs
Div. XII / Commission 46 PG	Exchange of Astronomers
Div. XII / Commission 46 PG	National Liaison on Astronomy Education
Div. XII / Commission 46 PG	Newsletters
Div. XII / Commission 46 PG	Public Education on the Occasions of Solar Eclipses
Div. XII / Commission 46 PG	Exchange of Books and Journals

TRIENNIAL REPORT 2006 - 2009

1. Introduction

Commission 46 continues its task in the triennium, which started in September 2006. It seeks to further contribute to the development and improvement of astronomical education at all levels all over the world through various projects initiated, maintained and to be developed by the Commission, and by disseminating information concerning astronomy education.

All members of the Commission contributed to the success of the Commission, as well as many others national liaisons. A new web site is kept at the Paris Observatory with the help of Chantal Balkowski and her team. The *Newsletter*s continue to be posted twice per year: in March and October, thanks to Barrie Jones. Commission 46 is working under Division XII *Union-Wide Activities*. The Commission Program Groups are listed above. Each Program Group has its own report. Here we mention only some generic aspects.

2. PG World Wide Development of Astronomy

John B. Hearnshaw, chair

PG-WWDA has as task to visit countries with some astronomy expertise at tertiary (i.e., post high school) level, which welcome further development of their capabilities in teaching and/or research in astronomy. One key target is to foster astronomy development in the 19 nations which currently have individual IAU members but which do not adhere to IAU. From time to time, PG-WWDA it organizes conferences or meetings on astronomy in developing countries. A longer term goal for PG-WWDA, in partnership with TAD, is to explore the possibilities of setting up a third-world institute for astronomy in a developing country, or to have a number of nodes to such an institute distributed through several geographical locations.

3. PG Teaching Astronomy for Development

Edward F. Guinan, Laurence A. Marschall, co-chairs

PG-TAD intends to enhance the astronomy in a country with currently underdeveloped astronomy: education or its astronomy research infrastructure in support of education. PG-TAD operates on the basis of proposals from national professional astronomy organizations or on the basis of a contract between the IAU and an academic or governmental institution, usually a university. In the triennium, countries benefiting from PG-TAD contributions were Kenya, D.P.R. Korea, Mongolia, Morocco, Nicaragua, Philippines, Trinidad & Tobago, and Vietnam.

4. PG for International Schools for Young Astronomers

Jean-Pierre de Greve, chair, Kam-Ching Leung, vice-chair

Initiated in the 1960s, ISYAs are intended to support astronomy in developing countries. They consist of three-weeks schools for $\sim$30 students at academic levels (BSc to PhD). They seeks to broaden the participants' perspective on astronomy through lectures from an international faculty on selected astronomical topics, seminars, practical exercises and observations, and exchange of experiences. There is a wide regional (multi-country) representation of both lecturers and students. In the triennium, ISYAs were organized in Malaysia, Kuala Lumpur and Langkawi Island, 6-27 March 2007, and in Turkey, Istanbul, 1-22 July 2008.

5. PG Collaborative Programs

Hans J. Haubold, chair

PG-CP is working on activities co-sponsored by UN, ESA, NASA, JAXA, UNESCO, COSPAR, IAU and others for the world-wide development of basic space science. In 2007, a number of major anniversaries occurred, among them the 50th anniversary of the *International Geophysical Year*, the launch of Sputnik 1, and the 50th session of the United Nations *Committee on the Peaceful Uses of Outer Space* (UNCOPUOS). Particularly, the International Heliophysical Year IHY 2007 was an opportunity: (*i*) to advance the understanding of the fundamental heliophysical processes that govern the Sun, Earth, and heliosphere; (*ii*) to continue the tradition of international research and advancing the legacy of IGY1957, and (*iii*) to demonstrate the beauty, relevance and significance of space and Earth science to the world. See <ihy2007.org>.

In preparation of IHY2007, UNOOSA in cooperation with ESA, NASA, JAXA, COS-PAR, IAU, and the IHY Secretariat, held international workshops in the United Arab Emirates in 2005, in India in 2006, in Japan in 2007, and in Bulgaria in 2008.

On the starting date of IHY2007, 19 February 2007, during the session of the *Scientific and Technical Subcommittee of UNCOPUOS*, the IHY kick-off included an IHY exhibit, press briefing, and an opening ceremony in the United Nations Office Vienna. IHY Regional Coordinators, Steering Committee members, and Advisory Committee members participated in the IHY kick-off event. The Austrian Academy of Sciences hosted a one-day symposium on IHY2007 in Vienna on 20 February 2007. The IHY workshop for 2009 will be hosted by South Korea. This workshop will also cover thematic areas as pursued by the *International Year of Astronomy 2009*.

6. PG Exchange of Astronomers

John R. Percy, chair, Kam-Ching Leung, co-chair

PG-EA is meant to provide travel grants, primarily for astronomers and advanced students from the developing countries, for research or study trips, of at least three months duration, to other institutes. A major goal is to enhance the opportunities for exchanges between developing countries and established astronomy centers elsewhere in the world. Guidelines and application procedures can be found on the web site <physics.open.ac.uk/IAU46/travel.html>.

7. PG National Liaisons on Astronomy Education

Barrie W. Jones, chair

In 2008 Barrie Jones has put considerable effort into updating the list and contact details of National Liaisons. See: <physics.open.ac.uk/~bwjones/IAU46>.

8. PG Newsletter

Barrie W. Jones, chair

Barrie Jones receives solicited and unsolicited material, edits it and constructs the *Newsletter*. He has been editor since late 1998 (*Newsletter* 49), since when the *Newsletter has appeared* in March and October every year. It is available in PDF form at <physics.open.ac.uk/~bwjones/IAU46> where there is every issue from Number 50 (March 1999) to Number 68 (March 2008). Jones notifies people via several email lists, including the Commission 46 Organising Committee, members of Program Groups, National Liaisons, and others. This amounts to about 110 people. The Newsletter would benefit from more and clearer links from elsewhere in the IAU.

9. PG Public Education at the Times of Solar Eclipses and Transit Phenomena

Jay M. Pasachoff, chair

The PG offers timely advice to countries that will experience a solar eclipse. It also offers advice about transits of the Sun by Venus and Mercury. It consults with local astronomers and newspapers. It has its own web page `<www.eclipses.info>`.

The PG takes advantage of solar eclipses to provide a variety of astronomy education to countries from which the eclipse is visible. The work is coordinated with that of the WG *Eclipses* of Division II (`<www.totalsolareclipse.net>`). One of the PG members, Ralph Chou, professor of optometry, is expert on eye safety during eclipses and on the safety of materials used as solar filters. The third PG member, Julieta Fierro, is experienced on using the occasions of eclipses to bring widespread public education.

A distinction between the PG and the WG *Eclipses* is that the former largely deals with the general public and the latter largely deals with professional astronomers making scientific observations. The PG works toward providing a wide variety of public education in astronomy, not limited to the sun and eclipses, when the general public's attention is primed toward astronomy by an approaching eclipse.

Among our past successes, jointly with the WG *Eclipses*, is the distribution of material for tens of thousands of eye-protection filters after online consultation. A review article on eclipses will be published in *Nature* during 2009 as part of the *International Year of Astronomy* and a knol on 'eclipse' is on-line at `<knol.google.com>`

In spite of efforts on the part of many, most people do not understand that the so-called 'solar filters' or 'eclipse glasses' are only for the partial phases of total eclipses or for any part of annular eclipses of those eclipses that are never more than partial. It would be good if these objects could be titled 'partial-eclipse glasses'. It seems that many people miss the glory of the total solar eclipse because, even though they are in the zone of totality, they leave the glasses on. The PG prefers cards that have filters in them to glasses that fit over eyes and ears, in part because people are then not tempted to leave the cards on for the duration of the eclipse. Too many people worldwide mistakenly believe that there are additional radiations that come out of the sun during eclipses.

10. Commission 46 web site

After being hosted by Barrie W. Jones at The Open University (UK) for about 10 years, it was transferred to the Observatoire de Paris in the summer of 2007. It is now in the charge of `<chantal.balkowski@obspm.fr>` and the URL is `<iau46.obspm.fr/>`.

11. Astronomy for the Developing World. Strategic Plan 2010 - 2020

An informal brainstorm was held in Paris, 27-29 January 2008, chaired by IAU Vice-President George K. Miley, aimed to an IAU strategic decadal plan for astronomy development. Topics were achievements and visions for the future of relevant Commission 46 Program Groups, and coordination with complementary non-IAU programs.

12. Meetings

Two Special Sessions, proposed by members of Commission 46, were held at the XXVI IAU General Assembly in Prague, August 2006: SpS2 on *Innovation in teaching/learning astronomy methods*, chaired by Rosa M. Ros and Jay M. Pasachoff; and SpS5 on *Astronomy for the developing world*, chaired by John B. Hearnshaw.

Cosmos in the Classroom 2007, a hands-on symposium on teaching astronomy to non-science majors, was held 3-5 August 2007 at Pomona College, southern California, USA.

The meeting was sponsored by the Astronomical Society of the Pacific (ASP), with co-sponsorship by a range of astronomical and educational organizations.

JENAM-2007 was held in Yerevan, Armenia, 20–25 August 2007. Its SpS5 was devoted to *Astronomy Education in Europe*. The conveners were: Hayk Harutyunyan (Armenia), Areg Mickaelian (Armenia), and Magda Stavinschi (Romania). The main topics discussed were the international astronomy education and development, effective teaching and learning at the university level, development of astronomy curriculum for schools, education research, planetariums and science; amateur astronomers and astronomy education; and multicultural astronomy.

JENAM-2008 was held in Vienna, Austria, 8–11 September 2008. Its SpS on *Education and Communicating Astronomy in Europe*, in preparation for IYA2009, was co-chaired by Magda Stavinschi (chair IAU Commission 46) and Ian Robson (chair IAU Commission 55).

13. Astronomy Olympiads

Astronomy Olympiads are a new topic for Commission 46. Many national and some international Astronomy Olympiads are being organized in different countries at high school level. The interest for such contests is increasing and tries replacing the lack of the systematic astronomy education in high schools. Examples are:
- 12th *International Astronomy Olympiad*, 29 September – 7 October 2007, Simeiz, Crimea.
- 3rd *Asian-Pacific Astronomy Olympiad*, 21-29 November 2007, Xiamen, South China
- 1st *International Olympiad on Astronomy and Astrophysics* was held in Chiang Mai, Thailand, 30 November - 9 December 2007.

14. Closing remarks

As all scientific bodies of the IAU, Commission 46 is deeply involved in preparations for the International Year of Astronomy 2009, as well as in preparations for scientific meetings to be held during the IAU XXVII General Assembly, 3-14 August 2009, Rio de Janeiro, Brazil.

Magdalena G. Stavinschi
president of the Commission

Transactions IAU, Volume XXVIIA
Reports on Astronomy 2006–2009
Karel A. van der Hucht, ed.

© 2009 International Astronomical Union
doi:10.1017/S174392130802601X

DIVISION XII / COMMISSION 46 / PROGRAM GROUP
WORLD-WIDE DEVELOPMENT OF ASTRONOMY

CHAIR   John B. Hearnshaw
PAST CHAIR   Alan H. Batten
MEMBERS   A. Athem Alsabti, Julieta Fierro,
                   Edward F. Guinan (TAD), Yoshihide Kozai,
                   Hugo Levato, Hakim L. Malasan,
                   Laurence A. Marschall (TAD), Peter Martinez,
                   Jayant V. Narlikar, J. Pereira Osório,
                   Jay M. Pasachoff, D. Kala Perkins,
                   Jin Zhu (China)

TRIENNIAL REPORT 2006 - 2009

1. Introduction

The Program Group for *World-wide Development of Astronomy* (PG-WWDA) is one of nine Commission 46 program groups engaged with various aspects of astronomical education or development of astronomy education and research in the developing world. In the case of PG-WWDA, its goals are to promote astronomy education and research in the developing world through a variety of activities, including visiting astronomers in developing countries and interacting with them by way of giving encouragement and support.

The principal aims and objectives of PG-WWDA are: (*i*) to visit developing countries (often IAU non-member states) with some limited astronomical expertise, and which would welcome some development of their capabilities in astronomy; (*ii*) to give encouragement, and to explore the possible assistance of the IAU in developing astronomy in these countries; (*iii*) to discuss with astronomers in developing countries the available resources for astronomical teaching or research, and to promote international contacts and exchanges between astronomers in these countries and those elsewhere; (*iv*) to write reports on the state of astronomy in developing countries for the Commission 46 president and to send these reports to the IAU Executive Committee; and (*v*) if the conditions were deemed favourable, then to follow-up any report with involvement by TAD or other program groups of Commission 46, as may be appropriate.

2. Visit to Thailand and Laos

John Hearnshaw undertook an astronomical tour to Thailand and Laos, 14-31 January 2007, as part of the activities of the PG-WWDA.

Thailand joined the IAU as a national member in 2006. It is a strongly developing country, both economically and astronomically. The recent decision by the Thai government to establish the National Astronomical Research Institute of Thailand (NARIT) means that in the coming 5 to 10 years, Thailand can be expected to become a strong regional centre for astronomical research and education. By the end of 2009 a 2.4 m optical telescope should be installed on Doi Inthanon (2550 m), Thailand's highest mountain, near Chiang Mai in the north of the country. This will be equal to the largest optical telescope in Asia when it is completed. The new institute in Chiang Mai is directed by Professor Boonrucksar Soonthornthum, who hosted the visit to Chiang Mai.

Hearnshaw visited four universities in Thailand and gave a series of lectures. These were the universities of Chiang Mai, Naresuan (in Phitsanulok), Khon Kaen and Mahidol (in Bangkok). All of these employ astronomers in physics departments, and research interests are in optical stellar astronomy, cosmology and solar physics. Chiang Mai has the strongest involvement in terms of numbers, with several astronomers and a small observatory (Sirindhorn Observatory) which is operated just out of the city. Mahidol University has an active research group in solar physics and cosmic rays, headed by expatriate American, Prof. David Ruffolo.

Hearnshaw also visited the Physics Department at the National University of Laos in Vientiane, which is sited on a spacious campus in the north-east part of the city. His visit there was hosted by Dr Khamphouth Phomassone, a geophysicist and Assistant Dean of Science. Two Laotian astronomers with MSc degrees from Chiang Mai University in Thailand are employed to teach astronomy to physics students as part of the bachelors' program in physics at NUL. At this stage no graduate program in physics or astronomy exists, though one is planned in the next few years. The university would benefit from many more computers, and a small telescope would do wonders for the teaching of astronomy. Laos and Thailand share very similar languages, culture and ethnicity. But economically they are a long way apart. In spite of that, the biggest asset is, as in many developing countries, the students, who had a tremendous enthusiasm for astronomy and learning.

3. Visit to Bangladesh

Jayant Narlikar made a visit to Bangladesh, 12-20 March 2007. He visited Chittagong 13-19 March and Dhaka 19-20 March. In Chittagong, he attended meetings organized by Professor J.N. Islam, Director Emeritus, Research Centre for Mathematical and Physical Sciences (RCMPS), Chittagong University. He gave two lectures, the first being the Abdus Salam Memorial Lecture on 'The interaction between particle physics and cosmology'. The second was entitled 'Research, teaching and development programmes of the International Astronomical Union'. This presentation described the important role played by the IAU in encouraging teaching, research and development in astronomy all over the world, with a special emphasis on the developing countries. The advantages of becoming a member of the IAU, both nationally as well as individually, were outlined. In this context, a brief presentation about the Inter-University Centre for Astronomy and Astrophysics (IUCAA) at Pune, India was made.

A similar lecture was given in Dhaka to physicists and mathematicians from Dhaka University, with a similar response from the academic community there. There were present young students interested in astronomy and they were running a magazine on behalf of the Bangladeshi Astronomical Association. Narlikar had, in addition, extensive discussions with Professor M. Shamsher Ali, the President of the Bangladesh Academy of Sciences, which would be the adhering organization to the IAU, if Bangladesh decides to join. Professor Ali was already keen that Bangladesh should become a part of the IAU member nations. All in all, the mission was successful in the first instance of introducing the IAU to Bangladesh and secondly, by generating an interest amongst Bangladeshi scientists in the IAU.

4. Visit to Uruguay

Hugo Levato made a visit to the Department of Physics of the Universidad de la República in Uruguay, 16–21 April 2007. He gave one lecture there to students and scientists, and one lecture at the Planetarium in Montevideo City. He also visited the Los Molinos Astronomical Observatory. Uruguay is an IAU member country since August 1970. It has four individual members of the Union.

At the Universidad de la República (founded in 1849) he visited the Institute of Physics, where three astronomers with PhD degrees are working in astronomy. They are Dr. Julio Fernandez, Dr. Tabaré Gallardo and Dr. Gonzalo Tancredi. The University of the Republic is Uruguay's only public university. The number of university students is today around 75 000.

The Institute of Physics is part of the Science Faculty and is divided into three departments, one of which is Astronomy. Around ten people work in the Department of Astronomy. The Physics Institute has a PhD program in Physics that includes astronomy as a specialization.

The Los Molinos Observatory, belonging to the university, operates a 35 cm telescope used for astrometric studies of minor planets and NEOs and a Centurion Telescope, of 46 cm aperture (operational since 2002), used in the search for supernovae, and for observations of asteroids and comets. An SBIG camera (ST9) is on this telescope.

It is recommended that: (*i*) the IAU support the visit of visiting professors to Uruguay to help with the ongoing PhD programs; (*ii*) a small amount of money should help in improving the instrumentation at Los Molinos Astronomical Observatory; (*iii*) IAU Commission 46 should consider supporting the three astronomers with PhDs and some of the students for observing trips to the nearest major observing facility, Complejo Astronómico El Leoncito, in Argentina, to gather data for their programs; (*iv*) the IAU should help in providing access to the astronomical literature through the facilities available in Argentina (SECYT) or Brasil (CAPES). Access to the astronomical literature is a problem at present.

Uruguay has no site with the proper astronomical climate to develop a medium size astronomical facility, but internet connectivity is well developed and the usage of the Virtual Observatory concept should be encouraged as well as the usage of other international facilities located in South America.

5. Visit to Mozambique

Peter Martinez visited Mozambique for PG-WWDA 3-5 June 2008. On arrival he met with Mr. Claudio Moises Paulo of the University Eduardo Mondlane (UEM) and Mr. Faustino Armando Nhanombe of the Instituto Nacional de Meteorología (INAM) to discuss the programme of visits for the following two days.

On 4 June, Martinez visited the Department of Physics at UEM, and presented a two-part lecture on 'Astronomy and Space Science' and 'Space science and technology for sustainable development'. The first lecture provided an overview of our knowledge of the astronomy, from Earth outwards to the Hubble Deep Field. The second lecture was about the practical applications and benefits of space technology, drawing connections with astronomy where appropriate. The *International Year of Astronomy* was also briefly discussed. The audience comprised faculty and students from various departments at UEM. Afterwards, there was a comprehensive discussion in the faculty room in which Martinez had the chance to interact with senior staff members at UEM. He was then shown the satellite tracking telescope facility on the roof of the mathematics building.

The meetings at the University focussed on the Departments plans to introduce an astronomy stream in the undergraduate Physics programme. The Department established a meteorology stream in collaboration with INAM and would like to repeat this in the area of astronomy also. The Department has for a long time intended to introduce astronomy, but lacked the staff with experience to teach it. Mr. Claudio Moises Paulo, a junior lecturer in the Department, had been given leave of absence to pursue an astronomy MSc degree in South Africa. The intention was that, on his return to Mozambique, he would develop the undergraduate astronomy course. The Department aims to position itself as the focal point for astronomy in Mozambique.

Martinez also discussed the acquisition of a small telescope to act as a facility for student training and also for public outreach. The satellite tracking telescope is dated and in a very poor condition and would probably not be worth trying to refurbish for use as a general astronomy instrument. Acquisition of a modern computer-controlled telescope and some accessories, like a CCD camera, is recommended.

On the afternoon of 4 June, Martinez visited the Ministry of Science and Technology (MCT in Portuguese). He met there with Dr. Leon. The purpose was to pay a courtesy call to the government ministry responsible for funding science and technology activities in the country, to inform MCT about the IAU and its capacity building activities in astronomy, as well as to inform about the IYA in 2009.

The discussions at MCT focussed on the role of MCT in supporting activities at UEM and INAM. Dr. Leon emphasized that MCT is very interested to promote astronomy in Mozambique. In particular, the MCT looks to the University to provide experts to popularize astronomy; to support Mozambiques participation in the SKA bid, and to support the development of space applications in general. The MCT sees the IYA as a huge opportunity to raise awareness about astronomy and space science in the country and was fully prepared to incorporate astronomy in the calendar of outreach activities supported by the MCT.

On the morning of 5 June, Martinez visited the Instituto Nacional de Meteorología (INAM). He made two presentations to the staff on 'The South African Astronomical Observatory' and on 'The International Astronomical Union'. The latter presentation provided an overview of the IAU, but went into more depth on the activities of Commission 46.

6. Visit to Uzbekistan

John Hearnshaw visited the Ulugh Begh Astronomical Institute (UBAI) in Tashkent, Uzbekistan, 7-14 August 2008. Uzbekistan is not currently an IAU member, but with 11 individual IAU members and a well-established observatory at Mt Maidanak (belonging to the Ulugh Begh Institute) it is a non-member country with substantial astronomical infrastructure. His visit was hosted by Prof. Shuhrat Ehgamberdiev, the director of the Ulugh Beg Institute.

During his time in Uzbekistan, he gave three seminars at UBAI, (one of them was on IAU Commission 46 programs to support astronomical development and two were research seminars) and he visited the Samarkand State University where a 50-cm student telescope was recently installed. Further student telescopes are planned for Andijan (in the Fergana Valley) and Parkent (about one hour north-east of Tashkent).

Hearnshaw also visited the Mt Maidanak Observatory (2700 m) in the Pamir Mountains, where 1.5 m and 1.0 m telescopes are the principal instruments. Mt Maidanak is one of the world's best sites for image quality (0″.6 median seeing) and clear nights (over 2000 photometric hours annually). It is located some four hours drive south-east of Samarkand. The main need for Maidanak in the future is international collaboration in order to develop the site with modern instrumentation and a larger telescope. This need is especially acute since the departure of Russian astronomers in 1991; a recently signed collaboration agreement with Russian astronomers may see the return of Russian support for the Maidanak site in the future. Collaborations are already in place with Taiwan and the USA, and a future collaboration with South Korea is likely to take place.

Hearnshaw returned from Uzbekistan with an updated list of individuals who are interested in being nominated for IAU membership, most of them working at UBAI in Tashkent. In addition, there is a strong interest in Uzbekistan in IAU national membership, and the hope is that this can be negotiated in the near future. This is especially relevant, given the fact that about 50 professional astronomers work in Uzbekistan, mostly at UBAI, but also in at least five universities, and also because of the outstanding climatic conditions for optical and infrared astronomy in the country.

7. Visit to Peru and Ecuador

Hugo Levato made visits to both Peru and Ecuador in September 2008 on behalf of IAU Commission 46. In Peru (an IAU member country) he met with scientists and professors from the Universidad Mayor de San Marcos, the Instituto de Geofísica del Peru and the Comisión Nacional de Investigación y Desarrollo Aeroespacial (CONIDA), all of these being groups with astronomical activities.

He lectured at the Instituto Geofísico del Peru to students and scientists, and also gave a public lecture at the Biblioteca Nacional in Lima, where all the interested scientists were also invited. He was hosted in Lima by Mara Luisa Aguilar Hurtado one of the IAU members from Peru.

The Universidad Nacional Mayor de San Marcos (UNMSM) is the oldest university in Peru, originally founded in 1551. Today UNMSM has twenty faculties, among them the Faculty of Physics. The program for Physics studies may involve the selection of astronomy subjects, but there is no formal program of a PhD in astronomy. SPACE is a program inside the Faculty of Physics which gathers professionals working in astronomy and in outreach. They have the help and contribution of some researchers from abroad.

Levato met with Rafael Carlos Reyes who was the first PhD in Astronomy in Peru. He studied in Brazil and he is working part time at UNMSM and at the Universidad del Callao where there is also interest in developing some astronomy program.

In 1984 there was a program for visiting professors in Peru sponsored by the IAU and carried out at the San Marcos University. The heads and some of the staff members of the groups and institutions which have astronomy programs in Peru today were the products of that visiting professors' program.

The Comisión Nacional de Investigaciones y Desarrollo Aeroespacial (CONIDA) is the Space Agency of Peru and has a group working in Solar Physics under the leadership of Walter Guevara. He graduated in Physics and is a product of the former Program of

Visiting Professors sponsored by the IAU in the 1980s. CONIDA has a project under development for an astronomical observatory devoted to solar studies.

The Instituto Geofísico del Peru (IGP) has a good infrastructure for several areas like seismology and volcanism. It also has a group working in astronomy under the leadership of Dr Jose Ishitsuka. Hugo Trigoso is the other astronomer of the group. They have some instrumentation for solar observations in Huancayo and some dipolar antennas in Jicamarca and also several others facilities in different locations in Peru.

The main project is to convert an old antenna for satellite tracking and communication for centimeter radio-astronomy. Also they have some ambitious projects for outreach. They operate the most modern planetarium of Lima, a recent donation from Japan.

It seems clear that Peru has the conditions to boost its overall astronomical activity. The problems are the lack of human resources in astronomy, no coordinated program to train PhD students in astronomy, and a lack of coordination between the different groups existing in Peruvian astronomy. Support from IAU Commission 46 could help overcome some of these issues, though 5 to 10 years are probably required before there is a viable program to train students to PhD level.

In Ecuador, Levato visited the Escuela Politécnica Nacional (EPN) and the Observatorio Astronómico de Quito. It is noted that Ecuador is not a national member of the IAU. He gave one lecture at EPN. He was hosted by Ericson López, who is the Director of the Observatorio Astronómico and also a member of the staff of EPN. There are no additional PhDs in Ecuador with a training in astronomy.

EPN is one of twelve national universities in Ecuador. It has nine faculties, one in Science where courses of Physics and Mathematics are provided. It is not possible at present to study astronomy at a scientific level, but there is an astronomical facility at EPN with a Meade 16-inch telescope which was recently purchased and which can be accessed through the internet and used by students of high schools in an outreach program managed by EPN.

The Observatorio Astronómico de Quito was dedicated in 1873 and typical instruments from that time are still kept at the Observatory, like a Repsold meridian circle. Today the observatory is used practically only for outreach, and presently is being completely refurbished with money coming from the city of Quito. A good museum of old instruments, not only for astronomy, will be finished probably by the end of this year.

It is clear that poor human resources are the main problem in astronomy in Ecuador. The immediate need is to establish some courses on astronomical subjects so that students of physics can take these courses and be trained in astronomical research. This should be done at the beginning with the help of visitors from abroad.

8. PG-WWDA participation at Paris brainstorm meeting of Commission 46

John Hearnshaw, as the representative of PG-WWDA, took part in the IAU Commission 46 brainstorm meeting, which took place at the Institut d'Astrophysique in Paris 27-29 January 2008. This was the first step in the development of an IAU decadal strategic plan for global astronomy development from 2010-2020. The goal of the meeting was to exchange ideas and views about future IAU activities and strategies, with a focus on astronomy in developing countries. The meeting was called on the initiative of the IAU Executive Committee, and was chaired by Vice-President George K. Miley.

9. Closing remarks

PG-WWDA continues to have a successful program in a number of key developing countries. The strong development of astronomy in Thailand since it joined the IAU in 2006 is one positive story; the interest expressed in Bangladesh to progress to IAU membership in 2009 is also encouraging.

We will continue to focus on countries that are not IAU members but which have professional astronomers or some astronomical infrastructure (for example Uzbekistan), or on developing countries that need guidance in establishing core programs in astronomy education at tertiary level (such as Mozambique and Laos). In addition it is important not to overlook a number of IAU member countries that still need continuing support from the Union for their existing astronomical programs; Uruguay is a case in point.

Beyond the IAU XXVII General Assembly, 2009, much work still remains for PG-WWDA. There are many opportunities in Central Asia (for example Kazakhstan, Tajikistan), in Africa and in Latin America for closer involvement of astronomers and educators with the IAU. It is noted that our activities are constrained at present by the time that can be devoted to these activities by PG-WWDA members on a largely voluntary basis, more so than by financial constraints.

John B. Hearnshaw
chair of the Program Group

Transactions IAU, Volume XXVIIA
Reports on Astronomy 2006–2009
Karel A. van der Hucht, ed.

© 2009 International Astronomical Union
doi:10.1017/S1743921308026021

DIVISION XII / COMMISSION 46 / PROGRAM GROUP
TEACHING FOR ASTRONOMY DEVELOPMENT

CO-CHAIRS   Edward F. Guinan,
                    Laurence A. Marschall

ANNUAL REPORT 2008

1. Introduction

Annual reports of the IAU Commission 46 Program Group *Teaching for Astronomy Development* (TAD) for the years 2006 and 2007 have been published in IAU *Information Bulletin* 100, 47; and 101, 40. Here the 2008 report is presented.

The primary goal of TAD is to aid in "the enhancement of the country's astronomy education and astronomical research in support of education." The TAD program continues to vigorously support the development of Astronomy education, teaching and research in several countries. Since the time of the IAU General Assembly in August 2006, Larry Marschall (Gettysburg College) and Edward Guinan (Villanova University) have been co-chairs of the TAD Program. TAD programs supported during 2008 include the following countries: Columbia, Kazakhstan, Mongolia, Namibia, Nepal, DPR Korea, and the Philippines. Also discussed are preliminary plans for joint *International Year of Astronomy* (IYA2009) and TAD programs for Africa during 2009. Summaries of the TAD sponsored programs are provided here.

2. Kathmandu, Nepal, International School on Astronomy and Astrophysics

An International School on Astronomy and Astrophysics (ISAA) was organized in Nepal, 28 March – 6 April 2008, by the B. P. Koirala Memorial Planetarium, Observatory and Science Museum Development Board, Ministry of Environment, Science and Technology, Nepal, jointly with the International Astronomical Union in the framework of the TAD program, Commission 46. The Government of Nepal has established the B. P. Koirala Memorial Planetarium, Observatory & Science Museum Development Board in 1992 to establish Planetariums, Observatories and Science Museums in Nepal and conducted research activities in Astronomy, Astrophysics and Cosmology. Michèle Gerbaldi served as the IAU scientific advisor and resource person to the School.

The venue of this School follows preliminary contacts between Michle Gerbaldi and former Nepalese ISYA-participant (Thailand, ISYA-2001) Sanat Kumar Sharma who is presently Co-Executive Director of B.P. Koirala Memorial Planetarium, Observatory and Science Museum Development Board. This ISAA took place from March 28 - April 6, 2008, in Kathmandu, and was honored by the presence of Dr. Catherine Cesarsky, President of the International Astronomical Union upon the invitation of the Nepali organizers. This is not the first School in Astronomy in Nepal. In 2007, a Fifth

School on Astronomy and Astrophysics was organized (January 16–19, 2007) and targeted school teachers to give them a basic background on Astronomy & Astrophysics. The 2008 School was focused on high school teachers (grades 11 and 12) as well as university students in Master of Science programs who have taken cosmology as one of their elective courses. A total of 32 students attended the program – all but one was from Nepal.The announcement of that School and other information can be found at the URL <www.planeta-observatory.gov.np>. Many of the participating students presented poster papers or gave oral presentations focused on their Masters of Science level research projects.

The topics covered and the lecturers are listed below:
1. Infra-red astronomy – Catherine J. Cesarsky, ESO, Germany, CEA, France
2. Stellar Evolution – Michèle Gerbaldi, Institut d'Astrophysique, France
3. Neutrino Cosmology – Udayaraj Khanal, Tribhuvan University, Nepal
4. Astro-particles – Pierre Darriulat, Institut des sciences techniques nuclaire, Vietnam
5. Exo-Planets – Edward Guinan, Villanova University, USA
6. Early Universe and Particles – M. M. Aryal, Tribhuvan University, Nepal

Practical activities and laboratory exercises were also organized by Gerbaldi and Guinan. Each participant received a set of astronomical exercises (DVD and printed leaflet) from the ESO Educational Office. These exercises have been produced by the European Space Agency (ESA) and the European Organization for Astronomical Research in the Southern Hemisphere (ESO) for use in high school.

Long term programs of scientific development have been undertaken by the Nepalese Government. The first part of this development program was the installation of a 35 cm telescope, which is now operational. Also, a basic course in astronomy-astrophysics is under development at the university for Master Students. It should be noted that all the teachers in grades 11 and 12 classes must have a M.Sc. degree in Physics with Cosmology/Astrophysics as their most commonly taken elective courses.

Gerbaldi and Guinan also met with a group of B.Sc. and M.Sc. students from the Tribhuvan University who had not yet graduated and were too young to participate in this ISAA according to the selection criteria. A number of the most enthusiastic of these students organized a very active Nepal Amateur Astronomy Society and have been very active organizing star parties for children and the general public. They also publish regularly astronomical (once a month) notes in one of the Kathmandu Newspapers. Also Gerbaldi, at the request of this group, gave a talk on 24 April at the NAST (National Academy of Science and Technology). Gerbaldi also had a short meeting with the Rector of NAST to discuss the training of the school teachers in the domain of Astronomy/Astrophysics via distance learning. While in Nepal, Dr. Catherine Cesarsky discussed the advantages of the future membership of Nepal in the IAU and gave advice about how Nepal can join.

3. Mongolia

The First International Astronomical Summer School in the People's Republic of Mongolia was held in Ulaanbaatar, 21-26 July 2008 with the support of the National University of Mongolia (NUM), which hosted the event, and the IAU-TAD program. The Summer School was the result of several years of work in advanced. John B. Hearnshaw (PG-WWDA) visited Mongolia in 2004 and presented a report on astronomy there, followed in 2006 by Mongolia's joining the IAU at the Prague General Assembly in 2006. In May 2007, TAD sponsored a visit by Dr. Katrien Kolenberg to NUM, where she presented an intensive 2-week course on astronomy and cemented relationships with Mongolian scientists. Plans for the summer school followed directly from Dr. Kolenberg's 2007 visit, and involved, as well as Dr. Kolenberg, an organizing committee of Prof. Lhagvajav, Prof. Batsukh G., Dr. Tsolmon R. and Mrs. Dulmaa A., all of the NUM. A series of lectures over 5 days were planned, along with an excursion to the Ulaanbaatar Astronomical Observatory at Bogd Mount. Guest lecturers included (*i*) Prof. R. Jayawardhana (University of Toronto, Canada), who spoke on stars, star formation, extrasolar planets, and eclipses; (*ii*) Dr. K. Kolenberg (University of Vienna, Austria), who spoke on general astronomy, variable stars, and the Sun; (*iii*) Prof. K. Sekiguchi (National Observatory, Japan), who spoke on extragalactic astronomy and cosmology; and (*iv*) Prof. George K. Miley, vice-president of the IAU (Leiden University, Netherlands), who spoke on radio astronomy.

Lectures for the first three days of the school took place in the newly-built Japan-Mongolia Center neighboring the buildings of the NUM, which had very modern and convenient facilities. During the last two days the lectures took place in a very nice main lecture hall the NUM. All lectures were presented with laptops and LCD projectors, which seem to be standard methods, even in the developing world. Because few attendees were at ease understanding lectures in English, lectures were spoken in English and translated by Tuguldur Sukhbold, a Mongolian student who is currently in graduate school at the University of Arizona. Details of the program and images from the Summer School can be found at <www.astromongolia2008.org>.

About 50 people registered for and attended the school. Most were students at NUM, but there were also many school teachers and amateur astronomers. In addition, three astronomers from the Democratic People's Republic of Korea attended the school: Dr Nam, Dr. Kim and Dr. Jong, the new director of the Observatory of Pyongyang. (Expenses of the North Koreans were covered by the IAU). In addition, the lectures were attended by two students, one from Austria and one from Spain, who learned about the lectures by chance while traveling through Ulaanbaatar.

The summer school was well covered in the media, including visits by four Mongolian TV stations, who interviewed organizers and lectures for broadcast. In addition, Mongolian Parliament member Mrs. Dr. Munkhtuya Budee gave an inaugural greeting. Additional activities included a formal Mongolian welcome dinner, song and dance performances, a picnic to welcome IAU vice-president Dr. Miley, a closing dinner in a Mongolian Yurt – also known as a ger – and informal dinners with local hosts.

During the visit to the National Observatory at Khureltogoot, on the slopes of the Bogd Mountain, attendees (teachers and students) were welcomed by Prof. D. Batmunkh (National Academy of Sciences), who showed the visitors the observatory and explained equipment and data acquisition with the coronagraph. In addition, there is an amateur astronomical community in Mongolia, which numbers about 60-70 mainly young people, about 25 of whom are building their own telescopes. The amateurs organized a stargazing party on the 4th night of the summer school, which was attended by about 35 people, who viewed the sky through small telescopes. The summer school ended with the awarding of certificates of attendance and an informal discussion by Dr. Miley, during which students asked general questions on astronomers and on being an astronomer.

Following the close of the formal school, the four foreign astronomers who had lectured at the summer school, along with one foreign journalist, John Bohannon, traveled to western Mongolia, where, after many adventures, they viewed the August 1, 2008 total solar eclipse under clear skies. The eclipse trip resulted in several articles, including one in Science magazine: <www.sciencemag.org/cgi/content/full/321/5894/1297b>.

Outcomes: The school can be considered a success, as measured by the enthusiasm of attending students. When George Miley asked the students at the end of the week who wanted to become a professional astronomer, many students raised their hands. Currently, however, this would require study abroad, since only very introductory courses in astronomy are taught at the university. One of the main concerns of the students was that the salary of an astronomer is one of the lowest for Mongolian academics. Changes from higher up are needed in order to improve the professional status of astronomers in Mongolia. The support of Parliament member Dr. Mukhtuya Budee offers some cause for hope in this regard. Local astronomers work mainly in solar physics. Mongolia has excellent observing sites, comparable to those of the best professional observatories in Chile, but the only instrument regularly used is the solar coronagraph. A small telescope in a network could allow Mongolian scientists to get involved in international research projects, e.g. variable star light curves, with minimal expenditure and training. During his visit, Prof. Kaz Sekiguchi also talked to Mongolian scientists and academics about the installation of a telescope and planetarium offered by Japan. Installing larger telescopes in Mongolia presents problems with technical support; a high level of maintenance is required in a climate with such extreme temperature changes. Installing a radio telescope would also be a possibility.

Until several motivated students study abroad and bring back their knowledge, which is likely to happen in the near future, it is important to focus more attention on the teaching of astronomy in elementary school and high school. The warm response of these teachers to Dr. Miley's talks indicates that there is a very receptive audience for astronomy among Mongolian school teachers. Young astronomy teacher Dulmaa Altangerel (NUM) will organize a lecture series specifically for school teachers in November 2008 at NUM. She will base her lectures on the lecture notes and power points from the summer school.

Finally, it should be noted that Prof. Miley took advantage of the presence of the North Korean astronomers to discuss with them the status of astronomy in the DPRK, and their possible membership in the IAU. (See the report on the IAU/TAD program initiative for North Korea for more details.)

4. Philippines

During the last few years the TAD program has supported the travel of astronomers between Gunma Astronomical Observatory (GAO) in Japan and the visiting staff of PAGASA (Philippine Atmospheric, Geophysical, and Astronomical Administration) to give Philippine scientists experience in astronomical research and to attend lectures. Japan and GAO generously provide the training for foreign students as well as providing housing and sustenance during their visits to Japan. Several years ago Japan donated a GOTO 45 cm telescope to the Philippines. During March 2008 the TAD program supported the travel expenses of Dr. Hakim Malasan from nearby Indonesia to give an intensive 5 day tutorial on all aspects of modern CCD photometry to the PAGASA staff astronomers and students. The TAD program is delighted that Hakim Malasan had agreed to carry out this important work. The local contact person is Dr. Cynthia Celebre - chief, Astronomy Research and Development Section (AsRDS), PAGASA (<cynthia_celebre@hotmail.com>).

5. Kazakhstan: initial TAD program for August 2008

At the request of Dr. Emmanuil Ya. Vilkoviskiy, Director of Fesenkov Astrophysical Institute, Almaty, Kazakhstan, an TAD program was established to support the visit of

Dr. Marina Romanova and Prof. Richard Lovelace (Cornell University, USA) to Kazakhstan during 9-15 August, 2008. Romanova and Lovelace visited several institutions in Almaty, gave lectures and had discussions with local scientists. Historically, Kazakhstan has a strong scientific base primarily because of previous close ties with Russia in which many scientists studied at Moscow and other Russian Universities. The main Astronomy Institute, named after Prof. V. G. Fesenkov, was founded in 1941 and is well known. In recent years a new generation of astronomers was trained by the earlier generation of scientists. In addition, some young scientists go abroad for study, work and return, thus creating an advantage to the development of Astronomy in Kazakhstan.

Recently, an Astronomy Department has been created at the Al-Farabi Kazakh National University, so that Kazakh young people can study astronomy right in Almaty. In 2008 the government of Kazakhstan decided to unite three major astronomical institutions (Fesenkov Astrophysical Institute, Ionospheric Institute and Space Research Institute) to one "National Centrum of Space Researches and Technologies" (NCSRT), which is expected to enhance Kazakhstan's Space exploration program, though the process is not easy and demands special attention to support astronomical scientific researchers. There are several small observatories around Almaty which are used for observations of different astronomical objects, for training young scholars and also to show astronomy for the public.

The visit was very successful. Romanova and Lovelace gave four lectures at the Fesenkov Institute and the Al-Farabi State University, and visited the Tjan-Shan Observatory. Prof. Lovelace gave two lectures on "Modern American Telescopes" where he described a number of telescopes under construction. Dr. Romanova gave two lectures: "Modeling of Young Stars" and "Numerical Modeling in Astrophysics" in which she concentrated on modern numerical simulations and visualization procedures in theoretical astrophysics. Also they had many discussions about mutually interesting scientific problems and possible future collaborations. Dr. Vilkoviskiy expresses his sincere thanks to the IAU for this support of Kazakhstan Astronomy.

6. Democratic People's Republic of Korea

Some background: While attending the 29th ISYA in Malaysia (March 2007), discussions were held with Dr. Ri Jin Yong, the science and technology Counselor at the embassy of the Democratic People's Republic of Korea (DPR Korea) in Kuala Lumpur. These discussions were lead by Michèle Gerbaldi and Ed Guinan (ISYA chair and vice-chair of ISYA at that time).

Previously, Dr. Gerbaldi has established cordial contacts with Dr Yong; he has been very helpful in arranging for two young astronomers from Pyongyang Astronomical Observatory (PAO) to participate in the ISYA-2007 in Malaysia. At the ISYA, Gerbaldi and Guinan met the PAO astronomers to discuss the help that the IAU could provide to North Korean Astronomy. The PAO astronomers requested technical astronomy books and many of these have already been sent and received.

Also Dr. Karel A. van der Hucht, General Secretary of the IAU, has kindly agreed to send additional spare copies of the Cambridge University Press (CUP) IAU symposia publications and other available books to the PAO. The IAU also has agreed to add PAO to the list of libraries receiving complimentary publications from the IAU. Additional books were sent from France by Michèle Gerbaldi during 2007-08. The shipment of additional books is planned to continue. The Director of PAO, Dr. Sok Jong, extends his sincere thanks for the help provided by the IAU.

6.1. *Support for astronomers from Pyongyang Astronomical Observatory (PAO) to participate in the Pacific Rim Conference in Stellar Astrophysics (PRCSA) in Phuket, Thailand, May 2008*

As a follow-up to the cooperation between the IAU and North Korea with the Malaysia ISYA, and with the continued support of Dr. Ri Jin Yong, the TAD program sponsored two visits of PAO astronomers to regional IAU astronomy meetings during 2008. The TAD program sponsored two young astronomers from PAO to attend and participate in the 8th Pacific Rim Conference on Stellar Astrophysics (PRCSA-2008) that was held in Phuket, Thailand during 5-9 May 2008. The DPR Korea Councilor, Dr. Ri Jin Yong also attended the meeting. TAD Co-chair Ed Guinan and Dr. Yong worked together to organize the participation of the PAO astronomers. Support was provided for the DPRK astronomers to attend the conference and to meet with international astronomers and IAU representatives. This was a unique opportunity for the DPRK astronomers to meet with IAU Commission 46 program representatives including the chair of the ISYA program Dr. Jean-Pierre De Greve (Belgium), and the vice chair of the IAU Commission 46 *Exchange of Astronomers Program*, Dr. Kam-Ching Leung – co-organizer of the conference. Moreover, this venue was a good opportunity for the PAO astronomers to have fruitful discussions and exchanges of ideas with Dr. Boonrucksar Soonthornthun of the National Astronomical Research Institute of Thailand. The IAU appreciates the great help that Dr. Soonthornthun provided in expediting the visas for the PAO astronomers to attend the PRCSA also for generously waiving the conference registration fees for the attendees from North Korea.

The PRCSA-2008 meeting was an excellent opportunity for more extended discussions with the DPRK astronomers and government representative about the DPRK renewing its membership in the IAU and participating in IAU sponsored programs such as TAD and ISYA. As a follow-up, it was proposed that three PAO astronomers, including the new PAO Director, attend the TAD sponsored Astronomy School in Ulaanbaatar, Mongolia during late July. Dr. Karel van der Hucht and IAU Vice president Dr.George Miley kindly agreed to support this program as did the DPR- Korea and the organizers of the Astronomy School in Mongolia.

6.2. *Participation of PAO astronomers in the Mongolia Astronomy School, in July 2008*

With the continued strong support of Dr.Ri Jin Yong, the TAD program supported the participation of three PAO astronomers in the IAU-sponsored Mongolia Astronomy School held in nearby Ulaanbaatar, Mongolia. One of the main reasons for the choice of this venue was that visas are not required for the North Korean astronomers to travel to Mongolia via China. Also, the new director of PAO, Professor Jong Sok, was able to attend. In addition, PAO staff astronomers, Drs. Kim Mun-Song and Nam Sok-Chon also attended. The PAO astronomers held fruitful discussions with Mongolian astronomers and the IAU-sponsored lecturers and School organizers Drs. Katrien Kolenberg and Tsolmon Retchin.

6.3. *International collaboration*

Dr.George Miley, IAU Vice-President and Coordinator of the IAU C-46 Education Programs, attended to the School to meet with the Mongolian and PAO astronomers and students. Dr. George Miley had substantive discussions with Dr. Sok and the PAO astronomers. Three main topics were covered - (*i*) the present situation of astronomy in the DPRK; (*ii*) future needs and plans of the PAO and (*iii*) the possibility of the DPR Korea rejoining the IAU. Also discussed were the possibilities of supporting visits from the DPR Korea to China. This would be a cost effective program since PAO has an agreement with NAO China in Beijing. However, the DPR Korea would need to rejoin the IAU before consideration could be given for funding of this type program.

Dr. Sok and PAO astronomers discussed other several additional aid priorities with Dr. Miley. These include: broadening the interests of the PAO into new areas, e.g., stellar astrophysics. Acquiring a small solar telescope and acquiring an adequate library of text books and access to modern journals. Also discussed was a proposal for the organization of a national astronomy school for interested physics students and school teachers, similar to the TAD school in Mongolia and hosting lectures in the DPR Korea by IAU sponsored astronomers. Dr. Miley recommended, on the longer term (next few years), the PAO should consider writing a strategic plan for the development of DPR Korea astronomy. Such a plan could help in acquiring outside funds. Dr. Miley offered to assist with the preparation of such a plan. Although help from outside for the building up DPR Korea astronomy is possible, significant internal support would also be required. Several sources of external support for the development of world astronomy require internal matching funds.

Representing North Korean Astronomy, Dr. Sok indicated that the DPR Korea astronomers wish to join the IAU as soon as possible. However, the annual dues would be a substantial burden for the PAO at this time, even those for interim membership. Dr. Sok already has had talks with the DPR Korea Academy about the matter. In that regard Dr. Sok agreed to inform the Academy in the DPR Korea about this meeting and will raise with them the matter of facilitating DPR Korea membership of the IAU as soon as possible. Also Dr. Miley agreed to explore possible European sources for funding visits by DPR Korea astronomers and investigate the possibility for future support by the IAU Commission programs.

7. Columbia. Support for student participation in the Columbian Congress of Astronomy & Astrophysics Meeting in Medellin

The TAD program assisted in supporting Columbian students and to attend and participate in the Colombian Congress of Astronomy and Astrophysics (COCOA) held in Medellin, Colombia during August 2008. The purpose of the meeting was to convene for the first time professional astronomers and astronomy students from Columbia to meet with Astronomy and Astrophysics researchers from the USA, Europe and South America to exchange ideas and establish collaborations. The meeting consisted of over a dozen invited speakers, half of which were from foreign institutions, covering a broad range of topics that include star and planet formation, astronomical instrumentation, galaxy formation and solar system astronomy. The long range goal of this meeting is the development of a comprehensive roadmap for the development of sustainable Astronomy and Astrophysics education and research programs for Columbia. TAD support was requested by Juan Rafael Martinez Galarza (PhD student at Leiden Observatory). More information about the Congress can be found the web site <urania.udea.edu.co/cocoa2008/index.php>.

8. Islamic Republic of Iran. New initiatives

TAD co-chair Guinan was invited to visit Iran after the 2008 ISYA in Istanbul, to discuss future cooperation with the IAU. He visited Iran from 22-29 July 2008. The invitation was extended by Dr. Yousef Sobouti, Director and President of the Institute for

Advanced Studies in Basic Sciences (IASBS), Zanjan. During this visit Guinan represented the TAD program and exchanged ideas for developing Astronomy education and training of Iranian students. In particular he discussed areas in which the TAD program can assist. In Tehran he met with Prof. Reza Mansouri, Head of the Astronomy School at the Institute for Theoretical Physics and Mathematics (IPM: <www.ipm.ac.ir>), and also met with Drs. Habib Khososhahi and Sepehr Arbabi. They discussed the development of the Iranian National Observatory (INO) and possible assistance and advice for this project from the IAU.

Guinan visited IASBS in Zanjan for three days and met with Dr. Sobouti and other members of the institute and held discussions about the development of Astronomy education and training in Iran. Dr. Sobouti and Guinan discussed possible future IAU cooperation and assistance for astronomy education and training of Iranian students within the framework of the TAD Program. As a result of these discussions a formal proposal was made by Dr. Sobouti and the IASBS to hold a one week TAD/IAU supported Astronomy and Astrophysics School at the IASBS during the spring 2009. This School would focus on "Recent Advances in Observational Astrophysics" and is expected be attended by 35-40 students from all over Iran. This Iran Astronomy School would be aimed at advanced undergraduates and graduate MSc students as well as new PhD students and thus would be similar to TAD supported schools held recently in Hanoi, Nepal and Mongolia. If this school is approved, the IAU would provide partial support for outside lecturers while the IASBS would provide the support for the housing and living costs of all attendees. It is hoped that this School could be continued annually during the next 2-3 years. While at the IASBS, Guinan presented an Astrophysics seminar and also met with graduate students and post-doctoral students to give advice and to discuss research projects and career opportunities.

9. Sub-Saharan Africa. Possible initiatives

The TAD program this year took steps to carry out major initiatives in sub-Saharan Africa during the IYA-2009, in the spirit of the Cornerstone 11 mandate for global development of Astronomy. The TAD program advisors communicated with Kevin Govender - the Manager of the SALT Collateral Benefits Program of the South African Astronomical Observatory (Cape Town), and coordinator of various astronomy development efforts. He has established contacts in several sub-Saharan countries and serves as an advisor to the TAD program for Africa. Kevin Govender also leads a working group on Astronomy in the region (currently focused on the IYA) which has put together a plan for the development of Astronomy in Africa themed "Astronomy for Education".

Govender also has arranged for representation at the International Council for Science (ICSU) General Assembly in Maputo, Mozambique in October 2008 and leads a survey of the state of Astronomy in each of the African countries they are working with. Also he is attending the *International Heliophysical Year* (IHY) meeting in Nigeria during November where, among other things, he will gauge interest in organizing a sub-Saharan astronomy education meeting. At the IHY meeting he will run training workshops on basic astronomy outreach which have proven extremely popular so far (based on feedback from similar workshops in Kenya, Namibia and a number of locations across South Africa). A consolidation of astronomy education resources mainly from South Africa have already been compiled and sent to coordinators in a number of African countries.

In response to requests from Kevin Govender, and with the help of TAD co-chair Larry Marshall, the TAD arranged to have instructional astronomy materials from Project

CLEA (Gettysburg College) sent for distribution to interested astronomy teachers at regional meetings. The TAD program also is looking forward to working closely with Dr. Peter Martinez (SAAO.) in organizing TAD programs in the region. Dr. Martinez is very knowledgeable of Astronomy in Africa and has worked over the years helping to develop programs. One plan of action being considered is an TAD sponsored one week Astronomy School to be held in the region during the IYA-2009. This school would target undergraduate Physics and Astronomy students and high school science teachers as well as Mathematics and Science outreach programs such as science centers and promotional activities by Departments of Science and Technology.

10. North Africa and Angola. Possible initiatives

The TAD program continues to work closely with Dr. Hassane Darhmaoui, Al Akhawayn University in Ifrane, for the development of Astronomy education, teaching and outreach programs in Morocco. During 2008 ongoing discussions and exchanges have been carried out with Dr. Khalil Chamcham (University of Oxford) with the help of Dr. Michèle Gerbaldi about a possible TAD program for Morocco. Plans for a five day TAD Astronomy School that will be primarily focused on the training of Moroccan high school science teachers (but also open to university students) are being developed for 2009.

Discussions with Dr. Roger Ferlet (IAP) have been carried out about possible help and advice from the TAD program for the establishment of the first "Ecole doctorale" in Astronomy (students who will begin PhD studies) in Algeria. Also Dr. Ferlet may be requesting IAU help for an Annual National Astronomy Festival for amateurs and others interested in Astronomy. Also in Tunisia there is an interest in working with the TAD program about the training of high school mathematics teachers in Astronomy. With French assistance, Dr. Ferlet has also expressed advice and help in the promotion of elementary Astronomy courses for high school students in Tunisia. He also reported on the establishment of an Astronomical observatory in Angola that could use some possible TAD aid and advice for related public outreach Astronomy programs in that country. Many of these programs in 2009 would be closely tied to IYA2009.

Edward F. Guinan & Laurence A. Marschall
co-chairs of the Program Group

Transactions IAU, Volume XXVIIA
Reports on Astronomy 2006–2009
Karel A. van der Hucht, ed.

© 2009 International Astronomical Union
doi:10.1017/S1743921308026033

DIVISION XII / COMMISSION 46 / PROGRAM GROUP
INTERNATIONAL SCHOOLS FOR YOUNG ASTRONOMERS

CHAIR	**Jean-Pierre De Greve**
VICE-CHAIR	**Kam-Ching Leung**
PAST CHAIR	**Michèle Gerbaldi**

TRIENNIAL REPORT 2006 - 2009

1. Introduction

The IAU Commission 46 Program Group *International Schools for Young Astronomers* (ISYA) was created in 1967 (Gerbaldi 2008). During the period 2006 till August 2008 two ISYAs took place, one in Malaysia and a second one in Turkey.

2. 29th ISYA, Malaysia, 2007

The 29th International School for Young Astronomers (ISYA) was held in Malaysia (Selangor and Langkawi Island), 5-23 March 2007. ISYA 2007 was organized by the IAU with Universiti Kebangsaan Malaysia (UKM) and with the cooperation of National Space Agency of Malaysia (ANGKASA), Ministry of Science, Technology and Innovation (MOSTI) and the Universiti Malaya (UM).

The School was held at the University Kebangsaan Malaysia (UKM), Selangor, 5-9 March 2007, and subsequently at the MARA Junior Science College, Langkawi Island, 10-23 March 2007, where the National Observatory is located.

2.1. *National Organizing Committee*

The members of the Organizing Committee in Malaysia were:
- Prof. Dr. Mazlan Othman, Director General ANGKASA, Secretary General of Academy Sciences of Malaysia, Chairman of National Committee on organizing ISYA2007
- Prof. Dr. Mohd Zambri Zainuddin, Head Space Science Laboratory and Deputy Dean of Malaya University
- Prof. Dr. Baharudin Yatim, Director Space Science Institut, National University of Malaysia
- Mr. Kassim Bahali, Head Astronomy programme, Al-Khawarizmi Observatory, Malacca
- Mr. Mhd Fairos Asillam, Science Officer ANGKASA, Secretary National Organizing Committee ISYA2007.

2.2. *ISYA 2007 participants*

The number of selected participants were 38 (female 9, male 29): 10 Malaysian and 28 foreigners from 11 countries: China (2), DPR Korea (3), India (3), Indonesia (7), Nepal (1), New Zealand (1), Philippines (4), Shri Lanka (1), Taiwan (1), Thailand (3) and Vietnam (2). The students had very mixed academic backgrounds and experience ranging from a MSc degree to having finishing their PhD.

2.3. *ISYA academic programme*

The faculty members and lectures were:

- Dr. Chenzhou Cui, China , National Astronomical Observatory, *Virtual Observatory*
- Prof. Jean-Pierre De Greve, Belgium, Brussels University, *Binary star evolution with massive components*
- Assoc. Prof Mamoru Doi, Japan, University of Tokyo, *Galaxies*
- Prof. Michele Gerbaldi, France, Institut d'Astrophysique de Paris, *Stellar atmosphere*
- Prof. Edward Guinan, USA, Villanova University, *Binary stars*
- Prof. K.R. Lang, USA, Tufts University, *The Sun*
- Dr. Hakim L. Malasan, Indonesia, Institut Teknologi Bandung, *Stellar observations*
- Assoc. Prof Mark Rast, USA, University of Colorado, *Astrophysics of the Sun*
- Prof. N. Udaya Shankar, India, Raman Research Institute, *Radioastronomy*
- Prof. Mohd Zambri Zainuddin, Malaysia, Universiti Malaya, *Astronomy*

The ISYA students were encouraged to describe their current research. Six sessions were organized for the 27 talks given, each talk was of 15 minutes plus 5 minutes for the questions. Emphasize has been put on the Virtual Observatory and data base concepts by Dr. Chenzhou Cui. For that purpose a network of 22 computers under LINUX/Windows was set up and image processing software installed. The practical classes were based on access to real data.

The participants also had to conduct optical observations. Six observing sessions took place at the Langkawi National Observatory for imaging and spectroscopy with the robotic 0.5 m telescope. These observing sessions were directed by Dr. Hakim L. Malasan with M. Ridwan Hidaya, Mrs. Lau Chen Chen and M. Karzaman Ahmad. Reduction of the images and spectra taken were done under the guidance of Dr. Chenzhou Cui and Dr. Hakim L. Malasan. One half-a-day session was devoted to solar observations at the Langkawi Solar Observatory. Solar images with Hα and Ca II-K-line filters were taken under the direction of Assoc. Prof. Mark Rast and Prof. Edward Guinan.

Profs. J.-P. De Greve and Edward Guinan provided information and advice on:
- how to present results and to give a talk
- how to apply to PhD programme, etc.
- how to apply for jobs and writing curricula vitae / resumes, etc.
- how to write science proposals, etc.

2.4. *Closing remarks*

Besides the academic programme of the ISYA, the lecturers gave several conferences at the National Science Museum and National Planetarium, at Kuala Lumpur, at the MARA Science College, Langkawi, and at the Terengganu University.

This ISYA is the starting point for the development of new programmes in Malaysia, in particular related to the use of the Langkawi National Observatory in both stellar and solar domain. Further developments are considered in the framework of the TAD programme.

3. 30th ISYA, Turkey, 2008

The 30th International School for Young Astronomers (ISYA 2008) was held in Istanbul, Turkey, 1-21 July 2008 It was organized with the Turkish Astronomical Society (TAD) and with the cooperation of the Yüzyil Isil School (YIS), a private primary and high school near Sariyer, Istanbul, the Istabul University (IU), and the Istanbul Kültür University. The lectures were held at the YIS (some 30 km from the center of Istanbul).

3.1. *National Organizing Committee*

The members of the organizing committee in Turkey were:
 - Prof. Ali Alpar, Turkish Astronomical Society and Sabanci University
 - Prof. Zeki Aslan, Istanbul Kültür University
 - Prof. Prof. Dursun Koer, Istanbul Kültür University
 - Sinan Alis, Istanbul University, secretary

3.2. *ISYA Participants*

Out of 46 candidates, 37 were selected for participation. Five of them didn't come (2 from Egypt, 1 from Nigeria, 2 from Turkey), two additional students were added later on, bringing the actual number of participants to 35, from 12 different countries from the region. The gender distribution was 40% female, 60% male. The students had very mixed academic backgrounds and experience ranging from a MSc degree to having finished their PhD.

3.3. *ISYA academic programme*

The faculty members and lectures were:
 - Prof. Ali Alpar, Turkey , Cibanci University, *Endpoints of stellar evolution, structure of neutron stars, observations of neutron stars*
 - Prof. Zeki Aslan, Turkey, Istanbul Kültür University, *Introduction to astrometry, CCD astrometry*
 - Prof. Jean-Pierre De Greve, Belgium, Brussels University, *Structure and evolution of stars, Structure and evolution of Binaries*
 - Dr. Antonio Frasca, Italy, Catania Observatory, *Stellar spectroscopy and data analysis with IRAF, spectroscopy of binaries*
 - Prof. Ed Guinan, USA, Villanova University, *Binary stars as astrophysical labs, extrasolar planets and life, stellar dynamos and effects on hosted planets, research with small telescopes*, and *Binary stars*
 - Prof. Kam-Ching Leung, USA, Nebraska University, *Photometry, novae, close binary observations*
 - Prof. Robert Williams, USA, Space Telescope Science Institute, *Hubble Space Telescope science, Hubble Deep Fields, novae*
 - Dr. Sinan K. Yerli, Turkey, Middle East Technical University, *CCD reduction techniques using IRAF with hands-on computer training on real/archive data*

In the second week the students had to give the topics for their presentations and a schedule was developed within the slots foreseen in the third week. Most of the students consulted Ed Guinan and/or Jean-Pierre De Greve for the content of their powerpoint presentation. The presentations took place on 5 slots of 1.5 hour from Tuesday till Saturday. For each student 15 minutes were foreseen for the presentation and the discussion. The lecturers agreed that the presentations were well prepared and of high quality.

Attention in the program was devoted to the development of observing competencies. Thus, an important aspect of the first week was the preparation of the observations to be carried out in the second week. With the support of Dr. Paul Roche, ISYA had access to the 2 m robotic Faulkes telescope on Maoui, Hawaii. Three slots of 2 hours were offered from 3.30 pm to 5.30 pm (Istanbul time) on Monday, Wednesday and Friday of the second week (through the appreciated support of Paul Roche). In the first week the students were asked to organize themselves into teams and to develop and propose a feasible observing project. In the second week slots of ten minutes were given to each team within the two-hour slots. In the lectures the students gained insight in the IRAF software, to be able to carry out proper reductions of the data. In the third week, students could reduce their data.

Profs. J.-P. De Greve and Edward Guinan provided information and advice on:
- how to present results and to give a talk
- how to apply to PhD programme, etc.
- how to apply for jobs and writing curricula vitae / resumes, etc.
- how to write a scientific paper and getting it published.

Besides the academic programme of the ISYA, Jean-Pierre De Greve assisted to a Saturday morning television programme on one of the Turkish television channels.

The cultural programme for the students consisted of a two trips to the center of Istanbul, to visit the Aya Sofia , the Imperial Palace, a visit to the Sultan's Palace, the Cistern and the Blue Mosque, and a sightseeing boat trip through the Bosphorus. A third visit and fourth visit offered a day at the grand bazaar and a visit to Istanbul's modern shopping area.

3.4. *Closing remarks*

During the ISYA in Turkey the young participants discussed future plans among each other and with the lecturers, and developed an intercultural network of young astronomers to help or advise each other in the future.

4. News Release - IAU0803: The International Astronomical Union teams up with the Norwegian Academy of Science and the Kavli Prize

June 5, 2008, Oslo

In an important move, the Norwegian Academy of Science and Letters and the International Astronomical Union have joined forces in support of both the Kavli Prize and the IAU International Schools for Young Astronomers.

The International Astronomical Union (IAU) and the Norwegian Academy of Sciences and Letters (NASL) have united in support of the Kavli Prize and the IAU International Schools for Young Astronomers (ISYA) after finding common ground in promoting astrophysics research and training young scientists in the latest astronomical research techniques.

The Kavli Prize Committee is composed of distinguished scientist based on recommendations from leading international academies and other equivalent scientific organizations. Beginning in 2010, the Norwegian Academy will seek advice by the International Astronomical Union in order to establish a balanced prize committee with respect to the various fields of Astrophysics. They will work together on publicising the events. From 2009 NASL will donate a yearly sum to the IAU ISYA programme, allowing the programme to expand. Furthermore, Kavli Prize winners will be invited to speak at the ISYA alongside the usual experts. The agreement was signed by the President of NASL, Prof. Ole Didrik Laerum, and the President of the IAU, Dr. Catherine J. Cesarsky.

The Kavli Prize, which is given in the areas of astrophysics, neuroscience and nanoscience, encourages future research directly through incentive by providing winners with $1,000,000, a medal and a diploma. Every two years the Kavli Prize in Astrophysics is awarded to one or more scientists who have demonstrated outstanding achievement in advancing human knowledge and understanding of the origin, evolution, and properties of the vast Universe. The Prize is open to a wide range of fields, including cosmology, astrophysics, astronomy, planetary science, solar physics, space science, astrobiology, astronomical and astrophysical instrumentation and particle astrophysics.

The first Kavli Prize in Astrophysics was awarded jointly to Maarten Schmidt, of the California Institute of Technology in the United States, and to Donald Lynden-Bell, of Cambridge University in the United Kingdom, on 28 May 2008.

Since 1967, the IAU ISYA programme has promoted the development of professional astronomers around the world by facilitating interaction between young scholars and established professionals. It also provides them with the practical knowledge essential to access the data archives that have become the basis for astronomical research. Individuals with a Masters degree in science can participate in the 21-day conference, which occurs at a different location each year to perpetuate international cooperation in astronomical research. Participants do everything from listen to lectures given by experts to becoming involved through hands-on practicals and computer sessions where they learn how to use essential astronomical tools.

This year the ISYA meeting will take place in Istanbul, Turkey in July. A diverse array of topics will be covered, from the structure of neutron stars to a discussion of the science completed using the *Hubble Space Telescope*.

– End of Press Release –

Jean-Pierre De Greve & Michèle Gerbaldi
chair and past-chair of the Program Group

References

Gerbaldi, M. 2008, in: K. A. van der Hucht (ed.), *Transactions IAU* XXVIB (Cambridge: CUP), p. 238

Transactions IAU, Volume XXVIIA
Reports on Astronomy 2006–2009
Karel A. van der Hucht, ed.

ⓒ 2009 International Astronomical Union
doi:10.1017/S1743921308026045

DIVISION XII / COMMISSION 46 / PROGRAM GROUP COLLABORATIVE PROGRAMMES

CHAIR   Hans J. Haubold

MEMBERS   Johannes Andersen, Christopher J. Corbally,
David L. Crawford, Julieta Fierro,
Arak M. Mathai, Margarita Metaxa,
Dale Smith, James C. White,
A. Peter Willmore

TRIENNIAL REPORT 2006 - 2009

1. Introduction

The working group is pursuing activities co-organized and/or co-sponsored by UN, ESA, NASA, JAXA, UNESCO, COSPAR, IAU and others for the world-wide development of basic space science.

2. COSPAR Workshops for Capacity Building

The IAU is pleased to have been co-sponsor of the following COSPAR Capacity Building Workshops.

2.1. *Sinaia, Romania, 4–16 June 2007. Solar-Terrestrial Interactions (STIINTE)*

This Workshop, organized by Joachim Vogt and Octav Marghitu, was dedicated to the analysis of data from multi-satellite space missions such as *Cluster*. It was attended by 24 very motivated PhD students and post-docs coming from Central and Eastern Europe: Romania, Hungary, Bulgaria, Czech Republic, Poland, Ukraine, Russia, Armenia, and Georgia. The scientific programme focused on various aspects of multi-satellite missions, ranging from data analysis and instrument design to kinetic modeling, the analysis of boundaries, and the analysis of auroral processes. After a first and already quite intensive week, the work culminated in the preparation of scientific projects by five teams. Each team had to address a specific problem in magnetospheric physics, leading to very intensive team-work and providing an excellent opportunity for the students to interact, put together their competences and apply what had been learned. On the last day of the school, each team defended its project to a panel of senior scientists who were participating at the simultaneous *Solar Terrestrial Interactions from Microscales to Global Models* (STIMM-2) meeting. The Workshop took place in the framework of the *Plan for European Cooperating State* (PECS) agreement between Romania and ESA, which is designed to prepare for full Romanian membership of ESA.

2.2. *Montevideo, Uruguay, 23 July – 3 August 2007. Planetary Science*

In recent years, the community of scientists working in planetary sciences in Latin America has been growing. Collectively, it was decided to pursue a new step in the academic development of this community by promoting the use of data from planetary space missions in its every day research. This Workshop, organized by Gonzalo Tancredi, the first in the series related to planets and small bodies of the Solar System and devoted to the use of planetary missions data bases, was designed to meet this objective. It focused on missions to comets, asteroids, satellites, rings and planets, like Mars. The participants worked for two weeks with real data from the missions and ideally started a research project with that data that they could continue after returning to their home institute.

2.3. *Alexandria, Egypt, 19 January – 1 February 2008.*
X-ray Astronomy using Chandra, XMM/Newton and Swift

This Workshop, organized by Alaa Ibrahim and Peter Willmore, took place in the prestigious new library of Alexandria, the first activity of this nature to take place in the region. The workshop activities were attended by the science minister, Prof. Hany Helal, the deputy minister Prof. Maged El Sherbiney and the president of Egyptian Academy of Science Prof. Tarek Hussein who chaired the opening session and the closing ceremonies, reflecting a recent decision to strengthen astronomy in Egypt. 31 participants from Egypt, Morocco, Turkey, Nigeria, South Africa, Ukraine, and India were admitted to the full programme while 5 more from local institutions were offered two-day participation. There was a public outreach day during which 100 high school students from Cairo and Alexandria attended a planetarium show and three public lectures. The program consisted of morning lectures and afternoon practical computer sessions. After the first few days the students started to work on research projects and they presented their results in a poster session at the end.

2.4. *Kuala Lumpur, Malaysia, 1–14 June 2008. Space Optical and UV Astronomy*

The programme of the Workshop, organized by Martin Barstow, included lectures on a range of topics in UV and optical astronomy, on-line data products and data processing pipelines, data reduction and analysis and analysis software, data archives, search for targets and display quick-look data for inspection purposes. Participants carried out a major project and also produced a draft guest observer proposal. A total of 30 student participants and 8 lecturers took part. The former included 9 from Malaysia, 9 from India, 4 from China, 3 from Indonesia 2 from Iran, 2 from Vietnam and 1 from Pakistan. One of the IAUs *International Schools for Young Astronomers* was held in Malaysia in 2007. The COSPAR workshop was intended to build on this as part of a general plan to strengthen Malaysian astronomy. A number of the Malaysian participants were in the early stages of research degree programmes, so their supervisors took part in the formulation of their project topics with the workshop lecturers and in discussion of possible follow-up activities after the workshop.

3. International Heliophysical Year 2007: astrophysics and space science

In 2007, a number of major anniversaries occurred, among them the 50th anniversary of the International Geophysical Year (IGY 1957), the launch of Sputnik 1, and the 50th session of the United Nations Committee on the Peaceful Uses of Outer Space (UNCOPUOS). Particularly IHY 2007 is an opportunity to: (*i*) advance the understanding of the fundamental heliophysical processes that govern the Sun, Earth, and heliosphere; (*ii*) continue the tradition of international research and advancing the legacy of IGY 1957; and (*iii*) demonstrate the beauty, relevance and significance of space and Earth science to the world (`<ihy2007.org>`, `<www.unoosa.org/oosa/en/SAP/bss/ihy2007/index.html>`).

In preparation of IHY 2007, UNOOSA, in cooperation with ESA, NASA, JAXA, COSPAR, IAU, and the IHY Secretariat, held international Workshops in:
- United Arab Emirates in 2005,
- India in 2006 www.iiap.res.in/ihy/>
- Japan in 2007 <solarwww.mtk.nao.ac.jp/UNBSS_Tokyo07/>
- Bulgaria in 2008 <www.stil.bas.bg/UNBSS-IHY/>

The starting date of IHY 2007 was 19 February 2007. On that date, during the session of the Scientific and Technical Subcommittee of UNCOPUOS, the IHY kick-off included an IHY exhibit, press briefing, and an opening ceremony in the United Nations Office Vienna (<www.lesia.obspm.fr/IHY/kickOFF/index.html>). IHY regional coordinators, Steering Committee members, and Advisory Committee members participated in the IHY kick-off event. The Austrian Academy of Sciences hosted a one-day symposium on IHY 2007 in Vienna on 20 February 2007.

The 2009 Workshop will be hosted by South Korea (<ihy.kasi.re.kr/meeting.php>). This workshop will also cover thematic areas as pursued by the *International Year of Astronomy 2009* (<www.astronomy2009.org/index.php/?option=com_content&view=article&id=112>).

Results on all above workshops have been made available, free of charge, on websites, in proceedings, UN documents, and reports in the international literature.

4. Centre for Mathematical Sciences, India: space science and mathematics

The Working Group is pursuing activities co-organized and/or co-sponsored by India and other nations for holding schools on the application of mathematical methods to astrophysics and space science. Such Schools have been organized since 1995 at the Centre for Mathematical Sciences, India. In 2006 and 2007, the cycle of five SERC Schools was successfully concluded. In 2009, a new cycle of five SERC Schools will start providing methods of mathematics applied to astrophysics and space science.

Centre for Mathematical Sciences (CMS) was established in 1977 and registered in Trivandrum, Kerala, India as a non-profit scientific society and a research and training centre covering all aspects of Mathematics, Statistics, Mathematical Physics, Computer and Information Sciences. Since 1977 CMS had executed a large number of research and training projects for various central and State governmental agencies.

CMS has a Publications Series (books, proceedings, collections of research papers, lecture notes etc; current number is 36), a Newsletter of two issues per year (current number is Volume 11, Number 1, 2008), a Modules Series (self-study books on basic topics; current number is 6) and a Mathematical Sciences for the General Public Series (current number is 2). The latest book from CMS is: A. M. Mathai & H. J. Haubold (2008), *Special Functions for Applied Scientists* (New York: Springer).

In 2002 CMS Pala Campus was established in a one-floor finished building donated to CMS by the Diocese of Palai in Kerala, India. In 2006, Hill Area Campus of CMS was established. The office, CMS library and most of the facilities are at CMS Pala Campus. The other campuses, namely, the South Campus (or Trivandrum campus) and the Hill Area Campus have occasional activities and libraries are being developed at South and Hill Area campuses also.

Starting from 1985, Professor Dr. A. M. Mathai of McGill University, Canada, is the Director of CMS. After taking early retirement in 2000, Professor Mathai is spending most of time at CMS and directing various CMS activities in an honorary capacity. CMS library is being built up by using the books and journals donated by Professor Mathai and colleagues, friends and well-wishers in Canada and USA. CMS has the best library in Kerala, India, in mathematical sciences.

In 2006 - 2007 the Department of Science and Technology, Government of India (DST), gave a development grant to CMS. Thus, starting from December 2006 CMS is being developed as a DST Centre for Mathematical Sciences. DST has similar centres at three other locations in India.

From 1977 to 2006 CMS activities are carried out by a group of researchers in Kerala, mostly retired professors, through voluntary service. Starting from 2007, DST created full time salaried positions of three assistant professors, one full professor and one Liaison Officer. They are at CMS Pala Campus. DST approved up to 15 junior research fellows (JRF) and one senior research fellow (SRF). They are the current PhD students at CMS Pala Campus. They will receive their PhD degrees from Mahatma Gandhi University (MG University) in due course, after fulfilling the residence requirement of three years. All the JRFs and SRF have got papers accepted/published in international refereed journals within one and a half years.

CMS conducts a five-week research orientation course called SERC Schools every year. The main theme for the first sequence of five schools was Special Functions and Functions of Matrix Argument and Their Applications. The theme for the second sequence of five schools is multivariable and matrix variable calculus, statistical distributions and model building. The second school in the second sequence will run in 2009 from 20th April to 22nd May 2009 at CMS Pala Campus. The total number of seats in each School is 30. International participation is allowed provided the candidates come with their own return air tickets and give an undertaking to attend all lectures, all problem-solving sessions and take all examinations (usually one examination in each week, thus five written examinations and one quiz). The local hospitality and study materials are given free of charge by CMS. All expenses are paid by DST for the nationally selected 30 participants in each School.

CMS has another activity at the undergraduate level. There are four courses in each year covering all undergraduate mathematics. Thirty motivated students are selected by CMS from among the names recommended by the college Principals in Kerala. Each course is of 10-days duration of around 40 hours of lectures and 40 hours of problem-solving sessions. Again, all expenses of the selected participants are met by DST. Apart from these two regular activities there are lecture series of three days and six days duration by international visiting faculty.

There are three categories of visitors: distinguished international visitors, distinguished national visitors, faculty and students from other institutions, colleges and universities in India. There is no provision to pay for international air tickets of the visitors. For all visitors, their local hospitality is provided free of charge by CMS. Those who wish to visit CMS Pala Campus need to write to the Director of CMS, giving the approximate dates and time that they would like to visit. The Director will then issue a formal invitation as per the availability of local accommodation. The recent, 2008, international visitors include Dr. A. A. Kilbas of Belarussian University, Belarus, Dr. Hans J. Haubold of the Office for Outer Space Affairs of the United Nations, Vienna, Austria, Dr. Serge B. Provost of the University of Western Ontario, Canada, and Dr. Peter Moschopoulos of the University of Texas at El Paso, USA.

5. Distribution of information material to the general public

The Working Group was selected as a focal point for world-wide distribution, through UN information dissemination channels, of the DVD titled *Hubble: 15 Years of Discovery*, produced by the European Space Agency (`<www.spacetelescope.org/projects/anniversary/index.html>`), IHY 2007 information material produced by the United Nations Office for Outer Space Affairs, and the IYA 2009 information brochure, produced by the International Astronomical Union, `<www.astronomy2009.org/resources/brochures/detail/brochure03_en>`.

Hans J. Haubold
chair of the Working Group

Transactions IAU, Volume XXVIIA
Reports on Astronomy 2006–2009
Karel A. van der Hucht, ed.

© 2009 International Astronomical Union
doi:10.1017/S1743921308026057

DIVISION XII / COMMISSION 46 / PROGRAM GROUP EXCHANGE OF ASTRONOMERS

CHAIR	**John R. Percy**
VICE-CHAIR	**Kam-Ching Leung**
PAST CHAIR	**Charles R. Tolbert**

TRIENNIAL REPORT 2006 - 2009

1. Introduction

The Commission 46 Program Group *Exchange of Astronomers* (PG-EA) provides travel grants to astronomers and advanced students for research or study trips of at least three months duration. Highest priority is given to applicants from developing countries whose visits will benefit them, their institution and country, and the institution visited. This program, if used strategically, has the potential to support other Commission 46 programs such as *Teaching for Astronomical Development* (PG-TAD) and *World Wide Development of Astronomy* (PG-WWDA). Complete information about the program, and the application procedure, can be found at <physics.open.ac.uk/IAU46/travel.html>.

2. Awarded grants

From January 2006 to July 2008, 15 grants were awarded. Recipients came from: Brazil (1), China (1), Egypt (6), Honduras (2), Nigeria (2), Peru (1), Sweden (1), Uzbekistan (1). The visits were for a variety of useful purposes, including beginning a graduate program or post-doctoral fellowship, visiting a vibrant research centre such as International Centre for Theoretical Physics (Trieste) or Inter-University Centre for Astronomy and Astrophysics (Pune), visiting an institute to learn new techniques useful for research and education in the home country, and taking up a summer research assistantship.

3. Improving the program

We note, as previous reports of this PG have done, that the number of applications has been decreasing, despite the fact that the program is well-publicized, both on the IAU web page, and through the Organizing Committee of Commission 46. We also note that, with some exceptions, the grants do not directly support other IAU and Commission 46 development programs such as TAD and WWDA. The role of and rules for this PG should therefore be re-assessed. The rules could be broadened, while focussing the program more sharply on the IAU's development priorities. Indeed, a broad discussion and reorganization of IAU education and development activities is presently underway.

Acknowledgments: We thank the IAU General Secretary and the IAU Secretariat for their advice and assistance, and the host institutions and astronomers who have made these visits possible.

John R. Percy
chair of the Working Group

Transactions IAU, Volume XXVIIA
Reports on Astronomy 2006–2009
Karel A. van der Hucht, ed.

© 2009 International Astronomical Union
doi:10.1017/S1743921308026069

COMMISSION 50

PROTECTION OF EXISTING AND POTENTIAL OBSERVATORY SITES

PROTECTION DES SITES DES OBSERVATOIRES ACTUELS ET DES SITES POTENTIELS DES OBSERVATOIRE FUTURS

PRESIDENT
PAST PRESIDENT
ORGANIZING COMMITTEE

Richard J. Wainscoat
Malcolm G. Smith
Carlo Blanco, David L. Crawford,
Margarita Metaxa,
Woodruff T. Sullivan

COMMISSION 50 WORKING GROUP
Div. XII / Commission 50 WG   Controlling Light Pollution

TRIENNIAL REPORT 2006 - 2009

1. Introduction

The activities of the Commission have continued to focus on controlling unwanted light and radio emissions at observatory sites, monitoring of conditions at observatory sites, and education and outreach. Commission members have been active in securing new legislation in several locations to further the protection of observatory sites.

During 2007 there were two landmark conferences related to light pollution. In February, the *International Dark Sky Association* organized a multi-disciplinary conference on light pollution in Washington DC, USA. This conference highlighted the many other problems that light pollution causes, including effects on wildlife, human health, and energy waste. It is clear that the need for protection of the night sky from light pollution extends far beyond astronomy, and astronomers can and should forge alliances with other groups to better protect the night sky.

In April 2007, the *Starlight* conference in the Canary Islands led to a "Declaration in Defence of the Night Sky and the Right to Starlight." The IAU was one of the participating organizations for this meeting. The meeting brought together many of the world's astronomers who are working on light pollution issues. In cooperation with UNESCO, the meeting has led to a draft proposal for *Starlight Reserves* which is described below.

2. Controlling light pollution

Population growth continues in regions close to the major astronomy sites in Hawaii, Chile, Arizona, and the Canary Islands. This growth, and its associated lighting continues to pose a threat to the observatory sites. Retrofits of existing lights are needed to maintain dark skies over these observatory sites as the nearby population grows.

Good progress has been made in the retrofit of outdoor light fixtures in the astronomy regions of Chile. In the Antofagasta region, approximately 70% of the fixtures have been replaced, in the Atacama region, approximately 65% of the fixtures have been replaced, and in the Coquimbo region, approximately 45% of the fixtures have been replaced. Money is available to complete the retrofit in the Antofagasta region, and further progress is expected in the Coquimbo region in the upcoming year.

Imaging from the *International Space Station* showed that the airports on the island of Hawaii are the major source of broad-spectrum light affecting Mauna Kea. State legislation was passed in 2007 that requires new lighting at the airports and harbors in Hawaii to conform to county lighting ordinances. A program to retrofit airport lights on the island of Hawaii is now underway. A new lighting ordinance for the island of Maui was passed in 2007 and will reduce light pollution over Haleakala observatory by requiring retrofitting of non-conforming light over the next 10 years.

Elizabeth Alvarez serves as Commission 50 representative to the *International Commission on Illumination (Commission International d'Éclairage* – CIE). This is an important engineering group setting standards for lighting.

Many activities related to light pollution are planned for the 2009 *International Year of Astronomy* (IYA 2009). Connie Walker at NOAO has been leading these efforts, and *Dark Skies Awareness* has been selected as one of 11 cornerstone projects for IYA 2009. Planned activities include *Globe at Night*, astronomy nights in National Parks, educational programs about good lighting, a night sky photography contest, a (radio) quiet skies program, dark skies discovery sites, *Dark skies week* and *Earth hour* when people around the Earth are encouraged to turn off their lights for 1 hour. These programs will have tremendous educational and outreach value.

A feature documentary entitled *The City Dark* is presently being produced by Wicked Delicate Films. It is scheduled to be released in 2009. It will highlight efforts being made to protect major observatory sites, including Mauna Kea and Arizona.

Wide interaction and cooperation with the *International Dark Sky Association* (IDA) is continuing, and the help that IDA provides is gratefully acknowledged.

3. Monitoring light pollution

As populations around observatory sites grow, there is a growing need for monitoring light pollution levels at observatories. Duriscoe, Luginbuhl, and Moore (2007) have developed a CCD camera system to measure light pollution in US National Parks. Systems such as this will be used in the future to quantify sources of light pollution and to measure trends in light pollution levels at observatories.

Dan McKenna and the *International Dark Sky Association* have been developing night sky brightness monitors. These are expected to be deployed to many observatory sites during 2009. These monitors will also help to measure long-term trends at the observatory sites.

Handheld Sky Quality Meters have been developed, and are affordable (approximately US $120). They deliver V-band sky brightness in magnitudes per square arcsecond. These have become invaluable tools for outreach and education.

A concept for a satellite that would observe light sources on the Earth has been developed by Elvidge *et al.* (2007). Such a satellite would allow astronomers to precisely pinpoint and measure sources of light pollution near observatory sites, monitor changes, and greatly help efforts to control light pollution at observatories.

4. RFI protection

Interaction with the *International Telecommunication Union* (ITU) is primarily made via the Scientific Committee on *Frequency Allocations for Radio Astronomy and Space Science* (IUCAF). This committee is composed of representatives from USRI (*International Union of Radio Science*), IAU, and COSPAR. Wim van Driel served as Commission 50 liaison to IUCAF.

An ITU *World Radiocommunication Conference* (WRC) was held in 2007 in Geneva. The main goal of these conferences is definition of the world-wide framework for spectrum management, including criteria for protection of radio astronomy from unwanted emission. WRCs are held every three to four years, and the agenda items are adopted at the previous WRC.

Of greatest relevance to astronomers, was an agenda item on protection of the radio astronomy service and Earth exploration-satellite (passive) service from unwanted emission in adjacent bands. Of particular importance to IUCAF was the case of the 1610.6 - 1613.8 MHz band that contains important spectral lines of the OH molecule. It was decided that the protection of this band is ensured.

The preliminary agenda adopted for the WRC in 2011 includes use of the radio spectrum from 275 to 3000 GHz. Although no allocations will be made at the WRC meeting in 2011, the radio astronomy community must identify a list of bands of interest.

International engineers and scientists have been studying the concept of solar power satellites. These typically have large $(10 \, km^2)$ solar panels in geostationary orbit, and transmit the power they collect to Earth in the form of microwaves. These have potentially disastrous consequences for both optical astronomy and radio astronomy. These satellites are not imminent, and many technological problems would need to be solved. However, they are so damaging to astronomy that careful attention is needed. URSI, IUCAF, and Commission 50 are continuing to monitor developments.

The major radio telescopes under development are presently ALMA and SKA. Commission 50 will provide help in efforts to protect these telescopes from radio interference when needed.

5. Developments within the past triennium

The emergence of light emitting diodes (LEDs) as a form of outdoor lighting is a new threat to astronomical observatories. For lighting near observatory sites, low-pressure sodium (LPS) lamps are the preferred light source. This is because the light they emit is nearly monochromatic. It can, therefore, be filtered out in some cases (e.g., *B* or *g* filter imaging); there is already natural emission from the atmosphere in the sodium line, so it is a wavelength that is already partially compromised. LPS lamps are highly energy efficient compared to other forms of lighting, and this has helped astronomers to get LPS lighting adopted around observatory sites, such as the Island of Hawaii, and La Palma.

White LEDs are now approaching, or in some cases exceeding the light output of LPS lamps in terms of lumens per Watt. They may require less maintenance, offer better color rendition, and the LED light fixtures may become less expensive than the LPS fixtures. White LEDs are essentially blue LEDs that have a phosphor that converts some of the blue light into redder light. They typically have an emission spike around 460 nm – a region of the sky that is very dark. Additionally, Rayleigh scattering of 460 nm blue light is 2.7 times stronger than for the yellow-orange 589 nm light of LPS lamps. The impact of LED lighting will be a topic for discussion within the Commission 50 Working Group for *Controlling Light Pollution.*

The Canary Islands *Starlight* conference was conducted in cooperation with UNESCO. A further meeting between IAU representatives and UNESCO was held in October 2007, and a draft proposal for *Starlight Reserves* is being widely circulated. The specific types of starlight reserves proposed include starlight heritage sites, starlight astronomy sites, starlight natural sites, starlight landscapes, rural and urban starlight oases, and mixed starlight sites. The most relevant to astronomers is the *Starlight Astronomy Site* designation. This designation would relate to exceptional observation sites for optical, infrared, and radio astronomy, and include potential future sites.

6. Closing remarks

The tragic accidental death of Hugo Schwarz was an enormous loss to Commission 50, and to the international effort to protect observatories from light pollution.

Our Commission continues to believe that more professional astronomers need to become involved in efforts to reduce light pollution and to reduce radio interference.

Richard J. Wainscoat
president of the Commission

References

Elvidge, C. D., Safran, J., Tuttle, B., Sutton, P., Cinzano, P., Pettit, D., Arvesen, J., & Small, C. 2007 *GeoJournal*, 69, 45
Duriscoe, D. M., Lugnibuhl, C. B., & Moore, C. A. 2007 *PASP*, 119, 192

Transactions IAU, Volume XXVIIA
Reports on Astronomy 2006–2009
Karel A. van der Hucht, ed.

© 2009 International Astronomical Union
doi:10.1017/S1743921308026070

COMMISSION 55

COMMUNICATING ASTRONOMY WITH THE PUBLIC
PARTAGER L'ASTRONOMIE

PRESIDENT — Ian E. Robson
VICE-PRESIDENT — Dennis R. Crabtree
SECRETARY — Lars Lindberg Christensen
ORGANIZING COMMITTEE — Oscar Alvarez Pomares,
Augusto Damineli Neto,
Richard T. Fienberg, Anne Green,
Ajit K. Kembhavi, Birgitta Nordström,
Kazuhiro Sekiguchi,
Patricia A. Whitelock, Jin Zhu

COMMISSION 55 WORKING GROUPS

Div. XII / Commission 55 WG	Washington Charter
Div. XII / Commission 55 WG	VAMP
Div. XII / Commission 55 WG	Best Practices
Div. XII / Commission 55 WG	Communicating Astronomy Journal
Div. XII / Commission 55 WG	New Ways of Communicating Astronomy with the Public
Div. XII / Commission 55 WG	CAP Conferences

TRIENNIAL REPORT 2006 - 2009

1. Introduction

Commission 55 was approved at the IAU XXVI General Assembly in Prague, 2006 ,following the great success of the Working Group *Communicating Astronomy*, which had been set up in 2003. It resides within Division XII and the mission statement of the WG has been incorporated into the Commission:

- to encourage and enable a much larger fraction of the astronomical community to take an active role in explaining what we do (and why) to our fellow citizens;
- to act as an international, impartial coordinating entity that furthers the recognition of outreach and public communication on all levels in astronomy;
- to encourage international collaborations on outreach and public communication; and
- to endorse standards, best practices and requirements for public communication

The first point is particularly important as, in principle; it is something over which we have some control or influence. It is now widely recognised that there are a number of barriers to communicating astronomy. Firstly, a number of professional astronomers do not feel comfortable with the very concept of talking with the public. Secondly, many of the employing organisations do not regard communication and outreach as a real part of the 'job description'. Hence, the time taken for public communication may not only go unrewarded for the researchers, it may well go against the researcher in effect if outreach is not counted as a merit in the same way as grants, refereed papers, etc. The final hurdle is that a number of organisations (especially those outside the USA) have not yet integrated public communication (or 'science and society') into their own organisational structure by providing the necessary support funding, training, infrastructure, personnel, etc.

To forward the work of the Commission, an extensive web-page has been set up: `<www.communicatingastronomy.org>`. This charts the history of the WG leading to the Commission and lists its organisation and officers. Indeed, this portal holds the key to the success of the Commission and all the information relating to the work is contained therein. Because of the nature of the work of the Commission, we are anxious to encourage as many people as possible to become active and spread the word. In that light we have set up a section where enthusiasts can sign-in as 'supporters', on whom we can call for general or specific tasks and propagation of information.

The actual work of the Commission has been split up into six Working Groups:
- *Washington Charter* (chair: Dennis Crabtree)
- *Virtual Astronomy Multimedia Project* – VAMP (chair: Adrienne Gauthier)
- *Best Practices* (chair: Lars Lindberg Christensen)
- *Communicating Astronomy Journal* (chair: Pedro Russo)
- *New Ways of Communicating Astronomy with the Public* (chair: Michael West)
- *Communicating Astronomy with the Public conferences* (chair: Ian Robson)

2. The Washington Charter

One of the key activities of the original Working Group *Communicating Astronomy* was the promulgation of the Washington Charter, which arose from the 2003 meeting in Washington, DC, USA. The Charter outlines *Principles of Action* for individuals and organizations that conduct astronomical research and that have a compelling obligation to communicate their results and efforts with the public for the benefit of all.

Although well received and heavily endorsed, a few organizations were hesitant in their endorsement of the original version because of the strong language used in the Charter. The Charter was subsequently revised in terms of softening the language without altering the strong message contained and the ownership passed to the WG *Communicating Astronomy* and hence to Commission 55 and its sub-group.

The preamble to the Charter clearly states the rationale for the priority that those of us involved in astronomical research need to give to communicating with the public: As our world grows ever more complex and the pace of scientific discovery and technological change quickens, the global community of professional astronomers needs to communicate more effectively with the public. Astronomy enriches our culture, nourishes a scientific outlook in society, and addresses important questions about humanity's place in the universe. It contributes to areas of immediate practicality, including industry, medicine, and security, and it introduces young people to quantitative reasoning and attracts them to scientific and technical careers. Sharing what we learn about the universe is an investment in our fellow citizens, our institutions, and our future. Individuals and organizations that conduct astronomical research – especially those receiving public funding for this research – have a responsibility to communicate their results and efforts with the public for the benefit of all. The success of the wording was that the AAS endorsed the revised Charter at their June 2006 meeting in Calgary, Canada. At the moment, the Charter has been endorsed by 31 organisations, but work in pushing this forward has taken a back-seat to the major efforts of Commission 55: preparations for the *International Year of Astronomy 2009* (IYA 2009).

3. Virtual Astronomy Multimedia Project- VAMP

Good progress has been made on VAMP and much of this will feed into the *Portal to the Universe* cornerstone project for IYA 2009 - which will have huge benefits when it is finally launched. The VAMP project web page is: `<www.virtualastronomy.org/`.. One of the key results of this work is a now globally accepted standard for metadata for PR imagery called *Astronomy Visualization Metadata 1.1.*

4. Best Practices

Unfortunately there has been little progress on this topic but some of the work will find its way into the IYA 2009 activities and some has been reported in the *Communicating Astronomy Journal.*

5. Communicating Astronomy Journal

The new journal for *Communicating Astronomy with the Public*, named CAP Journal, was launched in October 2007. The first issue, a 36-page, glossy production was a great success and this has now been followed up by issues 2 and 3, the last being in May 2008. Editor-in-Chief is Pedro Russo, supported by Lars Lindberg Christensen and Terry Mahoney as Executive Editor and Editor, respectively. The first edition had a broad range of articles and the success was demonstrated by the latest edition, which attracted a much wider range of authors; a trend that we expect to continue. The printing of the first issues of the *Journal* were sponsored by ESA/ST-ECF and can be found at `<www.capjournal.org/>`. New sponsorships for individual issues are sought.

6. New Ways of Communicating Astronomy with the Public

As with topic 3, this has taken a bit of a back-seat although the work on preparing for IYA 2009 has meant that the theme has been well and truly taken up with bloggs, webcasts and podcasts all becoming more 'mainstream'. This was also one of the themes for CAP2007 (see below)

7. Communicating Astronomy with the Public Conference

The *Communicating Astronomy with the Public* (CAP) conference has now become a brand name of Commission 55, following the highly successful meeting held in ESO-HQ in 2005. CAP2007 was held in Athens, Greece, 7-10 October 2007. This was a huge success with well over 200 participants (the largest CAP meeting so far) from many countries around the world. The focus was on IYA 2009 and it proved to be an extremely successful and useful meeting with lots of cross pollination of ideas from around the world. Athens proved to be a marvelous setting and visits to sites and the conference dinner in the shadow of the Parthenon were jewels to remember.

8. International Year of Astronomy

By far the greatest focus and involvement during 2007 and 2008 has been in the preparation and support for the *International Year of Astronomy 2009* (IYA 2009), and this has consumed most of the activity of the key activists. The key members of Commission 55, Ian Robson, Lars Lindberg Christensen and Dennis Crabtree are all members of the IAU Executive Committee Working Group overseeing the Year and the monthly teleconferences have meant that everyone has maintained their noses to the grindstone.

Commission 55 organised the kick-off meeting for the Nations Single Point of Contacts (SPoCs) at the ESO-HQ in Garching, Germany, 3-4 March 2007. This meeting was a tremendous success, with 26 out of the 63 identified SPoCs attending, coming from places as far a field as China, Japan, South Korea, Australia, USA, Canada and Mexico (31 countries represented in total). The webcast of the meetings and the presentations can be found at `www.communicatingastronomy.org/iya_eso/programme.html>`. During the Saturday evening, four attendees conducted a web- and skype-cast of the total eclipse of the moon, linking two schools from Germany and South Africa and many other callers freely joining in. Out of this kick-off meeting came a number of ideas and themes that are being taken forward for IYA 2009. One of the key successes was the agreement of the logo and slogan for IYA 2009, and these are now being spread through the communities by web downloads. The number of SPoCs who have managed to set-up their own national node pages is extremely gratifying.

Overall, it is clear that there is a lot of activity going on, but the global economic downturn has hindered the fundraising. Nevertheless, the IYA 2009 secretariat, hosted by ESO, Garching, Germany, is now fully staffed and the web page is proving to be a huge success (`<www.astronomy2009.org>`) with an amazing hit-rate. A fabulous trailer (video) has been produced by the secretariat, this and lots of other information can be found on the web. The video can be seen at:
`<www.astronomy2009.org/index.php/?option=com_content&view=article&id=378>`.

For the future, IYA 2009 again points to the main focus. A session at the JENAM 2008 in Vienna, September 2008, was devoted to the IYA 2009 in terms of education and outreach. Support will also be provided for the IAU XXVII General Assembly in Rio de Janeiro, Brazil, 2009, and it has been agreed that the biennial CAP (2009) will be foregone and instead will be held in the Spring of 2010 in Cape Town. It will be an extended meeting so that the evaluation and legacy aspects of IYA 2009 can be fully appreciated and documented.

8.1. *News Release - IAU0702: The United Nations declares 2009 the International Year of Astronomy*

Dec 20, 2007, Paris

Early this morning (CET) the United Nations (UN) 62nd General Assembly proclaimed 2009 the *International Year of Astronomy*. The Resolution was submitted by Italy, Galileo Galilei's home country. The International Year of Astronomy 2009 is an initiative of the International Astronomical Union and UNESCO.

The *International Year of Astronomy 2009* (IYA 2009) celebrates the first astronomical use of the telescope by Galileo a momentous event that initiated 400 years of astronomical discoveries and triggered a scientific revolution which profoundly affected our world view. Now telescopes on the ground and in space explore the Universe, 24 hours a day, across all wavelengths of light. The President of the International Astronomical Union (IAU)

Catherine Cesarsky says: "The *International Year of Astronomy 2009* gives all nations a chance to participate in this ongoing exciting scientific and technological revolution".

The IYA 2009 will highlight global cooperation for peaceful purposes - the search for our cosmic origin and our common heritage which connect all citizens of planet Earth. For several millennia, astronomers have worked together across all boundaries including geographic, gender, age, culture and race, in line with the principles of the UN Charter. In that sense, astronomy is a classic example of how science can contribute toward furthering international cooperation.

At the IAU General Assembly on 23 July 2003 in Sydney (Australia), the IAU unanimously approved a resolution in favour of the proclamation of 2009 as the *International Year of Astronomy*. Based on Italy's initiative, UNESCOs General Conference at its 33rd session recommended that the UN General Assembly adopt a resolution to declare 2009 the *International Year of Astronomy*. On 20 December 2007 the *International Year of Astronomy 2009* was proclaimed by the United Nations 62nd General Assembly. The UN has designated the United Nations Educational, Scientific and Cultural Organization (UNESCO) as the lead agency for the IYA 2009. The IAU will function as the facilitating body for IYA 2009.

The IYA 2009 is, first and foremost, an activity for the citizens of planet Earth. It aims to convey the excitement of personal discovery, the pleasure of sharing fundamental knowledge about the Universe and our place in it, and the merits of the scientific method. Astronomy is an invaluable source of inspiration for humankind throughout all nations. So far 99 nations and 14 organisations have signed up to participate in the IYA 2009 – an unprecedented network of committed communicators and educators in astronomy.

For more information on the *International Year of Astronomy 2009* please visit the website at `<www.astronomy2009.org>`.

– End of Press Release –

9. Other

The web now includes two new topics: 'training' and a 'jobs bank'. The former provides links to specific courses on science communications, while the latter relates to positions in the outreach area.

Ian E. Robson, Dennis R. Crabtree & Lars Lindberg Christensen
president, vice-president & secretary of the Commission

Transactions IAU, Volume XXVIIA
Reports on Astronomy 2006–2009
Karel A. van der Hucht, ed.

© 2009 International Astronomical Union
doi:10.1017/S1743921308026082

EXECUTIVE COMMITTEE   PRESS OFFICE

PRESS OFFICER   Lars Lindberg Christensen

TRIENNIAL REPORT 2006 - 2009

1. Introduction

While expecting a huge increase of interest in the IAU as a result of the planet definition issue on the agenda of the IAU XXVI General Assembly, Prague, Czech Republic, August 2006, the Executive Committee appointed Lars Lindberg Christensen as Press Officer in June 2006. He will stay on the job till at least the IAU XXVII General Assembly in Rio de Janeiro, Brazil, August 2009.

Below the IAU Press Release issued between June 2006 and November 2008 are listed below. The full text of those IAU Press Releases can be found on: <www.iau.org/public_press/news/>.

2. Press releases

- News Release IAU0809, Nov 21, 2008:
The International Year of Astronomy 2009 Opening Ceremony: Save the Date
- News Release IAU0808, Oct 30, 2008:
UNESCO and the IAU sign key agreement on Astronomy and World Heritage Initiative
- News Release, IAU0807, Sep 17, 2008:
IAU names fifth dwarf planet Haumea
- News Release IAU0806, Jul 19, 2008:
Fourth dwarf planet named Makemake
- News Release IAU0805, Jun 12, 2008:
Bond receives $500,000 Gruber Cosmology Prize for major theoretical insights into the origin of the Universe
- News Release IAU0804, Jun 11, 2008:
Plutoid chosen as name for Solar System objects like Pluto
- News Release IAU0803, Jun 5, 2008:
The International Astronomical Union teams up with the Norwegian Academy of Science and the Kavli Prize
- News Release IAU0802, Jun 5, 2008:
The first International Year of Astronomy 2009 Cornerstone Project brings the stars to Liverpool, UK
- News Release IAU0801, Jan 17, 2008:
The International Year of Astronomy 2009 welcomes the 100th participating country
- News Release IAU0702, Dec 20, 2007:
The United Nations declares 2009 the International Year of Astronomy

- News Release IAU0701, Oct 26, 2007:

New journal for astronomy communicators goes live!

- News Release IAU0700, Jul 17, 2007:

PPGF Cosmology Prize Recipients 2007

- News Release IAU0607, Dec 21, 2006:

IYA2009 web page public, National Node chairs appointed

- News Release IAU0606, Oct 27, 2006:

The International Astronomical Union announces the International Year of Astronomy 2009

- News Release IAU0605, Sep 14, 2006:

IAU names dwarf planet Eris

- News Release IAU0604, Aug 28, 2006:

The International Astronomical Union elects Catherine Cesarsky as new President

- News Release IAU0603, Aug 24, 2006:

IAU 2006 General Assembly: Result of the IAU Resolution votes

- News Release IAU0602, Aug 24, 2006:

The Final IAU Resolution on the definition of "planet" ready for voting

- News Release IAU0601, Aug 16, 2006:

The IAU draft definition of "planet" and "plutons"

3. IAU web site and data base

The IAU Press Office worked closely with the IAU General Secretary and the IAU Secretariat on developing the IAU web site and the IAU membership database. In addition, frequent requests from public and press were answered.

Lars Lindberg Christensen
IAU Press Officer

AUTHOR INDEX